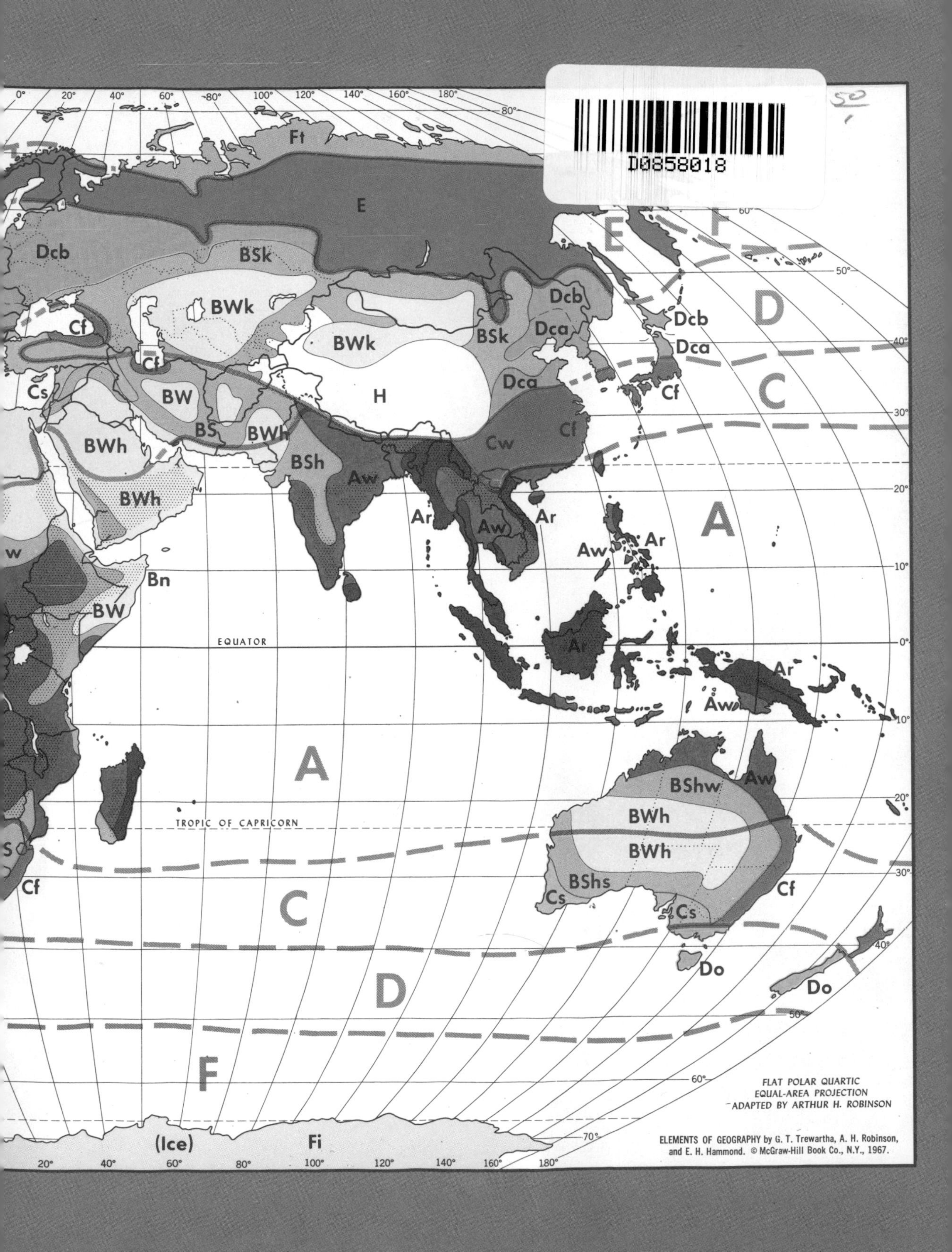
D0858018
Ft
E
Dcb
BSk
BWk
Cf
Cs
BW
BS
BWh
BSh
Aw
H
Cw
Dca
Ar
Bn
EQUATOR
A
TROPIC OF CAPRICORN
BShw
BShs
Do
C
D
F
(Ice)
Fi
FLAT POLAR QUARTIC
EQUAL-AREA PROJECTION
ADAPTED BY ARTHUR H. ROBINSON
ELEMENTS OF GEOGRAPHY by G. T. Trewartha, A. H. Robinson,
and E. H. Hammond. © McGraw-Hill Book Co., N.Y., 1967.

THE EARTH'S PROBLEM CLIMATES

Second Edition

wartha

Press

Published 1981

The University of Wisconsin Press
114 North Murray Street
Madison, Wisconsin 53715

The University of Wisconsin Press, Ltd.
1 Gower Street
London WC1E 6HA, England

First edition printings 1961, 1962, 1966, 1970
Second edition printing 1981

Printed in the United States of America
ISBN 0–299–08230–X

Contents

List of Tables, v

List of Figures, vi

Preface to the Second Edition, xi

Introductory Statement, 3

PART I *Latin America*

Chapter

1 Pacific South America: Colombia and Ecuador, 11
2 Pacific South America: The Dry Climates of Northern Chile and Peru, 23
3 Atlantic South America: Middle Latitudes and Subtropics, 41
4 Atlantic South America: Tropical Latitudes, 47
5 Caribbean South America and Middle America, 66

II *Australia, New Zealand, and the Equatorial Pacific*

6 Australia, New Zealand, and the Equatorial Pacific, 87

III *Africa*

7 The Sahara, Sudan, and Guinea Coast, 105
8 Tropical Central Africa, 125
9 Tropical East Africa, 134
10 Southern Africa, 159

IV *Southern and Eastern Asia*

11 Monsoon Circulation and the Indian Subcontinent, 173
12 Tropical Southeast Asia, 199
13 Eastern Asia, 207

V *Europe and the Mediterranean Borderlands*

14 Western and Central Europe, 235
15 Mediterranean Borderlands, 255

VI *Anglo-America*

16 Temperature and Precipitation, 281
17 Regional Rainfall Types: The West, 297
18 Regional Rainfall Types: The Interior, 311
19 Regional Rainfall Types: The South and East, 326

Appendix, 343

References, 349

Index, 367

List of Tables

1.1 Rainfall distribution at Andagoya, in the Choco area, 5°4′N, 76°55′W 17
2.1 Monthly percentages of total annual rainfall in Chile 40
3.1 Temperature and precipitation in Santa Cruz and Sarmiento′ 41
4.1 Wind distribution at Quixeramobim in Northeast Brazil during the rainy season 62
5.1 Average monthly pressure at San Salvador 78
6.1 Frequency of occurrence of the first four weather types for the area bounded by meridians 130°E and 180°E and parallels 20°S and 50°S 94
6.2 Percentage of days under major controlling weather situations in New Zealand 96
7.1 Percentage frequency of above-surface inversions and isothermals at Lagos, 1943 and 1944 119
7.2 Percentage frequency of inversions and of dry air at Lagos in August 120
9.1 Low-level wind direction and humidity at Dar es Salaam, 1960 to 1963 142
10.1 Frequency of different weather types at Pretoria 163
11.1 Average number of western disturbances per year affecting northern India 180
11.2 Average number of thundersqualls per month in Delhi and Calcutta 182
12.1 Surface and upper-air mean monthly relative humidity, Djakarta, Java 204
13.1 Seasonal distribution of precipitation in China in percent of the annual total 223
14.1 Rainfall distribution in the Netherlands 246
14.2 Frequency of *Grosswetterlagen* for Central Europe 253
15.1 Average number of depressions in the Mediterranean by season 263
15.2 Rainfall frequency over the Mediterranean Sea 266
15.3 Statistical analysis of Mediterranean weather situations 268
19.1 Frequency of occurrence of certain synoptic types over the state of Tennessee for the period 1925–1935 332
I.1 Köppen's climatic groups and types 343

List of Figures

A	General pattern of total annual precipitation on a hypothetical continent	4
B	General pattern of seasonal precipitation concentration on a hypothetical continent	4
C	Arrangement of principal types of climate on a hypothetical continent	5
1.1	Average annual rainfall in western Colombia	14
1.2	Annual march of rainfall in Pacific Colombia	15
1.3	Climatic arrangement in the equatorial eastern Pacific	16
1.4	Hourly distribution of rainfall at Quibdo, Pacific Colombia	17
1.5	Belted arrangement of the diurnal occurrence of heaviest rainfall in Pacific Colombia	18
1.6	Climatic arrangement (Köppen) in Pacific Ecuador	20
1.7	Profile of annual rainfall across western Ecuador from the coast toward the interior	20
1.8	Important climatic boundaries in western Ecuador	21
2.1	Pattern of the coastal currents along the coast of Peru	24
2.2	Vertical profiles of temperature and humidity for July and January at Lima, Peru	27
2.3	Temperature inversion and stratus layer along the Peruvian littoral	27
2.4	Annual precipitation curves plotted against latitude for 5 coastal deserts in tropical-subtropical latitudes	30
2.5	Percentage of annual rainfall in western Peru occurring in November–April	31
2.6	Climatic regions of the north Chilean desert	36
2.7	Pressure and wind patterns during an El Niño rainy period in the northern Peruvian desert	37
3.1	Arrangement of climatic types in Argentina according to the Köppen classification	42
3.2	Moisture regions of Argentina	42
3.3	Mean resultant winds over Patagonia in winter at 6 km	44
4.1	Northward advancing polar front invading tropical-subtropical Brazil	49
4.2	Time of maximum precipitation in eastern-central Brazil	51
4.3	Time of precipitation minimum in eastern Brazil	51
4.4	Stages in the development of a cold front along the Brazilian coast south of Recife	54
4.5	Showing stages in the southward advance over Brazil in spring and summer of the equatorial westerlies and the *ITC*	55
4.6	Average annual rainfall in northeastern Brazil	56
4.7	Location in eastern Brazil of the *BS* and *As* climates	57
4.8	Typical locations of the *ITC* and the trade winds of the Southern Hemisphere	58
4.9	Dry area in eastern Brazil occupying a position between the winter-rainfall area to the east and the summer-rainfall area to the west	60
4.10	Annual rainfall profiles of three stations in the Amazon Basin	64
4.11	Inferred arrangement of climatic types within the Amazon Basin	64
5.1	Rainfall characteristics in Caribbean South America	68
5.2	Composite annual rainfall profile of 17 stations in the Netherlands Antilles	69
5.3	Stress-induced divergence along the Caribbean coasts of Venezuela and Colombia in spring and fall	72
5.4	Percentage frequency of the trade-wind inversion in the Caribbean region	79
5.5	Average annual rainfall of the Yucatan Peninsula	80
5.6	Average annual rainfall profiles of three stations in Yucatan, Mexico	81
6.1	A meridional front crossing Australia from west to east	88

6.2 Precipitation distribution associated with a meridional front in southwestern Australia 90
6.3 Showing the location of the dry belt and the associated tongue of cool water in the equatorial eastern and central Pacific 97
6.4 Coefficient of rainfall variability over the equatorial Pacific Ocean 99
6.5 Mean monthly positions of the maximum brightness axes in the equatorial Pacific Ocean 99
6.6 Models of the Hadley and equatorial circulation cells 100
6.7 Longitudinal distribution of mean surface temperatures in the equatorial Pacific Ocean for February and August 101
6.8 Means of equatorial Pacific island rainfall and sea surface temperature anomalies 101
7.1 Moisture regions of Africa 108
7.2 Structure of the *ITC* in Africa north of the equator in summer 109
7.3 General locations and approximate boundaries of West Africa's 5 weather zones 113
7.4 Profiles of mean annual rainfall at Accra 116
7.5 Frequency of occurence of rainfalls exceeding 5 mm in the Sudan-Guinea Coast 117
7.6 Profiles of average annual rainfall at three stations along the Ghana coast 118
7.7 Tephigram for Lagos, Nigeria 119
7.8 Annual rainfall amounts along the Gulf of Guinea coast in Ghana, Togo, Benin, and western Nigeria 120
7.9 Annual profiles of monthly rainfall differentials between Axim and Accra, and between Lagos and Accra 121
7.10 Annual march of temperature at Accra 123
8.1 Distribution of average annual rainfall in the Congo Basin region 126
8.2 Depth of penetration of the southwest "monsoon" and associated weather in the Congo Basin 128
8.3 Position of two line squalls in the southwest "monsoon" and associated weather in the Congo Basin 129
8.4 Vertical structure of the atmosphere at Kinshasa in winter 131
8.5 Vertical structure of the winter atmosphere at Luanda and Mossämedes, Angola 131
8.6 Total annual rainfall over 25 years for 2 coastal stations in Angola 133
8.7 Isolines of equal coefficient of precipitation variability in Angola 133
9.1 Average annual rainfall in equatorial East Africa 135
9.2 Average January and July rainfall in equatorial East Africa 136–37
9.3 Principal climatic types in equatorial East Africa 138
9.4 Mean circulation at lower levels for eastern and southern Africa in July and January 139
9.5 Equatorial westerlies in the realm of the Indian Ocean easterlies in East Africa 141
9.6 Vertical structure of the air at Nairobi in August 141
9.7 Main large scale weather systems in equatorial East Africa 144
9.8 January and July isohyets in tropical East Africa north of the equator 151
9.9 Representative low-level pressure arrangement over the eastern Sahara and Sudan in winter 152
9.10 Representative low-level pressure arrangement over the eastern Sahara and Sudan in summer 152
9.11 Low-level winds and ocean currents along the Somali coast and over the adjacent sea in summer 154
9.12 Isotherms of surface sea temperature in July along the Somali coast 154
9.13 Frequency of fog and mist along the Somali coast in summer 155
9.14 Illustrating the low reliability of the annual rainfall in East Africa 156
9.15 A cold front from higher latitudes advancing equatorward over eastern Africa south of the equator 158
10.1 Five important weather types of South Africa 162
10.2 Vertical structure of the air at Durban in summer 166
10.3 Vertical structure of the air at Durban in winter 167
11.1 Main features of the low-level circulation over eastern and southern Asia in the cool season 174
11.2 Characteristic features of air flow over eastern and southern Asia at about 3,000 m, November to March 175
11.3 Principal elements of the low-level circulation pattern of eastern and southern Asia in the warm season 176
11.4 Position of the monsoon pressure trough and streamlines of the resultant winds over southern Asia in summer 176

11.5 Schematic profile of the planetary wind belts along meridian 78°E over India 177
11.6 Average rainfall in the Indian subcontinent for a winter, spring, and fall month and for the year as a whole 179
11.7 Mean positions of the two jet streams in July–August and mean July rainfall over South Asia 185
11.8 Schematic model of the Indian summer monsoon circulation 186
11.9 Percentage of the annual precipitation that falls over India during the southwest summer monsoon 186
11.10 Distribution of rainfall over India during an active monsoon condition and during a break monsoon condition 187
11.11 Normal dates of the onset and withdrawal of the summer monsoon over the Indian subcontinent 191
11.12 Moisture regions of the Indian subcontinent 193
11.13 Mean annual precipitation in the northwestern part of the Indian subcontinent 194
11.14 Mean resultant surface winds over the northwestern part of the Indian subcontinent 195
11.15 Precipitation subdivisions of the Indian subcontinent based on the time of rainfall maximum 196
12.1 Diagrammatic representation of the three principal air streams and their zones of confluence in Southeast Asia 200
12.2 Isohyetal arrangement showing the general location of the subhumid zone to the east and south of Java 203
12.3 Seasonal circulation patterns for January and July in the western South Pacific 203
12.4 Distribution in Southeast Asia of regions where the rainiest month occurs in winter and the driest month occurs in summer 205
13.1 Representation of air masses and fronts over eastern Asia in winter 208
13.2 Mean circulation and frontal zones over East Asia during the Baiu season 210
13.3 Mean specific humidity at the 700 mb level over China in July 211
13.4 Distribution of the precipitation over China in July 211
13.5 Frequency of cyclone formation in the Far East 216
13.6 Annual number of days with thunderstorms in eastern Asia 219
13.7 Distribution of mean west wind speed at 12 km over the Far East in winter, and mean winter precipitation over China 222
13.8 Composite profiles of 5-day means of rainfall and pressure for 5 stations in southern Japan, and composite of 5-day means of rainfall for 9 stations in southern China and Taiwan 224
13.9 Composite annual profile of 5-day means of rainfall for 5 stations in subtropical Japan 228
13.10 Showing the degree of synchronization of the first appearance of the southwest monsoon in India and the beginning of the Baiu rains in East Asia 228
13.11 Showing a typical location of the Baiu front over southern Japan 229
13.12 Five-day mean cloudiness and water-vapor transport during a recent Baiu season in East Asia 230
13.13 Average 10-day precipitation amounts for Shionomisaki in Japan 231
14.1 Major climatic boundaries in Europe according to the Köppen classification 236
14.2 Frequency of blocking action by a quasi-stationary high in the eastern North Atlantic 239
14.3 Annual frequency of cyclones in Europe 240
14.4 Annual frequency of deep cyclones with central pressure less than 1,000 mb 240
14.5 Frequency of deep cyclones with central pressure less than 1,000 mb in winter 240
14.6 Frequency of deep cyclones with central pressure less than 1,000 mb in summer 240
14.7 Mean European surface air temperature anomalies in winter during blocking action by an anticyclone in the easternmost Atlantic and adjacent western Europe 241
14.8 Mean European surface air temperature anomalies in summer during occasions of blocking action by an anticyclone in the easternmost Atlantic and adjacent western Europe 241
14.9 Showing the distribution of areas in Europe with very modest amounts of annual precipitation 244
14.10 Mean European precipitation anomalies in winter during occasions of blocking action 245
14.11 Mean European precipitation anomalies in summer during occasions of blocking action 245
14.12 Regions in Europe having a fall maximum of precipitation 248
14.13 Regions in Europe having a spring minimum of precipitation 249
14.14 Profiles showing the number of occurrences by month of 3 important weather situations in the British Isles 250
14.15 Frequency curves for the daily occurrences of 2 common large-scale weather situations in the British Isles 250

15.1 Three-kilometer pressure chart showing surface fronts under conditions of a northwest flow pattern in Europe 258
15.2 Common routes by which cold northerly air enters the Mediterranean Basin 258
15.3 Frequency of cyclonic *Grosswetterlagen* and frequency of precipitation in Italy and the Balkans 259
15.4 Three-kilometer chart with surface fronts under conditions of a southwest-steering situation in Europe 260
15.5 Frequency of cyclogenesis in the Mediterranean Basin 262
15.6 Main routes followed by cyclonic disturbances in the Mediterranean region 264
15.7 Temperature and rainfall contrasts between the east and west sides of the Mediterranean peninsula 265
15.8 Average number of days with rainfall in summer in the Mediterranean Basin 269
15.9 Average number of days with rainfall in winter in the Mediterranean Basin 270
15.10 Average number of days with rainfall during the entire year in the Mediterranean Basin 270
15.11 Showing the proportion of the annual rainfall received during the winter season in the Mediterranean Basin 272
15.12 Showing the proportion of the annual rainfall received during the summer season in the Mediterranean Basin 272
15.13 Showing seasons of maximum rainfall in the Mediterranean Basin 273
15.14 Showing those parts of the Mediterranean Basin having an annual march of rainfall typical of Mediterranean climates 274
15.15 Showing those parts of the Mediterranean Basin having a complicated annual march of rainfall 274
15.16 Showing the approximate southern limits of, and the area covered by, cyclonic winter rains in southwestern Asia 276
15.17 Showing the northern limits of, and the area covered by, summer rains associated with the southwest monsoon in southwestern Asia 278
16.1 Isolines of equal continentality for Anglo-America 284
16.2 Interdiurnal variability of daily minima in January for Anglo-America 285
16.3 Interdiurnal variability of daily minima in July for Anglo-America 286
16.4 Standard deviation of mean monthly temperatures for January for Anglo-America 287
16.5 Standard deviation of mean monthly temperatures for July for Anglo-America 287
16.6 Isolines of annual rainfall which show the general location of the Canadian Dry Belt in Saskatchewan and Alberta 290
16.7 The grasslands of Anglo-America east of the Rocky Mountains, showing the location of the "Prairie Wedge" 292
16.8 Average number of months per year with a mean transport of air from the eastern base of the Rocky Mountains 294
16.9 Showing the location of areas in the United States with a preponderance of night rainfall during April–September, inclusive 295
17.1 Precipitation regions of Anglo-America based on characteristics of the annual march of precipitation 298
17.2 Season of primary-maximum rainfall in the western United States 299
17.3 Generalized isohyets showing the location of areas of atypical modest annual precipitation in the northern Puget Sound region 303
17.4 Annual march of precipitation, San Francisco, California 303
17.5 Annual march of precipitation at a station located at the southern edge of Puget Sound 304
17.6 Annual march of precipitation at a station located on the southern side of the Strait of Juan de Fuca 304
17.7 Annual march of precipitation at a station located at about 50°N on the coast of British Columbia, Canada 304
17.8 Annual march of precipitation at a coastal station in Alaska at about 58°N 305
17.9 Annual march of precipitation at a desert station in southern Arizona 306
17.10 Contrasting mean air-flow patterns at 5,000 m for June and July over the southwestern United States 307
17.11 Annual march of precipitation in Salt Lake City, Utah 308
17.12 Annual march of precipitation in Redmond, Oregon 308

17.13 Annual march of precipitation in Pomeroy, Washington 308
17.14 Annual march of precipitation in Cranbrook, British Columbia 308
17.15 Change in precipitation amounts from June to July 309
17.16 Annual march of precipitation in Dease Lake, British Columbia 310
18.1 Annual march of precipitation in Albuquerque, New Mexico 312
18.2 Annual march of precipitation in Colorado Springs, Colorado 312
18.3 Annual march of precipitation in Lubbock, Texas 312
18.4 Average summer isentropic chart for the United States 313
18.5 Percentage of total annual precipitation occurring during summer in the United States 314
18.6 Average number of thunderstorms within the United States in August 315
18.7 Annual march of precipitation in Omaha, Nebraska; Bismarck, North Dakota; and Butte, Montana 316
18.8 That part of the mid-North American region where June's precipitation exceeds July's 316
18.9 Annual march of precipitation in Madison, Wisconsin; Burlington, Iowa; Grantsburg, Wisconsin; Chapleau, Ontario; and East Lansing, Michigan 317
18.10 Mean 7-day composite precipitation profile for 5 stations in southern Wisconsin 319
18.11 Precipitation patterns in the vicinity of Lake Superior and Lake Michigan during February 1948 322
18.12 Annual march of precipitation in Churchill, Manitoba, and Fort Vermillion, Alberta 324
19.1 Annual march of rainfall in Indianapolis, Indiana; Sandusky, Ohio; and Huntington, West Virginia 327
19.2 Showing the location of an extensive region in the subtropical southeastern United States where winter precipitation exceeds that of summer 328
19.3 Showing the location of an area within the subtropical southeastern United States where less than 50 percent of the annual precipitation occurs in the warmer half year, April–September, inclusive 328
19.4 Annual march of precipitation in Birmingham, Alabama; Chattanooga, Tennessee; Alexandria, Louisiana; and Memphis, Tennessee 329
19.5 Annual march of precipitation in Little Rock, Arkansas 330
19.6 Annual march of precipitation in Forth Smith, Arkansas 330
19.7 Standard deviation of the daily value of the south-north component of the geostrophic wind in m/sec at the 500 mb level in the eastern United States 332
19.8 Annual march of precipitation in Pensacola, Florida 334
19.9 Annual march of precipitation in Orlando, Florida 334
19.10 Annual march of precipitation in Lumberton, North Carolina 336
19.11 Annual march of precipitation in Trenton, New Jersey 336
19.12 Showing the location of areas in the Atlantic seaboard and Great Lakes regions where winter precipitation exceeds that of summer 337
19.13 Annual march of precipitation in New Haven, Connecticut, and London, Ontario 338
19.14 Annual march of precipitation in Annapolis Royal, Nova Scotia 338
19.15 Annual march of precipitation in Botwood, Newfoundland 339
19.16 Annual march of precipitation in Beauceville, Quebec 339
I.1 Köppen system of world climates 346–47

Preface to the Second Edition

In the nearly two decades that have elapsed since the initial publication of this book, new information as well as new climatic data have become available concerning some of the earth's unusual climates. There has been an attempt to incorporate these more recent concepts into the textual materials of the 1980 edition. For those regional climates where the recent literature yields little or nothing new, the original text has not been greatly altered. And this remains true even though the author may have developed some doubts concerning the complete validity of the earlier explanation.

The more extensive modifications of the earlier edition are to be found in the discussion of the following regions: Pacific Colombia, the Chilean-Peruvian desert, the drought region of Northeast Brazil, the southern Caribbean dry region, the dry-subhumid belt of the eastern and central parts of the equatorial Pacific, equatorial East Africa, the Indian subcontinent, East Asia, and western Anglo-America.

I wish to acknowledge the expertise of the Cartographic Laboratory at the University of Wisconsin-Madison; their help in refining the original maps and figures and in the preparation of additional cartography was indispensable.

THE EARTH'S PROBLEM CLIMATES
Second Edition

Introductory Statement

The World Pattern of Climates

As a consequence of the way the earth revolves about the sun and rotates on its axis, the low latitudes or tropics receive more solar energy than do the middle and higher latitudes. Energy is also lost from the earth to space by radiation processes at different rates in the different latitudes. But in the mean, the low latitudes receive more energy than they lose, while the reverse is the case in middle and higher latitudes. To correct this energy imbalance there is required a large-scale transfer of heat poleward, which is accomplished through the agencies of atmospheric and oceanic circulations. Earth rotation from west to east, and the frictional effects of the earth's surface upon air flow, cause these circulations to be relatively complex. Nevertheless, since solar energy and atmospheric and oceanic circulations are distributed in an organized fashion over the earth, it follows that these controls, operating jointly and in unison, will produce recognizable world patterns of temperature and precipitation distribution, the two most important climatic elements.

Were the earth's surface homogeneous (either land or water) and lacking in terrain irregularities, it may be presumed that atmospheric pressure, winds, temperature, and precipitation would be arranged in zonal or east-west belts. An approach to such a zonal arrangement actually prevails in the more uniform Southern Hemisphere. The fact that the earth's surface is not uniform, but instead is composed of continents and oceans, whose heating and frictional effects upon winds are quite in contrast, and that the lands have mountain ranges which are serious obstacles to horizontal air flow, result in a breakdown of the zonal belts of pressure and winds into cells, a fact which greatly complicates temperature and rainfall distributions.

In spite of these complications imposed by a nonhomogeneous earth's surface, and though the great land masses have a variety of dimensions, shapes, and terrain features, there is, nevertheless, a recognizable world pattern in each of the climatic elements, and, as a consequence, of the climatic types which are composites of these elements. The operation of the great planetary controls is sufficiently regular and dependable so that it tends to produce relatively similar climatic conditions in similar latitudinal and continental locations on most of the great land masses, even though the latter are separated by great linear distances. It is this repetition of climatic conditions that makes it possible to classify the numerous regional, and multitudinous local, climates into relatively few types.

Since this book is intended for a professional audience, the author considers it unnecessary to present here a detailed exposition of how the great planetary controls of solar energy distribution and the general circulation of the atmosphere and ocean water, modified by the differential effects of continents and oceans, act to produce a generalized world pattern of climatic distribution. This theme of the world pattern provides the core content for most introductory courses on climate. A discussion of the developmental aspects of the world pattern, supplemented by diagrams and maps, is readily accessible in a number of works written on climatic classification.

Since world temperature distribution is essentially zonal in character, showing a general decrease between tropical and polar latitudes, it presents little problem in the development of a rudimentary world pattern of climates. Precipitation distribution is more complex. Figures A and B adjacent purport to present in schematic form the distribution of the two main precipitation features, annual amount and seasonal concentration, as they might appear on a hypothetical continent of fairly low and uniform elevation extending from about 80°N to 60°S. Both are greatly simplified. Figure C is intended to represent the distribution on a similar continent of the main types of climate, involving *both* temperature and precipitation. Figure C in turn should be compared with the actual distribution of the earth's climatic types and regions as represented on the map located inside the front cover. This map represents a greatly modified version of the Köppen system of climates. A brief outline and map of the standard Köppen system

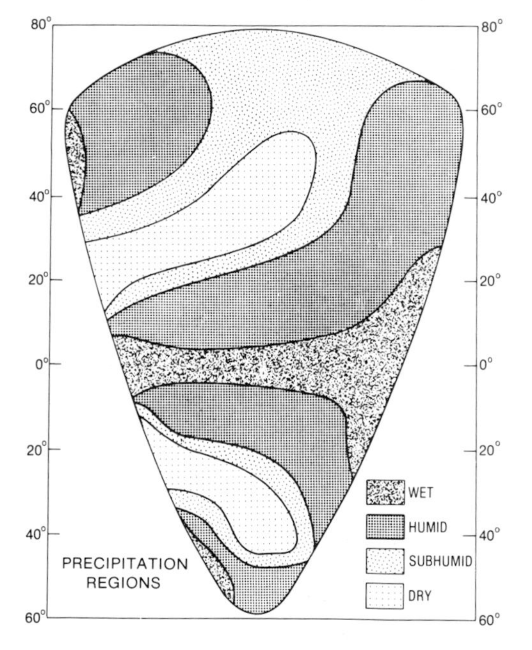

Fig. A. Illustrating the general pattern of *total annual precipitation* distribution as it likely would develop on a hypothetical continent of relatively low and uniform elevation. (After Thornthwaite, 1933, and others.)

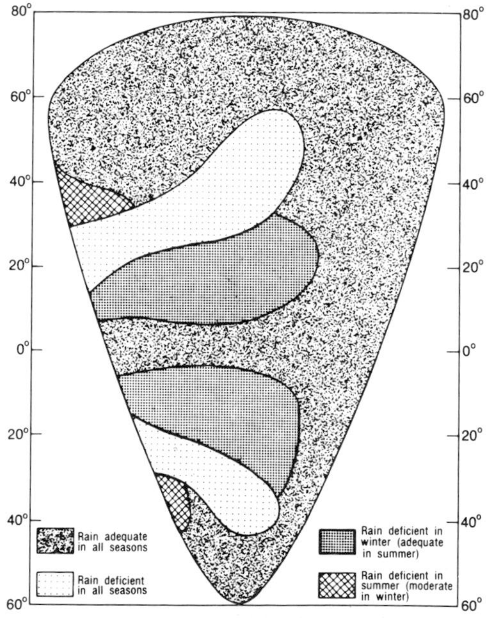

Fig. B. Illustrating the general distribution pattern of *seasonal precipitation concentration* as it likely would develop on a hypothetical continent of relatively low and uniform elevation. (After Thornthwaite, 1933, and others.)

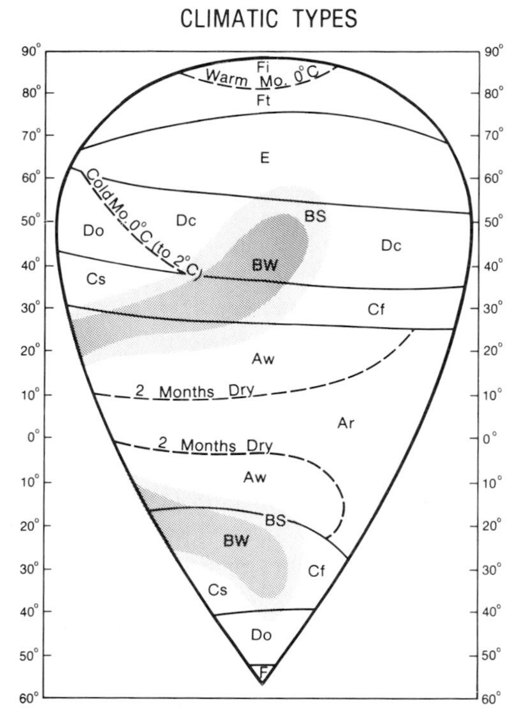

Fig. C. Illustrating the arrangement of the principal *types of climate* as it likely would develop on a hypothetical continent of relatively low and uniform elevation.

of world climates is to be found in the appendix.

Nature of the Book

But, while each of the continents exhibits much in its climatic arrangement which is in harmony with the general world pattern as determined chiefly by the great planetary controls, and some much more than others, there are at the same time numerous departures from the expected arrangement or intensity. These departures represent the unusual, the atypical; they are the climatic anomalies. In many instances they are the result of regional or local controls. Some of them, no doubt, result from geographical uniquenesses in the land areas themselves, and are associated with shape and size, terrain height and alignment, trend of coastline in relation to air flow, and the like. Others appear to be related to exceptional and local features of the atmospheric or oceanic circulations. It is very likely that some climates which at present appear to be anomalous will, upon a more complete understanding of atmospheric controls, lose much of their unusual or enigmatic character and seem to be more a part of the normal climatic pattern.

A major objective of the present book is to make an inventory of the more important of the earth's problem climates and to offer a brief and precise description of their characteristics. One of the most common types of anomalous climates, and also one of the most difficult to explain, is the localized dry area existing in a type of location normally characterized by humid conditions. Many of these areas are coastal in situation so that the supply of water vapor should be abundant. Still another type of climatic anomaly appears in the form of a seasonal or diurnal deficiency or excess of a particular climatic element, most commonly precipitation. Or the deviation may be one of intensity in heat or cold, wetness or dryness.

A further objective of this book, and a much more difficult one to attain, is to suggest explanations for at least some of the problem climates. A number of them have received serious research attention from climatologists but many still remain conundrums. Most climatological treatises written by geographers are heavily weighted on the side of description, but the practice can scarcely be defended. Certainly the rapid advances in atmospheric science during the past few decades have provided the climatologist with more effective tools for solving the climatic enigmas than was the case previously. Moreover, the meteorological literature is rich in findings and concepts which are applicable to the climatologist's regional problems. Obviously the author cannot have investigated personally the variety of problem climates here made note of. Much of what has been included as explanation is the result of gleanings from published sources, the original purpose of which may not have been climatological, but whose contents have here been made to serve a climatological end. The result of such a method is not so much the establishment of firm conclusions concerning the origin of the numerous regional climatic aberrations as it is to provide a threshold for subsequent and more focused investigations. While the author disavows any attitude of iconoclasm, it has been his intent to bring to bear upon climatic interpretation certain useful new concepts, the result of which scarcely can help weakening some long-established notions as to climatic genesis. In the present book there has been an attempt to make genesis an integral part of the climatic analysis, to the end that the treatment shall be dynamic, with its roots in the physical processes of the atmosphere. Unfortunately, in the present state of climatology, much of the treatment must of necessity be qualitative and speculative.

But while the description and explanation of the climatic anomalies are the principal subjects of the present book, it does not entirely neglect the more usual climates. In the literature, ordinarily these have been described in terms of averages of the climatic elements, while the weather types, which are the more realistic units comprising climate, have been slighted. An attempt is here made to inject into the descriptions and explanations of the more usual, as well as of the problem, climates a treatment of atmospheric perturbations, large-scale synoptic situations (*Grosswetterlagen*), and weather types whose combined effects produce a climate. It is especially in the tropics, where the weather element is most commonly neglected, that atmospheric disturbances are emphasized. However, a comprehensive and quantitative synoptic type of climatology is not at present possible of realization, for much remains to be discovered about the kinds and distribution of the atmospheric disturbances which affect weather in different parts of the earth, and likewise about the distribution of the weather elements associated with these perturbations. Moreover, every classification of weather types is open to question since there is no single parameter which delineates the totality of weather conditions. As a consequence, statistical averages of the elements are by no means abandoned as a method of climatic description, but on the other hand, this method is supplemented wherever possible with a treatment of the various types of synoptic situations and their perturbations which comprise climate.

No short title adequately describes the content of the present treatise since it attempts to do more than one thing. While its focus is the regional climates of the earth's larger land areas, it is not a general regional climatology in the usual meaning of that title. Rather it is a specialized treatment of climatic differentiation which emphasizes (1) the earth's problem climates, (2) the perturbation element in climate, and (3) genesis, or the dynamic processes in which climatic differentiation is rooted. It is designed to meet the needs of those interested in the professional aspects of climate rather than of laymen. A methodical description of all the earth's climates is not attempted, for many areas are climatically so normal or usual that they require little comment in a book which professes to emphasize the exceptional. This endeavor is meant to supplement the existing books on regional climates and not to cover the same ground. It is assumed, also, that the reader is already acquainted with the rudiments of meteorology and of physical climatology, and that he has been introduced to climatic classification and is familiar with the world pattern of climatic types. As descriptive tools, the modified form of the Köppen classification as outlined in the author's book *Introduction to Climate* and Thornthwaite's basic elements of classification are both employed. On occasions the standard Köppen system is referred to (see appendix).

It has not been considered essential that the organization of the materials on each of the major land areas should always be similar. The method of approach and treatment has been determined in part by what was thought to be most appropriate, considering the kinds of climatic problems involved. The inclusions and exclusions were likewise strongly influenced by the fact that our knowledge of climatic distribution and atmospheric processes varies greatly for different parts of the earth.

It is not presumed that this treatise will provide all of the answers to the conundrums posed by the earth's anomalous climates. Without doubt some problem areas have been unintentionally overlooked. Moreover, the reasons for many of the climatic enigmas will remain inadequately explained. Nevertheless, the book should cause to stand out in bold relief some of the regional climatic problems which demand further investigation. This in turn may stimulate regional climatic research. In addition, it is hoped that the book may serve to reorient climatic research by professional geographers toward a more dynamic and genetic approach, made possible by a better use of meteorological techniques.

It is the custom for an author to acknowledge his debt to those who have contributed to the development of his publication. While extremely conscious of the indebtedness, unhappily I am unable to make this awareness specific by naming the many persons who have given assistance. Particular gratitude, however, is felt to those men in the meteorological offices of Pan Air do Brasil in Rio de Janeiro and Recife, and in the government weather bureaus in Rio de Janeiro, at Accra in Ghana, Leopoldville (Kinshasa) in the Belgian Congo (Zaire), Nairobi in East Africa, and Pretoria in the Union of South Africa who took time from their crowded work schedules to discuss with me regional climatic problems. To the meteorologists at Wisconsin, I have gone on numerous occasions for advice on technical matters involving air physics. Financial aid for research has been generously contributed by the Geography Branch of the Office of Naval Research, Washington, D. C., and the University of Wisconsin.

PART I
Latin America

CHAPTER 1

Pacific South America: Colombia and Ecuador

Atmospheric Circulation Features of South America

The topography of the Southern Hemisphere strongly favors a vigorous zonal circulation of the atmosphere. Antarctica, although excessively frigid, has a convex surface profile which facilitates a downslope seaward movement of the cold air almost as rapidly as it forms, so that there is no build-up into well-developed cold anticyclones such as are generated over subarctic Eurasia and North America in winter. As a result the middle and lower latitudes of the Southern Hemisphere are not invaded by cold anticyclonic surges of *cP* air of a magnitude and intensity comparable to those of the Northern Hemisphere. Weak anticyclones that do develop along the periphery of parts of Antarctica tend to be carried rapidly downstream in the strong zonal westerly flow, while the few stronger ones that may penetrate the westerlies to a greater or less degree are greatly modified over a broad ocean surface before reaching the continental areas.

Only South America extends far enough poleward, and offers sufficiently high terrain barriers, to produce dynamic and thermal effects of a magnitude to disturb greatly the strong zonal circulation. The minor effects produced by southern Africa and by Australia-New Zealand are limited chiefly to the immediate vicinities of those regions and these effects are quickly damped downstream. Consequently, it is in the South American-South Atlantic sector, where the zonal circulation is most disrupted, that the greatest meridional exchange of air between high and low latitudes in the Southern Hemisphere occurs. Surges of sea-modified Antarctic air do move northward in the western South Atlantic and penetrate well into the tropics, causing disturbed weather along the east coast of South America and in the interior as well. Such cold-front disturbances provide an important element in the weather in that part of South America which lies east of the Andes and south of the equator. Throughout most of the Southern Hemisphere, however, long unbroken fronts, oriented in an east-west direction, reflect the strong and relatively uninterrupted zonal circulation.

Over the fairly homogeneous Southern Hemisphere the subtropical high-pressure system is more like a continuous ridge or belt, and less fragmented into individual cells, than is true in the Northern Hemisphere where large land masses with their obstructing cordillera exist. Moreover, the Southern-Hemisphere subtropical belt is, over extensive longitudes, essentially a statistical average of moving high-pressure systems carried eastward by the vigorous westerly circulation, while permanent cells, fixed in position, are less obvious. Here again the South American area is the exception, for the high Andes so obstruct and divert the zonal flow that relatively permanent anticyclonic cells are characteristic of both the South Pacific and South Atlantic on either side of subtropical South America. The blunt eastern end of the South Pacific cell terminates abruptly at the mountainous coastline in northern Chile and in Peru where it remains longitudinally fixed in its position. As a consequence, the circulation, both atmospheric and oceanic, around the eastern end of this cell is strong and remarkably persistent, so that the aridifying effects of the anticyclone are well developed west of the Andes. The fact that the subtropical belt of high pressure is not continuous across South America, but instead there exists a continental corridor between the Pacific and Atlantic cells, permitting a relatively free meridional exchange of tropical and polar air, has remarkable climatic consequences. Unlike most parts of tropical-subtropical southern Africa and Australia south of about 15° or 20°S, and broad Africa north of the equator in similar latitudes, which are dry, comparable latitudes in South America east of the Andes are, for the most part, humid, and extensive deserts are lacking. It is significant in this respect that the principal area of cyclogenesis in the Southern Hemisphere is located over the ocean east of Patagonia between about 40° and 60°W.

By contrast, the western end of the South Atlantic high is not as positionally stable with respect to South America as is its Pacific counterpart, for during the low-sun period it pushes westward deep into interior Brazil, while at high sun it retreats seaward. Nevertheless, because of the far eastward projection of the "hump" of Brazil, that part of the continent lies closer to the strongly subsiding center of a subtropical cell even in summer than is true of comparable latitudes in eastern North America which lie 4,000 to 5,000 km west of Cape São Roque in easternmost Brazil. Thus, the temperature inversion base is lower in the western tropical South Atlantic than in the western tropical North Atlantic, resulting in important climatic contrasts between the two regions.

Because of the more vigorous westerly circulation in the Southern Hemisphere, stemming from the powerful Antarctic cold source and reduced continental friction, the longitudinal axes of the subtropical highs in the eastern South Pacific and western South Atlantic lie closer to the equator than do their counterparts in the Northern Hemisphere. Accordingly, over the oceans, the equatorial pressure trough and its wind convergence (*ITC*) are asymmetrically located to the north of the geographic equator at all times of the year. Even along the coast of Brazil the *ITC* never has a seasonal position more than three degrees south of the geographic equator, and along the Pacific side of the continent it scarcely reaches the equator. This hemispheric asymmetry in atmospheric circulations (also oceanic) is reflected in contrasting rainfall patterns to the north and south of the equator and in the general northward displacement of climatic zones.

For a continent that reaches from about 10°N to 55°S, with its greatest breadth in tropical latitudes and rapidly tapering in width south of 15° or 20°, South America exhibits a climatic arrangement which in most respects conforms to that represented on a hypothetical continent (Figs. A, B, and C). Thus, along the Pacific coast west of the Andes the latitudinal succession of climates is precisely what might be expected from the operation of the great planetary controls. Accordingly, *Ar* climate in western Colombia is replaced by *Aw* in Ecuador, *B* climates prevail in Peru and northern Chile, *Cs* in central Chile, and *Do* still farther south. East of the Andes as well there is much in the major climatic arrangement which follows the expected pattern. Accordingly, an extensive *Ar-Am* area in the Amazon valley is flanked by *Aw* climates to the north and south, while farther poleward to the south *Aw* merges into *Cf (w)* (see front endpaper map and appendix).

Still, there are important departures from the standard pattern which need to be noted and accounted for. Examples of these are the excessive wetness of the Pacific side of Colombia; the unusual equatorward displacement and intense aridity of the Chilean-Peruvian dry climates; the low-sun rainfall maximum (*As*) along the Brazilian coast south of Cape São Roque; the drought area in northeastern Brazil inland from Cape São Roque; the dry littoral in northern Venezuela-Colombia, and arid Patagonia. The above are illustrative only, and are not meant to be a complete listing of the anomalous climates of South America.

South America West of the Andes

It has been pointed out above that the general latitudinal *succession* of climates in Pacific South America west of the Andes is approximately what might be expected. But while the north-south sequence of climates appears normal, there are, nevertheless, important instances where climates depart from their expected latitudinal or continental locations, or where the intensity of one or more of the climatic elements presents an unusual condition.

One such major aberration is the fact that all of the tropical climates appear to be displaced northward of their usual latitudinal positions along the Pacific littoral of South America. Striking is the unusually wide latitudinal spread (25° to 30°) of dry climates to the south of the equator in northern Chile and Peru, so that drought conditions are characteristic of the coastal strip northward almost to the equator. By contrast, dry climates to the north of the equator along the Pacific littoral of Mexico reach southward only to about 23°N. Just north of the equator in coastal Ecuador precipitation increases very rapidly, so that the latitudinal rainfall gradient is one of the steepest of the earth near sea level. The consequence is that a very narrow transition belt of tropical wet-and-dry climate (*As-Aw*) some 2° wide, and mostly in the Northern Hemisphere, is wedged in between the equatorward-displaced dry climates on the south, and the excessively wet *Ar-Am* climates which begin at about the Ecuador-Colombia boundary, or 1½°N. Rainy *Ar* climate continues to prevail northward to somewhat beyond the Panama-Colombia boundary (7° or 8° N). This unusual latitudinal positioning of climates in tropical Pacific South America, in marked contrast to that in Pacific North America, seems to point to exceptionally potent drought-making controls south of the equator, which result in a northward displacement of all of the tropical climates.

The Pacific Lowlands and Slopes of Colombia

The narrow crescent-shaped region of Pacific Colombia, terminated on its land side by the Cordillera Occidental, is the renowned first contender for the earth's rainiest locality. An added anomaly is the fact that some three-quarters of this region's annual rainfall occurs at night, which is unusual for tropical latitudes. The region mainly consists of a low coastal plain some 60–150 km in breadth, which is bordered seaward by a range of low coastal hills and landward by the west-facing lower slopes of the Cordillera Occidental.

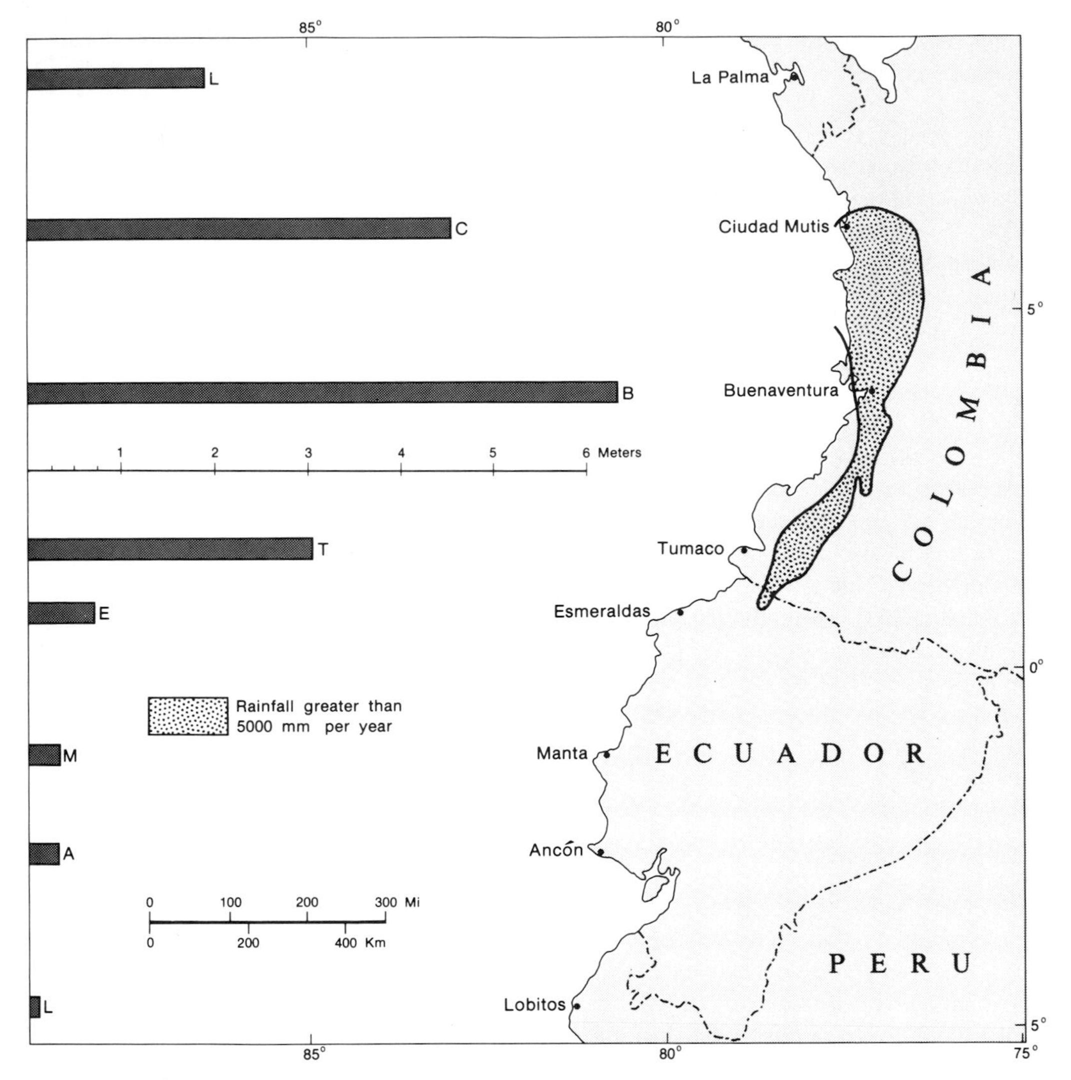

Fig. 1.1. Average annual rainfall amounts for eight stations in western Colombia. (After West, 1957; Snow, 1976; and others.)

Based upon data covering the period 1931 to 1960 the average annual rainfall of Colombia's Pacific coast crescent is about 5,500 mm/yr. The amounts range from nearly 3,000 mm/yr at the northern and southern extremities of the region to over 10,000 mm/yr in the central part near the headwaters of the Atrato River (Fig. 1.1). At Lloró (5°31′N), the mean for 1952 to 1954 inclusive was 13,473 mm. At Quibdó (5°42′N), probably the rainiest locality in the Americas, in 1936 19,839 mm were recorded (Snow, 1976).

In general the rainfall distribution pattern is one of belts running parallel to the Cordillera Occidental, with the band of maximum amounts located over the inland hilly portion of the lowland near the base of the mountain slope. Here the elevation is only 50–250 m. Night rain predominates. All gradients out-

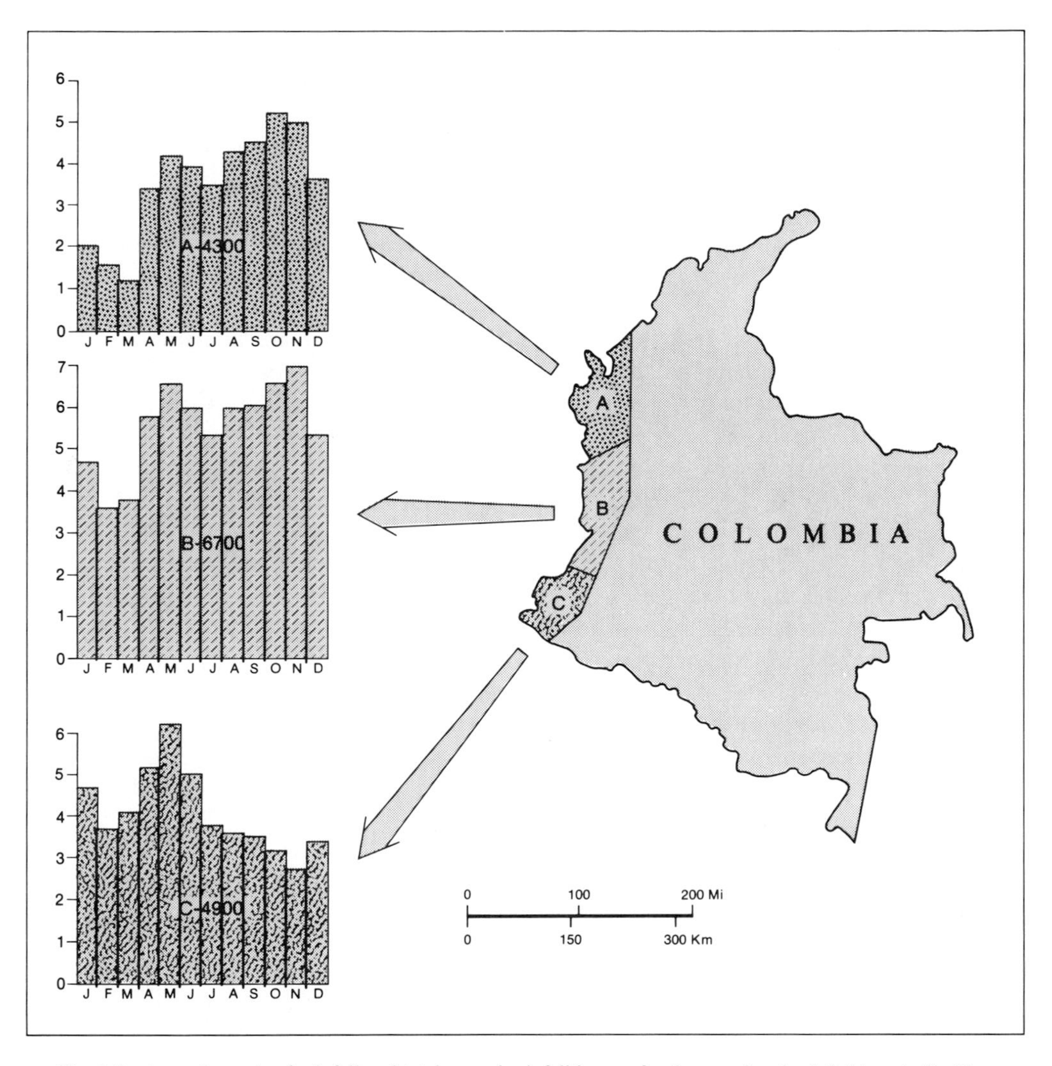

Fig. 1.2. Annual march of rainfall and total annual rainfall in mm for three regional subdivisions in Pacific Colombia. (After Snow, 1976.)

ward from the band-maximum (upward along the mountain slopes, over the lowland toward the coast, and coast-parallel over the ocean) are steep. Seaward from the band-maximum the rate of decrease is about 50–100 mm/km. The coast-parallel gradient is not so large in Colombia's central part, but steeper at the crescent's northern and southern extremities, and more so in Ecuador than in Colombia. Along the mountain's western slope the vertical rate of rainfall decrease is typically 5 mm/m between elevations of 500 and 2,000 m (Snow, 1976). So while the band of maximum is at a low elevation, normally <250 m, it is clear that the mountain slopes in general also have abundant rainfall.

Annual March of Rainfall.—Within the super-wet crescent annual rainfall does not exhibit strong seasonality; typically most months have at least 350 mm. It is mainly at the northern extremity where rainfall drops to less than 200 mm during two months, February and March, that seasonality is marked (Fig. 1.2). A fairly consistent but modest double maximum is evident throughout most of the crescent. But beyond, to the north and south in Panama and Ecuador, seasonality of rainfall is much stronger, as might be expected.

Figure 1.3, taken from Alpert (1948), shows the seasonality of rainfall by means of the climatic arrangement in the equatorial eastern Pacific, *employing Köppen's climatic system of classification*. Thus there is a broad midbelt of heavy rainfall, which along the Colombia coast spreads over some 5° of latitude (2°N to 7°N). In it precipitation is well distributed throughout the year (*Afw″*). Even here there is evidence of two wetter periods and two that are somewhat less wet, these coinciding with the slight seasonal shifts of the *ITC* and its disturbances. Toward the southern margins of the *Af* belt, however, the *w″* symbol changes to *s″*, which signifies that there the stronger of the two minima is in summer-fall and the weaker one in winter-spring. This is evidence that the climatic equator, on the average, is asymmetrically located some 3° north of the geographic equator and that a Southern Hemisphere rainfall regime prevails along the coasts of northern Ecuador and southern Colombia, even though these areas are in the Northern Hemisphere (see appendix).

Flanking the *Af* climates both to the north and south are narrow belts of *Am*, indicating a more pronounced dry season, but with double maxima and minima still prevailing. The *Am* belt to the north significantly bears the *w″* symbol, while that to the south has *s″*, reflecting the Southern Hemisphere type of annual rainfall variation. As rainfall declines in amount still farther to the north and south, and the dry season becomes increasingly emphatic, the *Am* changes to *Aw″* on the north and to *As′* to the south. The latter symbol indicates a single dry season in summer. Where this climatic belt extends into coastal Ecuador south of the equator the symbol changes to *Aw′*. It bears re-emphasizing that the *s* symbol in the tropics is relatively rare since there is a strong tendency for rainfall to follow the latitudinal shifting of sun and *ITC*. Hence one singles out the tropical climates of coastal southern Colombia and northern Ecuador as unusual, in that their wetter periods are winter-spring or low sun, and their drier periods summer-fall or high sun. It is evidence that the strong drought-making controls located south of the equator are able to extend their influence north of the equator at the time of the maximum northward migration of the sun.

The number of days with rain within the crescent is large, characteristically over 200 per year. Close to the Pacific coast this figure increases to at least 300 days, which is

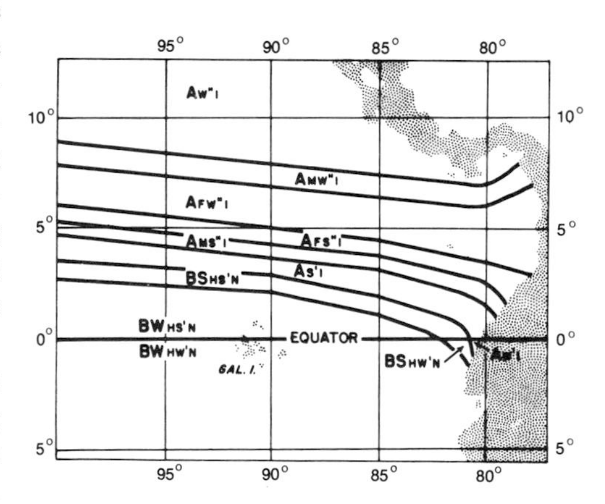

Fig. 1.3. Climatic arrangement in the eastern equatorial Pacific employing the Köppen scheme of classification. (See appendix.) Note that the climatic equator lies to the north of the geographic equator, so that the *s* (summer dry) symbol is conspicuous in southern Colombia and northern Ecuador. (After Alpert, 1948.)

more numerous than in the band-maximum rainfall farther inland in the foothill zone (see Table 1.1).

The Diurnal March of Rainfall.—In the low latitudes rainfall characteristically follows the sun not only in its seasonal cycle, but likewise in its diurnal course. Hence, Knoch's (1930) statement that close to 80 percent of the coastal rain falls at night, between the hours 1900 and 0700, indicates a major departure from the rule (Fig. 1.4). Knoch is also responsible for the statement that the lesser day rains along the coast are concentrated between the hours 0700–0900 and 1700–1900. At Andagoya, where nearly 79 percent of the rain falls in the period between the 19th and 7th hours, on the average there are 303 rainy days per year, 158 with daytime rains and 277 with night rains. Murphy (1926; 1939) describes the night rains as relatively continuous and states, "At night there were few hours without steady rain, *xxxx*." By contrast most of the day rains "*xxx* come as heavy showers or cloudbursts, and a constant succession of black squalls marched up the coast." The day rainfall he describes as much more violent, but less continuous, than that at night and associated with spectacular thunderstorms.

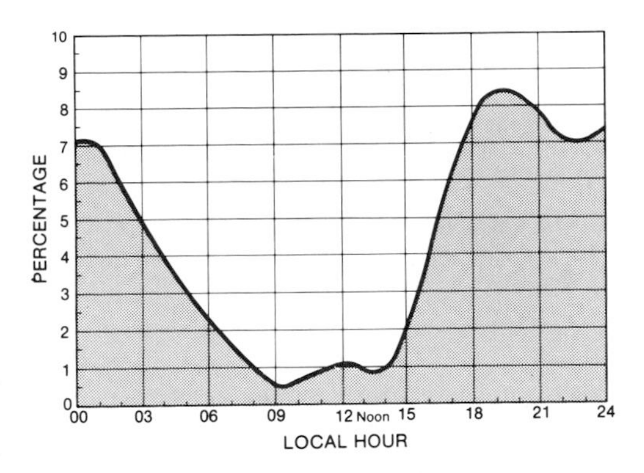

Fig. 1.4. Hourly distribution of rainfall at Quibdo, Pacific Colombia. (After Snow, 1976.)

Robert West (1957) differentiates three parallel belts of contrasting diurnal rainfall variation, representing a temporal progression from the slopes to the seacost (Fig. 1.5). In Zone 1, comprising the western slopes of the cordillera, the heaviest rains occur in the afternoon, with light rains continuing through the night. Farther seaward in the interior lowlands and the lower slope lands (Zone 2) the heaviest rains are concentrated in the evening hours, roughly 1900–2400, with occasional heavy showers continuing throughout the night and ending in drizzle rains falling from a gray overcast by 0700 or 0800 in the morning. The coastal rains (Zone 3) are very largely nocturnal, often beginning as heavy downpours after midnight and ending in drizzles in midmorning. Thick cumuli overlie the coastal lowlands at night, but during the midday hours there is usually some clear sky.

Table 1.1. Rainfall distribution at Andagoya in western Colombia, 76 m elevation, in the Choco area, 5°4′N, 76°55′W. Based on 12 years of observation.

Amounts of precipitation (in mm)												
J	F	M	A	M	J	J	A	S	O	N	D	Year
601	553	470	671	606	647	589	643	667	540	595	512	7089
Number of days with precipitation												
J	F	M	A	M	J	J	A	S	O	N	D	Year
26	21	23	25	26	24	26	27	27	24	27	27	303

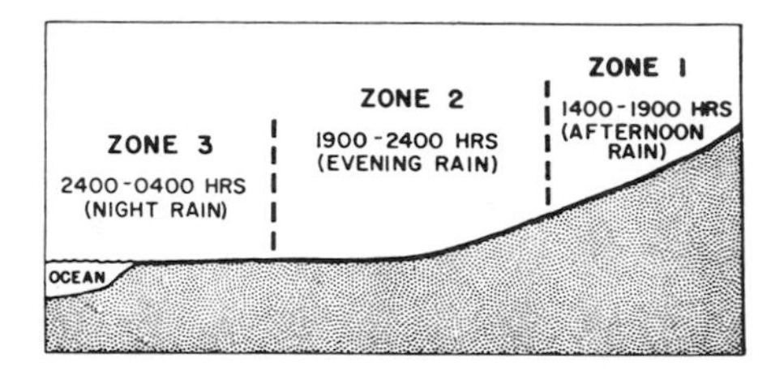

Fig. 1.5. Belted arrangement of the diurnal occurrence of heaviest rainfall in Pacific Colombia. Night rains greatly predominate along the coast, while day rains prevail interior along the mountain slopes. (After West, 1957.)

The origin of this belted arrangement of the diurnal march of rainfall has not been adequately explained. It is possible that it is related to land- and sea-breeze phenomena in which the developing cumuli are carried inland by the sea breeze during the warmer hours, to drift seaward again during the evening and night hours as the sea breeze weakens and the land breeze sets in. Another suggestion relates the night rains along mountainous tropical coasts to local convergence caused by katabatic winds descending the highland slopes (Brückner, 1951).

Causes for the Excessive Rainfall

To be sure, the equatorial latitudes are on the average the rainiest of the earth, but the usual reasons given for this phenomenon, such as high temperature and humidity, potential instability, and convergent atmospheric circulation, are scarcely adequate for interpreting the super wetness of the Colombia crescent. Admittedly its origin is still somewhat obscure. But almost certainly the obstruction provided by the Cordillera Occidental to the prevailing onshore westerly air flow must be a factor. Here the inflow of equatorial air is boxed in, somewhat as the southwest monsoon from off the Bay of Bengal is boxed in by the highlands in super-wet northeastern India. In addition, evidence from satellite cloud pictures and from ship observations indicates that the equatorial pressure trough lingers continuously in that sector of the eastern Pacific situated between the equator and the Gulf of Panama. Likewise, in or near the equatorial trough one can observe that important feature of atmospheric convergence, the *ITC*, that shifts somewhat north and south following the course of the sun and is responsible for the weak double maximum in the annual rainfall profiles of several, but not all, weather stations.

Snow (1976) points out that the large number of rain days and its small annual variation suggests that the causes of the excessive rainfall may have some association with the semidiurnal pressure oscillation, which in turn is linked to the semidiurnal convergence-divergence cycle of the tropics. Thereby convection during the warmer midday hours is suppressed, while it is stimulated during dawn and sunset periods. Still, in most of the tropics this daily pressure oscillation of the atmosphere is generally ineffective in rain production. Why should it be so effective in Pacific Colombia?

Probably the decisive factor in producing the extreme wetness of the Colombia crescent is the wind, but regrettably, station data for that particular climatic element are seriously deficient; one must turn to ship observations and travelers' accounts. It appears from these sources (not wholly reliable) that a southwesterly circulation, in the nature of equatorial westerlies, prevails over the ocean near the southern half of the crescent almost year round. Over the ocean to the north of the equator, the evidence suggests that the prevailing circulation is from the northwest, but closer to the coast both wind and water appear to move from the southwest. Thus, potentially unstable southwesterlies appear to prevail near the coast. Evidence indicates a low-level convergence exists over the Pacific coast of Colombia during much of the year. Also, over the sea directional or speed convergence is present most of the time. But this all adds up to the conclusion that no complete and definitive answer to the super wetness of the Colombia crescent exists at the present. Can it be that some cause-and-effect relationship exists for the striking fact that on the Pacific side of tropical South America what may be the earth's wettest climate prevails a few degrees north of the equator, while an equivalent distance south of 0° there is the beginning of the earth's driest desert? Here, on either side of the equator, are to be found the earth's wettest and driest parts—in close juxtaposition. The straight-line distance of less than 1,000 km separates the ultimate wet climate near the headwaters of the Atrato River in Colombia from the super-arid Lobitos (only 51 mm of annual rainfall) in northernmost coastal Peru.

The Transitional Climates in Western Ecuador

Speaking of the Pacific lowlands of Ecuador, the naturalist H. B. Guppy (1906) has written, "I suppose there can be no region of the globe where there are so many climatic anomalies as interesting to the meteorologist." Because of the general northward displacement of all climates in western tropical South America, Ecuador, while located astride the equator, occupies the abrupt transition zone between the excessive rainfalls slightly to the north and the intense aridity just to the south. Within a latitudinal spread of only about 5° along the coast, the littoral of northernmost Ecuador (1°–2°N) experiences an annual rainfall of 2,500 mm or more, while a semiarid coastal climate prevails at about 1°S, and at 2° or 3°S, desert conditions with 250 mm of rainfall or even less are the rule. The transitional character of the climate is still further denoted by the fact that this area experiences seasonally, and in modified form, the extreme rainfall conditions flanking it on the north and south, for the seasonal rhythm of alternating wet and dry seasons is striking. The climatic pattern in western Ecuador is further complicated by the fact that there is a rapid climatic change not only in a north-south direction, but likewise at right angles to the seacoast. Thus the cooler, drier coastal climates contrast strikingly with the warmer and wetter climates a few score miles inland.

Instrumental records of rainfall are entirely too few to permit the construction of a satisfactory map of annual precipitation for the Pacific lowlands and slopes of Ecuador. Consequently the few and inadequate records must be supplemented by the landscape observations of field scientists and the reports of travelers if any semblance of precipitation and climatic patterns is to be deduced (Ferdon, 1950). Figure 1.6 represents an attempt to integrate the gleanings from a variety of sources and to combine them into a single map showing generalized rainfall and climatic types. There appear to be no reliable instrumental rainfall records for northwestern Ecuador north of about 1°S. Tumaco in southernmost coastal Colombia, only a short distance from the Ecuador boundary, records 2,750 mm and rainfall increases inland. From the above facts, and also from the observations of plant scientists concerning the rainforest of this northern region, one concludes that there is a continuation of the heavy but declining rainfall of southern Colombia southward into adjacent Ecuador (Svenson, 1946). The decrease southward appears to be much more rapid along the coast than on the lowlands and slopes to the rear, where a meridional belted arrangement is conspicuous. An east-west profile of precipitation at about 2°S, based upon the records of six stations illustrates the increase in rainfall inland from the seacoast (Fig. 1.7).

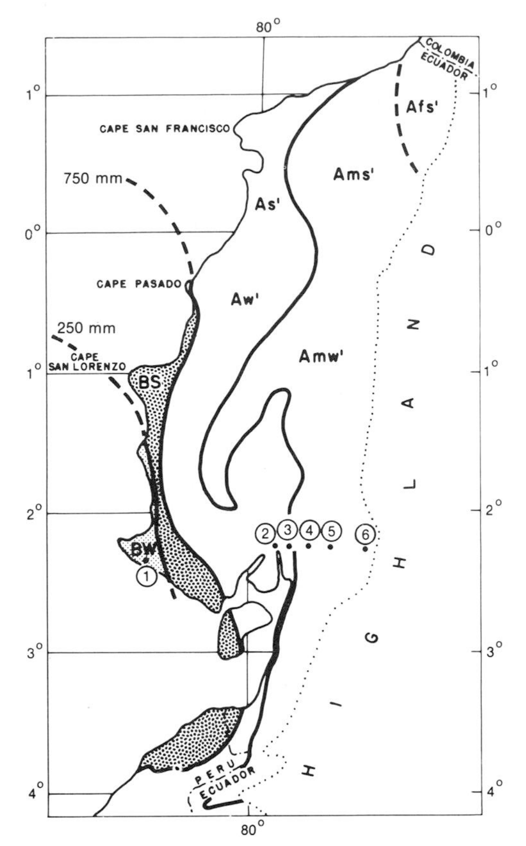

Fig. 1.6. Climatic arrangement (Köppen) in Pacific Ecuador. Circled numerals 1 to 6 across southern Ecuador refer to the location of stations employed in Fig. 1.7. (After Troll, 1930; Alpert, 1948; Ferdon, 1950; and others.)

Among the climatic features which appear to be somewhat unusual for an equatorial latitude are these: the dry but overcast littoral, together with a marked increase in rainfall inland from the coast; the excessively steep rainfall gradient along the littoral in a north-south direction; the strong seasonal variation in precipitation with a well-developed dry season; and the prevalence of a Southern Hemisphere type of annual rainfall variation for several degrees to the north of the geographic equator. Most of these climatic irregularities have their origins in the strength and average northward positioning of the South Pacific anticyclone, features which have been commented upon in earlier sections.

Without doubt the most unusual climatic feature of this equatorial region is the dryness of its coastal margins. The dry littoral in Ecuador appears to be a northward extension, in less extreme form, of the intense drought which prevails farther south in coastal Peru. It is logical to assume, therefore, that the drought controls which operate so strongly and consistently farther south, appear in these equatorial latitudes in somewhat weakened form and also somewhat more seasonally. Precipitation statistics for the dry coastal belt are too meager to be of great use in precisely defining the boundaries of, or describing the regional differences within, the dry area. Consequently recourse

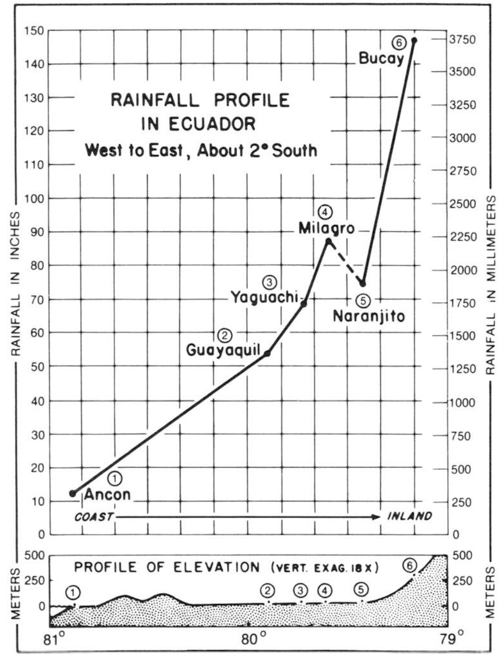

Fig. 1.7. Profile of annual rainfall across western Ecuador from the coast toward the interior at about 2°S. The locations of the stations here used are shown in Fig. 1.6.

must be had to indirect geographical information. During the dry season in coastal Ecuador, from June to December, Murphy (1939) observed a sharp and dramatic change in weather in the latitude of Cape San Francisco, or about 1°N. There seems to be extraordinary agreement in the literature on the region, ancient as well as modern, concerning the above phenomenon. Throughout the dry season "the rains and squalls are in full progress from Cape San Francisco northward, whereas south of that cape the traveler encounters only an occasional flurry of the mist known as *garúa*" (Murphy, 1939). The vegetation also undergoes its sharpest transition at this point, and the appearance of the sky shows equally rapid change. To the north, the clouds are changeable in form, feature, and color and the "canopy is brilliant and filled with movement, as the heavy cloud layers swirl and mingle and the cumuli develop into afternoon thunderheads" (Murphy, 1939). To the south of Cape San Francisco the dull, gray, featureless stratus layer stretches from horizon to horizon, at least during the daylight hours.

In the wet season, from January to June, the northern boundary of the dry coastal belt is presumed to be slightly farther south, but only slightly. Murphy locates the boundary separating regions having annual rains from those having rains only at intervals of several years at about Cape San Lorenzo or 1°S. Troll (1930) places the northern boundary of regions with a short rainy season (up to 3 months) at Cape Pasado (20′S). Alpert (1948) draws the 750 mm isohyet as reaching the coast at about Cape Pasado and the 250 mm isohyet at about Cape San Lorenzo. Svenson's (1946) map of native vegetation indicates that the xerophytic coastal forest begins at about Cape Pasado or just slightly south of the equator (Fig. 1.8). It cannot be greatly in error, therefore, to show the dry climates as beginning at about Cape Pasado

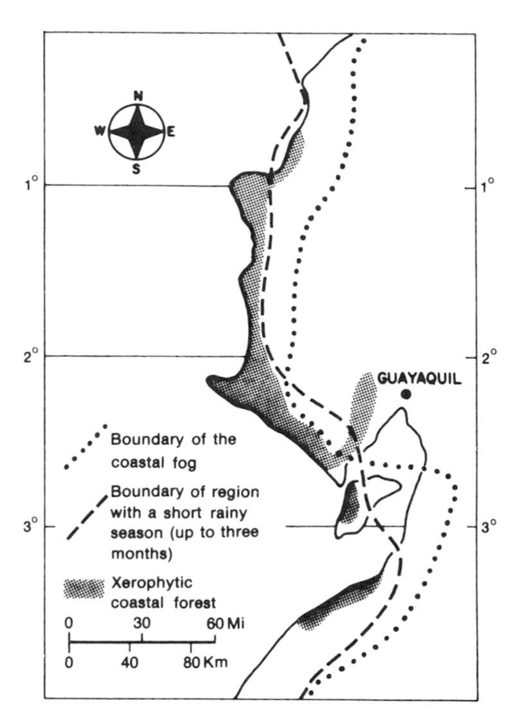

Fig. 1.8. Important climatic boundaries in western Ecuador. (After Troll, 1930, and Svenson, 1946.)

only 30–40 km south of the equator. Between capes Pasado and San Francisco is a stretch of coast where gray skies and drought prevail from June to December, but a modest rainfall during the period January to June prevents its being classed as prevailingly dry.

The depth to which the dry climates penetrate inland is also uncertain, but the maps of Troll, Murphy, and Svenson, as well as fragmentary descriptive evidence from other sources, suggest that the dry zone is relatively narrow, widening somewhat as the coast projects westward into the open ocean, and shrinking as the coastline retreats eastward into protected bays. In other words, the aridity seems to be in direct proportion to the extent to which the coast is exposed to the effects of the cool current.

But although the coast lands of middle

and southern Ecuador are dry in terms of the amount of precipitation which is caught in a rain gauge and can be measured, fog and drizzle are prevalent and the natural vegetation cover is not as xeric in appearance as one would expect in view of the slight amount of measured rainfall. The cool coastal current, fed by upwelled water, and a part of the main Peru current, is felt throughout the year northward to about Cape Pasado (20′S), and as far north as Esmeraldas (about 1°N) during the Northern-Hemisphere summer when the southerly winds are at their maximum development. This cool water, in addition to being partly responsible for the general aridity of the coastal belt, also accounts for two other climatic peculiarities, (a) the relatively low air temperatures for equatorial latitudes, and (b) the coastal fogs and drizzle (*garúa*). At Ancon on the exposed coast at about 2°20′S the average temperature in August when cool water prevails is only 21.4°C (70.5°F), while at Guayaquil in approximately the same latitude, but back from the open ocean at the head of Guayaquil Bay, it is 24.6°C (76.3°F). Since the average temperatures of the warmest month at the two stations are almost identical, the annual range at coastal Ancon reaches 5° or 6°C (10°F), a relatively high figure for an equatorial station, compared with about half that at Guayaquil.

Still more unusual are the gray overcast and the fog-drizzle which are characteristic of the cool-water littoral of Ecuador. On flattish coastal lands there is a uniformly gray overcast or high fog that provides practically no surface moisture. Where the land is somewhat more elevated, and especially along southwestern or windward exposures, the fog commonly is at ground level, and under favorable conditions genuine drizzle occurs. The lower level of the fog layer lifts rapidly inland so that drizzle is rarely experienced more than 30 km from the coast. Where the foggy moisture is moderate it results in epiphytic vegetation forms which envelop the giant cacti and the branches of the trees. Where the fog-drizzle is abundant it may actually produce a unique kind of rain-forest. In areas where the summer rains are very light and low-sun *garúa* abundant, the coastal inhabitants are inclined to reverse the concept of *verano* and *invierno*, with low sun being thought of as the wet season. Winter then becomes the period of maize harvest, of maximum fever epidemics, and of impassable roads.

Since Ecuador's dry coastal climate is apparently a consequence of those same drought-making controls which are performing at full strength just to the south in coastal Peru, a full analysis of the operation of these controls is postponed until a later section where the Peruvian desert is under consideration. It may be noted at this point, however, that two drought-makers, the subtropical anticyclone, and the cool ocean current which the anticyclone's southerly circulation generates, tend to operate in unison, so that it is difficult to differentiate their effects. In northernmost Peru, where the coastline turns abruptly eastward, the main body of cool water leaves the coast, but a branch of the Peru coastal current continues northward for some distance along the coast of Ecuador. It is debatable, therefore, whether the dry littoral of Ecuador is primarily the result of the stabilizing effects of the cool water, or whether the stabilizing influence of subsidence in the subtropical anticyclone likewise continues to function this far north. The fact that the dry climates in Ecuador are confined to such a narrow coastal strip, and that aridity is most intense on the westernmost extremities of land, suggest that cool water here may be the more important of the two controls.

CHAPTER 2

Pacific South America: The Dry Climates of Northern Chile and Peru

The Dry Climates of Northern Chile and Peru

Nowhere on the earth is there a coastal desert of such intense aridity that spans so many degrees of latitude as the Chilean-Peruvian desert. Deserts in somewhat similar geographical locations are found along the western littorals of tropical-subtropical North America, northern Africa, southern Africa, and Australia, but none equals in aridity or latitudinal spread this one in western South America. The extreme dryness of the Chilean-Peruvian desert is all the more unusual when one considers that no part of this arid strip is far from a tropical ocean, and that the desert is never more than a few score kilometers in width. That atmospheric humidity is high is attested by the presence of much low stratus cloud, fog, drizzle, and heavy dew in many parts. There is no lack of precipitable water. Still it does not rain. Obviously the dynamic controls making for the suppression of convection and the near absence of precipitation are unusually well developed. Also, this desert is terminated with extreme abruptness on both the north and south, especially the former, so that transitional steppe climate (*BS*) is only meagerly present.

In addition to intense aridity, the Chilean-Peruvian desert has other climatic anomalies: (1) the high degree of atmospheric humidity throughout the entire year; (2) the unusual uniformity of the weather; (3) the uncommon northward extension of desertic conditions to within a degree or two of the equator, giving rise to a desert of remarkable latitudinal dimensions; (4) a relatively large number of hours with rainfall that add up to only a trivial total amount; (5) low annual temperature; (6) the relatively large annual, but small diurnal ranges of air temperature, features unusual in tropical climates; (7) the strong persistance but shallow depth of the trade winds; (8) a strongly prevalent stratus deck that exists over much of the coastal desert from northern Chile to about 8°N in Peru; and (9) the brief periods of heavy rainfall and high temperatures which very occasionally affect chiefly the northern parts of this desert area, temporarily revolutionizing its weather and converting it to wet, tropical conditions (Prohaska, 1973).

Geographical Setting

Throughout Peru and northern Chile, the high Andean cordillera are never far removed from the coast, thereby restricting the width of the desert, and also exerting an important influence upon both general and local atmospheric conditions. However, the coastal terrain shows considerable variation in its different parts. In northern Peru between about latitudes 5° and 7°S, lowlands border the coast and extend inland for 100 to 160 km at a maximum. Southward from about 7°S the Andes gradually approach closer to the sea, narrowing the coastal lowland, so that it is finally pinched out just north of the mouth of the Rio Santo (about 9°S). Southward for about 240 km the precipitous slopes of the Andes rise abruptly from tidewater, and thence to Pisco (about 14°S) the coast is alternately bordered by small alluvial fans and by spurs of the nearby mountains. South from Pisco, in southern Peru and northern Chile, the coast is bordered by a low coastal range which in places reaches elevations of 900 m. Back of the coastal range, and separating it from the Andes proper, is a slightly lower desert surface, the pampas, which in Peru is an intricately dissected, rock-strewn surface, and in Chile a series of smoothly floored basins or bolsons. Naturally variations in terrain character are likely to produce some localisms in climate, so all parts of the elongated Chilean-Peruvian desert are not climatically identical.

It is the southerly circulation of air and water around the stable eastern end of the subtropical anticyclone in the eastern Pacific that so dominates the dry climate of Peru and Chile. And the extraordinary persistence and intensity of that anticyclone is thought to be caused, at least partly, by the blocking effect of the Andes cordillera upon the eastward drift of that system. The steady southerly coastal winds are one segment of the Southern Hemisphere trades; they approximately parallel the coast line. In conformity with the general wind flow, the movement of ocean water is from south to north, so that its surface temperatures are relatively cool.

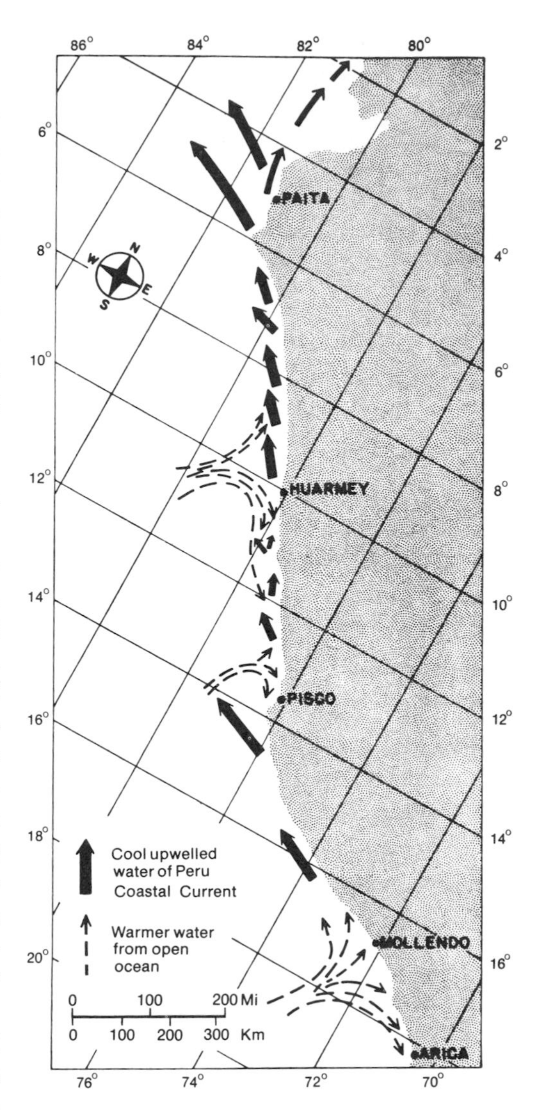

Fig. 2.1. Pattern of the coastal currents along the coast of Peru. (After Schweigger, 1949.)

This is the well-known Humboldt or Peru current system, which in turn may be subdivided into several parts (Fig. 2.1). The *oceanic current* flowing from about 40° to 5°S, the main element of the circulation, is several hundred kilometers wide and has a mean depth of more than 500 m. Its relative coolness reflects its high latitude region of origin. The second component, the *Peru Coastal Current*, also from the south, flows next to the coastline from central Chile to within a few degrees of the equator (Fig. 2.1). Its average depth is about 200 m and its width is much less than that of the oceanic current. Between the oceanic and coastal currents, near to the equator, a wedge of warm water, the *Humboldt Countercurrent*, thrusts southward between the two cool currents for variable distances.

The coolest surface-water temperatures are along the land margins of the coastal current where there is an upwelling of cool subsurface water caused by the persistent S-SW winds. The surface stress produced by the shoreline-paralleling winds and related Coriolis effects, causes a net mass transport of water away from the coast, known as the offshore Ekman drift. This offshore movement of surface water is compensated by a replacement with cooler water upwelling from moderate depths of 50 to 300 m. Accordingly, water temperatures increase toward the open sea, the inshore waters having an average negative temperature anomaly of 4°–6°C (9°–13°F). Eighty to 160 kilometers from the coast the water temperatures increase as much as 3°C (5°F). The main Peruvian current swings away from the land at about 5°S in northern Peru where the coastline changes direction and bends abruptly to the north, although a small branch of the main current, with some upwelling, continues northward along the coast of Ecuador nearly to the equator (Fig. 2.1).

Deserving of emphasis is the fact that the Coriolois force and the Ekman drift gradually weaken northward as latitude decreases and finally change sign at the equator. This change helps to explain the northern limit of the arid zone in the proximity of the Peru-Ecuador border. It also makes clear the abrupt transition from northern Peru's superaridity, related to cooi-water upwelling, to the wet coastal climates of northern Ecuador and Colombia, which are bordered by a warm ocean.

The description of surface waters and winds given above is a greatly simplified version of thc prevailing conditions. Actually the patterns of winds and currents are complicated and variable both areally and temporally. One might expect the waters to gradually increase in temperature to the north, and so they do away from the coast in the Peru Oceanic Current. But such is not always the case in the coastal current, where the volume of upwelling and the water temperatures as well are variable (Fig. 2.1). The most active upwelling and lowest sea temperatures appear to be concentrated along certain sections of the coast, and between these are other stretches in which upwelling is weaker and less-cool water from the open ocean approaches, or even reaches the coast. There is not complete agreement, however, regarding the location and extent of these areas of cooler and warmer coastal waters. Sverdrup (1942) indicates that there are four coastal regions of strong upwelling between 3°S and 33°S. The approximate location of the two most important of these regions is at 7° or 8° and 15° or 16°, with upwellings of lesser importance located at about 22° or 23° and again in the vicinity of about 30° (Schweigger, 1949). Between these focal centers of strong upwelling and ocean currents which set away from the land, are other areas where less-cool water moves toward the coast from the open sea, often accompanied by winds that are more westerly and northwesterly. The most striking of these coastal areas receiving water from the open ocean is in the vicinity of latitude 9° to 11°, a lesser one occurs around 13°, and a very marked one in latitudes 17° to 19° (Schweigger, 1949). Perhaps these alternating coastal stretches of cooler and less-cool water indicate the presence of major oceanic swirls, with cold upwelled water in the branch that moves away from the coast, and less-cool oceanic water in that part of the eddy moving toward the coast in another branch.

One additional area of warmer water, which is quite different from the others mentioned, needs to be noted. In extreme northern Peru in the vicinity of latitude 6°S and northward, are felt the effects of the warm Equatorial Counter Current, and the minor El Niño emanating from the Gulf of Guayaquil. These currents from the north are genuinely tropical in origin and have a much lower salt content and much higher temperatures (as much as 7° or 8°C or 12° to 15°F) than those that invade the coast from the open ocean farther south. The coastal region of Ecuador and Peru between about the equator and 6°S is the zone of conflict between the geniunely warm waters from the north and the cool Peru Coastal Current from the south, with the warm waters annually advancing farthest south in the summer season of the Southern Hemisphere. At irregular intervals of several years, however, the warm Equatorial Counter Current advances unusually far south along the coast of Peru, displacing the cool coastal waters normal to these areas. Such abnormal southward invasions of the equatorial waters appear to accompany large-scale dislocations of the atmospheric circulation.

Another striking phenomenon of this coastal desert is the strength and regularity of the diurnal land and sea breezes, the latter not uncommonly reaching almost gale force by midafternoon. This diurnal wind reversal has important effects upon the local temperatures, cloud, fog, and even precipitation.

The Climatic Pattern

Regrettably, the meager and poorly distributed instrumental climatic data for this desert region allow only the broader outlines of the regional climates to be portrayed with any assurance. Many of the suspected regionalisms and localisms of climate growing out of terrain and ocean-temperature variations cannot be verified from instrumental records of several years' duration. Frequently the notes of travelers must be resorted to. Unfortunately, nearly all of the official weather records are from coastal stations, so that the suspected important climatic differences along a profile at right angles to the coast are difficult to confirm.

Temperature.—Since cool waters and a long stretched band of associated cool air gird this whole coast, it is not unexpected that isotherms, both of sea-surface and air temperatures, should approximately parallel the coastline, indicating considerable temperature homogeneity along the more than 3,200 km of desert coast. The lowest air temperatures are along the immediate coast in closest proximity to the upwelled cool water and the relatively persistent deck of stratus cloud.

Surface air temperatures along the coast depend almost completely upon the cool marine air being advected inland by the onshore component of the trades and by the sea breeze. Where a coastal range is present, temperatures promptly increase inland. Under the prevailing cloudless skies interior from the coastal range in southern Peru and northern Chile, air temperatures at an elevation of a thousand + meters may be as high as or even higher than those along the coast. A normal latitudinal increase in temperature equatorward is lacking over long distances. Thus the average January temperature at Taltal in Chile—between 25° and 26°S—is very similar to that at Callao, Peru, some 1,700 km closer to the equator. Indeed, it is not unusual for normal latitudinal temperature gradients even to be reversed along certain stretches of the dry coast, with stations in Peru recording lower temperatures than those in northern Chile. Thus, Callao in Peru at about 12°S shows an average January temperature of 22°C (71°F), while Arica in Chile at 18°30′S records 23°C (73°F). With the data available, it is not possible to present a detailed regional analysis of air temperatures along the entire desert coast so as to reveal the suspected localisms. In the latitudinal temperature profiles of both the warm-month and the cold-month there are peculiar irregularities, the most striking one being the higher temperatures which prevail in the vicinity of Arica in northernmost Chile. This positive temperature anomaly is coincidental with a weakening of the cool coastal current and its actual displacement by less cool waters arriving from the open ocean to the west. If, and to what extent, the other two coastal stretches of reduced upwelling (10°–11°S and 13°S) are also areas of higher than normal air temperatures cannot be determined. Since one effect of the warmer water is to disperse the stratus deck which is so prevalent, at least in winter, the stronger insolation reaching the surface, plus the direct effect of the warmer waters, should act to increase the local air temperature.

There is evidence to indicate that departures from the normal seasonal temperatures along this desert coast are often associated with variations in wind velocities. Thus, air temperatures along the coast from Antofagasta in Chile to Esmeraldas in Ecuador were below normal from March to December, 1954. Similarly, water temperatures showed a negative anomaly for the same period, but the speed of the prevailing southerly winds was above average, the latter condition acting to increase the amount of upwelling (Graves, Schweigger, and Valdivia, 1955).

Employing Lima airport (Callao) as representative of the northern part of this elongated coastal desert, some further details on temperatures of that region can be provided. A year-round feature is the temperature inversion, but its effects differ somewhat in winter and summer (Fig. 2.2). During the entire low-sun or *winter* season a stratus cloud deck that develops underneath the inversion persists day and night (Fig. 2.3). Its effect in that season is to intercept and absorb some 80 percent of the incoming solar

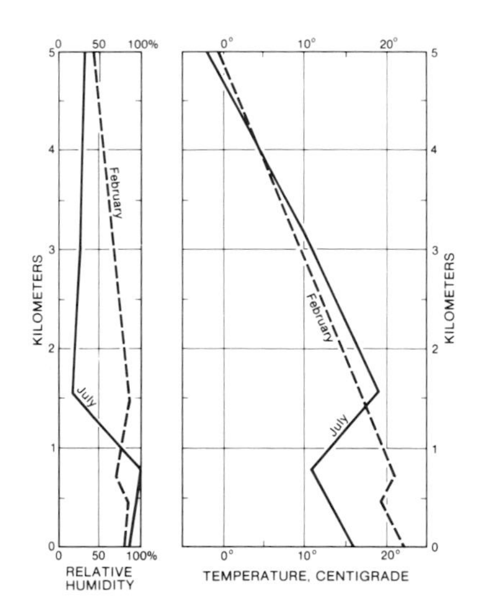

Fig. 2.2. Vertical profiles of temperature and humidity for July and January at Lima, Peru. (After Prohaska, 1973.)

radiation, which therefore has little effect on surface air temperatures. So along the coast surface air temperatures depend almost entirely on marine air advected inland by the onshore component of the southerly trade winds (Prohaska, 1973). During the four winter months, June through September 1967, the mean maxima temperature was 17°C (63°F), the mean minima 14.4°C (58°F),

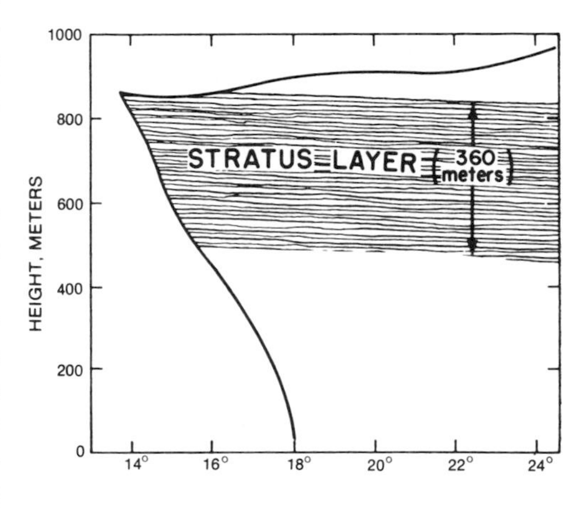

Fig. 2.3. Temperature inversion and stratus layer along the Peruvian littoral. (After Schweigger, 1949.)

the range 2.8°C (5°F), and the average 15.6°C (60°F). The mean for the ocean surface was 15°C (59°F) indicating that air temperature was a function of sea temperature (Prohaska, 1973). In Lima the winter temperature is 9°C (16°F) below the latitude average.

Typical *summer* conditions prevail from January through April when inversion, cloud, radiation, and temperature conditions are quite different from what they are in winter. The summer inversion, which is primarily a consequence of the cool ocean, starts closer to the surface, is shallower, less strong, and varies more from day to day in intensity, depth, and altitude. The fog or stratus layer in this season forms offshore over the cool upwelled water and is carried onshore by a morning type of sea breeze. It lasts only until about noon at which time the southerly trade-wind flow takes over again. Any inland cloud is dissolved, but the humidity remains high. Coastal temperatures in summer, then, reflect the combined effects of oceanic, radiative, and inversion influences. The inversion, because it accounts for the moist marine layer, inhibits rapid nighttime cooling, which is characteristically intense in most arid climates. The consequence is high minimum temperatures, some 3°C (5°F) above the water temperature offshore. Daily range of temperature in summer is double what it is in winter but still it is only 7°C (13°F). Daily maximum temperatures in summer, because of clearer skies, are fairly high, as much as 10°C (18°F) higher than the ocean and more than that where the plain widens. Actually the summer temperatures are only 3°–4°C (5°–7°F) lower than the norm for these tropical latitudes. Moreover, since the inversion stifles the upward movement of water vapor, the heat is a humid variety, resembling that of the wet tropics rather than the desert. The diurnal temperature range in summer is about twice that in winter.

Farther south, in northern Chile, temperature conditions within the humid coastal marine layer are not greatly different than they are in Peru. Weather conditions are monotonously uniform. Daily mean temperatures at Antofogasta are 18°–20°C (64°–68°F) in summer and 13°–14°C (55°–57°F) in winter; the daily range is 7°–8°C (13°–14°F) in summer and 6°–7°C (11°–13°F) in winter. Humidity is high. In summer stratus occurs along the coast on only about half of the days, normally breaking up during early morning with the rest of the day clear. Stratus is almost a daily occurrence in the winter months, but it usually disperses for a few hours in the afternoon. Fog drip is common on the coastal range in winter. The interior back of the coastal range, where climatic data are largely lacking, is the most barren region imaginable. Temperaturewise it is more severe than the coastal strip.

Cloud and Precipitation.—The concept of temperature inversion has been touched upon previously in order to clarify certain thermal characteristics. At this point it is referred to again in order to make clear its functions relative to cloud and precipitation. An inverted temperature profile, of variable intensities, is the normal type of atmospheric structure along the desert coasts of Peru and Chile from about Salinas at 2°S to Serena at nearly 30°S, and even beyond, especially in summer. The inversion is of dual origin. In the main it is simply the normal trade-wind inversion and is the result of subsidence and heating aloft in the stable eastern end of an oceanic subtropical high. This control operates year round but functions more exclusively in the cooler low-sun months. The second instigator of the inversion is the cooling from below of the *mT* trade-wind air mass as it moves over the cool upwelled water of the coastal current. This latter control is relatively most effective as a stabilizing agent in the high-sun summer months, at the time when subsidence aloft is weakest. The general effect of the temperature inversion is to stifle any upward movement of air, and hence to suppress the precipitation processes. Throughout the depth of the very humid, nearly saturated air, continuous mixing below the inversion results in a low and stable stratus cover.

Because of the general sparseness of weather stations in most arid regions, it is almost impossible to indicate with any degree of certainty where the absolutely driest part of the earth is located. Without fear of contradiction, however, as previously stated, the Chilean-Peruvian desert is a strong contender for the distinction of being the earth's most arid region over such a large range of latitude. From Piura in northern Peru at about 5° or 6°S, to Coquimbo in northern Chile at about 30°S, the highest average annual rainfall over a six-year period at any coastal station was under 50 mm and a number of stations in northern Chile have recorded no rainfall over periods of one or more decades. (Over the recent decade, 1961 to 1970, Lima airport recorded an average annual rainfall of 10 mm, the driest months being July-August-September [winter] and January [summer]. At Antofogasta in northern Chile over a recent 10-year period, the average annual rainfall was only 7.7 mm, with nearly 70 percent falling in June-July-August [winter].) In the Chilean desert behind the coastal range even the drizzle is absent; there an intensity of aridity prevails that, so far as is known, is not equaled in any other desert of the earth. To be sure, annual rainfall averages are of practically no significance in such a region, for along this entire coastal strip genuine rains occur only at intervals of several years. At Piura in northernmost Peru at about 5° or 6°S a very slight amount of summer rain associated with the southward-shifted *ITC* occurs almost every year, although the total is only about 20–30 mm annually. But southward from Piura the zone of annual rains is found only at elevations of several thousand meters along the slopes of the Andes, and, if winter drizzle is omitted, the coastal zone is rainless except for the very occasional brief summer showers separated by periods of several years. Such infrequent showers arrive from the east and represent a spillover from Andean cumuli developing in air from Atlantic moisture sources. A plotting of the annual rainfall against latitude is shown in Figure 2.4. It is clearly evident that the extreme aridity remains almost constant throughout nearly 24° of latitude, and also that there is a sharp increase in rainfall at both the northern and southern ends of the desert. The inversion, in conjunction with the advection produced by the trade-wind onshore component, makes for a very humid marine layer. Steady trade winds maximize the sea evaporation, while the inversion inhibits upward transport of water vapor. Consequently the moisture content of the marine layer is very high. It is capped by a low and stable stratus layer. At the Lima airport the average relative humidity was 88 percent during the four winter months of 1967, and 90 percent of the days were overcast with a stratus deck. Ceilings were usually between 150 and 300 meters and the stratus layer was only a few hundred meters thick.

Fog, Low Stratus, and Drizzle.—While low stratus and fog are a characteristic feature of this coastal desert, there are important seasonal and regional variations. Most frequently the fog is of the high-inversion type, so that it is in the form of a low, uniform stratus deck whose base averages between 350 and 550 m above sea level (Fig. 2.3). It is the result of the combined effects of the strong anticyclonic subsidence and the cool coastal waters. At times the fog reaches down to the land-sea surface, but more often it is at variable elevations above sea level. Turbulence near the surface usually is sufficient, especially during the day when the sea breeze is strong, to prevent the fog from de-

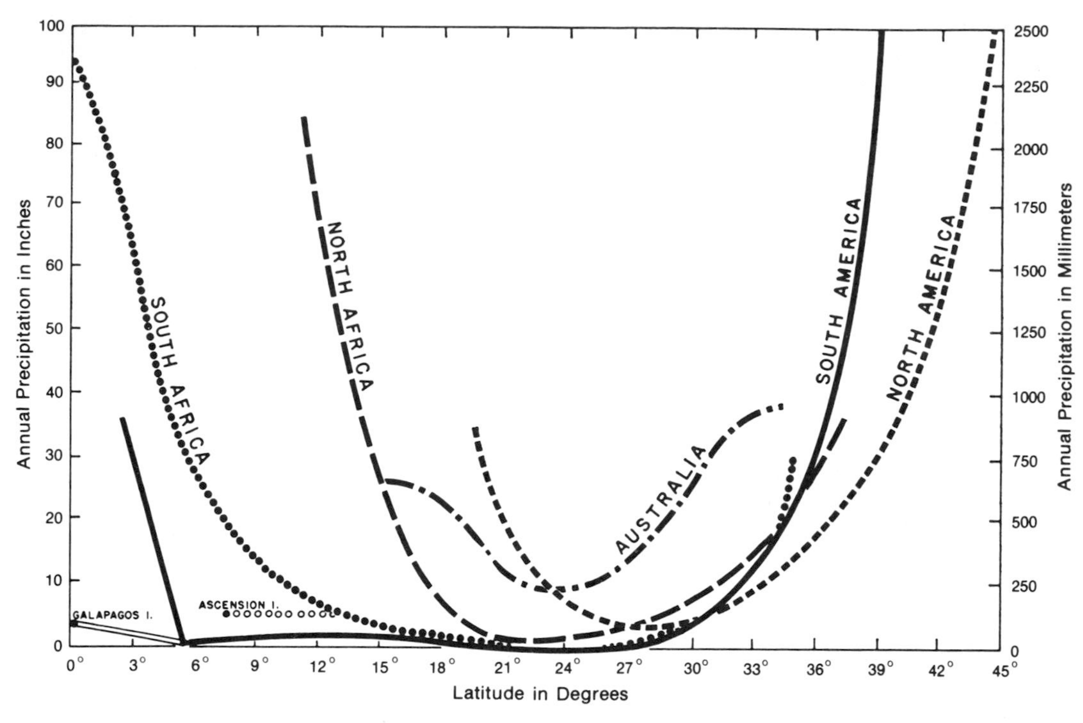

Fig. 2.4. The annual precipitation curves plotted against latitude for 5 coastal deserts in tropical-subtropical latitudes. Note that the Chilean-Peruvian desert exhibits the highest intensity of aridity, maintained throughout the largest span of latitude. (After Lydolph, 1973.)

scending to the ground. This explains why the slopes of the coastal range record more fog and drizzle than the immediate coastal lowlands. In summer the stratus deck is higher and less persistent so that sunshine is abundant. It lowers, thickens, and is more continuous in the winter months, especially at night, and in August and September it may persist unbroken for weeks on end. For this reason most of the Peruvian coastal desert has a maximum of drizzle in the cooler seasons. The stratus fog does not extend inland to the dry pampas beyond the crest of the coastal range.

Real rain rarely if ever falls from the stratus deck, but only a fine penetrating drizzle, the so-called *garúa*. The drizzle is most common between 2100 and 0900 and is least frequent in the afternoon between 1200 and 1700. This pronounced diurnal pattern, typical of the Lima region, is especially characteristic of the coastal strip. Farther inland toward the mountains, at elevations of 150 to 750 m where some orographic lifting of the humid marine air takes place, there is almost continuous drizzle along the zone where the stratus deck makes contact with the slopes. Here 100 to 200 mm of drizzle-rain may be received during the winter season, which is sufficient to induce a vegetation cover adequate to support grazing. The resulting landscape consists of a clearly visible green pasture zone, dotted with grazing cattle, which is limited above and below by tawny dry belts. While drizzle is common on the slopes and coasts of Peru and Ecuador, reportedly it is less common in coastal northern Chile.

Some drizzly weather appears to be associated with the passage of cold fronts arriving from the Antarctic cold source. During the period April to July inclusive, 1956, 31 cold fronts were recorded as passing the Chilean island of Juan Fernandez, 2,400 km south of Lima. Nothing is known of the fate of these disturbances after leaving Juan Fernandez, but there is some evidence that they move deeply into the tropics and are responsible for the periods of disturbed weather with drizzle at Lima. Passage of a front is commonly denoted by a change in the elevation of the stratus deck, an altering of the cloud form from flat stratus to fractostratus or fractocumulus, a drop followed by a rise in pressure, and a slight wind shift. During the period of 122 days studied, there were 97 days at Lima when there was drizzle precipitation amounting to 0.1 mm or more, and a total of 10.1 mm was recorded (Rudloff, 1956).

The geographic distribution of rainfall seasonality is very striking in the Chilean-Peruvian coastal desert. South from northern Ecuador to about 11°S, and including the northern third of Peru, as much as 90 percent of the meager annual rainfall comes in the six summer months, November through April. But from there southward, and including northern Chile, some 80–90 percent arrives in winter or between May and October (Fig. 2.5). The transition summer-to-winter rainfall gradient, situated in about the latitude of Lima, is excessively steep. Thus the northern part of the desert, which occasionally feels the effects of a high-sun southward shift of the *ITC*, and an infrequent spillover of convective rains from east of the Andes, has a rainfall regime resembling that of the wet tropics. But southward from about Lima what scanty drizzle and rain occurs may be mostly generated by northward-moving Antarctic fronts in the cooler low-sun months.

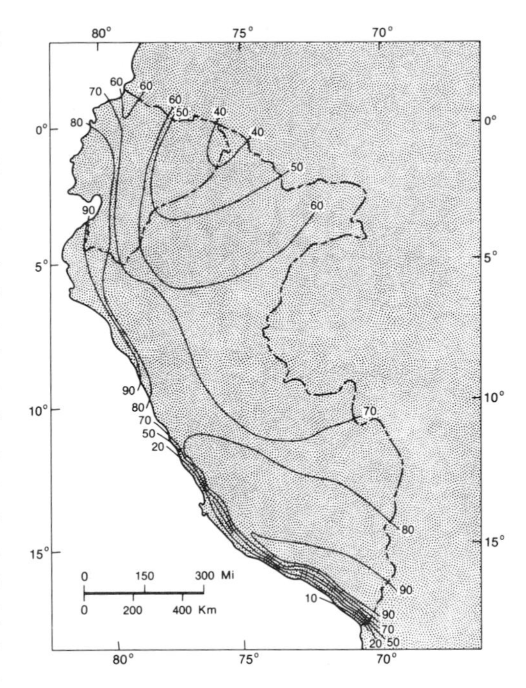

Fig. 2.5. Percentage of the annual rainfall in western Peru occurring in the 6 summer months, November–April. (After A. M. Johnson, 1976.)

Causes of the Intense Aridity.—What the controls are, and which, acting in combination, produce such intense and latitudinally extensive aridity have been analyzed by Lydolph (1957; 1973) and, more recently by Lettau (1978) and others. Of primary importance is the fact that this region is prevailingly under the influence of the strong subsidence associated with the eastern limb of a positionally stable anticyclone. It has been noted earlier that over extensive longitudes in the Southern Hemisphere the zonal subtropical ridge of high pressure is composed of migratory individual anticyclones. Only South America and its Andean cordillera are able to seriously impede and divert the zonal flow pattern, with the result that the South Pacific cell is a highly permanent feature, which on the east is terminated abruptly by the Andes, and whose position fluctuates only slightly throughout the year. A persistent and low inversion, frequently reaching down to nearly sea level, is the result. No doubt this strong and locationally fixed eastern end of the cell is by itself capable of producing more than moderate aridity. Western Australia may be cited as an example of a dry tropical-subtropical west coast which is dominantly under anticyclonic control, but where, instead of a permanent cell, individual migratory cells separated by cols with zones of convergence are the rule, and where upwelled cool water and high bordering mountains are lacking. Here minimum rainfalls of nearly 250 mm occur along the coast through only a few degrees of latitude, so that the conclusion might be drawn that the migratory anticyclones, without the assistance of supplementary controls tending to depress precipitation, are capable of reducing annual rainfall only to about 250 mm and even that for only a few latitude degrees.

Along most of the earth's dry western littorals the meager data available indicate that the most intense aridity is concentrated in a narrow strip along the coast and that rainfall increases both inland and seaward. An increase in rainfall inland might be expected along those coastal sections of Peru and Chile where elevation increases abruptly back of the coast. But even in lowland Ecuador, in the vicinity of the Gulf of Guayaquil, rainfall rapidly increases inland from the dry coast. Moreover, the coastal rainfalls seemingly are lowest on those points of land which project farthest westward into the ocean. All of this evidence points to the fact that one or more controls favoring drought, other than the anticyclone, reaches its maximum development in the vicinity of the coastline.

One such factor making for intensified aridity along the littoral is the presence of the unusually cool upwelled water of the Peruvian Coastal Current. Its obvious effect is to chill and stabilize the surface air, thereby intensifying and lowering the temperature inversion. In this respect reference may be made again to the dry coast of Ecuador, which is marginal with respect to the anticyclone, so that there a minor branch of the cool current or local upwelling would appear to be a principal drought-maker. Yet the more exposed parts of this coast have as little as 250 mm of annual rainfall. Farther south, where stronger anticyclonic subsidence and colder water combine their effects, the aridity is greatly intensified.

It is remarkable that the cool upwelled water and the associated aridity continue so far equatorward. This appears to be a consequence, in part at least, of the fact that the coastline in Peru maintains a NW-SE direction to within 4° or 5° of the equator. Just as long as the Peruvian coast continues to bend westward into the main oceanic circulation, so that cool upwelled water is concentrated near the coast, the intense aridity is continued equatorward, with the result that even at Piura, at about 5°S, annual rainfall is still only 250 mm, although, in contrast to regions farther south, some rain falls each year. But as soon as the coastline bends northward and eastward, and therefore away from the oceanic circulation around the eastern end of the anticyclone, the main cool current leaves the coast and precipitation increases rapidly northward. It is significant that the general dry zone of the equatorial eastern Pacific, as well as the 250 mm isohyet as drawn by Alpert (1948), leave the coast somewhat to the south of the equator and follow a WNW course, approximately coinciding with the route of the cool current which continues westward in equatorial latitudes. The arid Galapagos Islands, almost on the equator, and about 1,100 km west of the mainland, might be considered an extension of the Peruvian coastline along a path of constant curvature from Pt. Parina in northern Peru. What seems remarkable is the fact that the minor branch of the cool current that continues northward along the Ecuador coast can be so effective as a drought-maker. It appears likely, therefore, that the westward bending of the Peruvian coastline in the lower latitudes is an important factor in carrying the cool current and its aridifying effects so far equatorward, and consequently in greatly extending the latitudinal spread of the desert.

It might be suggested that the case for the effectiveness of the cool water as an aridifying agent is weakened by the observation that those localities along the desert coast where upwelling weakens and warmer waters from the open ocean move toward the land, do not seem to be appreciably less arid than those other areas where upwelling is at a maximum and coastal-water temperatures at a minimum. The observations of sea and atmosphere are still too meager to enable one to make a judgment on this problem. Moreover, the waters approaching the coast from the main Peruvian Current are by no means characterized by tropical warmth. They are only less cold than those of the Peruvian Coastal Current. Schweigger (1959) indicates that while no striking climatic localisms are at present known to be associated with these invasions of the coastal waters by currents from the open ocean, there has been observed in such localities a dispersion of the coastal stratus deck in winter, and on occasions in summer the appearance of cloud types which suggest the possibility of rainfall. Obviously this is a problem for further study.

In addition to the strong and positionally fixed anticyclone, the cool coastal waters, and a coastline which bends westward into the main oceanic circulation, there may be supplementary aridifying factors, chiefly concentrated near the coast. It is significant in the case of several of the subtropical oceanic highs, and more especially that of the eastern South Pacific, that there is a strong coincidence between the eastern margins of the subtropical cells and the western coastlines of the respective continents. This feature provides for maximum atmospheric subsidence over the coastal area. The high degree of coincidence between atmospheric circulations and the coastlines strongly suggests that surface characteristics along the coast exert some influence in determining the positions of the eastern margins of the high-pressure cells. Such an influence must result from the temperature and frictional contrasts that sea and land exert upon the atmosphere, which induces a flow of surface air approximately parallel to the coast. This in turn tends to sever the subtropical high-pressure belt, breaking the zonal flow into distinct cells, whose circulations are closed on the east and semiisolated from the circulation over the adjacent land. In South America the high wall of the Andes rising abruptly back of the coast joins its effects to those of the other controls mentioned in stabilizing the position of the eastern end of the South Pacific cell and abruptly terminating the anticyclonic circulation at the coastline. Once the cellular flow is established, surface temperature contrasts between land and water are further magnified by the upwelling of colder water caused by the slightly offshore paralleling winds, so that as long as the coastline continues to bend into the main oceanic circulation and cool water flanks the coast the drought effects continue even into equatorial latitudes.

Still another control of rainfall along littorals is the differential resistance to air flow over land and water surfaces resulting in some vertical motion. In surface air this involves chiefly the factor of frictional drag variation when wind blows across a coastline. At mid-troposphere levels there is the effect of greater convective turbulence over land surfaces. Thus, when surface air moves across a coastline, some upward or downward motion of the air should result from the greater drag upon the atmosphere's friction layer over land than over water. When there is an onshore component the air is slowed up over the rougher land and a piling up or stowing effect with vertical upward movement results. Conversely, with an offshore component, the surface air speeds up over the smoother water surface and subsidence occurs in the vicinity of the coastline. When the air flow is approximately parallel with the coastline there exists a stress differential between the current over the land and that over the adjacent sea, which operates to produce surface divergence and subsidence if the land is on the side toward lower pressure, but convergence and ascent if the land is on the high-pressure side. Accordingly in a wind which parallels a coastline there will be divergence in the friction layer if the land is to the left of the wind in the Northern Hemisphere and to the right in the Southern Hemisphere. In Peru and northern Chile, where the circulation around the eastern end of the subtropical high creates southerly winds that approximately parallel the coast, the stress differential over the adjacent land and water surfaces induces coastline divergence and subsidence within the friction layer (Bryson and Kuhn, 1961).

It seems, therefore, that the coastline itself becomes a climatic control, which intensifies the aridity along the littoral of northern Chile and Peru by (1) producing (in conjunction with the Andes) a closed anticyclonic circulation with paralleling winds, a cool ocean current, and a strengthened upwelling of cool water, and (2) generating coastal divergence and subsidence in the surface winds which approximately parallel the coast.

Further littoral subsidence may result from the strong sea breeze which prevails along much of this desert coast. The cool maritime air moving inland during the day produces a strong temperature inversion at the upper boundary of the marine layer where it makes contact with the warmer land air. Moreover, since the air involved in the sea breeze does not originate far out at sea, the sea air undergoes horizontal divergence during the heat of the day when it spreads farthest inland, with resulting subsidence. Significantly it has been noted in California that the summer inversion base along the coast is usually lowest in late afternoon when the maritime air has reached its greatest horizontal extent (Lydolph, 1957).

Very recently Lettau (1978) presented a new and supplementary hypothesis which gives additional emphasis to the sea breeze as a prime cause of the superaridity of the Chilean-Peruvian desert. He reasons as follows. In these low latitudes intense solar radiation strongly heats the long coastal strip of desert. Even with some cloudiness, considerable solar radiation gets through to ground level. Because the sea has a lower albedo than the land, it absorbs even more solar energy than does the bare soil surface. But the addition of this energy to the sea has little effect in raising the temperature of the cool water. The usual reasons given for the minor warming of the surface water are mixing, the high specific heat of water, and the penetration of solar rays to some depth. In addition there is the important feature of the Ekman drift (produced by Coriolis force) which continuously carries the slightly warmed surface water out to sea and away from the coast. Cool upwelled water is thus maintained adjacent to the strongly heated coastal strip. This is an ideal situation for the development of a strong and durable sea breeze. Even at night the sea continues to be about as cold or colder than the land, so there is little tendency for a land breeze to develop, and the sea breeze is likely to prevail throughout most of the 24-hour period, although it is much stronger during the day. The vigorous and nearly continuous sea breeze circulation dominates, in which the heated air over the land rises, flows seaward at elevations of 0.5 to 1.5 km, and then sinks and gives up its heat to the coastal water. But there the effect of the Ekman drift is to carry the warmed surface water out to sea, away from the coast, thereby maintaining the upwelling of cooler subsurface water along the land margins. This extremely efficient mechanism for the transfer, at low elevation, of land heat to the sea beneath the subsidence inversion, operates to prevent any significant deep convection and associated precipitation over the land. It is this feature which intensifies the aridity of the coastal desert in Peru and northern Chile.

A low-level jet, paralleling the coastline, is generated by the large horizontal temperature gradient in the vicinity of the cool-water coast. This air moves from south to north in the form of a helical circulation. In this low-level jet, wind speeds are greatest along those stretches of the coastline where temperature gradients between land and sea are steepest. The northward moving air south of these coastal sections will undergo a downstream increase in speed, causing increased divergence, intensified subsidence, and a further diminishing of chances for rainfall. Hence, the greatest subsidence and aridity are likely to be found somewhat to the south of the latitude of the strongest southerly winds of the low-level jet.

In summary, the extreme aridity prevailing along the great extent of coast in Peru and northern Chile appears to be related to a unique combination of circumstances, all elements of which are closely interrelated with each reinforcing the others. A strong and positionally stable anticyclonic cell, abnormally cool coastal waters, an abrupt mountainous coast, a coastline which continues to bend westward into the circulation of the open ocean until near the equator, an abrupt termination of the anticyclonic circulation at the coastline, and an unusually strong sea breeze—these conjoin to produce maximum aridity (Fig. 2.6).

Thunderstorm Activity in the Subtropical Andes.—Resembling the high plateau of Tibet and its mountain ramparts in subtropical Asia, the elevated Altiplano of western Bolivia and adjacent southeastern Peru, with its flanking mountains, is a focus of intense convective activity during the summer months (Schwerdtfeger, 1976). It is most pronounced over the mountains. The heavy rain showers are accompanied by violent thunderstorms, hail, and, on occasion, snow. Summer's strong high sun intensely heats the highland surface and it in turn warms the lower atmospheric layers resting on it; this takes place especially during the morning hours before there is much cloud cover. Thus, a high altitude thermal anticyclone is created above the Altiplano. The lapse rate in the heated lower air becomes superadiabatic stimulating a strong upward flux of both sensible and latent heat, the latter as much as twice the former. Numerous thunderstorms accompanied by heavy rain showers are the result. It has been suggested that these strong convective systems of the Altiplano region, like their counterpart in Tibet, make an important contribution to the maintenance of the earth's electric field.

THE OCCASIONAL ABNORMAL SUMMER (THE SO-CALLED EL NIÑO EFFECT).—A unique feature of the northern third of the Peruvian coastal desert is the very occasional revolution in weather which occurs in this region. In an area where the normal condition is a succession of near-rainless years, there are occasions when one or more of the summer months (usually January and Feb-

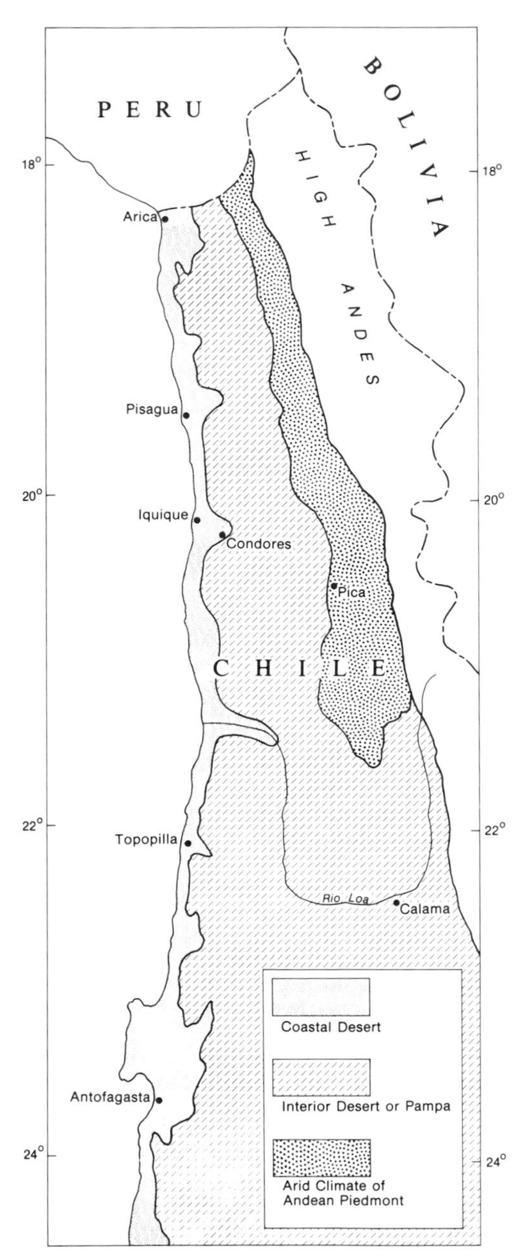

Fig. 2.6. Climatic regions of the north Chilean desert. (After Caviedes, 1973*b*.)

ruary) may experience spells of torrential rains which characteristically accompany severe electrical storms. On such occasions, what normally is one of the earth's most arid deserts, in short order takes on some features of a tropical-wet climate. In none of the other dry western littorals is there good evidence of similar catastrophic weather changes. Schott (1932) indicated that during the preceding 140 years there had been at least 12 years in northern coastal Peru when very heavy and destructive rains fell on land, and 21 additional years when there was a more moderate departure from the usual aridity. During about 100 of the years there was almost no rainfall except for drizzle. Since 1932 there has been a distinct moderating of the aridity at least in 1948, 1952, 1958, 1966, and 1973. These temporary cataclysmic reversals of weather are chiefly characteristic of the northern third of the Peruvian desert and decrease in frequency and duration toward the south. This suggests that whatever the controls are which bring the unusual rainy periods, they develop farther to the north and extend their influence southward.

The heavy rains seem to be due to a number of causally interrelated phenomena. The normal cool southerly and southeasterly winds of the subtropical anticyclone are replaced by northerly and westerly winds which cause a rapid increase in surface air temperatures, so that characteristic tropical heat prevails. The northerly winds, in turn, induce a southward flow of warm equatorial water which displaces the cool upwelled water of the Peruvian Coastal Current. This northerly water has temperatures up to 7°C (12°F) in excess of that which normally prevails, and its salinity is markedly lower. With the invasion of the warm equatorial water there is a great destruction of plankton and of fish, and dead fish in great numbers litter the beaches and befoul the air. In the decomposition, so much hydrogen sulphide is released that the paint of ships is blackened, a phenomenon known as the "Callao Painter." Much more serious is the effect on the guano birds, which because of the reduced supply of fish food either die of starvation or abandon their nests leaving the young to perish. The effects of the torrential downpours in what is normally an arid country are disastrous. In March, 1925, Trujillo at 8°S recorded about 400 mm of rain. Such deluges flood the irrigated valleys, inundating fields, destroying crops, roads, and bridges, and even ruining whole villages.

In normal years the coastal region for a few degrees south of the equator is an area of conflict and confluence between the warm northerly winds and sea currents and the cooler air and water from the south. But the average southern limit of *annual*, but very light, high-sun rains is about Piura, just poleward of 5°S. The meager high-sun rains, therefore, are a normal annual phenomenon in northernmost coastal Peru and southern Ecuador. However, it is when the controls which operate annually in summer at these very low latitudes extend 5° to 10° farther south along the coast that the catastrophic wet seasons occur. In the summer of 1925, which was a year of unusually heavy rains in coastal Peru, the duration of the period of warm water and heavy rains was as follows (Sverdrup, 1942):

4°20′S	76 days
7°40′S	63 days
12°20′S	15 days
13°40′S	8 days

While it is generally agreed that these striking episodes of increased rainfall along the desert coast of northern Peru are associated with the temporary displacement of the normal presence of cool water by warm water, the reasons for this radical change remain controversial. The increase in sea temperature clearly augments the air temperature, but a more dramatic result is the greatly increased rainfall that suddenly appears in the coastal desert. The abrupt change in the weather, both in temperature and rainfall, seems to arrive when the ocean temperature begins to equal or exceed the mean temperature of the atmosphere; the exceptional rainfall persists as long as the altered ratio between sea and land temperature prevails. It was formerly believed that the invasion of warm water mainly represented a displacement of the cold Humboldt current by a warm northerly current from beyond the equator; now the term El Niño is more likely to apply to *any* substantial increase in water temperature, no matter what the cause. There is significant evidence that El Niño happens when the trade winds of both hemispheres are weak, and when, simultaneously, the equatorial trough is displaced farther south in summer than normal, thus removing the usual southerly air flow which produces the cold upwelling (Fig. 2.7). This warming effect may occur as far south as about 14°S. During a prevalence of weak trades the transfer of heat, in both sensible

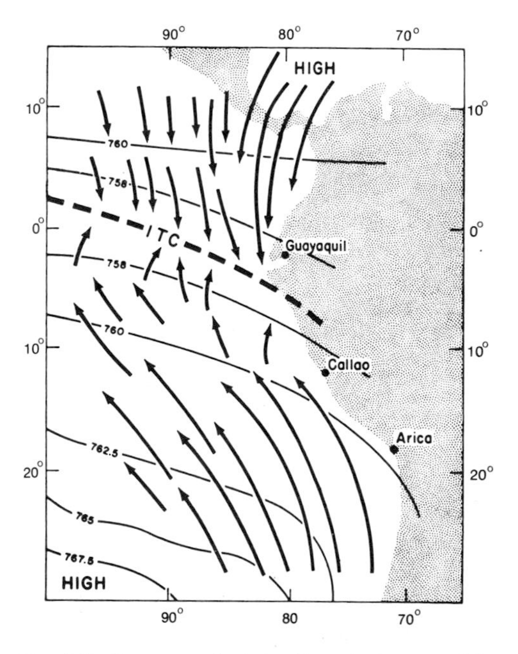

Fig. 2.7. Pressure and wind patterns during one of the infrequent El Niño rainy periods in the northern Peruvian desert. Note that the *ITC*, with its northerly winds and warm water, has advanced south of the equator along the coast of Peru. (After Schott, 1932.)

and latent form from ocean to atmosphere is below normal, and more than average amounts of heat continue to be stored in the ocean, raising its surface temperature. However, weak trades produce more spectacular effects on ocean temperatures through the halting of upwelling.

From the fairly abundant literature one must conclude that the origin of the El Niño phenomenon is still a controversial topic. The basic question is whether El Niño episodes have a local or a global explanation. Prevailing opinion seems to be that El Niño is a profound climatic event resulting from changes in the ocean-atmosphere system over the entire equatorial Pacific Ocean. But others contend that it is mainly associated with *regional* energetics and dynamics.

Wyrtki et al. (1976), Quinn (1974), and others choose to explain the onset of El Niño as the dynamic response of the Pacific Ocean to atmospheric forcing. Wyrtki et al. (1976) contend that during a period of strong southeast trades lasting longer than one year, the circulation around the South Pacific high is intensified, especially in the South Equatorial Current. This leads to an accumulation of water in the tropical western Pacific so that there comes to be an east-west slope of sea level. As soon as the southeast trades weaken, the water accumulated in the western Pacific tends to return eastward. The result of this backslosh in the equatorial Pacific is an accumulation of warm water in the region off Ecuador and Peru. So, according to this theory, the beginning of El Niño is a consequence of a weakening of the southeast trades resulting in an accumulation of warm water in the vicinity of northern Peru.

In contrast to the above planetary type of explanation, Lettau (1976) proposes that the onset of El Niño may have origins that are regional rather than global in dimension. As an example, he suggests that an occasional migrating disturbance from the winter circulation north of the equator might produce a fleeting spell of rainy weather in the normally dry region of coastal southern Ecuador and northern Peru, at about 4°–5°S. Even though the disturbance hovers in the vicinity only a few days, the soil moisture from its rains remains until it is evaporated. As long as evaporation continues, the portion of net radiation available for dry convection is reduced, as is the diurnal range of air temperature. This weakens the wind stress along the coast and with it the upwelling of cool water offshore. With the disappearance of upwelling, warm water from the north invades the area south of the equator, evaporation increases, and heavy local showers may occur. What becomes evident is that a limited intrusion of moisture from outside sources acts as a trigger, resulting in showers maintained by local moisture sources that can continue long after the original disturbance disappears. This emphasis on concepts of regional energetics and dynamics is in sharp contrast to the one that views the El Niño as an event resulting from changes in the ocean-atmosphere system over the entire equatorial Pacific and beyond.

While the heavy showers associated with a southward displacement of equatorial air, water, and the *ITC* are not felt along the coast much south of Callao, there were reports of heavy rains in the highlands of interior northern Chile in the summers of 1925 and 1952, seasons which were likewise wet within coastal Peru (Rudolph, 1953). Just what relationship exists between the rains on the Peruvian littoral and those in northern interior Chile is not known.

The Subtropical *(Cs)* and Temperate *(Do)* Climates of Central and Southern Chile

Poleward from the desert of Chile the arrangement and characteristics of the climates west of the Andes show few noteworthy anomalies, so that only brief comments are required. A very narrow area of semiarid climate (*BS*) is followed on the south by the dry-summer subtropical type (*Cs*) occupying approximately latitudes 30° to 36°, and this in turn is succeeded by the temperate oceanic (*Do*) climate at still higher latitudes. Poleward of about 50° there may be limited areas of tundra climate (*Ft*).

THE *Cs* CLIMATES OF CENTRAL CHILE. —Throughout the region of *Cs* climates in central Chile, the terrain layout resembles that of California, with a coastal range separated from the high Andes by the Vale of Chile whose altitude above sea level ranges from 350 to 700 m in the north to 100 to 300 m along the Bio-Bio River in the south.

A cool marine Mediterranean climate (*Csb*) prevails along the littoral, as one might expect. But unlike the situation in California where a distinctly continental Mediterranean climate (*Csa*) with high summer temperatures prevails inland from the coastal hills in the Great Valley, the Vale of Chile shows no such marked temperature contrast with the coast. This is because altitudes in the Vale tend to offset somewhat the effects of the inland and protected location. Still, while the annual range of temperature is only about 7°C (12°F) at Valparaiso on the coast, it is twice as great at Santiago in the interior at 500 m. But the average of the warmest month at inland Santiago is only about 20°C (68°F), so that it falls short by a few degrees of qualifying as a *Csa* climate. Although station data are lacking to prove it, one suspects that south of Santiago, as the Vale declines in elevation, summer temperatures may rise high enough to permit classifying it as a *Csa* climate.

Mark Jefferson's (1921) rainfall map of central Chile shows that the Vale has less precipitation than the coast, the coastal range, or the Andean slopes in similar latitudes. Thus, the isohyets loop southward in the Vale, with the 500 mm isohyet located 400 km farther south in the Vale than it is along the coast. Within the Vale there is consistently more rainfall on the eastern than on the western side which experiences the effects of lee location behind coastal hills.

THE MARINE *Do* CLIMATIC AREA OF SOUTHERN CHILE.—Southward of about latitude 38°S the dry summers typical of *Cs* climate become markedly wetter. At the same time the Mediterranean scrub forest terminates at about the Bio-Bio River and dense mixed forests take its place. Rainfall also increases markedly in total amount: 500 mm at Valparaiso and 3,700 mm at Valdivia. But while poleward of 38°S the summer months are not dry, as they are farther north, they are far less wet than the winter months. Thus, Valdivia at nearly 40°S receives 71 percent of its annual precipitation in the six winter months, indicating that in the summer period the Pacific anticyclone is stronger and cyclonic activity less marked. As Table 2.1 clearly indicates, with increasing latitude the discrepancy between rainfall in the summer and winter half-years wanes, until in latitudes poleward of about 50°S seasonal accent ceases to be conspicuous. In these higher latitudes there is little seasonal variation in the strength of the westerlies or in the cyclonic activity.

According to Köppen's classification extreme southern Chile, poleward of about 50°, is designated as having tundra climate, since the average temperature of the warmest month is slightly below 10°C (50°F). On the other hand, this region does not have a tundra vegetation, but instead is forested. It would appear that while Köppen's employment of the isotherm 10°C (50°F) for the warmest month is fairly satisfactory as marking the northern limit of forest in the continental subarctic regions of northern Eurasia and North America, it is less satis-

Table 2.1. Monthly percentages of total annual rainfall in Chile (Jefferson, 1921)

Stations	Lat.	J	F	M	A	M	J	J	A	S	O	N	D	6 Winter months
Santiago	33°27′	0	.5	2	2	22	25	26	13	4	3	1	1.5	90
Valdivia	39°48′	2	3	5	8	12	17	15	14	10	6	5	3	71
Puerto Montt	41°28′	6	6	7	8	12	10	16	9	7	6	7	5	62
Melinka	43°54′	4	5	6	8	13	11	13	10	9	7	7	7	61
San Miguel	53°43′	7	8	8	13	10	8	9	6	5	9	8	9	54
Bahia	58°58′	9	10	11	9	9	6	8	8	8	8	8	9	48

factory in its coincidence in a very marine area such as southern South America. In order to exclude such cool marine forested areas as southern Chile from being classed as tundra climate, it has been suggested that the definition of tundra climate be supplemented by stipulating also a mean annual temperature of 0°C (32°F) or below.

CHAPTER 3

Atlantic South America: Middle Latitudes and Subtropics

The Patagonian Dry Climate

Other than in Patagonia, nowhere does desert climate extend down to the open ocean on the eastern side of a continent in middle latitudes. To be sure, in northern China a subhumid or semiarid climate with around 500 – 600 mm of rainfall does reach tidewater along the western margins of the nearly enclosed Gulf of Pohai. But this is quite a different situation, for the Gulf of Pohai, although salt water, is far from the open ocean, and what is more, the total annual rainfall along the littoral in Patagonia is only about one-quarter of that which prevails in North China. Bearing a somewhat closer resemblance to the situation in Patagonia is the subhumid Canterbury Plain on the eastern side of South Island, New Zealand, situated to the lee of a great terrain barrier. But here too annual rainfall is usually four times and more that of most of Patagonia. What at first glance is even stranger is that the Patagonian desert lies in a part of the continent where the land mass is narrow, and marine influence consequently strong. The eastern margins of both North America and Asia, with their more severe continental temperatures, show far greater amounts of precipitation. Patagonia appears to represent a distinct climatic anomaly, for homoclimes for the two littoral stations of Santa Cruz and Sarmiento (see Table 3.1) are nonexistent on other continents. A middle-latitude lee-side desert with marine temperatures is unique to South America.

Throughout the Patagonian dry area the seasonal temperatures resemble those characteristic of the west sides of middle-latitude continents rather than of the east sides. Only

Table 3.1. Temperature and precipitation in Santa Cruz and Sarmiento in Argentina

	J	F	M	A	M	J	J	A	S	O	N	D	Year
Santa Cruz 50°01′S													
Temp. (°C)	14	14	12	9	5	2	2	4	6	10	12	13	8.5
Precip. (mm)	21	16	20	17	25	18	16	15	12	7	15	18	200
Sarmiento 45°35′S													
Temp. (°C)	17	17	14	11	7	4	4	6	8	12	14	16	10.8
Precip. (mm)	10	8	11	15	24	16	17	15	10	6	12	9	153

at the very northeastern margin, really beyond the desert proper, does the temperature of the warmest month exceed 21°C (70°F). Characteristic summer-month temperatures of 16°–18°C (60 – 65°F), winter-month temperatures of 1°–4°C (35–40°F), and annual ranges of about 14°C (25°F), bear close resemblance to their counterparts along the littorals of western Europe and western North America. Narrowness of the continent, the great strength and persistence of the Southern Hemisphere westerlies, and the narrower, lower, and more broken nature of the Andean barrier in these latitudes, all combine to permit a large-scale transport of maritime air from the eastern Pacific. There is too little longitudinal width to permit the development of any considerable degree of continentality. Thus, Patagonia is a cloudy desert, the average annual cloudiness being 60 to 70 percent, reaching over 70 percent in its southern parts.

Precipitation everywhere is low and the desert extends down to tidewater, about 125 mm of annual rainfall being characteristic of east-coast stations. Inland, toward the base of the Andes, rainfall increases somewhat, Col. de 16 Octubre at 560 m recording nearly 400 mm. The climatic map shows the desert to be bordered on its inland side by a belt of steppe paralleling the highlands (Fig. 3.1). Thus, there is the further anomalous situation of the precipitation increasing from the coast toward the interior (Fig. 3.2).

Rainfall in the Patagonian desert shows a seasonal concentration which roughly corresponds with the situation in comparable latitudes on the western side of the Andes in Chile. This would suggest that the rainfall controls operating on the Pacific side of the mountains are likewise effective to the east. Thus, Santa Cruz, which is representative of the southerly parts, shows no marked seasonal concentration of rainfall, thereby resembling stations in southern Chile. But far-

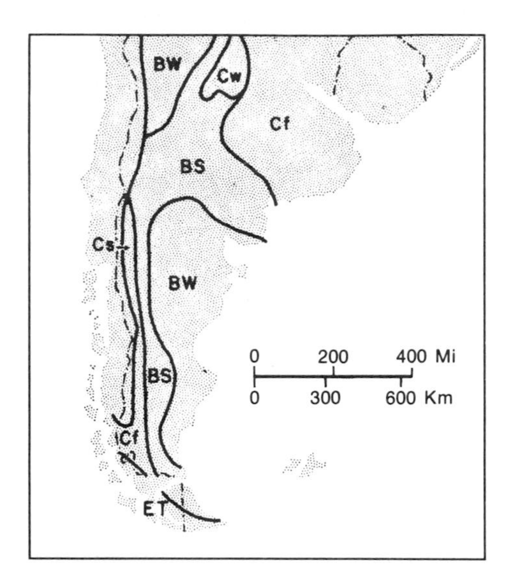

Fig. 3.1. Arrangement of climatic types in Argentina according to the Köppen classification. In Patagonia a desert climate prevails on the ocean side and the climates become less dry toward the interior. In subtropical latitudes farther north the east-west arrangement of dry and humid climates is reversed, with the dry climates located to the west, interior of the humid climates. (After Burgos and Vidal, 1951.)

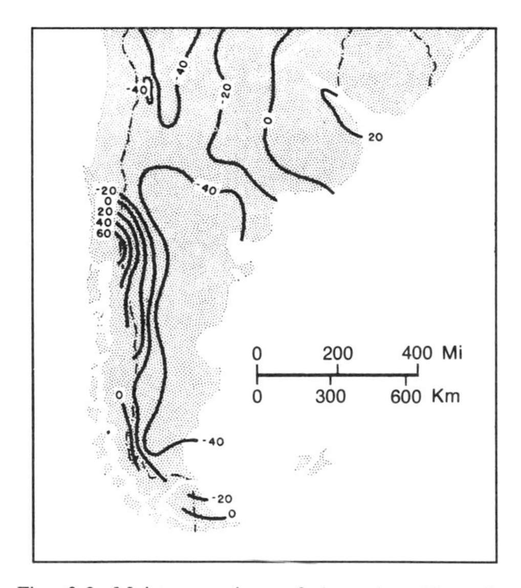

Fig. 3.2. Moisture regions of Argentina. Note the strong negative value over most of Patagonia, with an increase toward the interior. (After Thornthwaite, 1948.)

ther north at Sarmiento (45°30′S), and Col. de 16 Octubre (42°12′S), the winter months have three to four times the amount of the summer months. This seems to suggest that the South Pacific high makes its influence felt well into the middle latitudes during the high-sun period, when the routes of perturbations in the westerlies are displaced farther poleward.

The two principal problems of Patagonian climate requiring further analysis are (1) the causes for the aridity of this marine climate and (2) the reason why rainfall increases inland from the sea, so that steppe replaces desert in the submontane zone.

Origin of the Rainfall Deficiency.— The seemingly obvious reason for the general deficiency of rainfall is that the region lies on the lee side of the Andes, in the belt of strong zonal westerly winds. But in the middle-latitude westerlies most of the rainfall over lowlands is associated with disturbances of various kinds, including cyclonic and frontal systems. The question then becomes one of why disturbances in the westerlies produce so little rainfall in the maritime air over Patagonia. A partial explanation, at least, is related to the effects of the Andean barrier upon the westerly flow and its disturbances (Boffi, 1949). The broadscale air currents from the Pacific Ocean, since they cannot flow around the Andean barrier, are forced to surmount it. As they ascend the western slopes, the air columns are caused to shrink vertically, while on the lee side as they descend they are stretched. The blocking and the thermal effects of the highland barrier result in the development of an upper-level pressure ridge over the mountains, while a meridional trough is generated farther to the east over the ocean. This pressure ridge over the Andes, and the trough farther to the east, provide the anchor points for the high-level long waves in the westerlies of the Southern Hemisphere.

Schwerdtfeger (personal communication) notes that when *stable* air in the lowest 2,000 ± m of the atmosphere moves from the west toward the Andes in southern Chile, because of adiabatic lifting it becomes colder over the windward slopes of the mountains than it is farther out over the Pacific Ocean. As a consequence, a high pressure ridge will form over the western slopes, for the westerly winds over the ocean are deflected to northwesterlies near the coast of southern Chile; this continues as long as the little-disturbed flow of air above the level of the crest of the mountains is approximately from west to east. On the downwind, or lee, side over Patagonia, while the *surface* winds are still from the west, at several hundred meters above ground level, as friction fades, a southerly component must exist. This implies that there must be a flow of colder and absolutely drier air over Patagonia. Such an environment is scarcely conducive to condensation and precipitation.

Prohaska (1976) has pointed out that the climatic charts of the oceans reveal distinctly less cloud and precipitation frequency over the western South Atlantic east of Patagonia than in the eastern South Pacific west of Chile. Satellite photographs reveal a similar disparity in the cloudiness of these two oceanic regions. Such evidence appears to suggest the presence of contrasting atmospheric circulation patterns.

In conformity with the orographically induced pressure ridge described above, the winds over Patagonia east of the Andes are southwesterly and characterized by anticyclonically curved flow and associated subsidence (Fig. 3.3). This southwesterly anticyclonic flow gradually gives way to cyclonically curved flow farther to the northeast in latitudes 35° to 40°S, so that the southwesterly current by degrees becomes westerly and then northwesterly. Boffi (1949) states that there is no evidence at the surface

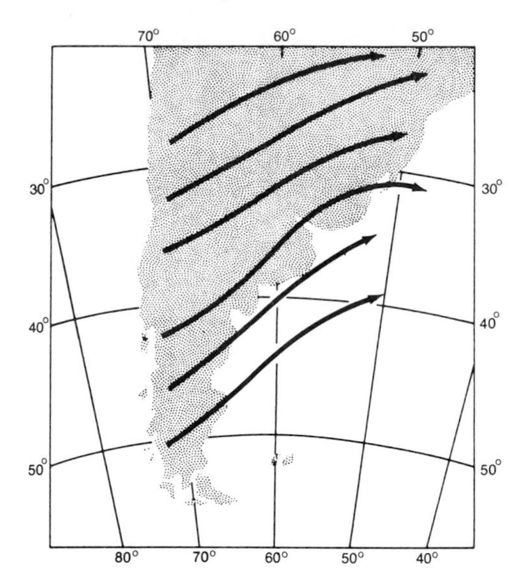

Fig. 3.3. Mean resultant winds over Patagonia in winter at 6 km. (After Boffi, 1949.)

that cyclones actually cross the mountains north of 50°, although one may assume that their high-level counterparts do. But these do not regenerate into active storms observable at the surface until they reach the longitude of the upper trough which lies well to the east of Patagonia over the ocean, but which approaches much closer to the coast at latitudes 30° to 40°S where the continent extends farther eastward.

The surface cyclones arriving from south of 50°S are relatively few, only about two a year in summer and three a year in winter (Boffi, 1949). These move from SW to NE toward lower latitudes, following the upper-level anticyclonic flow, and tend to decrease in intensity in this subsident air stream, so that they are poor rain-bringers. On the other hand, those cyclones originating farther north in subtropical Argentina and Uruguay, where they are affected by tropical air from the north, intensify as they move toward the southeast and higher latitudes, since they come under the convergent effects of the upper-air trough which here is closer to the coast.

A somewhat similar situation to that in Patagonia exists in the United States east of the Rockies where the above-surface cyclones which have crossed the mountains are weak and ineffective rain-producers, and do not regenerate as long as they are moving toward lower latitudes in the anticyclonically curved flow. Active regeneration of the disturbances begins only after their tracks revert from a southeast to a northeast direction coincident with the change from anticyclonic to cyclonic curvature in the flow as the trough is approached. Thus, the difference between the eastern coasts of middle-latitude North America and South America is that the orographically induced trough aloft is situated well east of the coastline in narrow Patagonia, while it is well inland from the east coast over the United States. It is chiefly over the semiarid Great Plains that cyclonic disturbances have attributes somewhat similar to those in Patagonia.

It appears, therefore, that the general effect of the Andes upon the westerlies is to create an upper-level pressure ridge over the narrow continent and a trough of low pressure well to the east over the sea. Hence, any disturbances in the westerlies remain weak, failing to regenerate and become effective rain-bringers until after they have left the continent and approach the longitude of the trough. Farther north where the continent is wider the east coast is brought nearer to the position of the trough aloft and hence the rainfall is greater. If the continent were broader to the south, so that the trough were positioned near the coast, it seems likely that its eastern margins would be wet as they are in North America and Eurasia. It must be admitted, however, in spite of the usefulness of the preceding explanation, that the abnormally low annual precipitation of maritime Patagonia, much lower than that over the

Great Plains in the interior of a great continent, continues to remain something of a puzzle.

Auxiliary controls acting to intensify the drought may be (1) the unusual strength and constancy of the westerly circulation in the Southern Hemisphere; (2) the lack of any monsoonal indrafts of tropical maritime air; and (3) the stabilizing effects of the cold Falkland Current which parallels the Patagonian coast. The vigor and constancy of the westerlies operate to maintain undiminished the anticyclonic subsidence over and just to the east of the Andes described previously. This same vigorous westerly flow, acting in conjunction with weak cyclonic activity and the narrowness of the continent which precludes the development of a thermally induced summer low, operates to exclude invasions of humid tropical air, such as occasionally affect the weather of the Great Plains in summer.

Cause of Rainfall Increase Inland.— The second unusual feature of Patagonia's rainfall requiring explanation is the fact that rainfall increases inland away from the coast, with the result that a belt of steppe borders the desert on its land side. Still closer to the mountains the rainfall is sufficiently plentiful to produce a narrow belt of humid mesothermal climates (Köppen's *C*). This somewhat more humid submontane belt, located well inland from the coast, is most conspicuously developed between about latitudes 35° and 50° (Figs. 3.1, 3.2).

Conceivably the cause of the increased rainfall of the interior steppe climate might be the upslope winds with an easterly component associated with the passage of the relatively few and weak cyclonic storms. However, the fact that these disturbances have a trajectory which carries their centers well to the east of the mountains, appears to reduce the likelihood of their generating an easterly flow in the steppe belt. This is verified by the fact that stations representative of this western Patagonian region experience surface winds with a direction other than SSW to NNW only rarely. Rainfall occurs at these stations in association with westerly winds or when there is little surface wind movement. It appears likely that most of the increased rainfall in western Patagonia is derived from air of Pacific origin and is a consequence of a spillover of precipitation from the orographically rising currents to the west of the Andean watershed. In the strong westerly flow, cumuli developed over the mountains are displaced eastward for some distance beyond the crest. There exist instances of rain produced by the orographic ascent of Atlantic air with winds from an easterly direction, but these are so rare that their effects on the annual total are almost negligible.

Subtropical South America East of the Andes

Much of this region, located between approximately 25° and 40°S, and including the northern half of Argentina, Uruguay, southern Paraguay, and southeastern Brazil, exhibits little in the way of important climatic anomalies.

With the greatly increased breadth of the continent north of about 40°S, its eastern side comes more and more under the influence of the trough aloft induced by the Andean barrier. Farther south where the continent is narrower and withdraws westward, the upper trough is positioned well eastward of Patagonia, over the ocean. In the latitude of the Argentine Pampa the anticyclonically curved southwesterly flow characteristic of the Patagonian region gradually changes to west and northwest and its curvature becomes cyclonic with resulting convergence. Cold fronts marking the leading edge of cold air masses advancing northward, here for the first time make contact with tropical air, with the result that this subtropical region becomes a major area of cyclogenesis and cyclonic regeneration. Disturbances moving eastward and southeastward in the cyclonically curved air stream deepen and intensify as they approach the longitude of the trough aloft, so that they are much better rain-bringers than those of Patagonia. As a consequence the humid climates (*Cf*) occupy a broad belt on the eastern side of the land mass.

In these subtropical latitudes rainfall decreases toward the interior so that the steppe is located east of the desert. This is the reverse of the climatic arrangement in Patagonia where desert climate lies on the eastern or oceanic side of the more humid steppe. The inverted locations of steppe and desert in the two regions suggest contrasting sources of moisture, with the westerlies from the Pacific providing the main source in the latitudes of Patagonia, while the Atlantic serves as the moisture source for the subtropical latitudes farther north. The higher and broader Andes effectively block the arrival of Pacific moisture in these somewhat lower latitudes. It has been pointed out previously why dry and subhumid climates are less extensively developed in subtropical South America than in either Australia or southern Africa in comparable latitudes.

As one might expect, summer is the rainiest season throughout most of subtropical eastern South America, the accent becoming more marked toward the north where contact is eventually made with tropical wet-and-dry climate (*Aw*). Summer rainfall likewise becomes relatively more important toward the interior.

A noteworthy feature of the Argentine Pampa is the fact that its modest-to-moderate rainfall arrives in the form of heavy showers concentrated in a relatively few rainy days. Thus, the average rainy day on the Pampa records 15 to 20 mm of rain. Based upon frequency of precipitation alone, this region might be classed as steppe. It has been suggested that this fact of the precipitation being in the form of heavy showers distributed on so few days may be the cause for the lack of trees on the humid eastern Pampa, in spite of an annual amount of precipitation which normally would permit forest growth (Köppen, 1923).

CHAPTER 4

Atlantic South America: Tropical Latitudes

Tropical South America East of the Andes

Over most of tropical South America east of the Andes the normal world pattern of climates is conspicuous. Thus, in the vicinity of the equator there is an extensive area of tropical-wet climate (*Ar*) with abundant rainfall and no prolonged drought season. To the north and south of the Amazonian *Ar* are extensive areas of more moderate precipitation with a marked dry season at low sun. These are the *Aw* climates of the Orinoco Lowlands in Venezuela and Colombia north of the equator, and of the grasslands and scrub forest of interior Brazil south of the equator. Such an arrangement is the expected one.

The more notable departures from the expected and normal pattern of climates are (1) the relatively modest rainfall in easternmost Brazil in the region of the "hump," culminating in the extensive semiarid area, chiefly interior, but reaching the coast just west and north of Cape São Roque; (2) the region of winter maximum precipitation (*As*) in the coastal area of Brazil east of the highlands and southward from Cape São Roque to about 13°S; (3) the dry area along the Caribbean coast in Venezuela and Colombia, including as well the adjacent islands of Aruba and Curaçao (see chapter 5). In addition to these three major climatic singularities there are other less striking ones, such as the spring minimum and fall maximum in equatorial northeastern Brazil south of the equator, and the unusually heavy rainfall of the interior western Amazon Basin.

Local Controls of Brazilian Climates

Since the several climatic pecularities of eastern tropical Brazil south of the equator appear to be genetically interrelated, it is judged an efficient procedure to consider some of the controls common to the general area prior to an analysis of each of the climatic anomalies individually. In an earlier section dealing with the general climatic pattern and its controls for the whole of South America, brief comment was made concerning certain aberrations in the climatic controls in tropical eastern South America. These require further elaboration at this point.

Of outstanding importance is the fact that the easternmost part of Brazil is dominated by divergent, subsident anticyclonic air to an extent that is not true of land areas in the western North Atlantic in similar latitudes. In part, at least, this appears to be the result of wedge-shaped tropical South America south of the equator thrusting so much farther eastward than do the tropical lands between 0° and 15°N, so that easternmost Brazil projects farther into the more stable parts of the South Atlantic anticyclone than do Northern-Hemisphere South America and the Caribbean lands into the North Atlantic high. Not only is the center of the South Atlantic cell closer to the western side of the ocean, but it likewise extends closer to the geographical equator than does its counterpart in the North Atlantic, with the result that the climatic equator, or equatorial trough of low pressure, is held in a position north of 0° throughout most of the year, and is able to extend its influence only a few degrees south of the equator along the northeastern coast of Brazil.

The greater prevalence of stable anticyclonic air in the westernmost tropical South Atlantic and adjacent eastern Brazil, compared to the situation in the westernmost tropical North Atlantic, is evidenced by the lower base of the trade-wind inversion in the former and a larger amount of temperature increase and relative humidity decrease from bottom to top of the inversion (Riehl, 1954). Additional evidence of prevailing atmospheric stability and the feeble penetration of the *ITC* in the vicinity of easternmost Brazil is the complete absence of hurricane development in the western tropical South Atlantic and the relatively few days with thunderstorms, only one-third to one-fourth as many as in the westernmost tropical North Atlantic. Studies of seasonal and daily inversion patterns are not available for easternmost Brazil, but the local meteorologists recognize that variations in the inversion, as related to both temperature and humidity, play a prime role in determining the current weather of that area. There appears to be something of a seasonal cycle, with inversions stronger and lower at Recife in spring and higher in fall. The drying effect of the anticyclone's easterly flow is carried deeply into central Brazil, even to the middle Amazon Basin in winter-spring. In summer the anticyclone withdraws eastward and southward so that a flow of humid unstable northerly and westerly air is able to cross the equator and enter deeply into central and southern Brazil. But the eastward withdrawal of the high is not sufficient to permit the easternmost "hump" of Brazil to escape its influence even in summer, when a divergent flow continues to prevail, thereby obstructing a normal southward migration of the *ITC* over that region.

Cold Fronts.—Still another weather control of first magnitude influencing the weather of much of eastern tropical South America south of the equator is the northward movement of masses of cold air from the subantarctic regions. In fact the northeastward progress of these cold fronts over subtropical and tropical South America east of the Andes and south of the equator constitutes the main weather activity with which the forecaster in this region is obliged to deal. It has been noted in an earlier section that because of the disrupting effects of the Andes upon the zonal westerlies, the longitudes to the east of the Andes are the principal quadrant within the Southern Hemisphere where there occurs an important meridional exchange of air between high and low latitudes. This exchange takes place through the break in the subtropical ridge of high pressure which is represented by the atmospheric corridor between the South Pacific and South Atlantic cells. As a surge of cold air moves equatorward a cyclonic circulation is likely to develop along the leading edge of the air mass at about latitudes 30°–40°S, either over the land or over the waters to the east. This temporarily slows down the northward movement of the cold air, but as the cyclone moves eastward there is a resurge of the polar air northward. The cool current of southerly air is retarded by the Brazilian Highlands, with the result that the cold front pushes northward more rapidly through the interior lowlands than it does farther east and along the coast (Fig. 4.1).

There are significant seasonal contrasts in the movement of the cold air, both over the continent and along the east coast. In the warmer seasons the depth, magnitude, and frequency of the cool air masses are less, as might be expected. Along the coast of Brazil they move farther north in the low-sun period, some even reaching the latitude of Cape São Roque (about 5°S), while Rio de Janeiro approximately marks the equatorward limits of the weaker fronts of summer. During the cool season the cold fronts follow each other in rapid succession, so that much cloud and disturbed weather prevails along the east coast. But, because there is no southward streaming of tropical air between the frequent cold fronts of winter, each new

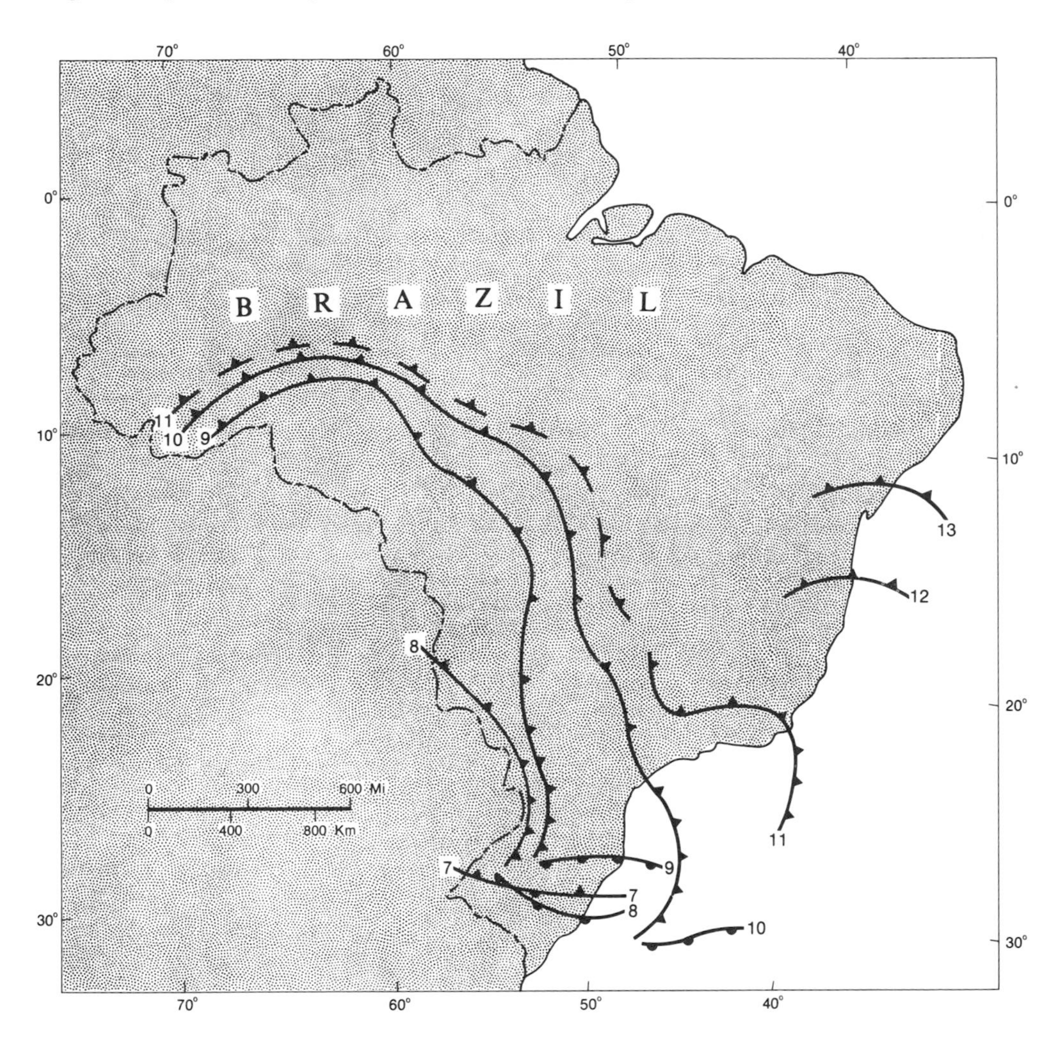

Fig. 4.1. A northward advancing polar front invading tropical-subtropical Brazil, both in the interior and along the coast, November 7–13, 1963, at 1200 GMT. (After Ratisbona, 1976.)

cold front encounters only the modified cool air of the preceding one, with the result that strong convective activity is less likely than in the warmer months of summer and the intermediate seasons when air-mass contrasts are greater (Coyle, 1940).

In the interior the situation differs somewhat. There during the low-sun period the cold fronts invade the realm of the stable land trades, occasionally reaching even to the equator, so that much less weather results than during similar synoptic situations along the coast, stratus clouds being the chief consequence. But in summer, when the interior, southward as far as the tropic or beyond, is flooded with humid and unstable northwesterly air, the weaker infrequent thrusts of cool air develop very active weather, with vigorous convection occurring along the zone of convergence with tropical air.

The disturbances which move northward in association with the leading edge of the polar air, do not, according to Coyle, show true cyclonic form with closed isobars, and warm, cold, and occluded fronts. Instead, the disturbance is in the form of a trough, with which cloud and rain are associated. Even after frontolysis has begun at the surface, the effects of the cold air continue to be felt in the form of disturbed weather farther north, resulting perhaps from continued cold-air advection aloft, or from the development of an induced trough. True vortex disturbances may develop along the cold front eastward from the continent over the ocean, and the strong southeast winds from these oceanic disturbances are able to affect the weather along the littoral, especially south of Rio de Janeiro.

In reality a considerable variety of local weather conditions may accompany the northward advance of the cold front, depending on the season, the depth and strength of the cold air mass, the terrain character, and the directional alignment of the coast (Coyle, 1940). Along the east coast of Brazil the weaker and shallower cold thrusts are to a greater extent retarded by the highlands. As a result the front may take up a position parallel to the coast with the cold air banked up on the seaward side of the highlands between Porto Alegre and Rio de Janeiro. The result is bad weather along this whole stretch of littoral. This may last for several days in succession. A weak remnant of the cold air may move farther north along the coast toward Salvador, producing only mildly unsettled weather. A deeper and stronger cold air mass, because it is less obstructed by the eastern highlands, is likely to proceed northward more rapidly and with the front maintained at nearly right angles to the coast. Under such conditions the bad weather associated with the front is of shorter duration, clearing is more rapid, and the disturbance is able to progress farther into the lower latitudes. Less is known about the variations in cold-front weather in the interior.

The Coastal Region of Summer Drought and Winter Rainfall Maximum (As) South of Cape São Roque

Between 10°S and 20°S along the Atlantic littoral of Brazil there occurs a very remarkable change in seasonal rainfall regime. Thus, in the latitude of Rio de Janeiro and Santos the summer half-year records 60 to 65 percent of the annual rainfall. This is relatively normal for the tropics. But between Cape São Roque and about 13°S the situation is reversed, and 70 to 80 percent of the year's rainfall comes in the winter half-year (Figs. 4.2 and 4.3). This is distinctly abnormal for these latitudes, where characteristically the annual march of rainfall is one in which low-sun is relatively dry and high-sun wet. In between these two east-coast areas of opposite annual rainfall varia-

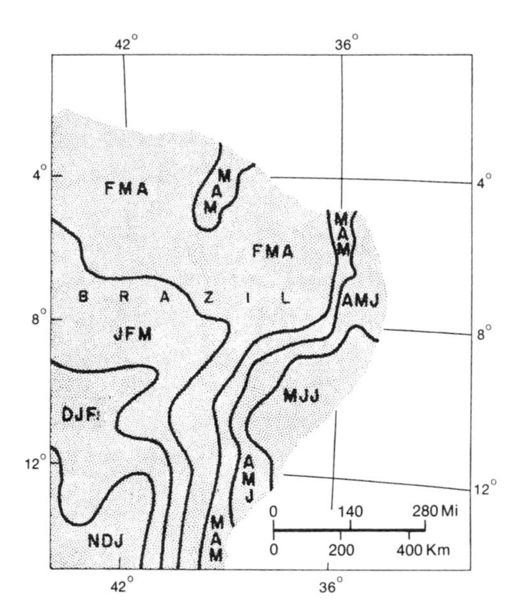

Fig. 4.2. Time of maximum precipitation in eastern-central Brazil, over a period of three consecutive months. The three-letter symbols are derived from the first letters of each of the three months included. (From *Atlas Pluviométrico do Brasil,* 1948.)

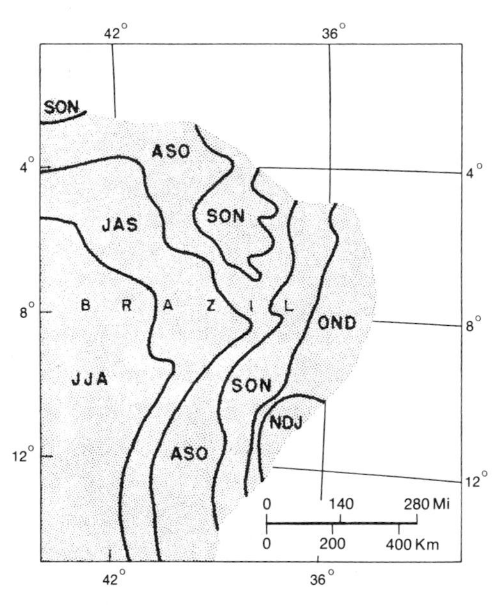

Fig. 4.3. Time of precipitation minimum in eastern Brazil during a period of three consecutive months. (From *Atlas Pluviométrico do Brasil,* 1948.)

tions, approximately latitudes 13° to 17° or 18°S, is a transition area where seasonal accent is not marked, but winter is still a rainy season. Attention here will be focused principally upon the coastal region of winter maximum southward from Cape São Roque. Here the annual rainfall variation of the coastal strip is not only in contrast to that farther south along the coast, but it is also opposite to that in the interior in the same latitude, where a strong summer maximum prevails. Moreover, the transition zone between the coastal winter maximum and the interior summer maximum is relatively narrow, for 120 to 250 km inland a strong summer maximum already prevails.

Annual rainfall amounts on the coastal lowlands and slopes immediately south of Cape São Roque are by no means large for a tropical littoral closely bordered by highlands and with onshore winds approximately normal to the coastline and highlands. The composite average annual rainfall for eight stations within this region is only 1,300 mm, and two of the stations show less than 875 mm. Similar windward coasts backed by highlands located north of the equator, such as in the Guianas, Central America, and the more rugged Caribbean islands, show annual rainfall amounts that are 50 to 150 percent greater. Clearly the precipitation is relatively modest considering the geographical arrangement of sea, winds, and highlands. If the onshore *mT* air were not divergent and moderately stable one would expect the precipitation to be much greater. It appears, therefore, that the relatively strong anticyclone and the frequent, if not necessarily low-level, inversions have a depressing effect upon rainfall over the whole northeastern region. Certainly the anticyclone operates to prevent a normal seasonal southward and southeastward migration of the *ITC* over the region of the Brazilian "hump," for only occasionally do heavy summer rains associated with the *ITC* extend southward of Cape São Roque. Here the moderate rainfall is partly of orographic origin, but with important contributions made by several types of perturbations including easterly disturbances in the deep trades, and troughs associated with cold fronts which move northward along this east coast.

Even more striking than the moderate amount of rainfall is the fact that the annual precipitation profile shows a dry season in spring-summer and a pronounced maximum in winter-fall. At Recife October rainfall is only 18 mm. North of about 8°S April-May-June are the wettest three months, but farther south this changes to May-June-July. Similarly, there is a slight change in the time of the driest period, October-November-December in the north, and November-December-January in the south (Figs. 4.2, 4.3).

In attempting to explain the seasonal rainfall distribution, it may be advantageous to focus attention as much on the strong high-sun minimum as on the low-sun maximum. Actually it is not unusual for tropical east coasts backed by highlands to have relatively abundant rainfall in the low-sun period. What is more unusual is for the high-sun period to be so dry.

One important rain-making element which operates at a maximum in the low-sun period, but is very much weaker or even absent, in summer and spring, is the cold-front disturbance described earlier. There it was pointed out that the effects of such disturbances are felt as far north as Cape São Roque in winter but ordinarily not much beyond the tropic in summer. Cold-front weather is not experienced north and west of Cape São Roque where the coastline bends sharply to the northwest. A considerable variety of weather types is associated with the cold fronts, depending on the kind of atmospheric environment which the cold air is invading. Where it converges with tropical air curving southward around the western end of the anticyclone, as is more likely in the lower latitudes, showery squally weather is usually the result and rains may be heavy.

Hunter (unpublished) is of the opinion that cool-air advection associated with polar pressure waves is the primary cause of the cool-season (April-August) rainfall maximum at Recife on the northeast coast of Brazil. He observes that during a cool-season rainy spell in this area the air is normally cool, moist, and unstable up to the 500 mb level, or somewhat beyond the icing level. As the rains begin there is both a temperature and pressure drop, while rising pressure and temperature are coincident with diminishing rains. The weather accompanying a northward advancing cold front depends upon the pressure and wind conditions that it encounters. In winter, despite their greater original intensity, polar fronts along the east coast of Brazil are subject to frontolysis somewhat farther south than is true in the warmer months. During the east-coast rainy period, April to August, the cold air invades the domain of a strong subtropical high over Brazil which usually causes frontolysis to begin between 15° and 25°S along the coast. On the surface weather map the cold front is shown as stationary or slowly moving with the cloud and rain diminishing in intensity. Significantly at this stage an important change in weather occurs some 650–1,300 km ahead of the cold front. Winds become gusty and veer to the south, the inversion disappears, and there is a drop in temperature and pressure at the 750 and 600 mb levels. At this stage rain showers begin. There is frequently a brief clearing period of 12 to 24 hours before the second, and often more active, period of the weather cycle commences. It begins with a fall in pressure poleward of about 30°S, winds back from southerly to easterly at Recife, the air flow aloft changes from anticyclonic to cyclonic, and precipitation starts in the form of heavy rain showers falling from lines of squall clouds advancing inland from the sea accompanied by strong gusty winds. Showery and even continuous moderate to heavy rains may continue for a day or two before the cycle of cold-front weather is gradually brought to a close, the whole cycle occupying perhaps 4 to 7 days.

It is Hunter's contention that frontolysis at the surface of the cold front farther south along the coast of Brazil is followed by frontogenesis aloft farther equatorward, and such a development is chiefly responsible for the winter rains south of Cape São Roque (Fig. 4.4). This explanation would not require the original surface cold front to advance deeply into the tropics in order to produce rain in the Recife area, although this is not uncommon. The result of frontolysis at the surface may be to produce a flow of cold air aloft northward of the stagnating cold front, or the action may be that of the polar front inducing a trough in the tropical easterly flow as described by Riehl (n.d.). During the periods of rainfall produced by such a pressure wave the trade-wind inversion at Recife normally is destroyed, either directly through cool-air advection aloft, or indirectly through the effects of large-scale pressure changes. If the temperature inversion persists, rainfall is meager or lacking. During the period November through February the diminished cool air along the east coast of Brazil, plus the lack of frontolysis action, precludes the development of much warm season rainfall along the coast just to the south of Cape São Roque.

To what extent easterly waves, or some modification of this type of disturbance, are responsible for the seasonal rainfall of the Recife region is not clear. There is published evidence that they do occur in this region (Berry, Bollay, and Beers, 1945) and the meteorologists at Rio de Janeiro and Recife with whom the subject was discussed agreed that they are a weather type of the tropical east coast of Brazil. This can only be inferred from the character of the local weather, however, for no weather reports are received from ocean areas to the east. Some synopticians locate easterly disturbances on the surface weather maps, but others are inclined to refrain. The feeling was expressed that while such disturbances may be present at most times of the year, they are significant as weather-makers chiefly in those seasons, or on those occasions, when the inversion is weakest and the air less stable. It is possible that some disturbances at present designated as easterly waves may be synonymous with the induced troughs resulting from invasions of polar air described previously.

It is something of a paradox that toward the end of the rainy season in late August and September, when the polar thrusts are still strong, there should be a marked falling off of the rainfall, so that October and November (spring) are two of the driest months. This same period, however, marks the beginning of the up-trend of the annual rainfall profile in central and western interior Brazil, where humid northerly air and the *ITC* are crossing the equator and invading the Southern Hemisphere. The fact that the *ITC* in September does not in a similar fashion start to move south of the equator along the east coast, appears to be related to the same cause that stifles the effects of strong polar thrusts from the south. Both rain-makers seemingly are held at bay in the Recife area by the subtropical high, which, while in spring it is weakening and retreating eastward in the western interior, has built up to its maximum over eastern Brazil in this transition season. Coyle (1950) refers to the development of a separate anticyclonic cell over northeastern Brazil during the period September to November inclusive and his air-flow maps for September and October

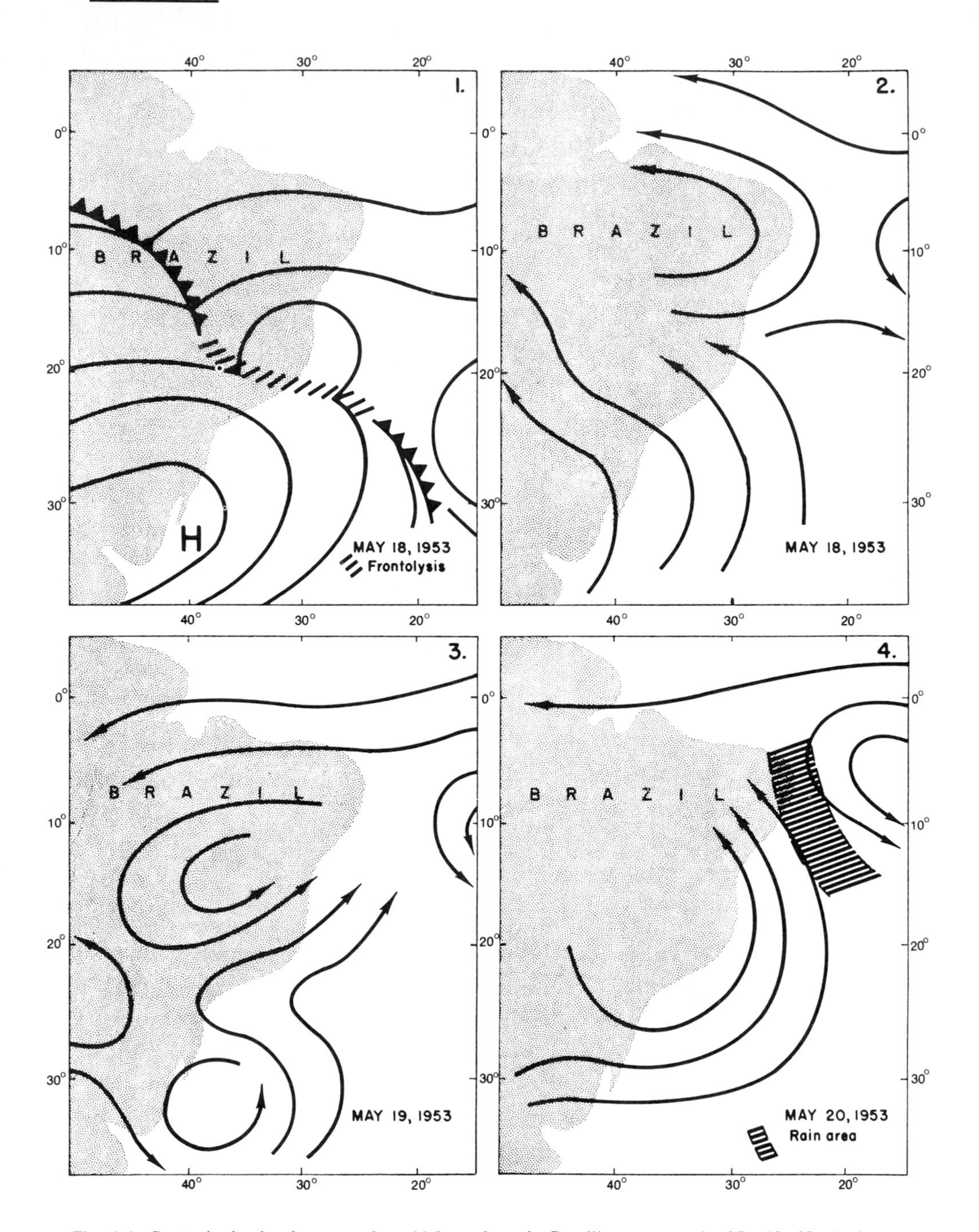

Fig. 4.4. Stages in the development of a cold front along the Brazilian coast south of Recife. No. 1 shows cold-front frontolysis in relation to the surface pressure pattern. No. 2 represents the streamline flow at 5,000 m on the same date, with frontolysis of cold front shown as located on the preceding surface chart. No. 3 shows the streamlines at 5,000 m on the following day, just prior to the beginning of rainfall associated with frontogenesis aloft developing equatorward of the surface frontolysis. No. 4 represents the streamline flow at 5,000 m in relationship to the rain area that has developed. (After Hunter [n.d.].)

show the stream of humid northerly air and the *ITC* invading the Southern Hemisphere as arriving from the west and northwest, but avoiding the eastern extremity of Brazil where the anticyclone continues to prevail (Fig. 4.5). Finally, by November the northwesterly equatorial air current reaches the east coast in the latitude of Rio via a land trajectory, but the northeast continues to escape its effects. It is, therefore, the strong build-up of an anticyclonic circulation over easternmost Brazil in September-October that stifles the effects of the cold fronts from the south and obstructs the advance of equatorial air and *ITC* from the north, so that spring is the driest season at Recife. It is not unexpected, therefore, that spring witnesses the lowest and most persistent inversions. In the following months of summer, which are still a part of the dry season at Recife, the weaker cool thrusts do not reach this far north, while the anticyclone, persisting in slightly weakened form, continues to repel the advance southward along the coast of the

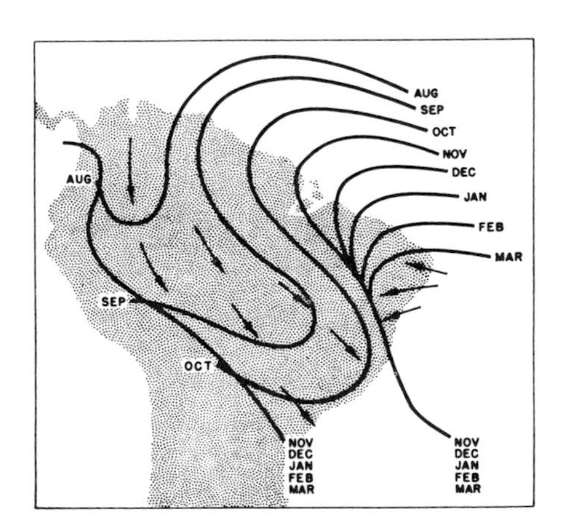

Fig. 4.5. Showing stages in the southward advance over Brazil in spring and summer (Southern Hemisphere) of the humid, unstable equatorial westerlies and the *ITC*. Note that the rate of southward advance of this northwesterly air stream is much more retarded along the north coast than it is in the interior, and that it normally does not succeed in overrunning extreme eastern Brazil. (After Coyle, [n.d., *e*].)

northerly air and the *ITC*. Nevertheless there is an appreciable increase in rainfall in January and again in February when the northerly equatorial air reaches its maximum southward development.

Still another factor making for a wetter winter-fall and a drier spring-summer has to do with the contrasting qualities in the opposite seasons of the easterly winds along this east coast south of Cape São Roque (Mintz and Dean, 1952). Thus, the magnitude of the resultant winds is greater in July (winter) and less in January (summer), so that more maritime air is crossing the coastline and being lifted over the coastal highlands in winter. At Recife higher wind velocities are indicated for winter-fall, the rainy season, than for summer-spring at both the 1,000 and the 2,000 m levels. The velocity of the resultant winds likewise is greater in July and less in January and the flow is more constant in July. Positive divergence is characteristic of the mean surface winds in January, the dry season, while a neutral condition as regards surface divergence is characteristic of July. The inversion base is likewise lower in summer than in winter (Vuorela, 1953). From the above observations of the seasonal surface winds, it is clear that in such properties as magnitude, velocity, constancy, and surface divergence the easterly air flow in July is more conducive to precipitation processes than is that of January.

In summary, the relatively dry spring and summer of the Brazilian coast south of Cape São Roque are attributable to the following conditions:

a. The relatively strong anticyclonic circulation continues to persist over easternmost Brazil in spring and summer, thereby preventing a normal southward and eastward migration of the unstable northwesterlies and the *ITC* at the time of high sun.

b. The cold-front disturbances are fewer and weaker in summer than in winter, and the same is true of easterly disturbances.

c. The characteristics of the general onshore easterly flow in summer are less conducive to the precipitation processes than are those of winter.

The Dry Region of Northeastern Brazil

That the atypically located dry climate in northeastern Brazil is difficult to define and bound is indicated by the variety of shapes and sizes that it assumes on various maps of rainfall and climatic types. To be sure, the area of drought varies so much in extent and location from year to year that the fluctuating *annual* boundaries of the dry climate add greatly to the difficulty of defining the outer limits of the normally dry region as a whole (Fig. 4.6). As defined by Serra's map (1941) (Fig. 4.7), it embraces roughly 400,000 sq km and it is considerably larger on Webb's (1954) map which employs Thornthwaite's formula for dry climates. Most of the dry region as outlined on Serra's map has an average annual rainfall of less than 760 mm, with fairly extensive areas having under 630 mm, and a few restricted spots with even less than 400 mm. Serra's map shows no arid or desert climate, but according to Webb (1954) both Köppen's and Thornthwaite's systems of classification indicate limited areas of arid climate. In reality Northeast Brazil is a region of alternating drought and flood. For about half the years the region experiences either disastrous drought or damaging flood, while during the other half, though free of climatic calamity, the region continues to suffer from the normal dryness.

For the most part the dry region lies well inland, but its northern extremity does extend down to tidewater along a limited stretch of coast westward from Cape São Roque. Here two stations record only 615 mm of rain, while farther to the northwest beyond Fortaleza the coastal station of Camocim likewise has an annual total of only 951 mm.

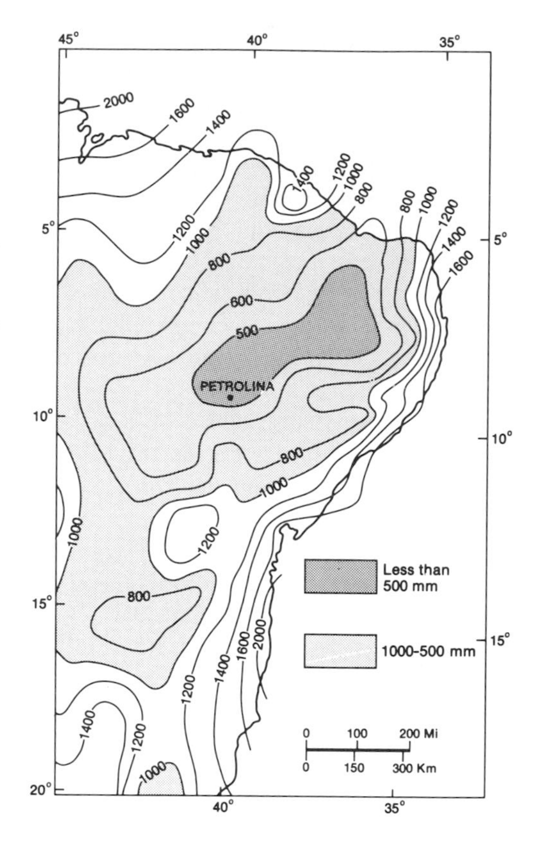

Fig. 4.6. Average annual rainfall in northeastern Brazil. (After Ramos, 1975.)

The average annual rainfall in this subequatorial dry region is not only meager (mostly under 750 mm, with the core area under 500 mm) but it exhibits a dangerous time and space variability as well. Such a climate is unusual in these tropical latitudes (5°–15°S), especially on the windward side of a continent not far removed from a warm ocean. Such an undependable climate causes serious economic and social problems, especially for a population mainly dependent on agriculture. Consequently, it has been a region of large-scale population out-migration. Most of the yearly rainfall over Northeast Brazil occurs in association with organized weather systems, or disturbances, rather than in random showers. Such distur-

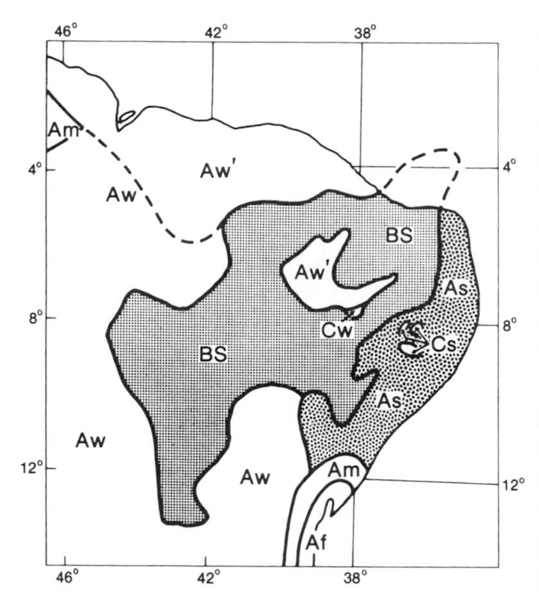

Fig. 4.7. Showing the location in eastern Brazil of the *BS* (dry) and *As* (cool season rainfall maximum) climates according to the Köppen system of classification. (After Serra, 1946, and Bernardes, 1951.)

bances move from east to west at slow speeds of 2°–3° of longitude (225–320 km) per day. During 1972, a representative year, Petrolina, located in the dry core region, had 533 mm of rain and 43 rain days, but only 13 days with precipitation of 10 mm or more per day. Any one day's precipitation is likely to occur in only a few hours. There was a total of only 10 rainy episodes (Ramos, 1974). Rainfall does not occur unless the relative humidity of the middle atmospheric layer is two to three times above the normal. The prevailing trade-wind inversion appears to play a major role in keeping rainfall amounts low, and in inhibiting isolated random showers. Only well-organized weather systems are able to weaken the prevailing temperature inversion sufficiently to permit deep cumulus convection with associated rainfall. Oddly, the northeast dry region has an amount of cloudiness which seems high considering the meager precipitation. However, most of this is either layered or small cumulus in type.

Over the entire dry area more rain falls in the high-sun than in the low-sun period so that in this respect there is sharp contrast with the winter-rainfall coastal region south of Cape São Roque. But within the dry area itself there are modest differences in the annual rainfall profile, the northern and northeastern parts, including the dry coast, experiencing a maximum in late summer and fall (usually March-April) and a minimum in spring and winter. By contrast, the more interior and southerly parts show a greater symmetry in rainfall profile, with summer the wettest season and winter the driest, although spring continues to be somewhat drier than fall. Consequently, much of the dry region, except its more northerly parts, has an annual rainfall variation which resembles that of the extensive *Aw* climatic region of interior Brazil south of the equator, in which the Northeast dry region is embedded. It is like the *Aw* except that it has only one-third to one-half as much total precipitation. The rainy season in Northeast Brazil, centered in late summer and fall, occurs at the time of the southernmost seasonal migration of a lower troposphere confluence axis (*ITC*) over the continent and the adjacent tropical Atlantic (Hastenrath and Heller, 1977).

A complex terrain characterizes the dry region. In a general way it may be described as an upland plain, most of it below 450m in elevation, above which rise prominent crystalline-rock massifs, and also erosion remnants of a former covering of sedimentary rocks. The local rainfall distribution within the general dry area is very complicated because of the differences in elevation. Unfortunately, most of the meteorological stations are on the lowlands, so that it is impossible to map accurately the localities of somewhat higher precipitation.

Rainfall is largely of the showery convective type. Freise (1938) writes, " . . . the rain pours down in heavy showers lasting from a quarter to three-quarters of an hour over an area of a dozen hectares or so to perhaps some two square kilometers. Rains lasting several hours and rains extending over five square kilometers or more are extremely rare. Rather frequently the whole rainfall of a month is concentrated in three to five showers not more than a day apart." The remarkable variations in rainfall both in time of occurrence and area covered are matched by the great variability in total rainfall from one year to another. This annual variability is at a maximum in the inland areas of greatest aridity where elevation is lowest. It is least along the dry northeastern coast, and at higher elevations in the interior.

The *ultimate* cause of the striking time variability of annual precipitation in Northeast Brazil is problematic. However, it does seem to be well established that the immediate cause for occasional wet years is an extra southward shift in summer of the equatorial trough and the *ITC* in the western tropical South Atlantic and adjacent South America. Among the most disastrous climatic hazards in South America are the infrequent El Niño heavy rains and floodings in dry northwest Peru, and the Sêcas, or droughts, in Northeast Brazil. These phenomena appear to occur simultaneously in particular years. Thus, a Sêca year in Northeast Brazil is typically matched by a wet El Niño year in far distant northwestern Peru, although the two regions are separated by a broad continent (Fig. 4.8). A definite link appears to exist between the two phenomena. In normal summers the *ITC* pushes only a few degrees south of the equator in both regions. This results in the usual subhumid condition in Northeast Brazil and the

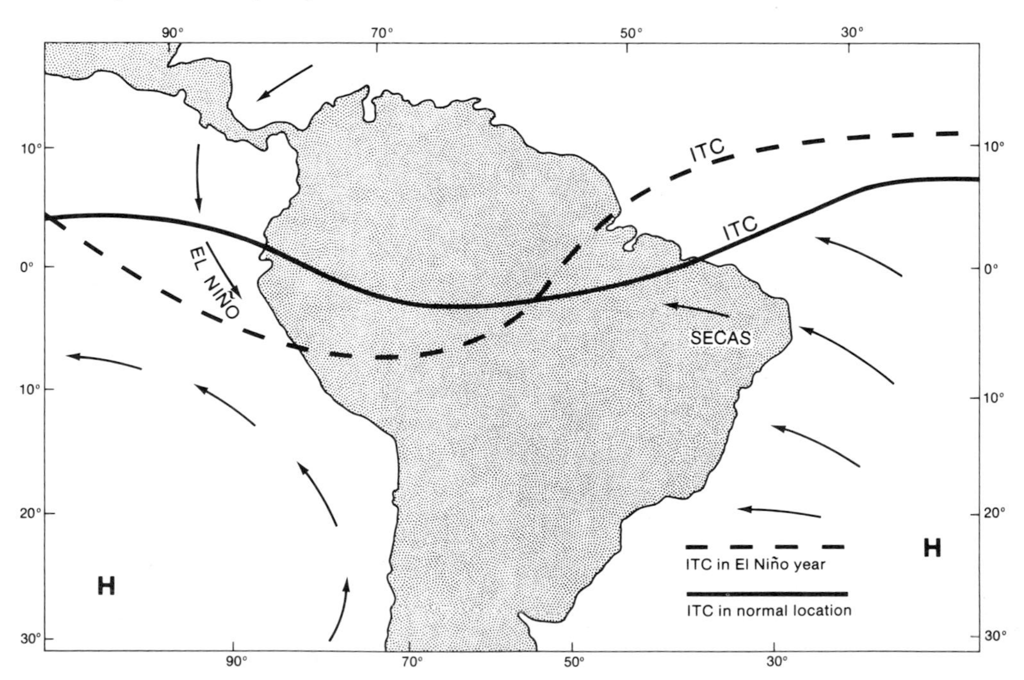

Fig. 4.8. Typical locations of the *ITC* and the trade winds of the Southern Hemisphere (1) during a normal year and (2) during a typical El Niño-Seca year. (After Caviedes, 1973*a*.)

normal (and desired) aridity in northern Peru, where dependence is upon irrigation. However, in abnormal summers, characterized by wetness and flood in Peru and intensified drought (Sêcas) in Northeast Brazil, the *ITC* is situated farther south of the equator (10° ±) in Peru and north of its usual position in the western South Atlantic, so it does not reach into Northeast Brazil. It is the unusual strength and persistence of the South Atlantic high-pressure cell that on those occasions dominates the situation and prevents the *ITC* from making its normal southward shift in Brazil. Simultaneously on the western, or Pacific side, a weakened and southward displaced subtropical high permits a farther-than-usual southward shift of the *ITC* and its accompanying rains (Fig. 4.8). The weakened Pacific high results in greatly diminished southerly wind and water circulation along the coast of northern Peru, and, therefore, an absence of cool upwelled water. It would seem, then, that the time coincidence of a plus anomaly rainfall in Peru with drought conditions in Northeast Brazil may have its origin in the counterbalance movement of the subtropical high pressure cells along the Pacific and Atlantic sides of tropical Southern Hemisphere South America, which in turn either attracts or repels the southward advance of the *ITC*. The likelihood of experiencing a drought year in Northeast Brazil when coastal northern Peru is experiencing a wet El Niño year is about 46 percent, and the probability of experiencing an El Niño year in Peru when Brazil is experiencing a drought is 60 percent (Caviedes, 1973*a*). These are high probabilities, given the relatively small number of occurrences of El Niño floods and Northeast Brazil droughts over a period of eight decades.

The causally related periods of wetness and drought in widely separated coastal regions, such as northwest Peru and Northeast Brazil, are thought to be linked to large-scale changes in the atmospheric circulation, not only south of the equator but in the Northern Hemisphere as well (Bjerknes, 1966). There is evidence that the strong interannual rainfall variation over Northeast Brazil is dependent on the degree of cyclonic activity or blocking in the New Foundland-Greenland area during the Northern-Hemisphere winter and spring (Namias, 1972). The linkage is traced through variations in the North Atlantic subtropical anticyclone, the northeast trades, and the responding low-latitude (Hadley) cell, which is obliged to change in both location and strength. Intense blocking action over North America and the North Atlantic is usually linked with ravaging drought in Northeast Brazil.

The Sêcas of Northeast Brazil appear to be related to an equatorward expansion of the South Atlantic high and a poleward retraction of its North Atlantic counterpart, with an associated northward displacement of the equatorial trough of low pressure. At the same time the zonal bands of maximum cloudiness and rainfall frequency are north of their normal positions. Also the North Atlantic trades are weaker and the South Atlantic trades are stronger than normal. During a wetter-than-usual rainy season in Northeast Brazil departure patterns are opposite to what they are in typical drought years (Hastenrath and Heller, 1977).

Since the semiarid coastal strip of dry Northeast Brazil, lying west and north of Cape São Roque, presents such striking contrasts in rainfall with the littoral just south of Cape São Roque, both in annual amounts and season of maximum, this coastal steppe warrants preliminary attention. Of great importance, as it relates to differing amounts of rainfall along these two coasts, is the contrasting alignment which they present to the prevailing southeasterly flow. Where the southeast trades meet the coast and its bordering highlands almost at right angles, as they do to the south off the Cape, the rainfall is heavier than to the north where they nearly parallel the coast or, in places, are even slightly offshore. In this respect it is significant that to the north of the Cape those stretches of coast which are more nearly east-west in direction, so that the southeast trades have a slight offshore component, are the driest. Further reducing the rainfall along the littoral north of the Cape is the fact that the highlands there are withdrawn farther from the coast, resulting in diminished orographic effects. In addition, the cold fronts and their induced troughs, which are to a considerable degree responsible for the cool-season maximum at Recife, do not follow the westward-bending coastline to the northwest of the Cape. Here the meager rainfall comes chiefly at high sun and is associated with northwesterly air flow and with disturbances of quite another variety, generated along the *ITC* farther to the north and west (Fig. 4.9).

We return now to the more general problem relating to the genesis of the dry region in its entirety. During the Southern Hemisphere winter almost all of Brazil south of the equator feels the drying effects of the greatly enlarged and intensified anticyclonic circulation over the continent. Throughout the period of this deep penetration of the subsident easterly flow, rainfall is at a minimum. But beginning in August or September the South Atlantic anticyclone weakens over the interior, and the *ITC*, accompanied by a flow of humid, unstable northerly and northwesterly air, crosses the equator and invades Southern Hemisphere Brazil. This invasion begins in the northwest and progresses rapidly across the country toward the southeast, thereby initiating the summer rainy season of the interior and reaching the southeast coast in the vicinity of Rio de Janeiro by October or November. Of unusual significance to the dry region of northeastern Brazil is the fact that while the anticyclone weakens and withdraws eastward in the interior as the sun advances southward, spring witnesses a maximum build-up of the cell of high pressure in easternmost Brazil, and in somewhat weakened form it continues to persist there throughout the summer. As a consequence, the southward and southeastward advance of the *ITC* and the equatorial westerly flow are greatly retarded in north-

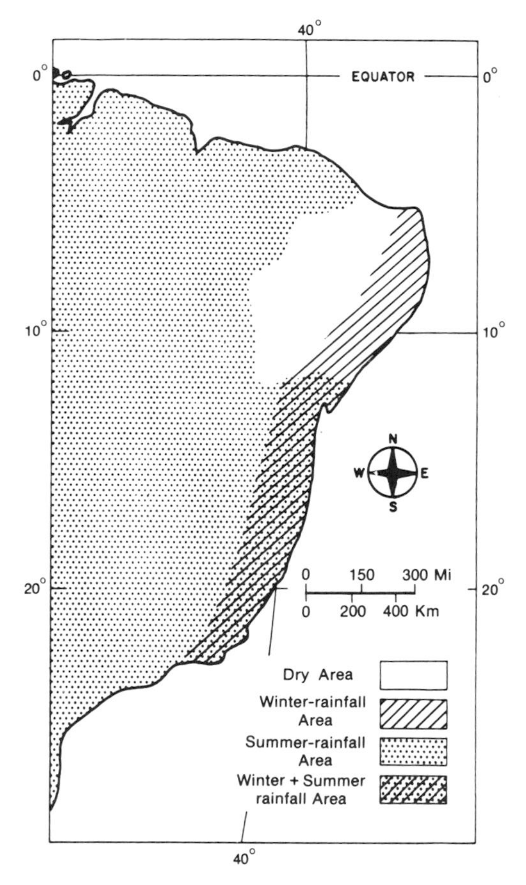

Fig. 4.9. The dry region in eastern Brazil occupies an intermediate position between the winter-rainfall (Köppen *As*) area to the east, and the strong summer-rainfall (*Aw*) area to the west, and normally escapes the full effects of the disturbances which produce these contrasting seasonal rainfalls.

eastern Brazil and along the coast north of Cape São Roque, so that whereas the equatorial air reaches the east coast at Rio de Janeiro by way of the interior in October or November, it does not appear at Fortaleza, nearly 20° nearer the equator on the northeast coast, until February or March (Fig. 4.5). Even at the time of maximum southward advancement of the equatorial air, in normal years it is not able to overrun all of easternmost Brazil, the anticyclone and its dry easterlies continuing to hold a bridgehead there throughout the summer. As a result, the summer isohyets in easternmost Brazil significantly assume a north-south alignment, whereas those of the interior trend east-west.

From the above description, it is clear that the somewhat later arrival of the rainfall maximum (February–April) in the northeastern part of the dry region, and especially along the coast, than in the interior, is a consequence of the much slower southward movements of the *ITC* along the coast (Fig. 4.5). That in this same northeastern part of the dry region spring is the season of most intense aridity, is related to the maximum build-up of the subtropical anticyclone over easternmost Brazil at this season.

During summer and early fall the *ITC*, which delimits the eastern and southern margins of the equatorial northwesterly current, is aligned in a NNE-SSW direction in the eastern interior of Brazil, its direction changing to NE-SW along the coast north of Cape São Roque. Significantly, the eastern and southern boundaries of the dry region have an alignment similar to that of the *ITC* on whose fluctuating eastern margins the drought area is located. As a result, Brazil's drought region in summer occupies a marginal position between the stable easterly flow of the persisting anticyclone, and the humid equatorial northwesterlies. The former dominates the dry region for much of the year, although in summer and early fall the fluctuating *ITC* approaches sufficiently close to provide modest amounts of local, showery precipitation, But years of flood and drought are well intermingled, depending on whether equatorial air or the anticyclone's easterlies temporarily are in the ascendancy in this transition zone. It is, therefore, the relative persistence of stable anticyclonic air over easternmost Brazil throughout the high-sun period, a feature unusual for such a location and season, which obstructs the normal southward and eastward displacement of the *ITC*, the consequence of which is the dry region. And, unlike the coastal region south of Cape São Roque which is also dry in summer and spring, the dry region does not benefit from winter rains. It has been suggested earlier that this general northward displacement of subtropical high and *ITC* in the equatorial Atlantic and adjacent parts of eastern Brazil may be a consequence of the stronger Southern Hemisphere circulation, which in turn is derived from the powerful and durable Antarctic cold source (Flohn and Hinkelmann, 1952).

That during the summer-fall period of maximum rainfall the dry spells are associated with easterly air flow is borne out by the following observations. At semiarid Quixeramobim during the period 1910 to 1925, and considering only the normal rainy period January to May, there were 11 unusually wet months (>260 mm) and 10 very dry ones (<20 mm). During the 10 driest months easterly winds prevailed 89 percent of the time, while during the 11 wettest months westerly winds and calms prevailed 43 percent of the time. Significantly the showers are described as coming from a westerly direction (see Table 4.1).

Serra (1946) has studied the synoptic contrasts prevailing in the dry northeast during abnormally wet and abnormally dry years, and his findings are significant, as well, in understanding the controls producing the general dry region. He points out that while the normal summer rains of this drought area are associated with the convective activity within the equatorial northwesterlies and along the *ITC* when they are displaced farthest into the Southern Hemisphere, this convergence zone reaches well into the dry northeast *only* when southern Brazil to the south, and the Gulf of Mexico to the north are crossed by strong cold-front invasions.

Table 4.1. Wind distribution at Quixeramobim in Northeast Brazil during the rainy season over a period of 14 years (Flohn and Hinkelmann, 1952)

	NE–S	SW–N	Calms
11 wet months (>260 mm)	57	14	29%
10 dry months (<20 mm)	89	3	8%

When this cold-front activity to the north and south is weak, the *ITC* remains north and west of the dry area and fair weather and drought are the rule. It would appear that a difference of 1° to 2° in the mean position of the *ITC* is sufficient to account for the wet and dry years. During the dry year of 1932 Serra (1946) found that over North America in January, the cold highs took a general easterly course and did not strongly invade the Gulf of Mexico and the Caribbean, while concurrently the *ITC* in South America remained somewhat north and west of its usual position and marginal with respect to the dry area. Simultaneously in South America the South Atlantic subtropical high and its stable easterly winds remained strong over the northeast region of Brazil, so that summer cold fronts from the south could not approach. By contrast, during the wet year of 1935 cold anticyclones characteristically moved farther southward over eastern North America and consequently invaded more deeply the tropical regions to the south, resulting in a greater southward and eastward displacement of the *ITC* over eastern Brazil. At the same time there was a weakening and displacement southward of the South Atlantic anticyclone and its easterlies, which in turn permitted the cold fronts from the south to advance farther into the northeastern region. Somewhat similar weather situations to those described above characterized other dry and wet years.

Conversations and correspondence with other members of the Brazilian Meteorological Service in Rio de Janeiro gave the impression that they were less impressed with the importance of summer cold fronts from the south as a factor in determining the occurrence of wet and dry years in the dry region. On the other hand, they were inclined to lay more stress upon the role played by the North Atlantic subtropical anticyclone, noting that when it was strong it acted to block or weaken the southward-moving cold fronts in the Caribbean and North Atlantic, with the result that the *ITC* over Brazil south of the equator remained north and west of its normal position.

A wet fall and dry spring are a characteristic feature of that northeastern part of Brazil's tropical wet-and-dry climate (*Aw*) which borders the coast west of the dry area. Fortaleza is representative of this region. The wet fall has its origin in the retarded southeastward advance of the equatorial northwesterly air and the associated *ITC* in these parts, as described earlier, so that they do not reach their farthest southeastward limits along the coast until the fall months of March and April. Subsequently the retreat of the *ITC* is relatively rapid. As noted earlier, the dry spring is a consequence of a strong build-up of the South Atlantic anticyclone.

The Tropical Wet Climates of the Amazon Valley

Although having world renown as the land area where tropical wet climate (*Af-Am*) is most extensively developed, the Amazon Valley is by no means a region of rainfall uniformity, either in annual amounts or in annual march. Thus, the lower Amazon within a few hundred kilometers of the coast has abundant precipitation of over 2,000 mm, and in close proximity to the sea it may exceed 3,000 mm. Likewise in the middle and upper Amazon, westward of about longitude 56° or 57°W, annual rainfalls exceed 2,000 mm, and the amounts appear to increase westward. Rainfalls greater than 3,000 mm are to be found in the lowlands of westernmost equatorial Brazil and adjacent parts of Colombia and Peru. But for some reason, rainfall drops below 2,000 mm in the western part of Para state (50°–56°W) where the Amazon lowland is narrowed by the encroaching highlands to the north and south. Two stations in this region, Obidos and Altamira, record rainfalls of slightly under 1,750 mm while several others have less than 2,000 mm. (Fig. 4.10). Just what the cause is of this decreased precipitation is not so clear. One may conjecture that the obstructing effects of the Guiana Highlands upon the northwesterly flow of equatorial air could produce a modest rainshadow effect. Supplementing this effect is the fact that the upper Amazon feels the invasion of the unstable humid northerly flow earlier than the middle parts and for a longer period of time.

It is likewise true that the heavy annual rainfall of most of the Amazon Basin has a distinct seasonal periodicity, with the least rainfall in spring and late winter (August–November). This situation emphasizes again the deep westward penetration of the South Atlantic subtropical high in spring. The core area of tropical wet climate with no dry season (Köppen's *Af*) is in the upper Amazon Basin westward from about longitude 60°–65°W. Elsewhere Köppen's *Am* (see appendix), characterized by a very short dry season, largely prevails, but with the possibility that there is a north-south corridor of *Aw* climate separating the *Am* both to the east and to the west (Fig. 4.11). The two stations, Obidos and Altamira, are definitely *Aw* in character. Even many of the stations in the State of Amazonas in the upper and middle Amazon which are classified as *Af* have a season of distinctly lower rainfall in the spring and winter of the Southern Hemisphere. Such a situation emphasizes again the unusual strength and deep penetration of the South Atlantic subtropical high and its stable easterly flow in the low-sun period. Along the coast a similar *Am* or *Aw* annual variation of rainfall prevails.

An unusual weather type for an equatorial region is occasionally experienced in the western Amazon Basin when a strong thrust of cool polar air from the south manages to penetrate deeply into the heart of the tropics. So far as is known, in no other tropical area does modified polar air reach so far equatorward. Such deep invasions, occurring only in the winter half-year, represent the western flank of extensive cool air masses which have followed a land trajectory northward between the Andean and Brazilian highlands. The cool air is greatly modified and very shallow by the time it reaches the western Amazon country; according to Coyle (1940) about 500 m in thickness, but still showing the features of a cold front which may be accompanied by some weather. With the passage of the front there is a sharp drop in temperature amounting to as much as 4° to 6°C (8° to 10°F). A minimum temperature of 1°C (34°F) has been recorded at Cuiaba at

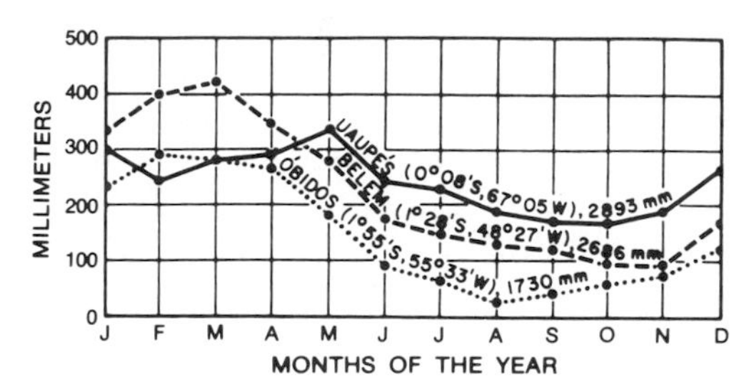

Fig. 4.10. Annual rainfall profiles of three stations in the Amazon Basin close to the equator. Uaupés is located in the extreme western part where rainfall is both heavy and well distributed throughout the year. Belém, at the eastern extremity, likewise has heavy annual rainfall, but there is greater seasonal contrast. Óbidos, situated between the other two stations, has distinctly less total rainfall and likewise a dry season. The Southern Hemisphere spring is normally the driest season.

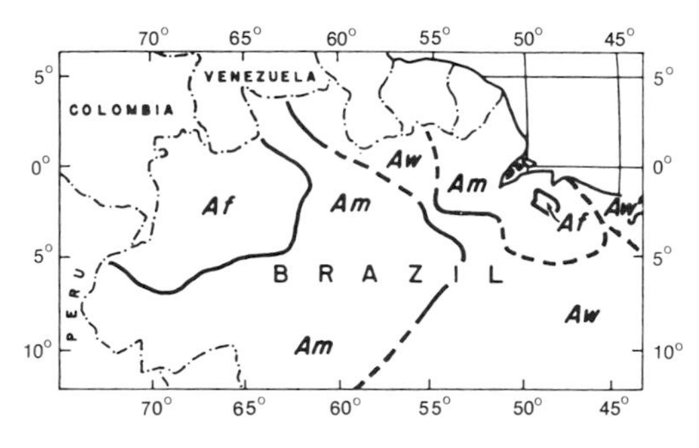

Fig. 4.11. Inferred arrangement of climatic types within the Amazon Basin. Note that a large part of the middle and lower Amazon has a dry season, either very brief (Köppen *Am*), or somewhat longer (Köppen *Aw*).

16°S on the occasion of one such invasion of polar air. Normally this unusual weather phenomenon, called the *friagem*, occurs in the equatorial areas only one to three times a year. Characteristically these cool thrusts are not experienced in the central and eastern equatorial region, but when the polar air does spread eastward its weather effects are not significant, for here the air being displaced is drier and more stable, so that stratus clouds are the principal evidence of the cool invasion. In summer the weaker invasions of cool air do not reach the Amazon country, but they do occasionally produce very active weather over the interior *Aw* region farther south.

Atmospheric Disturbances.—Because of the meagerness of upper-air observations in equatorial South America little is known for certain regarding the perturbations which are responsible for the heavy precipitation of that region. Palmer (1952) has suggested that disturbances developing from unstable long waves in the equatorial easterlies of the western Atlantic between latitudes 0° and 10°N, may be an important contributing element. These eddies, especially those near the equator, may recurve southward over Brazil and after crossing the equator serve to reinforce the extensive summer low, partly thermal in origin, which covers much of the interior southward as far as Bolivia. Thus, the heavy rains of the Amazon Valley and farther south in interior Brazil probably are not greatly different in origin from those of India and southeastern Asia where widespread rainfall accompanies the arrival of weak vortex and wave disturbances which had their origin far to the east over the western Pacific. Observations at Manaus suggest the westward passage of weak eddies of both clockwise and anticlockwise circulation over this region during the season of rains.

Coyle (1950) describes still another type of rain-bringing disturbance characteristic of equatorial and tropical interior Brazil during the Southern Hemisphere summer, which may or may not be related to the type mentioned by Palmer. These perturbations are described as being in the nature of wind surges representing speed convergences in the general northwesterly flow of humid unstable air which floods interior Brazil during the high-sun period. The surges travel toward the southeast in the fairly continuous northwest current. Seemingly, they bear close resemblance to the velocity surges in the Indian summer monsoon, and like them result in widespread instability with showers and thunderstorms. Such surges, while poorly defined at the surface where turbulence conceals them, can be observed as velocity increases aloft.

CHAPTER 5

Caribbean South America and Middle America

The Southern Caribbean Dry Zone

The dry littoral region of northern South America in the states of Venezuela and Colombia and including adjacent parts of the Caribbean Sea is one of the earth's most intriguing climatic anomalies. A xeric vegetation cover prevails. On the mainland this somewhat interrupted dry zone extends from Paria Peninsula on the east (62°W) to Cartegena (76°W) on the west, and for variable distances out over the adjacent sea. This dry zone, situated as it is only 10°–12° north of the equator, is abnormal in three respects: (1) its meager amount of annual precipitation for such a tropical latitude and marine location; (2) the light and undependable nature of the rains; and (3) their suppression during the high-sun months when ordinarily precipitation in these latitudes is at a maximum. This is that relatively rare climatic type designated by the symbol *As* in the Köppen classification. For given the latitude of the dry zone, its proximity to a tropical sea, its location at the western and less stable end of the North Atlantic subtropical high, and the presence of coastal mountains that act as obstructions to the obliquely onshore trade-wind circulation throughout the year, a humid, tropical type of climate, either *Ar* or *Aw*, should be created.

Defining the Venezuela dry zone as essentially that area having less than 1,000 mm of mean annual rainfall, the dry region widens from east to west, and in the latter part attains a breadth of 4° of latitude. Just how extensive it is can only be conjectured, but certainly its area is greater over sea than over land. Over the Caribbean it is probably a contiguous dry area; over the highly irregular land surface, there is a complicated intermingling of dry lowland and moister upland localities. The terrain within the land portion of the dry belt, which so greatly affects rainfall amounts and areal distribution, is one in which long stretches of dry highland-bordered east-west coasts are interrupted by shorter wet coasts oriented north-south. Within the highlands are pouch shaped valleys, the wetter ones opening to the windward east and the drier ones to the leeward west. The easterly trade winds blow obliquely onshore throughout the year, more northeast in winter, and east-northeast in summer. Some of the driest localities on land, receiving less than 500 mm of annual rainfall, lie along the east-west oriented stretches of highland coast, and extend inland for variable distances. Other dry localities are situated inland and are separated from the coast by wetter highlands. Significantly, constricted highland valleys opening eastward to the sea along a short stretch of north-south coast have large annual rainfalls, often 2,000–2,500 mm. The highly complex rainfall patterns of both areal and seasonal distribution make explanation difficult. Precipitation deficiency within the Venezuela dry zone results from both the infrequency of days with rain (under 50 per year) and the characteristic meager fall per rain day. The considerable size of the dry zone would indicate that this climatic anomaly is of larger than local scale. But over the coastal sections at least, where climates change greatly over very short distances, small-scale processes must also be operating (Fig. 5.1).

In attempting to explain the unusual rainfall features of the southern Caribbean dry belt, there are advantages in focusing on the low altitude, insular Netherlands Antilles (Aruba, Curaçao, and Bonaire), instead of the more complicated mountainous mainland coastal belt. These islands are near the heart of the dry belt, have relatively low relief, and are possessed of a goodly number of reliable weather stations. A composite annual rainfall profile for 17 stations on these islands shows an average of 569 mm, 75 percent of which comes during five low-sun fall-winter months, September to January inclusive (Fig. 5.2). Only 16 percent occurs in the five driest months, February through June, while July and August receive only slightly more than the driest months. Clearly, this regime is a reversal of what is normal for these tropical latitudes. For the same 17 insular stations the composite annual profile of rainfall occurrence, or number of days with rainfall, differs from that of rainfall amounts, in that June-July-August show a relatively strong secondary maximum of rain days (Fig. 5.2). The marked increase in the probability of rainfall over the Netherlands Antilles in June, July, and early August is not evident on the composite annual profile of rainfall amounts owing to the light and widely scattered nature of the summer showers, with only a few widespread general rains interspersed.

Causes for the South Caribbean Dry Zone and Its Unusual Annual Rainfall Profile

Dry climates owe their origin either to a deficiency of moisture in the air which makes rising and cooling unproductive of rain, or to the prevalence of atmospheric subsidence which suppresses convectional overturning. The marine tropical location of the South Caribbean dry zone permits dismissal of the first of these causes; the topic of subsidence therefore requires amplification. Subsidence on various scales induced by various mechanisms is surely the main cause for the unusual dry zone under consideration.

On the largest scale, but far from being the most intense and effective, is the subsidence associated with location within the western end of the North Atlantic subtropical high, and the upper level mass convergence that takes place there. But such subsidence at lower levels must of necessity be general rather than local, such as is required to explain the dry belt. Alone it is ineffective as a cause for the aridity; it may, however, be a supplementary cause, especially in certain seasons when the subtropical high is most developed.

Terrain.—Topography, operating in a variety of ways, also contributes to the drought-making process, and it is difficult to understand this dry anomaly without invoking the effects of the fragmented, but imposing, highlands that border the coast. Also, it should be noted that atmospheric circulations are compensatory in character, so that mass ascent of air in one locality must be counterbalanced by downward movement elsewhere, and that the action must occur as quickly and proximately as possible. Thus,

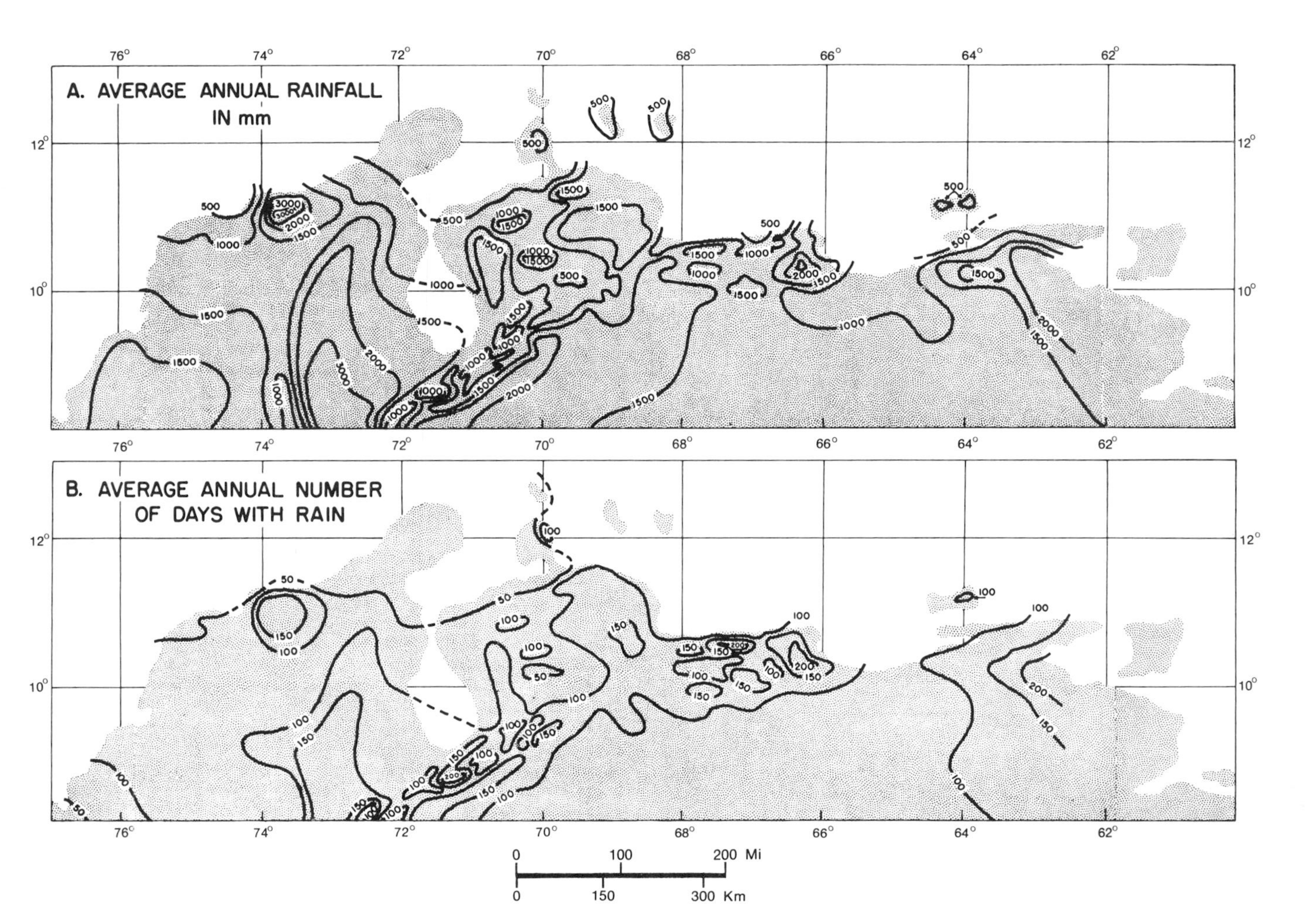

Fig. 5.1. Rainfall characteristics in Caribbean South America. (After Lahey, 1958.)

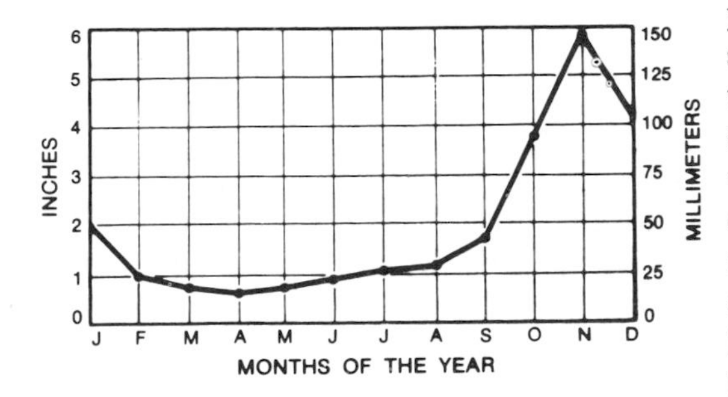

Fig. 5.2. Composite annual rainfall profile of 17 stations in the Netherlands Antilles. Average annual rainfall 569 mm. (After Lahey, 1958.)

during the high-sun period, the heated Llanos lowlands of interior Venezuela and Colombia experience vigorous convective lifting with attendant rainfall. Promptly compensating for this regional mass ascent there follows a downward movement of air elsewhere, at which time (March to October) the dry zone comes under the influence of the descending limb of the large-scale vertical circulation initiated by the heated Llanos (Snow, 1976). This particular subsidence acts to suppress rainfall, especially in the high-sun period, which normally is the time of precipitation maximum in the latitudes of the dry belt. The highlands separating Llanos and the dry zone function as the nodal zone for this regional-scale circulation. During the remaining months, or low-sun period, when the compensatory circulation could be expected to reverse its direction, the dry zone might be expected to experience its wet season, and so it could were it not for the fact that in winter there is increased subsidence in the subtropical high. Significantly then, the peak of dry-zone rainfall is not in midwinter but rather in fall and early winter.

Wind System, Cool Water.—Further elaborating the consequences of terrain upon the dry zone, the direct dynamic effects of the coast and the bordering highlands on the year-round, low-level trade-wind circulation should be mentioned. These winds flow into the dry zone from the east-northeast in all seasons. Over the Caribbean Sea the surface winds increase in velocity downstream, but decrease as they come onshore along the north coast of South America. Over the oceanic part of the dry zone seasonal streamlines are anticyclonically curved or straight, but they take on cyclonic curvature near to the coast. Thus the surface trade flow along the coast is northeasterly, while out over the sea it is east-northeasterly. Such flow patterns result in directional divergence in the windfield near the coast and speed divergence offshore (Lahey, 1973).

The altered trade-wind flow, with a seaward set, in turn leads to the upwelling of cooler waters along the east-west stretches of the coast and adjacent islands of the Netherlands Antilles. Subsidence then occurs over the cooler water, due to the presence of the heat sink below. Thus, year-round somewhat cooler water within the dry zone separates it from warmer waters in the rest of the Caribbean. The temperature difference is not large, to be sure, but is climatically significant. However, the upwelled cooler water can scarcely be considered a primary cause for the dry zone anomaly, but it probably does act to reinforce atmospheric subsidence. It seems unlikely, therefore, that Lettau's (1978) hypothesis for explaining the intense aridity of northern Chile is applicable to the southern Caribbean dry zone.

Lahey (1973) has found that in all seasons the resultant wind flow between about 900 and 3,000 meters is from land to sea over the dry belt. However, the velocities of these resultant winds increase downstream from the coast to the islands of the Netherlands Antilles. This increased downstream wind speed is pronounced at all seasons at the 900 meter level and is less evident at 1,800 meters only during the fall, which, significantly, is the wettest season. The seaward flow of air which has been slowed by the highlands and by the turbulence that develops over them, is speeded up over the smoother ocean surface, resulting in speed divergence and associated subsidence.

We turn now to an analysis of resultant flow patterns for the individual seasons and the vertical motions consequent upon those flows, in order to observe their effects upon *seasonal* rainfalls. In *spring*, the season of least rainfall, fewest number of rainy days, and greatest dominance of extended dry spells, a deep easterly flow prevails over the dry area. The surface easterlies show a downwind acceleration and are characterized by a directional and speed divergence up to at least 1,800 m, while at about 4,300 m convergence is indicated, so that subsidence prevails over the region in question. If this qualitative analysis of divergence based upon resultant air flow is supplemented by computing divergence and vertical motion using the Bellamy-Bennett method, it can be determined that in spring throughout the deep easterlies over the dry area there is strong downward movement from the surface up to at least 6,000 m. One suspects that this subsidence is relatively local to the dry area, although data are lacking which would permit computing divergence and vertical motion for most of the surrounding areas. However, within a triangle formed by Hato Field in Curaçao, Port of Spain in Trinidad, and Beane Field in Santa Lucia, enclosing an area which lies to the northeast of the dry zone and is distinctly wetter, the mean vertical motion is upward at elevations between 900 and 4,300 m.

Summer is still relatively dry, especially compared with the Llanos region of summer maximum to the south of the highlands, although compared with spring, the seasonal total is slightly higher and the number of rainy days strikingly so. Extended dry spells are still numerous, however, and most of the rainy days are characterized by light and widely scattered showers. Surface streamlines and isovels indicate directional and speed divergence. At 900 meters directional convergence and speed divergence make the situation indeterminate as far as vertical motion is concerned. Between 1,800 and 3,000 m there is a speed divergence and zero directional divergence, resulting in positive divergence. Convergence exists between 4,300 and 6,000 m. The above condition indicates summertime subsident motion over the Netherlands Antilles, although the vertical downward motion is not as strong as in spring. Computed subsidence substantiates the above conclusion based on qualitative analysis, and also indicates that the intensity of the subsidence increases toward the west. There appears to be an absence of *ITC* and equatorial westerlies (Lahey, 1973).

Fall is a season of increasing amounts and frequency of precipitation, with November the wettest month of the year in the Netherlands Antilles, and November and December having the largest number of days with rain. The qualitative evidence provided by resultant wind flow in the fall period is inconclusive as regards divergence. Computed subsidence in one of two triangles delineated within the dry belt indicates upward vertical velocities to 4,300 m and downward movement at the 6,000 m level. In the second triangle there is mean subsident motion at and below 900 m and again at 6,000 m, but upward vertical motion between these two levels. These simultaneous upward and downward movements at different levels in the atmosphere over the dry area in fall provide a condition which, while it is certainly more favorable for precipitation processes than is the more general subsidence of drier spring and summer, still does not represent a situation which is strongly conducive to abundant precipitation (Lahey, 1973).

Winter, like fall, is a relatively rainy season as compared with the dry spring and summer, although there is a rapid decline both in amount and occurrence of rain in February. Qualitative examination of the resultant winter windfield over the Netherlands Antilles indicates divergence at the surface, an indeterminate condition at 900 m and 3,000 m and convergence from 1,800 m, to 4,300 m. Convergence also prevails at 6,000 m. From the preceding observations it is difficult to evaluate the nature of the vertical motion over the dry area for the winter season as a whole. But when quantitative methods are employed for determining divergence, the calculated values of vertical motion are upward in all three of the triangles employed, which is in accord with the generally rainier conditions in this dry area in winter (Lahey, 1973).

In summary, it may be noted that the previous discussion of resultant atmospheric flow indicates that the dry area is a region in which vertical downward motion of the atmosphere is present in varying degrees of intensity throughout much of the year. This is in agreement with the region's general deficiency of rainfall. At the same time there are important seasonal differences, with subsidence being most prevalent at the various levels in the drier seasons of spring and summer. It is worth emphasizing that within the general dry area, where subsidence is so common, the rainfall deficiency is *not* characteristic of those coastal sections that have a north-south orientation and consequently feel the effects of directly onshore easterly winds.

EFFECTS OF LAND SEA DIFFERENTIAL FRICTION.—As a further step in the explanation of the climatic anomalies presented by the dry area it is required that the reasons for the localized divergence and subsidence previously described be investigated. Surface winds flow into the dry area from east-northeast during all seasons, thereby making an oblique angle with the east-west segments of coastline. Differential frictional effects of land and water surfaces upon this ENE flow result in directional divergence in the windfield in the immediate vicinity of the coast, with consequent subsidence. Over the Netherlands Antilles offshore speed divergence predominates. But while coastal divergence and subsidence are the rule along those sections of coast oriented approximately E-W, this effect is lacking along N-S coasts which the easterly air stream meets at nearly right angles, with a resultant stowing and lifting effect which produces abundant rainfall.

Stress-induced divergence for the coasts of the dry belt has been computed for each of the four seasons. In spring and summer, the two driest seasons, the divergence values are strong along all east-west coasts (Fig. 5.3). Only along coasts oriented north-south, which the wind flow meets almost at right angles, are values of convergence strong, and there rainfall is heavy even in these drier seasons. In the fall and winter, the wetter seasons, the windfield along east-west coasts is still divergent but only weakly so. The north-south coasts continue to remain convergent and also rainier.

Offshore Flow Aloft.—Throughout the year at 900 to 3,000 m the resultant wind flow is from land to sea, which is opposite to conditions prevailing at the surface. Since hills and mountains fringe the northern margins of South America, the effects of land surface friction must be carried up to considerable heights. Thermal daytime turbulence over the land supplements the retarding effects of the irregular land surface on

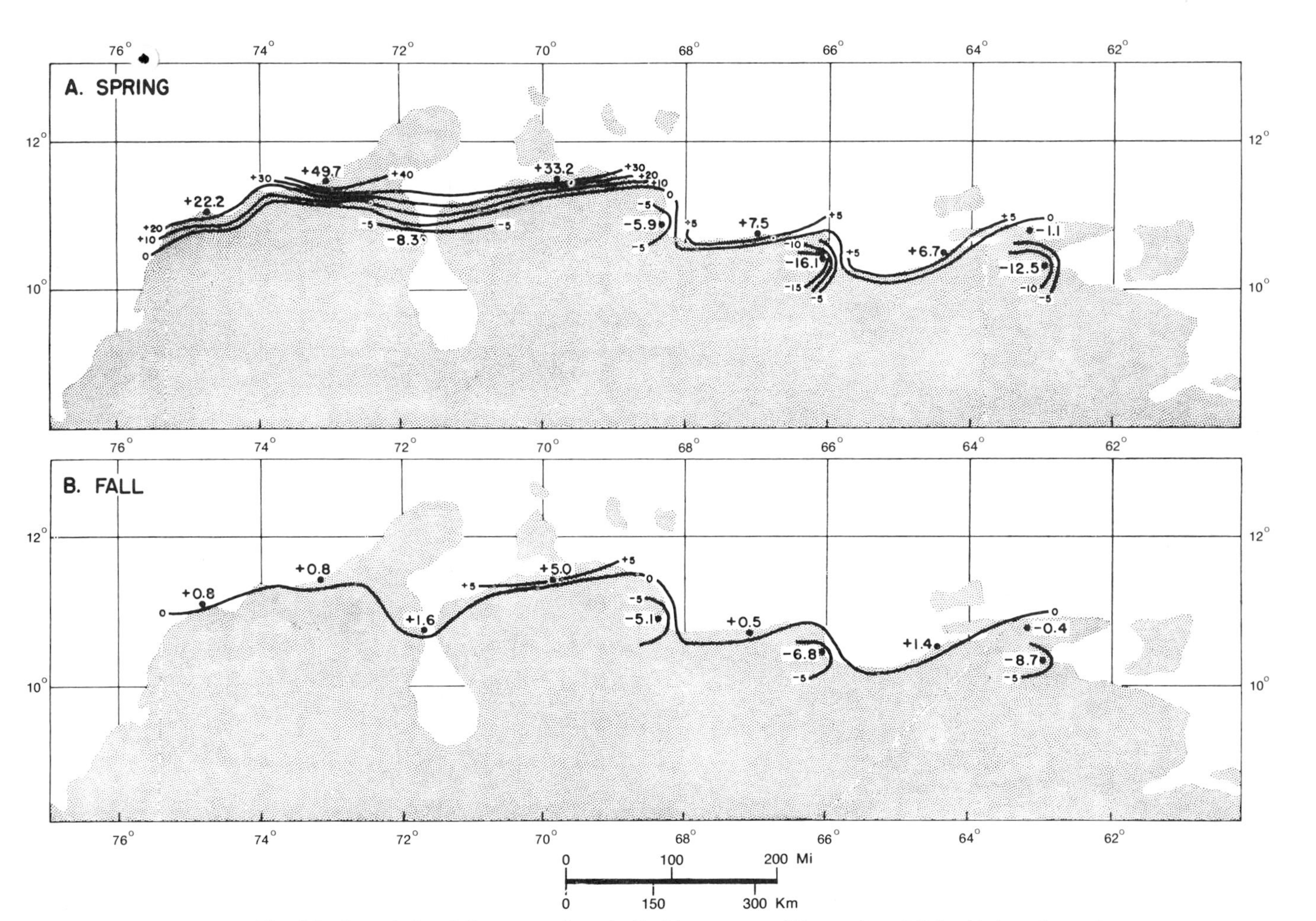

Fig. 5.3. Stress-induced divergence along the Caribbean coasts of Venezuela and Colombia in spring, a dry season, A; and fall, a wet season, B. (After Lahey, 1958.)

the middle-troposphere flow moving from southeast to northwest across the coastline. As this air moves from the land out over the cooler Caribbean the constraint imposed by terrain irregularity and thermally induced turbulence suddenly diminishes and the flow is accelerated. At these levels the observed increase in resultant velocities between coastal and island stations agrees with the proposed hypothesis. The result of the acceleration of the mid-troposphere air flow over the southern Caribbean is to cause a subsidence of air from above to replace the mass removed by speed divergence at lower levels. Subsident effects over the coast should likewise be intensified by downward wavelike motions on the lee, or seaward, side of the coastal highlands.

SEASONAL SYNOPTIC PATTERNS.—Lahey (1958, 1973) has sought further insight into the relationships between synoptic patterns and rainfall in the dry zone by constructing seasonal 500 mb charts of broadscale flow for northern South America and the Caribbean for five precipitation types and in addition an average seasonal map. The five precipitation types are (1) extended dry spells, (2) widely scattered light rains, (3) widespread moderate rains, (4) widespread locally heavy rains, and (5) widespread locally very heavy rains.

Spring, a dry season, is characterized on the average by deep easterlies over the southern Caribbean and an upper subtropical-high cell centered at about 16°N and greatly elongated in an east-west direction so that troughs and ridges can vary greatly in longitudinal position. Extended dry spells, the dominant weather type in spring, are associated with a 500 mb pattern resembling that of the mean chart just described. The major upper-level easterly trough system lies southeast of the dry area, so that easterly troughs over the dry area are either absent or they are fast-moving and weak. The occasional widespread moderate and heavy rains are caused by extended upper troughs which extend from the western Atlantic clear across the Caribbean Sea.

Summer, also a relatively dry season, is characterized by a marked increase of scattered light rains over the dry area. These are associated with a northward shift of the subtropical high and an increase in the number of transitory upper easterly troughs and accompanying low-level convergence lines over northern South America. Extended dry spells are still numerous. Widespread moderate and locally heavy rains are a result of temporary westward and southward shifts of the summertime trough and ridge system.

In fall, the primary rainy season in the Netherlands Antilles, a marked increase in precipitation begins after about October 10, for the probability of the occurrence of widespread rains increases from 5 percent to 30 percent between October 10 and November 16. The widespread moderate and locally heavy rains characteristic of fall occur when the normal seasonal flow pattern is shifted to the northeast, so that a low-level convergence lies over the Caribbean Sea, and deep extended trough systems extend southwestward into the western Caribbean.

Winter starts out rainy but ends up relatively dry. The humid climate of December and early January finds the subtropical high somewhat poleward of its normal position, with extended upper troughs similar to those of fall causing widespread heavy rains. Frequent extended dry spells in late winter are the result of a high-pressure cell aloft over northwestern South America, during which occasions the dry area lies on the subsiding southeastern flanks of this subtropical anticyclone.

MIDDLE AMERICA (MEXICO, CENTRAL AMERICA, WEST INDIES)

In spite of the great geographic variety which characterizes this region, resulting in a profusion of regional and local climates, there still persists a relatively strong and clearly observable broadscale pattern of climatic arrangement which reflects the great planetary controls. Located in the tropics and subtropics (about 6° to 33°N), the whole region is dominated by the seasonally shifting subtropical anticyclones of the North Atlantic and North Pacific, together with the tropical easterlies located on their equatorward margins. There appears to be no invasion of the Caribbean area in summer by a clearly defined *ITC* and accompanying equatorial westerlies, such as occurs in tropical Africa and Asia in similar latitudes, and in Brazil south of the equator. Here the circulation is easterly at all seasons, but in summer the trades are deep and the inversion either absent or confined to high levels. In winter and spring, by contrast, the inversion is lower and of more frequent occurrence. Except in the highlands where altitude offsets latitude to produce mesothermal (*C*) conditions, the climates of Middle America are either tropical humid (*Ar, Aw*) or dry (*BS, BW*).

Extensive areas of dry climate are limited to northern and northwestern Mexico, the latter, at least, representing a normal location for a subtropical western littoral. In the Mexican dry area the principal controls making for drought are (1) the stable eastern end of the North Pacific high in conjunction with the cool California ocean current, and (2) intermontane location. The dry, cool littoral (*Bn*) in tropical and subtropical Pacific North America is by no means as latitudinally extensive as that in Peru and Chile, and the intensity of drought is not so severe. Moreover, the point of minimum precipitation (80 to 100 mm) is here shifted about 5° poleward compared to situations along the dry western littorals in the Southern Hemisphere. Most of this difference is a consequence of the northward-displaced average position of the meteorological equator in the eastern Pacific Ocean and the attendant poleward shift of the North Pacific high. Ocean temperatures are not so low as in comparable locations in South America, since the coastal alignment in Mexico is such that upwelling is reduced, for the eastward bending of the coastline with decreasing latitude causes the divergence of the coast from the open ocean circulation to occur at relatively higher latitudes. The result is that the coastal dry climates extend equatorward only to about the tropic, whereas in South America they reach northward to with 3° or 4° of the equator.

Within the regions of tropical humid climates the larger amounts of precipitation are characteristic of the windward, or eastern and northern, sides of land areas. Leeward locations are less rainy. This contrast between eastern and western sides is greatly accentuated where highlands prevail, and is less conspicuous, or even lacking, where terrain barriers are absent.

Within the domain of tropical humid climates in Middle America, annual rainfall amounts show wide regional and local variations, many of them incompletely understood. As a rule larger amounts are characteristic of the windward, or eastern and northern sides of land areas; leeward western and southern sides are likely to be less wet. Consequently, annual rainfall amounts are usually higher on the windward Atlantic side than on the leeward Pacific side of Central America. But this differential is also due to the fact that most of the rain-bringing disturbances moving through the Middle America region approach from the east, or Caribbean side, as they are transported by the prevailing trade winds.

As a rule, summer and autumn are the seasons of maximum precipitation. This is to be expected since at those times the subtropical anticyclones are displaced farthest poleward and the prevailing *mT* air is more unstable than it is at low sun when the anticyclones are positioned more centrally over the region. Only in extreme northwestern Mexico is there a restricted area which has a winter maximum and this is normal for a subtropical western littoral. Throughout the areas with tropical humid climates, *Ar-Am* symbols tend to prevail along the windward sides and *Aw* on the leeward slopes and lowlands. As described and explained earlier, tropical humid climates extend much farther poleward along the western side of Central America and Mexico than they do in South America before they are terminated by drought conditions.

Circulation and Synoptic Elements in Relation to Annual Rainfall Profiles

One of the simplest annual rainfall profiles is that which characterizes the Pacific slopes and lowlands of western Central America and most of western Mexico. Here summer and fall are wet, and winter and spring are relatively dry (Bryson, 1957*b*). During winter the subtropical anticyclones are farther south, pressure is higher, the inversion lower, and steady northeasterly winds prevail even along the coast. At this season the rainshadow and leeward effects are at a maximum and disturbances, including the *nortes* which add appreciably to the winter precipitation on the east side, are here weak in their rainfall effects. But as the sun moves into the Northern Hemisphere again, the subtropical anticyclones move northward, pressure declines, the inversion lifts, and a variety of rainbringing perturbations, some of them mild in character such as easterly waves and those associated with the poleward displaced *ITC*, and others of a more boisterous nature like tropical storms and hurricanes, disturb the weather of the west coast (Portig, 1959).

Also of signal importance in increasing the summer precipitation of the Pacific lowlands and slopes as far north as Salvador (Botts, 1930), and at times even farther, are the unstable equatorial southwest winds and their westerly disturbances which prevail over the eastern equatorial North Pacific at the time of high sun. Along this southerly and easterly part of the Pacific coast of Central America the relatively strong and steady easterlies of winter are replaced in summer by equatorial westerly winds over the ocean just seaward from the coast, while along the immediate coast there is much calm and variable wind. Temporarily western and southern Panama and Costa Rica have a windward location. These southwesterly winds have had a long trajectory over equatorial waters and are humid and unstable, so that they are conducive to precipitation. When they converge against the western slopes, or with the tropical easterlies along the coast, they yield abundant precipitation. It is significant that those western peninsulas of Costa Rica and Panama which extend farthest out into the ocean, and hence into the domain of the equatorial westerlies, are the areas of heaviest summer precipitation. These westerly winds are farthest north in September and October and have begun to retreat in November. Sapper (1932) believes they are especially effective in producing the second of the two summer maxima, that of October-November.

The eastern or windward side of Central America has much more complicated annual rainfall profiles than does the Pacific leeward side. The high-sun period, when the trade inversion is weak, is still the rainier part of the year, but low-sun is less dry than on the west, and characteristically there is a short secondary minimum in summer. Here the lifting of the onshore trades by terrain barriers not only adds to the total annual precipitation, but also aids in continuing the rains in winter as well as summer. Consequently there is no genuinely dry season as there is on most of the Pacific side. Still another factor in producing the winter rainfall is those disturbances associated with cold fronts, here known as *nortes*, which move south and southeastward across the Gulf of Mexico and the Caribbean, accompanying expulsions of cold air from the North American continent. Along with strong winds and a sharp drop in temperature, such disturbances are associated with showery rainfall. The Isthmus of Tehuantepec experiences, on the average, about twenty northers each winter, while the Canal Zone may expect about three. Cuba, and to a less degree Jamaica, also feel the effects of this norther weather type, but rarely the islands farther to the east.

Summer rainfall of the eastern exposures, in addition to being of orographic origin, is greatly supplemented by that from the same types of disturbances mentioned previously for the Pacific side, more especially easterly waves. It is these disturbances, operating in the less stable trades, which account for much of the broad peak of precipitation in summer and fall. Most spectacular of the vigorous disturbances is the hurricane, whose heavy rains affect most of Middle America fronting on the Caribbean and the Gulf of Mexico, and are a significant weather element along the west and northwest coast of Mexico as well. The hurricane season begins in about June and continues through November, with a general maximum in late summer and fall. To an appreciable degree, the second of the two summer-rainfall maxima is a consequence of hurricane rainfall. Somewhat oddly, hurricanes are likewise a feature of the weather of the desert region of Baja California in northwestern Mexico in most years. During a 13-year period at least 15 hurricanes were experienced in this area (Meigs, 1955). Most of these violent storms appear to originate over the waters west of Mexico between 10° and 20°N. A single such storm may bring a rainfall equal to the average annual total.

The Principal Climatic Uniquenesses

Among the more striking climatic uniquenesses within the Middle American realm are the following: the midsummer secondary minimum of rainfall; the spring dry season at most east-coast stations of Mexico and Central America, and the November maximum; the winter rainfall maximum along the northern or Caribbean coast of Honduras; the dry climate in northern coastal Yucatan; and the subhumid and dry climate along the Gulf littoral in northeastern Mexico. Other local areas of subhumid or even dry climate appear to exist in certain coastal lee locations in the West Indies, as for example in the vicinity of Mole St. Nicolas in northwestern Haiti, Azua in southern Santo Domingo, and probably along the east and south coast of Cuba's Oriente Province from about Point Maisi to Guantanamo Bay where annual rainfall may drop below 750 mm (Grove, 1959).

Less publicized than the Sêcas of Northeast Brazil or the El Niño rains and floods of the normally dry littoral of northern Peru are the temporal variations in rainfall in large parts of Middle America. Yet throughout the present century at least, that region has exhibited an alternation of periods of predominantly wet years with others that were characteristically dry. Hastenrath (1976) points out that the quality of the rainy season in the Caribbean region has a large negative correlation with rainfall and sea-surface temperatures along the northern Peru littoral. Also he notes "During extreme dry years in CARIB the North Atlantic high expands equatorward, meridional pressure gradients steepen, and the trades are stronger, albeit in a somewhat more southward location; at the same time, the ITCZ over the eastern Pacific stays farther south and the South Atlantic high contracts on its equatorward side. For extreme wet years, the reverse departure patterns from the 1911–70 mean maps are characteristic. Sea surface temperature anomalies reflect a response to variations in the subtropical highs and major ocean currents: advection of cold waters in the eastern part of the oceans is favored by equatorward expansion of the subtropical high in the respective hemisphere; the wintertime Gulf Stream system has a distinct signature concomitant with departures in the equatorial and South Atlantic; and in the equatorial eastern Pacific departures reverse from the winter preceding toward the height of an extreme rainy season, warm waters in July/August being characteristic for drought in CARIB. In view of the strong spatial correlations, departure patterns constructed from stratification according to extreme events in CARIB are expected to have more general validity. Anomalous rainy season conditions are signaled in advance by large-scale departure patterns in January/February, thus offering the prospect of foreshadowing extreme rainy season behavior from the setup of low-latitude circulation during the preceding northern winter." This is of great economic importance.

THE MIDSUMMER SECONDARY RAINFALL MINIMUM.—A feature of the annual precipitation profile which is common to most of the Caribbean area, including Central America east and west, the southern Gulf coast of Mexico, the West Indies west of Puerto Rico, and the northern coast of Colombia, is the secondary minimum of mid- and late summer, the *veranillo* (Bryson and Lahey, 1958). Most stations show a rapid increase in rainfall in May-June as the sun advances toward its maximum northern position, and the subtropical high suddenly shifts poleward as well. Then, peculiarly, rainfall declines for one to three months (July, August, September), only to rise to a second, and usually higher, maximum in September, October, November (Schröder, 1955). It is not unusual for November to be the year's wettest month. This phenomenon of an appreciable falling off of rainfall in middle and late summer appears to coincide with a slight increase in pressure at that time. To be sure, for the year as a whole the maximum pressure is in winter when the subtropical anticyclones are displaced southward, and the minimum is in summer when the anticyclones shift northward. But the general summer minimum of pressure actually is composed of two depressions, a shallower one in May and a deeper one in October, separated by a slight increase in July-August (Table 5.1). This latter feature signifies that the North Atlantic Anticyclone is most extended in middle and late summer, resulting in a slight increase in pressure, with associated subsidence, and reduced convective overturning in the Caribbean area at this time (Ward and Brooks, 1934).

But recently Portig (1965; 1976) has pointed out certain incongruities between summer pressure and rainfall in the Caribbean region that diminish the validity of the simple pressure-rainfall hypothesis: (1) throughout the entire Caribbean region, summer pressure on the average is not fully in phase with the march of summer rainfall; (2) while pressure extremes occur throughout the entire region at the same time, such is not true of the summer rains; (3) moreover, in the years 1954–57 when the normal summer "dent" in the annual rainfall curve was absent, and instead a single July–August rainfall peak was present, atmospheric pressure reached its summer maximum in July precisely as in normal years.

Seemingly the hypothesis that a slight midsummer pressure maximum operates to

Table 5.1. Average monthly pressure at San Salvador (Sapper, 1932)

J	F	M	A	M	J	J	A	S	O	N	D	Year
704.3	704.4	704.0	703.6	703.4	703.3	703.7	703.8	703.2	703.1	703.9	704.0	703.7

reduce the rainfall in that season is not completely satisfactory. Something more is required, but up to this time that extra reason is not forthcoming.

Mexican climatologists (Mosiño Alemán and Garcia, 1976) have suggested that the modest falling off of rainfall in midsummer over the eastern and southern parts of their country may be associated with a change in the atmospheric circulation. This change involves the establishment of an upper air trough which extends from the North American Atlantic seaboard through Florida to Central America. When this situation prevails, easterly waves and tropical storms less frequently enter the Gulf of Mexico, and instead are likely to recurve to the north and take a course to the east of Florida.

The midsummer secondary pressure increase and associated secondary dry period probably do not have any connection with the latitudinal movement of the equatorial trough. Riehl (1947) relates them, rather, to a seasonal movement of the high-level western Atlantic polar trough. In midsummer this trough is displaced farther west and north into the Gulf of Mexico, so that its rain-producing effects are weakened over most of Middle America. The early-summer and the fall peaks of rainfall are correlated with the positions of the trough, which at those times are more nearly centered over the area. Proximity of the trough means more perturbations and hence more rainfall (Riehl, 1954).

Asymmetry of the Annual Rainfall Profile.—The spring rainfall minimum, or near minimum, is unusually conspicuous in a large majority of the Atlantic-coast stations of Central America and Mexico. It is also a moderately conspicuous feature of the Pacific slopes and of the West Indies. Even where spring is no drier than winter, it is still much drier than fall. Thus, on the average, the total rainfall of March-April-May is only one-fourth to one-tenth that of September-October-November. In searching for an explanation for this feature one needs to have in mind that the sun is vertical over much of this area twice during the year, once in late April to June and again in July to late August. Where marine influence is so strong, it is not unusual for the lag in precipitation to be as much as two months behind the vertical sun. With this lag in mind it does not seem so unusual for March-April to be drier than September-October, or even November. This climatic lag is further witnessed by the monthly distribution of hurricanes, for March and April are nearly devoid of those excessive rain-bringers, and in May they are likewise rare. On the other hand, they reach their maximum frequency in August-September-October, while November has considerably more than May.

Some insight into the reasons for the asymmetry of the rainfall profile is provided by noting the seasonal characteristics of the temperature inversion in the Caribbean area (Gutnick, 1958). Frequency of inversion occurrence is much higher in dry winter and spring than in summer and fall, which are the wetter seasons (Fig. 5.4). Moreover, except in the westernmost Caribbean, fall has a lower inversion frequency even than summer. The strength of the trade-wind inversion, measured by the difference in temperature between the base and the top of the inversion, is likewise greater in winter and spring and least in summer and fall, with fall

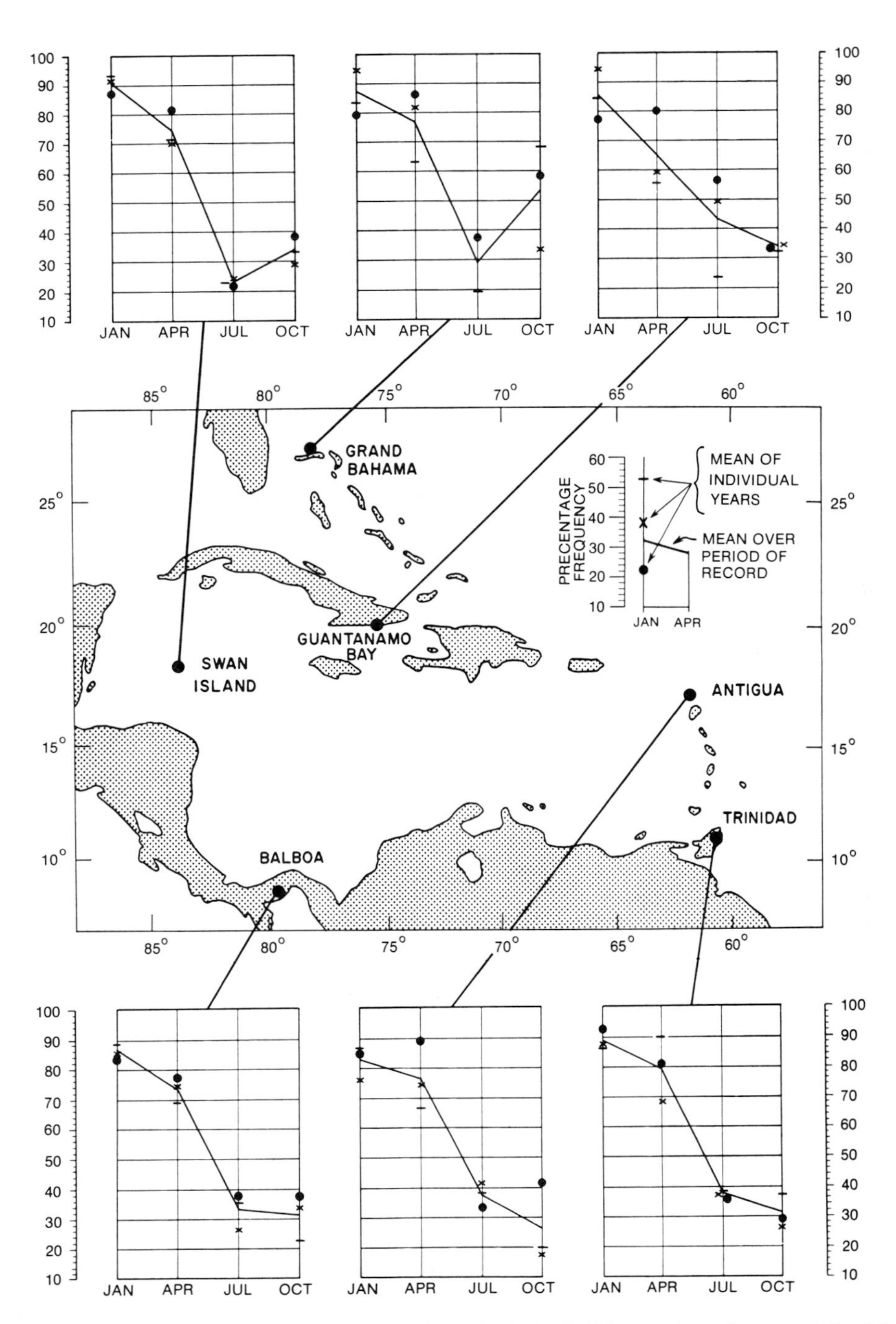

Fig. 5.4. Percentage frequency of the trade-wind inversion in the Caribbean region. (Courtesy of Gutnick, 1958.)

somewhat weaker than summer. Employing still a third criterion, viz., height of the inversion base, spring shows the lowest level, followed by winter, fall, and summer in that order. It will be noted from the above analysis that there is a strong correlation between frequency, strength, and height of the inversion and the seasonal rainfall amounts.

THE HONDURAS *Afs* REGION.—Along the Caribbean northern coast of Honduras the annual rainfall variation characteristic of tropical latitudes is reversed, so that the winter half-year records more precipitation than the summer half. For example, at Puerto Castillo, Honduras, the total rainfall for the six months October through March is between four and five times as great as that which falls during the period from April through September. Also the rainfall of the winter months December through February is more than three times as much as that which falls in June to August inclusive. Much the same situation prevails at La Ceiba and Trujillo as well. Sapper (1932) recognizes this unique climatic area where the reduced precipitation of the normally dry months of March and April is continued well into the summer period.

Along this Honduras littoral the drier summer, as compared with winter, is believed to be associated with the contrasting direction at which the easterly winds meet the east-west coastline and highlands in summer and in winter. In winter, when the resultant winds are northeasterly, they are obliquely onshore, so that there is a downstream convergence or stowing effect, and a forced elevation of the maritime air stream by the highlands. This results in a partial counteracting of the coastal divergence resulting from differential frictional effects of land and water upon the northeasterly flow. In summer, by contrast, the prevailing winds are more easterly so that they are approximately parallel with the coast and the highlands. As a consequence orographic convergence and lifting are reduced or absent, so that low-level coastal divergence and subsidence resulting from land-sea stress differential increase and rainfall declines. During June-July-August, the dry period, the measured divergence ($x10^{-5}$ sec $^{-1}$) along the north coast of Honduras is +1.6, while in wetter December-January-February it is only +0.6 (Bryson and Kuhn, 1961).

THE NORTHWESTERN COAST OF YUCATAN.—Because of its low elevation much of the Yucatan Peninsula has a relatively modest precipitation, in spite of its windward location (Fig. 5.5). The unexpected feature, however, is the marked decrease in annual rainfall from southeast to northwest, so that at Progreso on the northwest coast a meager 472 mm are recorded, while at Arenas Island about 160 km northwest of the dry coast, the annual total is only 416 mm. These rainfalls indicate that semiarid conditions prevail along the northwest littoral and for some distance out over the sea to the north, and possibly to the west as well. Precipitation increases rapidly inland away from the sea, for at Merida, 40–50 km inland and to the south of Progreso, annual rainfall amounts to 913 mm and at Valladolid some 110 km inland,

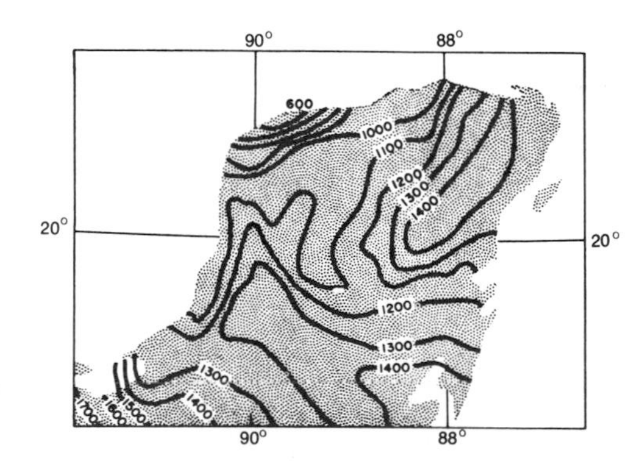

Fig. 5.5. Average annual rainfall of the Yucatan Peninsula. (Courtesy of the University of Mexico, Institute of Geography.)

but farther east, it is 1,181 mm. While the meager coastal rainfall is concentrated in the warmer months of May to October inclusive, it is very significant that the decrease, amounting to about 430 mm, between Merida and Progreso is very largely a matter of decreased warm-season precipitation. Whatever the aridifying control may be, two conclusions seem obvious; that rainfall damping is intensified along the coast and out over the adjacent sea, and it reaches its maximum development in the warmer half of the year, May through October (Fig. 5.6). During the drier winter months the principal decrease in rainfall is from east to west, so that the west coast represents the driest region, and isohyets are aligned almost at right angles with the north coast. But during the warmer months the isohyets approximately parallel the north coast and the rainfall gradient between that coast and the interior is relatively steep.

The available data are insufficient to permit drawing very valid conclusions as to the causes for the dry littoral of northern Yucatan. The Mexican meteorologists are inclined to ascribe the drought chiefly to the leeward location of this northwest coast in a circulation which is dominantly from ESE and SE (Arias, 1958). Such an interpretation, however, seems more applicable to the less rainy cooler months when the isohyets are aligned in essentially a north-south direction, than to the rainier summer when the isohyets parallel the north coast. The question still remains why there exists the large reduction in warm-season rainfall, both along the north coast and for some distance out over the adjacent sea as well.

During the warmer months surface wind flow along the north coast is either parallel to the shoreline or even slightly offshore, while in winter it is northeasterly and hence obliquely onshore. An offshore wind would result in coastline subsidence as a consequence of increased wind speed over the sea surface where frictional drag is reduced. At the same time paralleling winds would similarly produce surface divergence and subsidence near the coastline as a result of a stress differential, and this effect should extend some distance away from the coast out over the water. Unfortunately data are insufficient to permit a computation of the inferred divergence. These same paralleling and offshore winds likewise act to produce an upwelling of subsurface waters which creates a cool ocean surface to the north and west of Yucatan, a condition that is reflected in the more frequent stratus cloud and fog of this area. The fact that much lower annual rainfalls are to be found on islands as much as 150 km or more to the north and west of Yucatan (416 mm at Cayo Arenas and 608 mm at Faro de Cayo Arcas) suggests that the cool water may be an important factor in stabilizing the surface air and damping the rain-making processes.

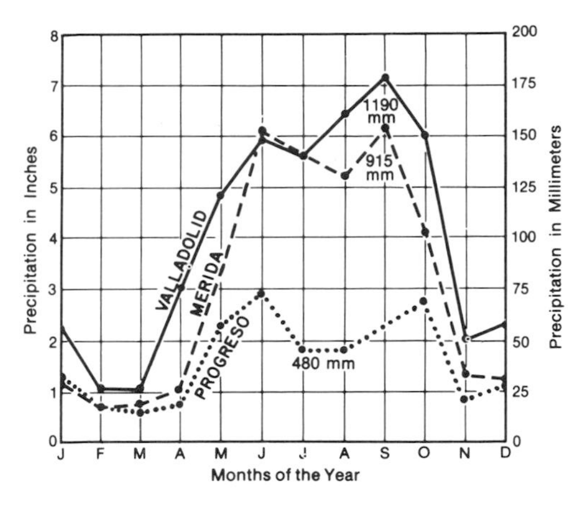

Fig. 5.6. Average annual rainfall profiles of three stations in Yucatan, Mexico. Progreso is located on the coast, while the other two are inland.

THE SUBHUMID LITTORAL OF NORTHEASTERN MEXICO.—Still another problem climate is the subhumid-semiarid condition along the western margins of the Gulf of Mexico in northeastern Mexico and adjacent southern Texas where the annual rainfall in places drops below 600 mm. The incongruous feature is that this part of the west-Gulf littoral should be so deficient in precipitation considering the fact that in summer it lies within a flow of surface easterlies with an extensive body of warm water to windward. Advected moisture seemingly should be abundant. The geographical layout would appear to indicate that it should experience abundant precipitation as does most of the Gulf coast of both Mexico and the United States. This limited section of the Gulf littoral exists as an area of deficient precipitation wedged in between two much rainier regions, one to the north and east along the Gulf coast of the United States, and the other to the south in Mexico. Actually it has only one-third to one-half the annual rainfall that the middle Gulf coast of Mexico receives, while there is a similar decline westward from the mid-Gulf coast of the United States to the Rio Grande. The rainfall gradient is matched by a comparable diminution in the number of summer thunderstorms; from 40 to 50 along the Alabama coast to under 15 in southernmost Texas. Along the lower Rio Grande the rainfall averages only 500 to 750 mm, but it increases rapidly southward, so that Tampico has 1,100 mm, Tuxpam nearly 1,300 mm, and Veracruz 1,600 mm. Some of this increase southward, but by no means all of it, may result from the fact that the highlands are closer to the seacoast. The striking contrast between the three Mexican stations mentioned and the Rio Grande area is in the amount of summer rainfall, the winter totals not being greatly different. Some control appears to be in operation along the northwestern Gulf coast which greatly depresses the rain-making processes of summer, and more especially midsummer, when a strong secondary minimum prevails.

A part of the explanation for this seemingly out-of-place area of deficient rainfall along the west-Gulf littoral is to be found in Wexler's analysis of the above-surface standing waves in the westerlies as they are positioned over the United States in the warmer months. The essential feature of the summer mid-troposphere circulation over the Mississippi Valley is the dominance of a large anticyclonic ridge and associated tongue of dry, subsiding, northerly air, extending in a northeast-southwest direction from Illinois-Indiana to the western Gulf coast (Figs. 18.4, 18.5, 18.6) (Wexler, 1943). Since summer rainfall depends largely on the presence of a deep moist current, the existence of the dry currents associated with the above-surface anticyclonic ridge tends to dampen the rain-making processes over the northern west-Gulf region in the warm season in spite of the humid lower atmosphere. For the northwestern Gulf of Mexico the layer of moist surface air, whose top is defined as that level where the relative humidity drops below 40 percent, averages only 2,500 to 3,000 m in thickness during the period June to October (Hosler, 1956).

Supplementing the explanation given above is the observation made by Riehl (1947) that the tropical easterly current over the Gulf of Mexico arriving from the Atlantic and the Caribbean, bifurcates in the western Gulf, one branch curving anticyclonically into the higher latitudes and the other branch turning southward toward Central America. Between these two diffluent branches subsidence develops over the western Gulf, which decreases precipitation over the lowlands to the north and south of the Rio Grande.

In addition, tropical disturbances of various kinds associated with the easterly circulation of summer are likewise much less frequent in the northwestern Gulf than they are farther east in the eastern Gulf and western Caribbean. Among these the easterly wave is the principal mechanism for producing rainfall along the northwestern Gulf coast in the warmer months June to October. Such disturbances are maintained only in a strong easterly flow associated with a well-developed North Atlantic subtropical high, and since the high and its easterlies weaken in the Gulf of Mexico it is to be expected that easterly waves will decline both in number and intensity toward the northwestern part of the Gulf (Hosler, 1956).

PART II

Australia, New Zealand, and the Equatorial Pacific

CHAPTER 6

Australia, New Zealand, and the Equatorial Pacific

AUSTRALIA

Australia approximates in nearly ideal form the climatic arrangement that one would expect on a hypothetical continent where the great planetary controls largely regulate the weather. Anomalous climates are few. Given a compact land mass of about 7 to 8 million sq km (3,000,000 square miles) located between 9° and 39°S, elliptical in shape, and generally lacking in important terrain obstacles except for a dissected plateau escarpment facing eastward not far inland from the east coast, one would assume a climatic arrangement which very closely fits the one that actually prevails. This is confirmed by the fact that the Australian weather service has been able to devise ingenious solar control models which adequately represent the annual march of the isohyets following the advance and retreat of the sun.

Much of Australia lies in the subsidence and divergence zone of the Southern Hemisphere subtropical anticyclones, so that dry climates occupy all but the northern, eastern, and parts of the southern margins of the continent. The isohyets of annual precipitation are in the form of ellipses which are open on the west and encircle the dry core of the continent. With the advance and retreat of the sun these isohyets, rotating on a pivot located in the southeast, advance southward with the sun in summer and northward in winter. Accordingly, the northern tropical margins have high-sun rain and low-sun drought, while exactly the reverse is true of the subtropical southern margins. The east side is transitional between these two extremes, there being no dry season, while at the same time there is a gradual shift in the time of rainfall maximum, from summer along the northeast coast, through fall in the southeast around Sydney, to a winter accent in Tasmania (Bryson, 1957*a*). Between the summer-maximum region to the north and the winter-maximum region to the south there is a poorly defined intermediate zone where the two regimes overlap, resulting in a rainfall profile without seasonal accent. This transition belt extends diagonally across Australia in a NW-SE direction from about 20°–25°S on the west to 32°–38°S on the southeast. Thus, extreme northern Australia has a tropical wet-and-dry climate (*Aw*); the south and southwest, dry-summer subtropical (*Cs*); middle latitude marine climate (*Do*) is characteristic of Tasmania and the southeast; and humid subtropical of much of the east coast (*Cf*). Desert climate (*BW*) prevails over most of the interior and extends to the west coast, while a belt of semiarid or steppe climate (*BS*) separates the desert from the humid climates lying to the north, east, and south. The arrangement is almost perfect in its simplicity. Minor departures and irregularities in the above-described arrangement of climates are imposed by the modest elevations lying just interior from the east coast.

The Weather Element

No very satisfactory identification and classification of Australian weather types, involving not only synoptic pressure and wind patterns, but also their associated rainfall features, is at present available (Foley, 1956). It has been noted earlier that the annual weather cycle over the continent closely follows the advance and retreat of the sun. The dominating element in Australian weather is the subtropical anticyclone, but in this part of the world instead of being a permanent feature the average seasonal high is a statistical composite of rapidly moving anticyclonic cells which cross the continent from west to east, moving at an average speed of about 800 km a day. Over the median latitudes of Australia the moving cells are so much in control that the weather element is consequently weak. But with increasing distance from the axis of the belt of moving highs, there is a progressive strengthening of the effects of atmospheric disturbances. In the tropical north these are principally perturbations of the high-sun period associated with the *ITC* and perhaps the deep trades. In the subtropical south they are mostly westerly disturbances, many of them in the form of troughs and cold fronts between the moving cells of high pressure. These are at a maximum in the cooler months.

The nearly perennial weather activity of eastern Australia, between the Coral and Tasman seas, is a consequence of the frontogenetic convergence field between moving anticyclones coming from the west across the eastern Indian Ocean and Australia and the more stationary South Pacific cell whose movement eastward is obstructed by the high Andes. The energy for the increased weather activity in eastern Australia and over the adjacent waters is derived chiefly from the heat and moisture in the easterly *mT* air which has been warmed and humidified over the equatorial waters east of the Coral Sea.

Disturbances of the Westerlies.—To those westerly disturbances so important to the weather of southern and eastern Australia, which are linked to the trough line between two eastward-moving anticyclonic cells, has been given the noncommittal name of meridional front (Troup, 1956). In western Australia the front separates warm, dry cloudless *cT* air on the east, from cooler, maritime air on the west derived partly from higher latitudes (Fig. 6.1). Thus, the moving cells are composed, at least in part, of modified *mP* air as it is being transformed into *mT* air in the subtropical system. Not all the weather disturbances of southern Australia are clear-cut examples of simple meridional fronts, some being more complicated derivations, while others are obscure in their origins. Palmer (1942) recognizes four types of disturbances crossing southwestern Australia:

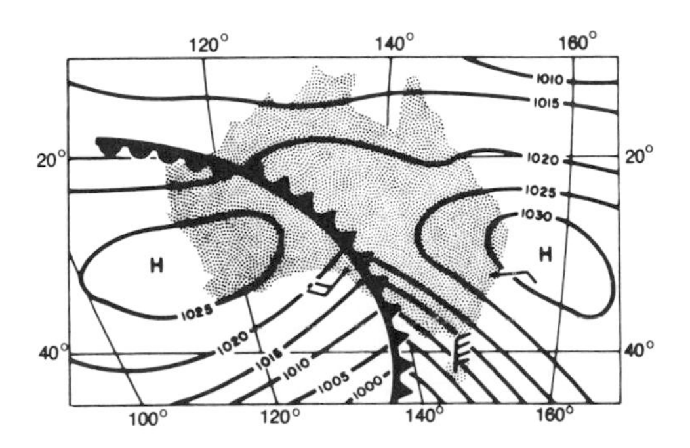

Fig. 6.1. A meridional front, located between two mobile anticyclonic cells, crossing Australia from west to east. (After Palmer, 1952.)

(1) simple meridional fronts without distortion or wave formation; (2) meridional fronts showing wave formation west of the 130°E meridian; (3) the "westerly type" which appears to be a meridional front followed not by an anticyclone but by westerly winds and low pressures, sometimes with secondary barometric minima; and (4) periods of low pressure in western and southern Australia due to incursions of tropical hurricanes from the north. Over a 12-month period southwestern Australia normally experiences 41 of type 1, 2 of type 2, 9 of type 3, and 1 of type 4. It becomes clear that the simple meridional front is by far the most common weather type in southwestern Australia.

Much controversy has developed concerning the structure of these troughline westerly disturbances situated between two migratory anticyclones. Most writers have considered them to be cold fronts. Others have likened them to occluded cold fronts, while still others have insisted that they are non-frontal. Troup (1956), in a study of meridional fronts passing between Perth and Kalgoorlie, stated as his conclusion, "The so-called 'meridional front' is as often as not non-frontal in character, the disturbance at the surface having in such cases many of the features of a cold front but being in fact associated with a rise or disruption of the subsidence inversion." Seemingly, however, the system of eastward-moving cells and troughs, some of the latter genuine cold fronts, is genetically associated with invasions of cool anticyclonic air moving in from higher latitudes. Thus, the invading cold air both feeds the subtropical highs and moves them eastward, and at the same time creates the troughs with their disturbed weather between the cells. As a rule the cool southerly air to the rear of the trough does not break out in strong surges toward lower latitudes as genuinely polar air does. However, a few of the meridional fronts can be traced from the south clear across northern Queensland and even to New Guinea (Hogan and Maher, 1943). From the Southern Hemisphere synoptic charts one gets the impression that westerly disturbances in northern tropical Australia, probably induced by cold fronts farther south, are of some consequence as rainbringers in that region.

The principal characteristics of the cool-season meridional-front disturbances over southern Australia are as follows. They move from west to east in line with the progress of the anticyclonic cells. With the frontal passage there is a sharp wind shift from N or NW to W or SW. The air to the east of the front is warm to hot (21° to 24°C [70° to 75°F] at 0600) with a low dew point, an absence of low convective clouds, and good visibility. The relatively cool air to the west of the front (about 18°C [65°F] at 0600) is modified *mP* in the process of being transformed into typical subsident tropical air. It has a high moisture content and a relatively high dew point. There is a moderate amount of convective cloudiness and showery precipitation in the maritime air behind the front, which gradually tends to disappear inland (Fig. 6.2).

Most meridional fronts maintain a simple and undeformed NW-SE alignment eastward to about mid-Australia or 130°E. Farther to the east the subtropical portion north of about 40°S tends to lag or become stationary, and it is at this stage that some of the fronts begin to lose their simplicity as a wave cyclone with a warm sector and warm front evolves. The location, rapidity, and degree of the deformation vary greatly among meridional fronts. In some fronts the deformation occurs over eastern Australia; it is very common over the Tasman Sea, with the wave cyclone moving southeastward across New Zealand; still other fronts pass eastward across New Zealand with no evidence of deformation.

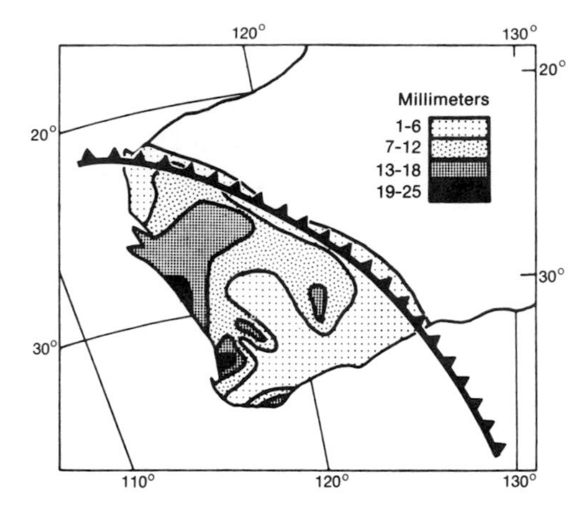

Fig. 6.2. Precipitation distribution associated with a meridional front in southwestern Australia. Note that nearly all of the weather is located west of the front within the maritime air masses. (After Palmer, 1952.)

It has come to be recognized, however, that the great variety of rainfall episodes in Australia are not always to be explained in terms of either simple or more complicated meridional fronts as they appear on the surface synoptic charts. Foley (1956) has shown that widespread, and often heavy, rains may be expected when certain patterns prevail in the 500 mb contours. One such is the upper trough with a meridionally oriented axis. To the east of the upper trough divergence prevails which is usually associated with falling pressure and low-level convergence making for ascending motion and rain. Evidence of the upper trough may or may not be present on the surface synoptic chart. A second 500 mb situation favoring widespread rain is the "cut-off low" which develops at the northern tip of an upper trough when anticyclogenesis farther south cuts off a cold pool from the general mass of cold air in higher latitudes. This likewise results in upper-level divergence, with convergence at lower levels. Not infrequently the "cut-off low" aloft is followed by the development of a strong low-level cyclone over eastern Australia as shown by both Foley (1956) and McRae (1955). Still a third upper-level situation favoring widespread rainfall is the presence of a deep upper low usually centered over the southern ocean but extending its weather effects well to the north. These three types of 500 mb disturbances appear to occur in almost all seasons, though the deep southern lows are distinctly most frequent in the cooler months. During the 3½ years from January, 1949, to June, 1952, Foley observed 24 occasions of upper troughs, 42 of "cut-off lows," and 54 of deep southern lows. While general rains associated with the above-described situations seemingly may occur in almost any part of the country, including the desert interior, the east experiences them most frequently and therefore benefits most from their rainfall effects.

Tropical Disturbances.—Almost no literature on disturbances over tropical northern and northeastern Australia has been written since the introduction of the current method of streamline analysis. Correspondence with the Bureau of Meteorology at Melbourne indicates a belief that, apart from rain associated with the hurricane type of storm and possibly with local convective clouds within the "monsoon" air streams, the summer rains over northern Australia are chiefly a product of streamline convergence in weak low-pressure systems in the equatorial trough which is located over tropical Australia at this time. Some rain may also result from the invasion of eastward-moving cold fronts from higher latitudes, or the inducing of disturbances by such cold invasions aloft. To what extent troughs in the easterlies, similar to the easterly waves of the Caribbean area, are responsible for rainfall in northern Australia is not known. The Australian Bureau of Meteorology considers them to be comparatively rare, although it is admitted that the synoptic network over tropical waters to the east may be too sparse to detect them.

There is no Australian literature on such easterly disturbances. The heavy rainfall along the northeast coast of Queensland no doubt is partly of orographic origin, but observations indicate that the summer precipitation occurs irregularly and is in the form of aperiodic showers and squalls moving in from the sea, suggesting some form of extensive disturbance.

There is no temporary retreat or weakening of the Australian *ITC* in mid-summer, with a secondary rainfall minimum corresponding to the Caribbean *veranillo*, for there is no revival of anticyclonic activity at high sun as a consequence of the ease with which anticyclones in the South Pacific can move toward the south and east as soon as the circulation index begins to rise.

Tropical Australia, more especially the northeast and the northwest, likewise occasionally feels the effects of disturbed weather associated with severe tropical storms of the hurricane variety. Few of the northeastern storms cross the coastline and move into higher latitudes over the land, but those of the northwest show a greater tendency to move southeastward into the continent. Their season of occurrence is the high-sun period December to April. Most of the tropical cyclones appear to originate in the region of the *ITC*, rather than as waves in the deep easterlies. Storms with surface wind velocities exceeding 34 knots have an annual frequency of occurrence amounting to 10.8 on the northeast coast and 8.5 on the northwest.

Climatic Anomalies

Omitting localisms, the only important departure from the usual climatic patterns found on the other extensive land areas in tropical-subtropical latitudes, is the absence in Australia of a cool and intensely dry western littoral (Köppen's *Bn*) with its relatively low temperatures and high frequency of fog and low stratus. Here there is nothing comparable to the cool coastal desert of Peru-Chile, or the Namib of southwestern Africa. An auxiliary fact is that nowhere in the Australian dry area, even though it is one of the earth's most extensive deserts, is rainfall as meager as in the other deserts of low latitudes. As far as can be determined from the sparse data, the minimum rainfall for Australia is about 150 mm and this is found in the deep interior and not along the western littoral as is the normal situation. Consequently in Australia, where the lowest rainfall along the west coast is about 230 mm, annual precipitation decreases inland, while in most of the other low-latitude deserts aridity is most intense along the west coast and rainfall increases inland. Figure 2.4, which represents annual precipitation of the five dry western littorals plotted against latitude, shows that not only is Australia's minimum coastal rainfall of about 230 mm considerably greater than those of the other dry western littorals, but even this modest aridity is maintained over a latitude of only a few degrees. In the four other dry western littorals, rainfall amounts not only are distinctly lower, three of them having less than 50 mm, but these smaller amounts are maintained along more extended stretches of coast, in Peru-Chile, the extreme case, through about 24° of latitude.

In a previous discussion of a comparable situation in western South America it has been noted that there are at least three factors controlling atmospheric stability, and therefore the precipitation processes, along that dry western littoral: (1) control of surface air temperatures by a cool ocean current and associated upwelling; (2) strong subsidence in the smoothly curved circulation around the eastern limb of a positionally stable anticyclone, resulting in a persistent and low inversion; and (3) intensified coastline subsidence where an anticyclone and its circulation are terminated abruptly at the coastline and surface divergence develops in the paralleling winds as a result of the differential frictional effects of land and water (Lydolph, 1957). All of these drought-making controls appear to be brought to a maximum intensity and latitudinal development where the coastline continues to bend into the main atmospheric and oceanic circulations even into low latitudes, so that the coast is held in contact with these circulations. Since Australia represents an exceptional case, it will be enlightening to discover whether in this instance any of these drought-making controls are lacking or weakly developed.

One factor making for reduced aridity is the lack of a strong and locationally fixed anticyclonic cell in the eastern Indian Ocean whose smoothly curved easternmost circulation approximately coincides with the Australian coastline. It is this absence of a persistent anticyclonic flow of southerly air along the Australian west coast which largely accounts for the absence of a cool ocean current and for associated upwelling. Probably the lack of an important terrain obstacle in Australia is one factor making for the absence of a strong stationary cell of high pressure along the western side of that continent. In the Southern Hemisphere, where only the South American Andes offer a serious obstacle to westerly air flow, the latter is much stronger than the easterly flow on the equatorward side of the subtropical pressure ridge. Consequently the subtropical highs over extensive longitudes in the Southern Hemisphere are essentially statistical averages of individual migratory cells of high pressure which travel from west to east. This is especially true in the eastern Indian and the western South Pacific oceans. The obstructing effects of the Andes are to be observed in the fairly stationary position of the cell along the west coast of South America and possibly of the cell bordering the Namib of southwestern Africa as well. In these locations, also, the atmospheric and oceanic circulations are relatively strong and persistent. But in those longitudes far removed from the effects of the Andes, as for example in the vicinity of Australia, there is a continuous procession of eastward-moving anticyclonic cells, the individual cells separated by troughs. Such a variable circulation pattern derived from the moving cells and the intervening troughs, not only discourages the development of a persistent cool current with upwelling, but likewise makes for a weakened and less enduring subsidence and inversion, especially along the west coast, but also over the interior as well. In the troughs between the cells of high pressure, surface convergence is present which lifts and weakens the inversion and occasionally results in a modest amount of showery rainfall. Vowinckel's (1953) maps actually show that the region of southwesternmost Australia and the waters to the south and west are an area of modest cyclogenesis, with the consequence that Perth has a greater frequency of frontal weather than does Cape Town and its lapse rate in the lower air is steeper. Along the Australian west coast, therefore, lapse rates are steeper on the average, and inversions weaker and less persistent than along the more arid west coasts of Peru-Chile and southern Africa.

In contrast to the situation along dry tropical west coasts of the other continents, Australia lacks a cool current and its stabilizing effects. In addition, there is no upwelled subsurface water. Such currents as do exist are weak, fickle, and appear to reverse their direction in the opposite seasons. The coastal waters are actually warmer than those in the open ocean farther west, where cool Antarctic waters do flow northward around the eastern end of the more stationary central part of the Indian-Ocean high. But closer to Australia where the invasions of cool maritime air from higher latitudes are more frequent, and moving cells of high pressure with intervening troughs characteristic, the atmospheric circulation is too variable in direction to permit the generation of a steady oceanic flow. Moreover, there is no prevailing set of the water away from the coast which would give rise to upwelling. Confirming the absence of cool water along the west coast of Australia is the fact that here the isotherms sag poleward along the littoral, in contrast to their marked equatorward bending along the west coasts of South America and southern Africa. In addition, coastal fog and stratus are largely lacking.

And finally, the west coast of Australia, lacking the smoothly curved wind flow around the eastern end of a positionally stable high, with paralleling winds along the littoral, is not subjected to a coastline divergence derived from the differential frictional effects of land and water. At the same time the convex coastline, instead of bending westward into the main atmospheric and oceanic circulations, equatorward of about 22°S recedes abruptly eastward.

New Zealand

New Zealand, situated as it is in the frontogenetic oceanic area eastward from Australia, is affected by complicated and variable weather types. Additional complexity in the weather and climatic pattern derived from these synoptic types results from the mountainous nature of the islands, with highlands oriented in various directions in different parts, and large valleys and straits serving as important wind and weather funnels. As a consequence a slight difference in the orientation of a front and its isobars may produce very different types of local weather, even when the overall synoptic situation is similar.

The basic synoptic patterns which in large measure determine New Zealand's weather are as follows (Watts, 1945; 1947):

1. Moving anticyclonic cells which cross northern New Zealand from west to east. Toward the centers of these highs, fronts are absent and light winds and clear skies prevail.
2. Undeformed trough lines or meridional fronts between two eastward-moving anticyclonic cells. The accompanying weather pattern resembles that described for similar disturbances in Australia, except that the northerly winds to the east of the trough are here usually humid as well as warm, since they have had a water trajectory and consequently are capable of yielding heavy precipitation on windward slopes, usually northern and northwestern. The humid, cooler southwesterlies on the rear of the trough are likewise potentially unstable and capable of producing heavy, showery rainfall. Kerr (n.d.) recognizes four subdivisions of this main type.
3. Meridional fronts which are deformed before reaching New Zealand. When the resulting wave cyclone and its warm and cold fronts pass over the islands, they cause bad weather over an extensive area. Along the warm front widespread precipitation falls from a slowly lowering overcast without cumulus forms. As a rule more precipitation falls in the forward sector of the depression than to the rear. Since in the former the low-level winds have an easterly component, the regions affected by the cyclone receive a larger share of their rainfall when the wind is from an easterly direction. Based largely upon the route followed, four subdivisions of this basic type are recognized by Kerr (n.d.).
4. Tropical cyclones, of varying degrees of strength, which originate well to the north, but in taking a southward course bring heavy rains to northern and eastern districts, though

they do not greatly affect the southern parts or the western littorals.

5. Although not recognized by Kerr (n.d.) as a distinct type, there is also the westerly-wave series, in which depressions traveling well to the south of New Zealand in latitudes 50°–60° cause sympathetic fluctuations in the air flow over these islands, with winds shifting to W-SW as the center passes their longitude. With this sequence showery weather prevails on the western side of the islands.

From Table 6.1 it is clear that simple and deformed meridional fronts (Types 2 and 3) together provide much of the weather of New Zealand. Moreover, their frequency of occurrence remains high throughout the year, only fall showing a very significant decline. Storms of tropical origin (Type 4), however, are concentrated in summer and to a less degree in fall. Without doubt the frequency indicated in the table for situations when fronts are lacking in the area and anticyclonic conditions prevail (Type 1), are not truly indicative of the amount of fair weather over New Zealand, for the area for which the tabulation was made is much more extensive than New Zealand itself.

Based upon the relative dominance of meridional fronts, New Zealand's climate appears to be a synthesis of a weather series, each series composed of four components: anticyclone, northwesterlies, frontal weather, and the conditions associated with the following air flow, the direction of the latter to a considerable degree determining the distribution of rainfall between eastern and western sides (Watts, 1947). Not infrequently, though, the weather sequence is complicated by the formation of cyclonic depressions along the meridional front.

Watts (1947) has classified New Zealand weather into 15 large-scale weather situations and shows the percentage of days on which each of these situations controls the weather in the northern, central, and southern parts. Anticyclonic fair weather with light winds is the dominant type among the 15, prevailing on 28 percent of the days in the north, 19 percent in the central part, and 20 percent in the south. Cold fronts control the weather on only 5 and 6 percent of the days in the north and in the center, but 10 percent in the south, indicating that about half of these fronts that cross South Island weaken or dissipate before they reach the center and north. Depressions, warm fronts, and stationary fronts combined control the weather on 9 percent of the days in each section.

In abbreviated Table 6.2, Watts (1947) has combined his 15 original weather situations into 5 major groupings, each of which supposedly is somewhat similar in its weather effects. When northwesterly, westerly, and southwesterly air components are joined, it becomes clear that undisturbed westerly weather is the most important and frequent controlling influence in all three parts—39 percent in the north, 41 percent in the center,

Table 6.1. Frequency of occurrence in New Zealand (in percent) of the first four weather types noted above, for the area bounded by meridians 130°E and 180°E and parallels 20°S and 50°S (Kerr, n.d.)

Type	Winter	Spring	Summer	Fall	Year
1	8.3	6.0	7.7	21.7	10.8
2	41.7	49.3	41.4	33.6	41.6
3	50.0	42.5	40.6	35.0	42.1
4	0.0	0.0	5.3	2.4	1.9
Uncertain	0.0	2.2	5.0	7.3	3.6

and 37 percent in the south. Southerly to easterly flow is somewhat more frequent in the center and south than in the north, while the reverse is true for moist northerly flow. That during about 40 percent of the year the weather is under the control of westerly flow does without doubt have a predominant influence upon the average distribution of rainfall, more especially in South Island, where high mountains extend in a general NE-SW direction. But at the same time, the 20 –25 percent of the weather situations with northerly, southerly, and southeasterly flow are at present considered to play a more important part in rainfall distribution than was formerly realized. Disturbed weather associated with fronts and depressions is common to all parts, but reaches a maximum in the south.

Problem Climates

Other than the multiple local climates of anomalous character not uncommon in such a mountainous island region, which must here be neglected, the problem climates of greater dimensions are few. Prominent among them are the conspicuous contrasts in rainfall distribution in North Island and South Island. In the latter there is a superhumid west side and a subhumid east. Over 5,000 mm are deposited annually on the western slopes and 3,700 mm on the lower coastal stretches of Westland. By contrast extensive areas east of the mountains receive only 500 – 750 mm and some localities less than 500 mm, suggesting subhumid conditions.

North Island, on the other hand, has a much more complex rainfall distribution and striking contrasts between east and west sides are lacking. Rainfalls exceeding 2,000 mm are to be found on highlands in both eastern and western parts, while the less rainy lowlands as well, on opposite sides, are not greatly in contrast.

The explanation of these regional rainfall contrasts between North and South Island is not simple. In part it reflects the relative contribution of orographic effects in the two areas, for while South Island is compact and has a high, steep, and unbroken wall of mountains separating Westland from Canterbury, which highlands are oriented at nearly right angles to the dominant westerly flow, North Island, by contrast, is much less compact, its outline is irregular, and its highlands lower, less continuous, and variously oriented in direction. Moreover, the east-west contrasts in rainfall are increasingly magnified in a southerly direction due to the fact that there is a steady increase southward over New Zealand in the mean strength of the westerly flow at lower levels.

With closer attention paid to the dynamic elements' contribution to rainfall contrasts between north and south, the problem is much more complicated. Due to uncertainties in the synoptic analyses, data are not available in suitable form on the frequency, intensity, and movement of rain-bringing synoptic situations, to permit a satisfactory regionalizing of their effects. Watts's (1947) classification of controlling weather situations, and his tabulation of their frequency of occurrence in the northern, central, and southern parts, provide only modest help. The fact that anticyclonic days with light and variable winds are more frequent (28 percent) in the north than in other parts, does suggest decreased windward-leeward rainfall effects in North Island. Likewise the fact that strong northwesterly flow is two to three times as prevalent in the central and southern parts as in the north has a bearing upon the problem. Depressions and fronts of all kinds are less frequent in the north (14 percent) than in the south (19 percent) but this difference is entirely accounted for by the fact that there are twice as many days dominated by cold fronts in the south as in the

Table 6.2. Percentage of days under major controlling weather situations in New Zealand (Watts, 1947)

Controlling situation	Northern Region	Central Region	Southern Region
Northwesterly, westerly, and southwesterly flow	39	41	37
Southerly to easterly flow	14	21	21
Anticyclone	28	19	20
Moist, northerly flow	5	4	3
Frontal weather (cold, warm, stationary) and depressions	14	15	19
(cold front separately)	(5)	(6)	(10)

north. However, cold fronts in the south are seriously dissipated by the high mountains, so that the same conditions of rainfall distribution occur as one might expect from an undisturbed westerly current.

In New Zealand wave depressions normally take a southeastward course, with much more precipitation falling in the forward sector where winds with an easterly component prevail. Prolonged and heavy cyclonic rains ahead of the advancing center are much more frequent on the east side of less elevated North Island than on the leeside Canterbury area of South Island. This is reflected in the fact that while the number of rainy days is similar on the eastern sides of both islands, the mean rainfall per rainy day is definitely higher in the north. There can be no doubt the high southern mountains exert a considerable effect upon the depressions themselves. It is a variable effect, to be sure, sometimes operating to steer them around the southern end of South Island, at other times causing them to become stationary and fill to the west of the highlands, after which a new center forms to the east over the ocean. It is possible, also, that the obstructing highlands act to induce anticyclonic flow curvature in the westerly current over and just to leeward of the mountains, a situation described by Boffi (1949) for Patagonia and elaborated in Chapter 3. Taken together, these influences operate to greatly reduce the cyclonic rainfall on the lee side of South Island. No equivalent effect on depressions is produced by the lower and more fragmented highlands on North Island.

Two additional controls are worthy of comment. In late summer and fall cyclones of tropical origin on occasions bring heavy rains to the northern and eastern parts of North Island without much affecting the western districts or the regions farther south. Likewise air masses of tropical origin with higher specific humidities are much more involved in the depressions that affect the northern parts, thereby contributing more rainfall to the eastern parts of North Island than to the Canterbury area farther south.

Problem climates associated with seasonal concentration of rainfall seem numerous at the local level, but few if the scale adopted is extensive (Garnier, 1958). Over the entire area there are almost no strong seasonal contrasts, as might be expected in such a marine location. Most of North Island shows a fairly consistent maximum in winter when the routes of the subtropical anticyclones are farthest north and therefore less centered in these latitudes, while summer, the time of maximum anticyclonic control, has relatively less rain. Winter is likewise the season of most frequent and most vigorous cold fronts. But though the latitude is subtropical, North Island scarcely has a Mediterranean (*Cs*) climate, for the annual total is too great (1,100± mm) and summers are far from dry. Along the western side of South Island winter is the least rainy season, a feature which may be associated with the greater prevalence of anticyclones formed in the cool air behind a cold front. Seasonal contrasts are not marked, however. The only parts of New Zealand which are distinguished by a warm-season maximum of any strength are the more continental parts of interior South Island where a few stations may experience twice as much rain in summer as in winter. But on the Canterbury Plain any seasonal concentration is slight.

The Dry-Subhumid Zone of the Equatorial Eastern and Central Pacific Ocean

It is a normal feature of the world pattern of climates for the eastern-central parts of subtropical oceans to be dry. Such oceanic dry climates are simply westward extensions of the subtropical deserts over the continents in similar latitudes and are derived from the same planetary controls. However, a most unusual feature is the existence, in the equatorial eastern-central Pacific Ocean, of an equatorial dry zone of immense east-west extent, which is characterized not only by low annual precipitation totals, but, in addition, high annual rainfall variability. This oceanic dry zone exists precisely where one would expect the moisture-laden converging northeast and southeast trades to produce a single intertropical convergence zone with a zonal cloud band representing strong convectional activity, frequent thunderstorms, disturbed weather, and heavy rainfall (Fig. 6.3). In summary, the main features of annual rainfall totals in the tropical Pacific are:

1. A zonal equatorial dry belt with less than 750 mm of annual rainfall extending westward in the form of a tapering wedge, from

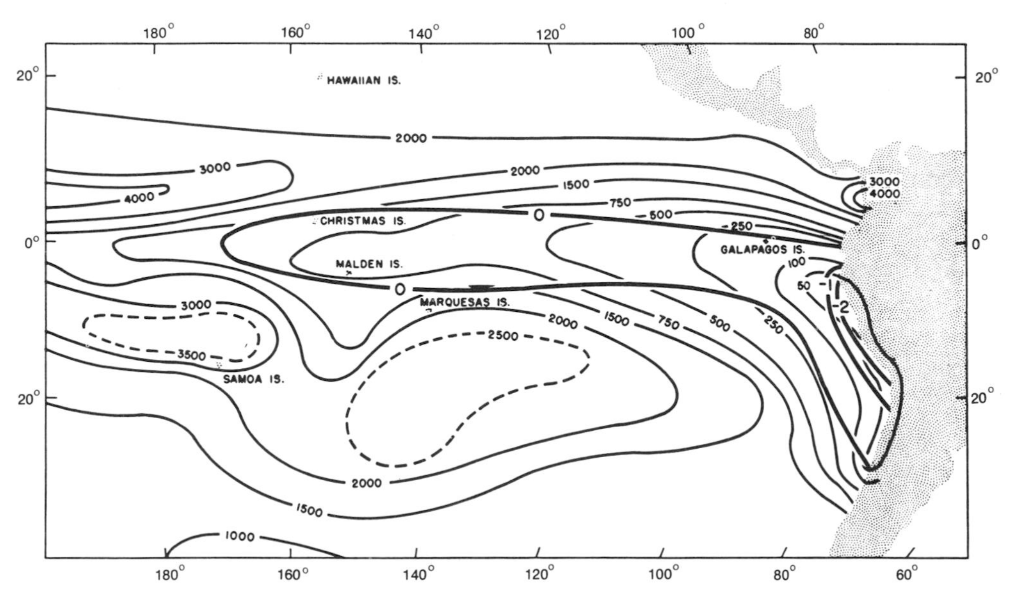

Fig. 6.3. Showing the location of the dry belt and the associated tongue of cool water in the equatorial eastern and central Pacific. The lighter lines are isohyets, with annual rainfall shown in millimeters. (After Sekiguchi, 1952.) The heavy lines show the temperature differences (in °C) between water and air in February. The area within the outermost heavy line has water temperatures lower than those of the air. (After Schott, 1935.)

the coast of South America to about 160°–170°E, a distance of 11,000 km or more. Undoubtedly this is one of the earth's most striking climatic anomalies;
2. Wet zones flank this linear equatorial dry belt on the north (7°–12°N) and south (5°S +) corresponding to the two zones of convergence in the double *ITC* that appear to prevail here; and
3. Rainfall decreases poleward from these two equatorial wet belts as the domain of trades and subtropical highs are approached.

A somewhat similar dry zone is to be observed in the equatorial eastern and central Atlantic, but there its longitudinal extent is by no means as great, and also, while equatorial in location, it appears to remain slightly south of the equator and is therefore confined wholly to the Southern Hemisphere. The discovery of these two equatorial oceanic dry zones somewhat weakens the concept of a continuous equatorial maximum of rainfall.

Since few observations of rainfall amounts have been made for the equatorial Pacific Ocean, recourse must be had to data from island stations which are complicated by orographic effects and consequently may not accurately represent conditions over the sea. Rainfall within the dry wedge increases westward. In the Galapagos Islands, on the equator and about 1,100 km west of South America, the annual amount is only about 100 mm. At Malden Island (155°27′W) and Christmas Island (157°27′W) in about mid-ocean there has been an increase to 710 mm and 970 mm, while at Banaba (about 170°E) the annual depth of rainfall is around 2,000 mm. It is impossible precisely to define the belt of subhumid precipitation, but in mid-Pacific it appears to be about 10° wide and it straddles the equator asymmetrically, with its northern boundary at about 2°–3°N and its southern limits at 6°–10°S. Along its northern boundary an excessively steep rainfall gradient separates it from a belt of very heavy precipitation coinciding with the locationally fixed *ITC* just to the north. Thus, while Christmas Island at 1°58′N has an annual rainfall of only 950 mm, Fanning Island at 3°54′N records over 2,000 mm. Except along the southwestern margins of the dry tongue, where there exists a shear zone between obliquely converging easterly air streams (Gentilli, 1952), the southern boundary is in the nature of a more gradual transition. Remarkably little latitudinal shifting of the dry belt occurs during the course of the year.

The subhumid zone is delimited not only by a reduced annual rainfall, but also by a strikingly higher annual variability than is the case to the north and south (Seelye, 1950). This is not surprising considering the close juxtaposition of controls making for rain and drought within this area, and the steep rainfall gradients on the northern and southwestern margins of the dry belt (Fig. 6.4). Isolines showing the percentage frequency of dry months for the year, and also for each of the seasons, likewise serve to demarcate this anomalous region (Seelye, 1950). In addition, the seasonal maps of dry-month frequency reflect how small is the latitudinal shifting of the dry belt, especially along its northern margins. Maximum annual rainfall variability occurs at the western end of the equatorial dry belt.

The synoptic network in the central-eastern equatorial Pacific is so sparse that recent weather satellite photographs have proved to be an invaluable tool for investigating the climatic patterns of that region. On these photographs (taken in 1967) the areas of maximum brightness, and, hence, strong reflection of sunlight, represent regions of persistent and deep cloudiness over the ocean (Fig. 6.5). In contrast, the relatively cloud-free equatorial dry zone is the conspicuous dark area of the satellite photographs. The

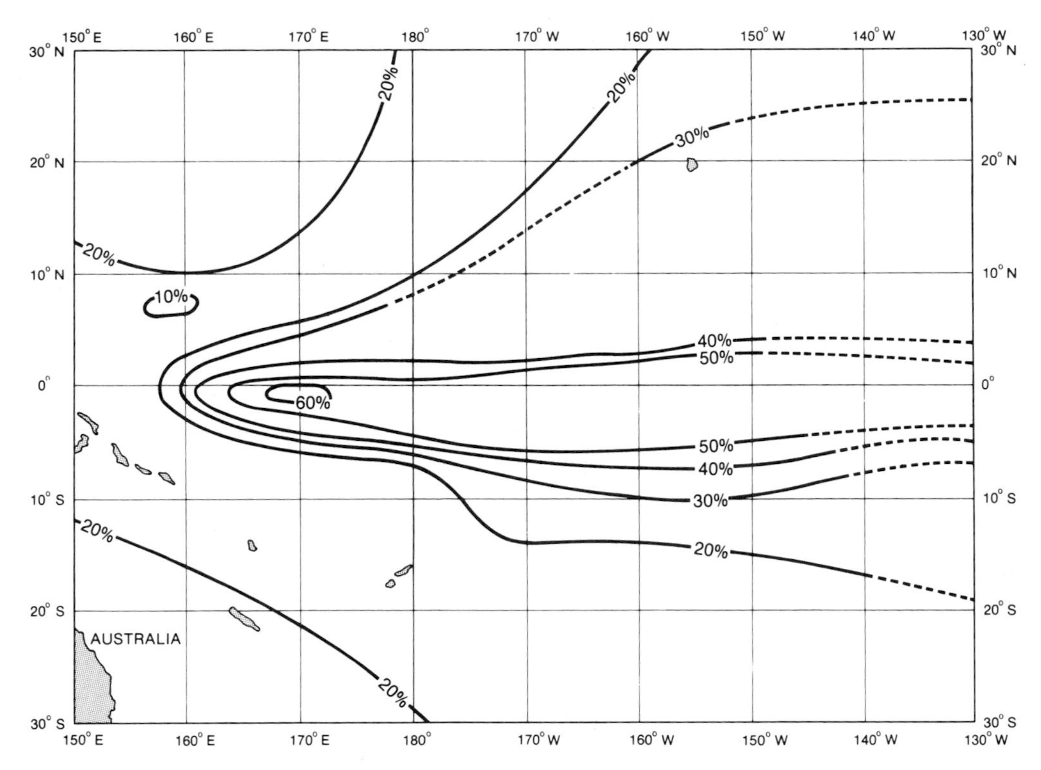

Fig. 6.4. Coefficient of rainfall variability over the equatorial Pacific Ocean, 1951–1960. (From Musk, 1976.)

zones where the *ITC* is persistent in its location, where convective activity and cloud are prevalent and rain-bringing disturbances are frequent are easily seen. The photographs clearly reveal that in the equatorial eastern and central Pacific there are two *ITC*'s with relatively persistent deep cloudiness, not the usual one. They show that the northern *ITC* is both enduring and well developed, and undergoes little shifting of position throughout the year, maintaining a location between about 7° and 12°N. The southern *ITC* is more variable spatially and temporally than its Northern-Hemisphere counterpart, is less well developed, shows great east-west variation in its cloud intensity, and reaches its maximum development at about 5°S. Coincident with the double *ITC*, the two cloud bands represent the lo-

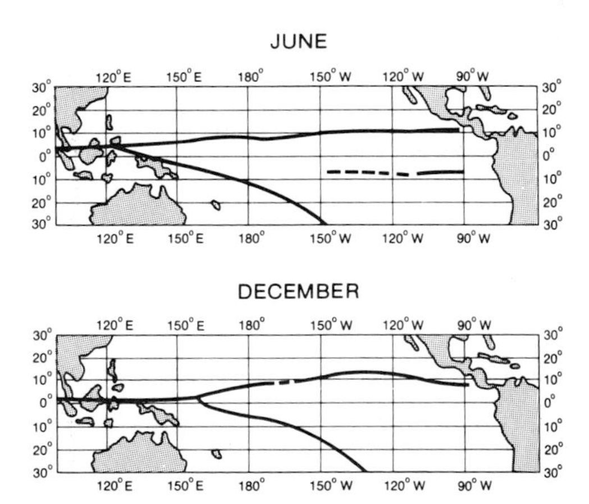

Fig. 6.5. Mean monthly positions of the maximum brightness axes (hence, cloudiness) in the equatorial Pacific Ocean; from ESSA 3 and 5 satellite evidence, June and December 1967. (From Musk, 1976.)

cations of cumulus cloud clusters associated with wave disturbances moving from east to west. The two roughly parallel cloud bands denote the ascending branches of the double *ITC* comprising the meridional circulation of the Hadley cell. Satellite photographs reveal a third belt of persistent cloudiness which joins the northern *ITC* belt in the vicinity of the New Guinea-Indonesia region, and from there extends southeastward well to the east of Australia and New Zealand.

Two major atmospheric circulations are recognized in the equatorial Pacific. One of these is the large scale zonal (E-W) Walker circulation, having a prime region of ascent in the warm western equatorial Pacific-Indonesian region, and a general region of subsidence over the eastern half of the equatorial Pacific, where sea temperatures are cooler, in the dry zone between the northern and southern limbs of the *ITC*. This subsidence zone intensifies and widens toward the eastern Pacific where the ocean water is increasingly cooler and the area of cool water is more extensive. Thus, the eastern half of the equatorial Pacific, with its cooler sea temperatures, acts as a major mass source, where the chilled atmosphere gives rise to local high pressure. The western half, by contrast, where ocean temperatures are higher, is a great mass sink where the atmosphere is being warmed with resulting lower pressure. Easterly winds prevail at lower levels.

The other major circulation, meridional (N-S) in character, is the so-called Hadley cell, one in each hemisphere (Fig. 6.6). Normally the circulation in the Hadley cell involves convergence and ascent of air near the equator and flanking subsidence belts in the subtropics. But in the central and eastern Pacific a double *ITC* structure appears to prevail with an equatorial cell of high pressure positioned over the cool water, thus separating the two Hadley cells from each other. Subsidence prevails in in the equatorial cell resulting in greatly diminished cloud and rainfall.

So the paradox of a dry zone in the equatorial Pacific appears to be due to several interrelated controls. Perhaps the most important factor influencing the rainfall deficiency in this region is the tongue of relatively cool

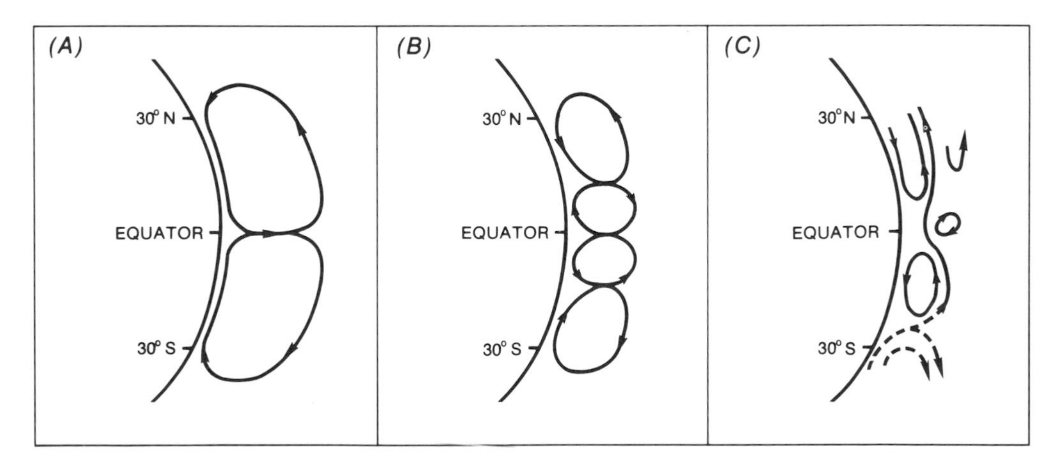

Fig. 6.6. Models of the Hadley and equatorial circulation cells: (*A*) simple classical model, with a Hadley cell in each hemisphere; (*B*) Fletcher's 1945 model showing equatorial cells between the two Hadley cells; and (*C*) Bunker's simplified model of the actual situation over the equatorial Pacific in April 1967. (From Musk, 1976.)

water originating off the west coast of South America and progressing westward along the equator to the central Pacific. The current flows westward owing to the wind stress of the low-level easterlies or trades as a part of the Walker cell. The coolness of the equatorial ocean increases eastward, for a part of its water temperature anomaly derives from the cool Peru current which flows northward along the west coast of South America (Fig. 6.7). Its effect is mainly to be felt east of the Galapagos Islands. But over most of its longitude, the coolness of the current is due to cool upwelling and the presence, at a depth of about 100 m, of a cool eastward-flowing equatorial undercurrent (Cromwell current); these are probably supplemented, to a diminishing degree westward, by some advection from the cool Peru current (Flohn, 1972; Musk, 1976). Dryness is further accentuated by a zone of active subsidence, local divergence, and high pressure between the two convergence lines of the double *ITC*.

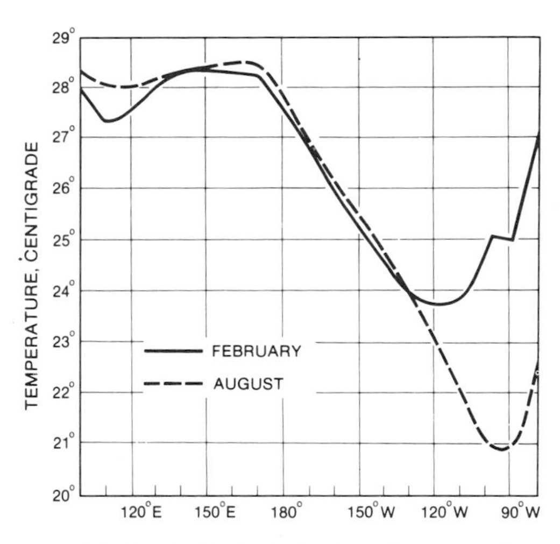

Fig. 6.7. Longitudinal distribution of mean surface temperatures in the equatorial Pacific Ocean for February and August. (After Saha, 1970.)

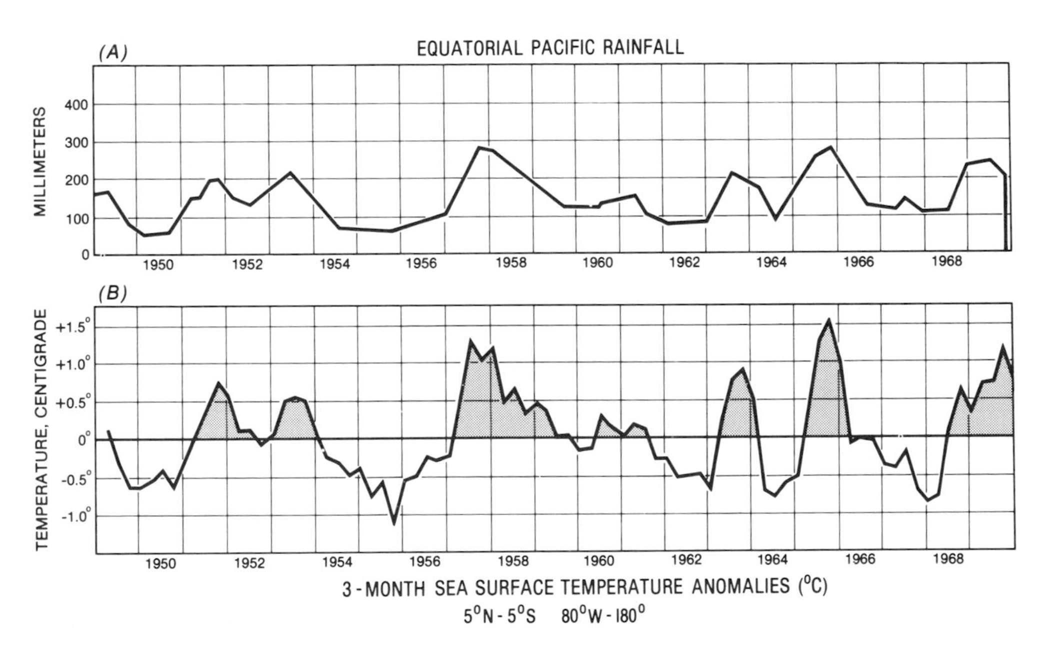

Fig. 6.8. (*A*) Twelve-month running means of equatorial Pacific island rainfall (mm), 5°N–5°S, 150°W–165°E, 1949–1969, and (*B*) three-month means of sea surface temperature anomalies (°C), 5°N–5°S, 80°W–180, 1949–1969. Note that rainfall increases as sea temperatures rise. (From Allison et al., 1971, as shown in Pyke, 1972.)

This divergence in turn stimulates the upwelling of cool Cromwell water, which further stabilizes the local atmosphere.

As a rule the driest years correspond with periods of undisturbed easterly winds and low ocean temperatures, or when the Walker cell is well developed. Wet years are characterized by above-average sea surface temperatures and a greater frequency of westerly winds, signifying a greater frequency of disturbed weather-associated traveling depressions (Fig. 6.8).

In summary, the dry zone of the central and eastern Pacific appears to correspond with a zonal equatorial belt of subsidence within the Walker cell, oriented along the axis of the cool easterly current. The regions of highest rainfall variability within the dry zone are those areas where the cool current experiences the greatest interannual variability. The lowest cool-water temperatures are centered along the geographical equator, from where the Ekman drift causes the water to diverge and move north and south, owing to the change in sign of the Coriolis parameter (Bjerknes, 1960); upwelling of cooler water from below results. The wet tongue, reaching from the Solomon islands to Samoa and farther east and lying to the south and west of the dry wedge in the central Pacific, may be a consequence of the shear zone developed between two branches of the easterly flow (Gentilli, 1952).

PART III
Africa

CHAPTER 7

The Sahara, Sudan, and Guinea Coast

No other continent is so symmetrically located with respect to latitude as is Africa, for its northern and southern extremities extend almost equidistant from the equator. Nearly the whole of the continent lies in tropical latitudes, only the extreme northern and southern subtropical parts reaching far enough poleward to feel the effects of the middle-latitude westerlies and their disturbances in the cooler seasons. It is unique among the continents in that it is strongly affected by the subtropical anticyclonic belts of both hemispheres, and hence has extensive areas of tropical-subtropical dry climate both to the north and to the south of the equator.

Africa is sometimes described as a continental plateau and it is true that much of its area is moderately elevated above sea level, so that altitude plays a significant role in reducing the otherwise tropical temperatures over very extensive areas. On the other hand, great cordillera like those in North and South America and Eurasia, which might function as climatic divides, are lacking.

For obvious reasons southern Africa can have no such important disrupting effects upon the strong zonal circulation of the Southern Hemisphere as those produced by the Andean barrier in the higher latitudes of South America. The continuity of the subtropical high-pressure ridge is, therefore, greater in the vicinity of South Africa, so that there is not the same important exchange of air between high and low latitudes in the South African quadrant as there is in the South American. Disturbed weather associated with the equatorial advance of surface cold fronts consequently appears to play a relatively minor role in the climate of eastern Africa south of the equator. An upper anticyclonic circulation continues aloft over both northern and southern Africa at all seasons, so that a high zonal index prevails, and in the Southern Hemisphere, at least, a procession of anticyclonic centers moves from southwest to east across the southern part of the continent.

The South Atlantic high which abuts against the southwestern margins of Africa is relatively strong and is somewhat more fixed in its position, especially in summer, than is true of the subtropical highs over the oceans to the east and west of Australia. Well-developed and persistent atmospheric and oceanic circulations around the eastern end of this cell are the result. As a consequence the cool Benguela current with its associated upwelled water is a prominent feature along the coast of southwestern Africa, so that the stabilizing effects of cool water are added to those of subsidence in the anticyclone to produce intense coastal aridity. Somewhat similar atmospheric and ocean circulations prevail along the northwestern coast of Africa which is paralleled by the cool Canaries current, but in this case the effects of the cool water do not extend so far equatorward.

As a result of the absence of extensive cordillera the climatic pattern of Africa in its larger aspects resembles that of the hypothetical continent which emphasizes the great planetary controls. In addition, there is a strong resemblance in the arrangement of climates to the north and to the south of the equator. Thus, along a Northern Hemisphere profile from Algeria to Zaire the succession of climatic types very closely matches that along a Southern Hemisphere profile from Cape Town to Zaire. The principal hemispherical contrasts seem to stem from the fact that Africa north of the equator is longitudinally much broader than its counterpart to the south, and also that it lacks an oceanic boundary along most of its eastern side. North of about 12°N, eastern north Africa is bordered by the great Asiatic land mass, and as might be expected, that continent's seasonal temperature, pressure, and wind systems have a strong influence upon adjacent Africa. Because there is no permanent oceanic anticyclone over the Arabian Sea, eastern Africa north of the equator, even south of 12°N where it does have frontage on an ocean, lacks the effects of onshore neutral *mT* air. Instead, dry continental air prevails and aridity is the result. This is not true to the same degree in eastern Africa south of the equator where a somewhat more normal arrangement of oceanic anticyclone, maritime air, and humid littoral prevails.

A further element of hemispherical asymmetry in Africa is associated with the fact that in tropical western Africa just a few degrees north of the equator, the coastline extends east-west for roughly 2,500 km, with a heated land mass to the north, an equatorial ocean to the south, and winds onshore throughout the year. This arrangement extends tropical humid climates far to the west along the Guinea coast of Africa. Equally important, the Gulf of Guinea moisture source extends the humid tropical climates farther north in West Africa than is true farther to the east, so that the trend of the isohyets across northern tropical Africa south of the Sahara is slightly west-northwest by east-southeast.

Among the noteworthy deviations from the standard world pattern of climatic distribution, which is so conspicuous in Africa, may be mentioned the following: (1) the unusual longitudinal breadth and remarkable equatorward extension of the dry climates across all of northern Africa, a feature which is quite in contrast to the situation south of the equator. As a consequence the transitional wet-and-dry climates have a much more restricted latitudinal spread to the north of the equator than to the south; (2) the dry littoral in Ghana-Togo on the Gulf of Guinea, in the central part of what is elsewhere on the Guinea coast a tropical wet climate; (3) the high-sun secondary rainfall minimum along the Guinea coast, a feature which is greatly intensified in the mid-sections of that littoral; (4) the generally below-normal rainfalls over much of eastern tropical-equatorial Africa whose situation on the windward side of the continent, with highlands rising abruptly interior from the coast, warrants the expectation of much wetter conditions. Here the greatest climatic abnormality is the desert of Somalia, eastern Ethiopia, and northeastern Kenya—an equatorial dry climate along an eastern littoral. Other climatic abnormalities less striking than those listed above will be noted in the general regional descriptions. A discussion of Mediterranean Africa will be postponed to a later section dealing with the Mediterranean Basin.

The Sahara and Its Southern Margins

This is the most extensive and the most continental of all the tropical-subtropical dry climates, so that, for the latitude, it is characterized by a maximum severity in temperature, both seasonal and daily, and in drought as well. Stations recording rainfall are few, so that no clear picture of the details of precipitation distribution is available. Everywhere it is low. At Bilma in eastern Niger the annual total is less than 25 mm, which is one of the lowest for the earth. Undoubtedly the Sahara represents the world's most extensive area of intense aridity, only three of the dry western littorals being able to match it in rainfall deficiency, and then only in restricted areas. Along the Atlantic coast of the Sahara where it is bordered by the cool Canaries current, maximum stability is reached in the vicinity of Port Étienne at about 21°N, where rainfall drops to 28 mm. For a few degrees of latitude, therefore, this littoral compares in dryness with the arid coasts of Peru-Chile and southwestern Africa. But because the Saharan coast abruptly changes direction south of Port Étienne and Cape Blanco, and bends away from the main cool-water circulation, the superaridity is quickly lost southward from 21°N, with the consequence that the littoral desert does not extend to such low latitudes as it does in western South America and southwestern Africa. Throughout the Sahara, subsidence associated with a stable subtropical anticyclone is in control throughout the year, for even in summer when a thermal low pressure appears at the surface over the southern Sahara and adjacent Sudan, the anticyclonic circulation and its drought-making effects continue to prevail aloft, so that deep convectional overturning is stifled.

In such a desert region it is meaningless to speak of seasons of maximum and minimum rainfall. Only along the northern and southern margins is the concept of annual rainfall variation meaningful, the poleward margins having their entire meager precipitation in the cooler months, while exactly the reverse of this is true along the southern margins.

One of the most remarkable features of aridity in northern Africa is its extensiveness (Fig. 7.1). In reality the Sahara is only one part of a still larger unit of tropical-subtropical dry climate which extends from the Atlantic Ocean across North Africa, Arabia, and western Asia to northwestern India-Pakistan. The east-west dimension of this arid area simply reflects the great longitudinal extensiveness of the land surface lying within the latitudes of anticyclonic subsidence.

More unusual is the latitudinal breadth of the Sahara, resulting from an abnormal southward penetration of drought conditions. In no other continent over a similar extent of longitude is there such a limited poleward migration of the summer isohyets following the sun. At this point Africa north of the equator may be compared with tropical South America south of the equator, both areas having great longitudinal breadth. Yet in the very latitudes where broad northern Africa has extensive dry climates, broad South America east of the Andes and south of the equator has mostly humid tropical and subtropical climates. No doubt the high Andean cordillera act to abruptly terminate the effects of the Pacific anticyclone in western South America, so that its aridifying effects do not extend eastward beyond the Andes, and only in winter does the South Atlantic

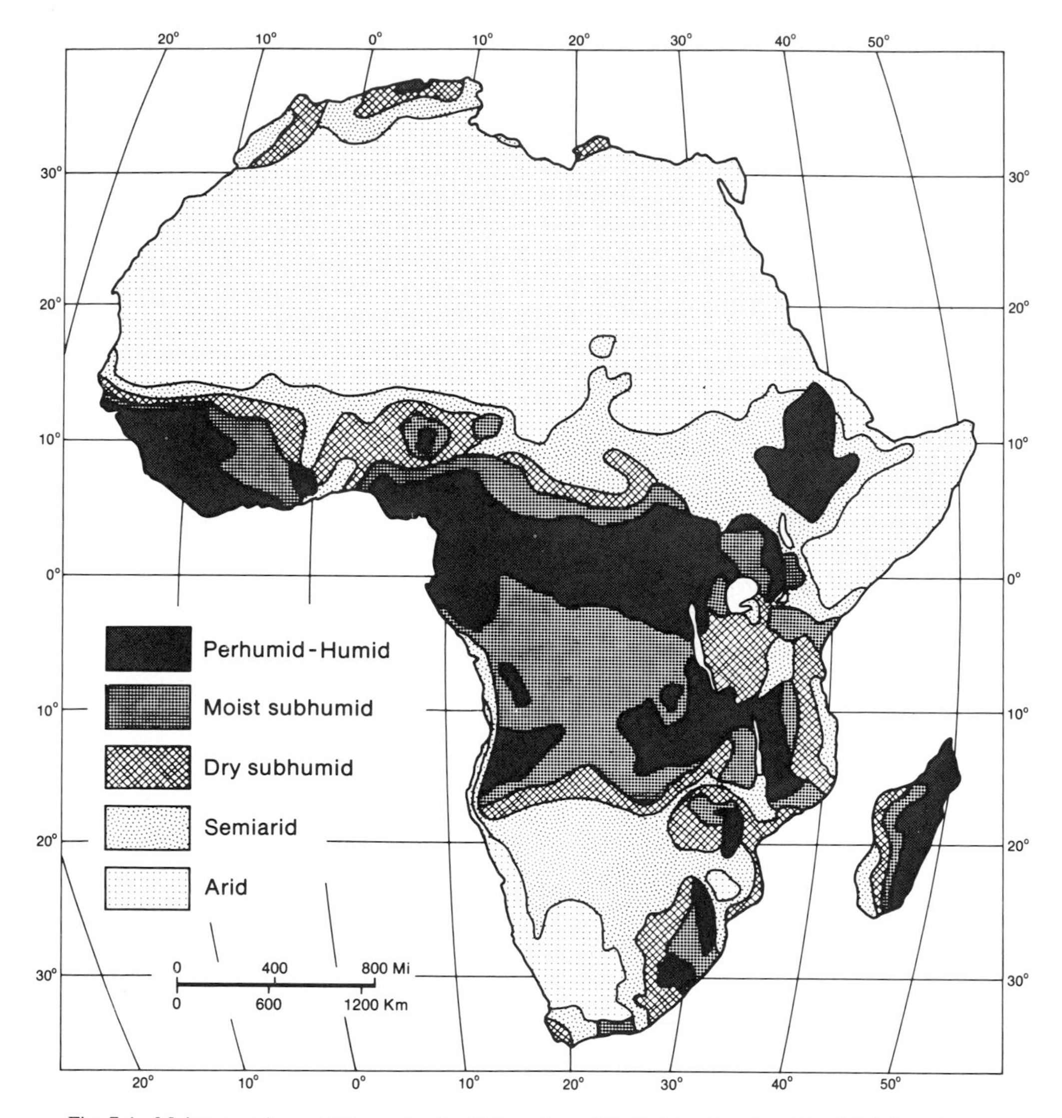

Fig. 7.1. Moisture regions of Africa after the 1948 system of C. W. Thornthwaite. (Simplified from Carter, 1954.)

high dominate the interior. Here there is a relatively free latitudinal exchange of air, with unstable equatorial air extending 20° and more south of the equator in summer. Because in Africa there is no such terrain barrier as the Andes, an anticyclonic circulation prevails across the whole of North Africa, with another similar cell over Arabia, and their subsident effects continue in summer as well as winter, even though a *surface* thermal low does exist in the warmer months.

Also in Africa south of the equator the tropical humid climates are able to reach to 17°–20°S, whereas in North Africa their northern limit is about 13°N in the west, and they scarcely reach to the equator in the extreme east. Thus, the dry climates of northern Africa broaden from west to east as the distance from the Gulf of Guinea and the Atlantic Ocean moisture sources increases.

The reasons for the unusual southward penetration of the Saharan dry climate are not entirely clear. Part of the explanation lies in the vertical structure of the *ITC* in that region at the time of high sun. In July and August the surface *ITC* reaches a mean position of about 20°–21°N, deep within the Sahara, so that if the summer rainfall were concentrated in the latitude of the surface convergence, or the northern limit of *mT* air, the desert area would be greatly shrunken on its equatorward margins. Actually the summer position of the *ITC* fluctuates greatly, not only aperiodically but also within the diurnal period. The *ITC* represents the boundary between the hot, dry Sahara air of anticyclonic origin arriving from the north, and the cooler, moister, maritime air from the south (Fig. 7.2). The latter originates in either of two source regions, the Atlantic Ocean and Gulf of Guinea in the west, and the Indian Ocean in the east. The northern continental current in summer is warmer than the southerly maritime air, and since the slope of the front is determined by the density contrast between the two air masses, the front must slope upward toward the south and at a very low angle of inclination. Thus, there is developed a situation in which hot, dry, stable air is ascending over less hot, moist, unstable air. The climatic implications of this atmospheric structure are profound, for therein lies the key to much of the climatic character of the southern Sahara and Sudan (Solot, 1943).

The dry and stable Saharan air precludes the development of any precipitation from its ascent along the very mildly inclined surface of discontinuity. Moreover, in its trajectory northward the maritime air has been undergoing convection, condensation, and precipitation, so that its moisture content has decreased. Concurrently its vertical depth has shallowed. Any afternoon cumuli developing in the thin wedge of maritime air are quickly dissipated when they penetrate through the above-surface front and enter the dry Saharan air. The effect of the frontal structure, therefore, is to inhibit the development of deep cumulo-nimbus and hence the occurrence of precipitation. Only at distances far enough south of the surface *ITC* so that the layer of moist, unstable surface air of maritime origin is sufficiently deep to allow a development of cumulo-nimbus clouds can precipitation occur. This does not take place until approximately 320 km south of the surface front, where a line drawn par-

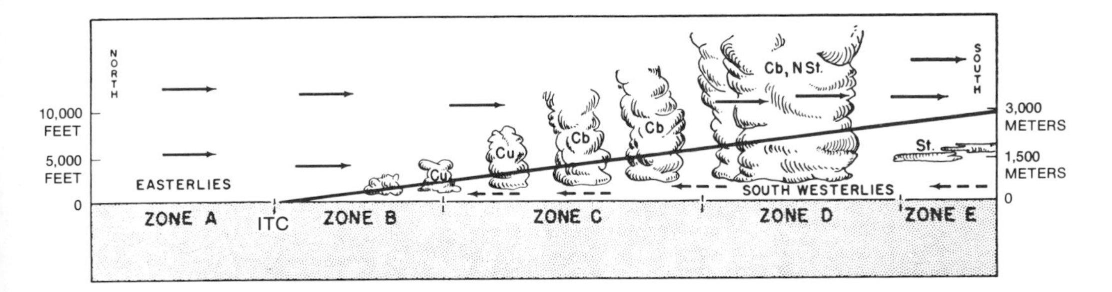

Fig. 7.2. The structure of the *ITC* in Africa north of the equator in summer. Here hot, dry easterlies from the Sahara overrun humid and less hot southwesterlies of maritime origin along a very mildly sloping surface of discontinuity. Five weather zones are indicated. (After Solot, 1950; Great Britain Meteorological Office, 1944*b*; and Walker, 1958.)

allel to the convergence marks the approximate northern limit of precipitation in summer (Solot, 1943). It is even much farther south that the maritime air is sufficiently deep and humid, and persists over a long enough period so that the high-sun rains are adequate to produce a humid climate. The surface *ITC*, therefore, is here a drought-maker in summer, and its July position is coincident with the southern Sahara.

Still another factor which may aid in extending the Saharan drought so far equatorward, especially in eastern Africa, is the subsidence on the northern flanks of a decelerating easterly jet stream, which in July is positioned at about 15°N over eastern Africa (Koteswaram, 1958).

According to Solot (1943) the summer rains in the transition zone along the southern margins of the Sahara may be of two kinds, one possibly associated with local random convection, and the other in the form of organized shower activity occurring in connection with extensive atmospheric disturbances. He notes, for example, that at Khartoum at about 16°N, where the annual rainfall is 150 to 180 mm, and strongly concentrated in July and August, there were in those two months in 1942 only 8 days with moderate or heavy rains, 20 with light rains, and the remainder with no rain. On the occasions of heavier showers it was found that the *ITC* at the surface was north of its normal position and there was an increased southerly flow of moist air near the ground which converged with a correspondingly strong northerly flow aloft at about 2,400 m. The spells of heavier widespread shower activity also coincided with increased horizontal convergence and cyclonic vorticity throughout a great depth of the troposphere. In other words, some sort of atmospheric disturbance was essential in order to cause widespread and heavy showers.

The semiarid Sahel savanna belt, situated approximately astride the 15°N parallel, and thus along the Sahara's southern borders, is the northernmost and the most arid of West Africa's agricultural zones. The region is notorious for its ravaging droughts, most recently those of the late 1960s and early 1970s. At present the causes for the large interannual variations in the July–September rains are not well understood, but it is recognized that if it were possible to foretell some months in advance the nature of the approaching summer rains, the effects of a future drought might be minimized by taking appropriate measures.

Efforts to explain the recent sub-Saharan droughts have been generally in the direction of the possible function played by anomalies in the large-scale atmospheric circulation. Schupelius (1976) has found a correlation between the Sahel's summer rainfall and the latitudinal position of the subtropical high over the Sahara. A more southerly location of the high correlated with dry years, while a northerly shift coincided with less dry years. Lamb (1978) found that a number of interrelated features of the tropical Atlantic, both atmospheric and oceanic, were situated 300 –500 km farther south in the drier sub-Saharan 1968 rainy season (July–September) than in the somewhat moister 1967 rainy season. Among these features he notes the following: "The kinematic axis between the Northern and Southern Hemisphere trades, the near-equatorial convergence zone, the near-equatorial pressure trough, the zone of maximum sea surface temperature (*SST*), the mid-Atlantic maxima of precipitation frequency and total cloudiness, and the center of the North Atlantic subtropical high."

A special weather type characteristic of the dry season in the Sahara and the climatic transition lands to the south of it, is the well-known harmattan. It is one of the few types of disturbed dry-season weather in this extensive region where air-mass conditions largely dominate during the period of low sun. The harmattan consists of a strong southward surge of fresh, dry, northerly air. Since the air has had a long trajectory over the Sahara, it is very dry, cool by night and warm by day, and is laden with dust. Thus, a dense and widespread haze is characteristic of harmattan weather and this condition may persist for several days.

The source of the harmattan air is twofold. It may be derived either from an intensification and southward displacement of the Azores high or from cold Eurasian sources. The Eurasian air, after crossing the Mediterranean, invades the African continent and underruns the normal Saharan air in the form of a vigorous cold front. Such *cP* air masses may advance southward to within 5° of the equator. Cool air from the first-mentioned source is localized chiefly in Nigeria and West Africa. It should be noted that winter conditions in northern Africa are particularly favorable for a deep and rapid southward movement of polar air, for unlike the situation in the Western Hemisphere where a large mass of opposing tropical maritime air must be penetrated, here there is no such opposition. Once the polar air has crossed the Mediterranean Sea it makes contact with Saharan air which likewise has a northerly component, so that the front advances southward with great rapidity. The harmattan cold front, although characteristically a dry-weather disturbance, may become a cause of rainfall in the low-sun period along the Gulf of Guinea.

Throughout much of the transition area along the southern margin of the Sahara, where a vertical sun is experienced twice a year, two seasonal maxima of temperature are characteristic. The first of these usually occurs in May, and is fairly coincident with the first vertical sun and a prevalence of *cT* air. But at the time of the second period of high sun in August the prevailing air mass is *mT*, and cloudiness and precipitation are at their peak. It is only after the retreat southward of the *ITC*, when *cT* air and cloudless skies again prevail, that the second or October temperature maximum occurs (Solot, 1950).

The Transitional Climates of the Sudan and Guinea-Coast Lands

As a textbook example of what tropical wet-and-dry climates (*Aw*) should be in terms both of fundamental climatic characteristics as well as of type location, the Sudan region of tropical Africa north of the equator probably has no peer. Throughout its great east-west extent it is bordered on its poleward side by the Sahara, while southward it merges into the tropical wet climates along the margins of the Gulf of Guinea and the Atlantic Ocean in the west, and the Congo Basin to the east. True, the north-south dimensions of the transitional *Aw* climates of the Sudan are narrower than might be expected, but this already has been explained as the consequence of an unusual southward expansion of Saharan climate.

Over the Sudan, and all but the southern margins of the lands bordering the Gulf of Guinea, there occurs a reversal of wind direction during the annual march of the seasons, with dry northerly Saharan air, or trade winds, prevailing at low sun, and maritime southerly currents at high sun. This seasonal reversal in the circulation is commonly designated as a monsoon, and so it is, provided a genetic element is not attached to the definition of a monsoon. But the seasonal wind reversal is not a consequence of the differential heating of land and water; it simply follows upon the orderly advance and retreat of the wind systems consequent upon the course of the sun. To be sure, the superheated Sahara does have the effect of displacing the *ITC* unusually far poleward in July (20°–21°N) when a surface thermal low is developed over the southern Sahara, but the shallow flow of maritime air in these latitudes, just to the south of the *ITC*, is not an effective rain-maker.

Circulation Patterns

Along the Guinea coast and for a distance of 150 to 300 km inland, there is no seasonal wind reversal and the prevailing air movement is onshore from the southwest throughout the year, although the vertical depth of the maritime current is much shallower at the time of low sun. However, because the coastal margins experience a shallow inflow of maritime air during the Northern Hemisphere winter when the Sudan just to the north is dominated by dry northerly Saharan air, the coast lands do have a very modest amount of winter rainfall. There appears to be some disagreement as to why southerly maritime air continues to cross the Guinea coast even in winter. The meteorologists at Accra believe that the *ITC* does not move southward over the cooler waters, but instead continues to have a mean winter location over the warmer lands slightly to the north of the coast and therefore north of the equator. Solot (1943) on the other hand, is inclined to view the southwesterly onshore air flow as partly in the nature of a perennial sea breeze.

Over the western Sudan and adjacent Guinea coast, and over the central African Sudan as well, the southerly flow of maritime air in summer has its origin over the tropical Atlantic and Gulf of Guinea. Only the easternmost parts feel the effects of southerly air derived from Indian Ocean sources. Apparently the southwesterly current of summer over the Sudan is only the western part of a longitudinally much more extensive (150° or more) equatorial southwesterly current which dominates not only tropical and equatorial Africa, but the Bay of Bengal and southern Asia as well, and even reaches eastward to China and Japan. In the latter regions it is commonly designated as the southwest monsoon. These summer westerlies of the low latitudes in Africa north of the equator and southern Asia are convergent and unstable, so that they provide an environment favorable for precipitation.

To understand the seasonal and geographical distribution of precipitation in the region of the Guinea coast and northward into the Sudan, it is necessary to comprehend the vertical structure of the atmosphere. Unfortunately that is a feature on which there are still differences of opinion. About the only matter on which there appears to be general agreement is that the humid maritime westerly and southwesterly current is overlain by air with an easterly component. But the structure of the discontinuity, and the nature and origin of the easterlies aloft, are still items of dispute. It has been suggested that the easterlies may exist in the form of a "nose" penetrating into the maritime air, as described by Sawyer (1947) for northwestern India-Pakistan, but this has not been verified. Certainly the easterlies which override the maritime current in its northerly parts are of anticyclonic and Saharan origin and consequently dry, but whether their origin and their humidity characteristics change farther south, as suggested by Hamilton and Archbold (1945), remains uncertain.

The Weather Zones.—Of signal importance in understanding seasonal weather and the related annual march of rainfall in tropical West Africa, is the fact that the region has four or five latitudinal weather zones (Fig. 7.3). These have been described by a number of writers, together with the climatic consequences of their north-south movements following the sun. The fivefold division here employed follows the classification suggested by Walker (1958). Reference is to the July period when the belts are displaced farthest north.

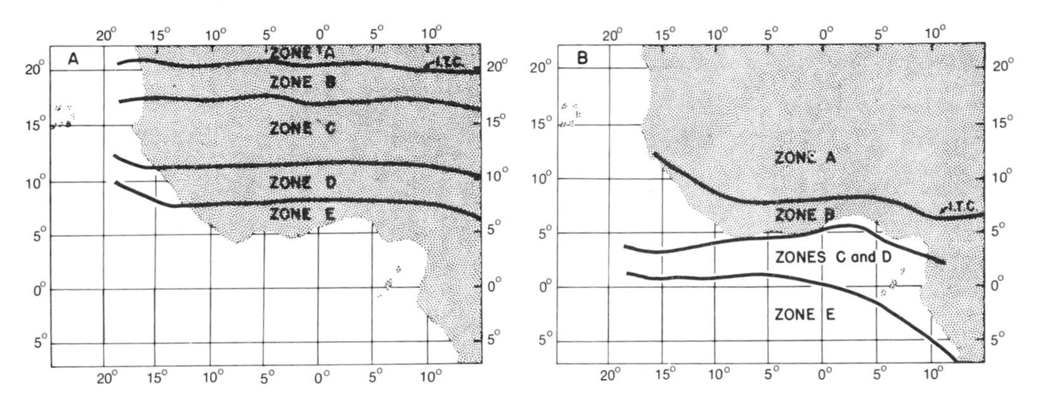

Fig. 7.3. General locations and approximate boundaries of West Africa's 5 weather zones as shown in Fig. 7.2, in summer, A, and in winter, B. (After Walker, 1958.)

Zone *A*, located farthest north, lies poleward of the *ITC* and hence is completely within the easterly flow of dry, anticyclonic, Saharan air. Cloudless skies, with hot days and marked night cooling, prevail. Harmattan periods, characterized by extensive dust haze, are frequent.

Zone *B* extends for 250 to 400 km south of Zone *A*, with its southern boundary poorly defined. Surface winds are WSW, but the depth of the maritime current is shallow. On a few days each month, usually not over five, thunderstorms develop in the afternoon or evening, which may or may not bring rain. The July or August rainfall usually is under 75 mm.

Zone *C* extends southward from Zone *B* for 500 to 650 km, but its width is variable and its boundaries are poorly defined. Here the maritime surface current is deeper, so that total rainfall is greater than in Zone *B*, but normally it does not exceed 125 mm a month. Much of the rain falls in heavy showers associated with disturbance lines oriented roughly north-south and moving in a westerly direction, as determined by the easterly circulation aloft.

Zone *D*, to the south of Zone *C*, averages about 300 to 500 km in width, but its boundaries are variable and ill-defined. Here the deeper maritime current, and perhaps the more humid easterlies aloft, create an increasingly favorable environment for precipitation, so that days with rain are the rule rather than the exception, and the rainfall is more prolonged and less intense. Cloud is abundant. A considerable part of the precipitation appears to be associated with disturbances that approach from the south or southwest, with rain belts that are aligned east-west.

Zone *E*, the most southerly of the weather zones, is located much of the year over the ocean and is able to shallowly penetrate the coastal lands for only a short period between July and September. Although the southwesterly flow is steady and strong, and cloud abundant, rainfall in it is usually modest and temperatures not excessively high. Significantly, a stratus deck is of frequent occurrence, suggesting the presence of an above-surface inversion.

Walker (1958) points out that the weather zones of tropical West Africa advance northward at a rate of about 160 km a month, while their southward retreat is about twice as rapid. Moreover, the description of weather in the different zones as given above fits better the conditions during the northward advance, for during the more rapid southward retreat the zones are less distinct and *C* and *D* are particularly difficult to differentiate.

Throughout the extensive region under consideration there is a strong maximum of rainfall at the time of high sun when the humid maritime southwesterlies are deepest over the land and penetrate farthest inland. Low sun coincides with a marked period of drought, for at that time the *ITC* has migrated farthest south and the dry easterlies dominate the weather except along the Guinea coast. But although summer is the rainy season, with the isohyets following behind the northward advance of the southwesterlies and the *ITC*, the precipitation is not associated with the surface convergence between easterlies and westerlies. The reason why this surface convergence is associated with drought instead of rainfall has been pointed out in a previous section dealing with the Sahara. As a consequence, and in spite of an abnormally large northward displacement of the *ITC*, the isohyets lag far behind and the belts of transitional climates remain narrow. Only where the maritime surface current is of adequate depth, and the atmospheric disturbances are sufficiently numerous and well developed, is precipitation moderately abundant.

Perturbations and Rainfall

In addition to rainfall caused by relief features, and possibly also by random thermal convection, although the latter is controversial, there is much which has its origin in extensive atmospheric disturbances. One generally dry weather type, the harmattan, has been described earlier, and the conditions under which it may produce cloud and rain in the coastal areas during the cooler months will be pointed out in a subsequent section. Two other types of disturbances, both prime rain-bringers to the Sudan and Guinea coast, require comment.

The first of these is the well-known disturbance line, which, because it has been described in detail by a number of authors (Hamilton and Archbold, 1945; Chancellor, 1946; Eldridge, 1957), requires only summarization here. Although the disturbance line is characterized by thunderstorms and squall winds, and thereby resembles a cold front, it is not frontal in origin and structure, but instead is embedded within a homogeneous air mass. This perturbation originates within the deep easterly flow located above the maritime southwesterlies and is carried by the easterlies in a westerly direction, although the accompanying cloud and rain are developed in the underlying humid air. West African disturbance lines probably belong to that variety of perturbation known as the easterly wave. Believed to be characteristic of the whole Sudan area, they can be traced from the longitude of Lake Chad westward throughout tropical West Africa.

A typical disturbance line has the following characteristics. The axis of the wave is oriented roughly north-south and it moves in a WSW direction at about 40–50 kph. Surface winds ahead of the trough are the normal southwesterlies. As it comes nearer, a dark, heavy bank of cumulus reaching up to 9,000 m or higher is observable to the east. A distinct roll of low cloud lies at the forward edge of the cloud bank, and as this roll cloud passes there is a sudden and strong squall wind from an easterly direction. A few minutes later the downpour of rain commences, usually accompanied by thunder and lightning. As defined by strong turbulence, heavy rain, and low cloud, the average disturbance line is roughtly 50 km wide, so that normally the rain does not last more than an hour or two. Not unlikely disturbance lines exist throughout the year, although their rainfall effects are most noticeable in those months when the surface maritime current from the southwest is relatively deep and extends well inland. They appear to be particularly characteristic of Walker's Zone *C*. Without doubt the disturbance-line weather type is a major cause of rainfall in the Sudan and the Guinea coast. At four widely separated stations in Ghana, Eldridge (1957) has computed the proportion of rainfall in each month of one year, 1955, which owed its origin to disturbance lines. The proportion varied from over 90 percent at Navrongo in northernmost Ghana, to about 30 percent at Axim on the southwest coast.

A second type of disturbance which is an important cause of rainfall is one concerning which little is known at present and still less has been written (Walker, 1958; Swan, 1958), and to which no specific name has been assigned. About all that is known concerning it is that it is an extensive area of bad weather within the deeper parts of the maritime southwesterly flow and is considered to be an important weather element especially of Walker's Zone *D*. Radar at Accra first picks up the southwesterly perturbations as bands of disturbed weather over the ocean to the south, where their longer east-west dimension may be roughly 150 km and their width about half as great. As they move onshore, carried by the southwesterly air flow, they are accompanied by weather which resembles warm-front conditions and hence stands in contrast to that associated with the more vigorous disturbance lines. There are no significant wind shifts and squall lines, and the rainfall is less intense, but usually of longer duration. By the West African meteorologists they are described as bringing the "monsoon rains." Little is known about their structure although in conversations it has been suggested they may resemble the surges in the southwest monsoon of southern Asia and in the northwesterly current over interior Brazil at high sun. Such speed convergences may not be marked by an increase in wind speed near the surface where turbulence is relatively strong.

If the annual rainfall profile of Accra is constructed from 5, 10, or 15-day means instead of from monthly averages, it is revealed that there are three periods of maxima instead of two, a primary one in late May, June, and early July, and two secondaries, an earlier one in March and early April, and a later one in October and early November (Fig. 7.4). The meteorologists at Accra are of the opinion that the earlier rains in March–April, which occur when the southwesterly flow is less deep (Zone *C*), are probably more a result of disturbance lines, while the later rains in June and early July are to a greater extent associated with those other disturbances which come in from the south and are characteristic of the deeper parts of the southwesterly current (Zone *D*). Since the retreat southward of the weather belts is more rapid than their advance northward, the third maximum in October probably combines the rainfall effects of both types of disturbances. It is not to be inferred that the rainfall effects of the two disturbance types can be completely separated even in the spring and summer maxima, but only that the maximum frequency of disturbance lines tends to precede that of the southwesterly perturbations.

This concept of a differentiation in the time of occurrence of the two types of disturbances has been studied in some detail by Swan (1958). Unfortunately his data are organized by monthly means only, instead of by shorter periods. Nevertheless, Swan has been able to show that the earlier rains of March to May have a relatively high frequency of occurrence on occasions with (1) wind velocities in excess of 40 kph and (2) easterly winds (Fig. 7.5). Both of these features are common to disturbance lines but

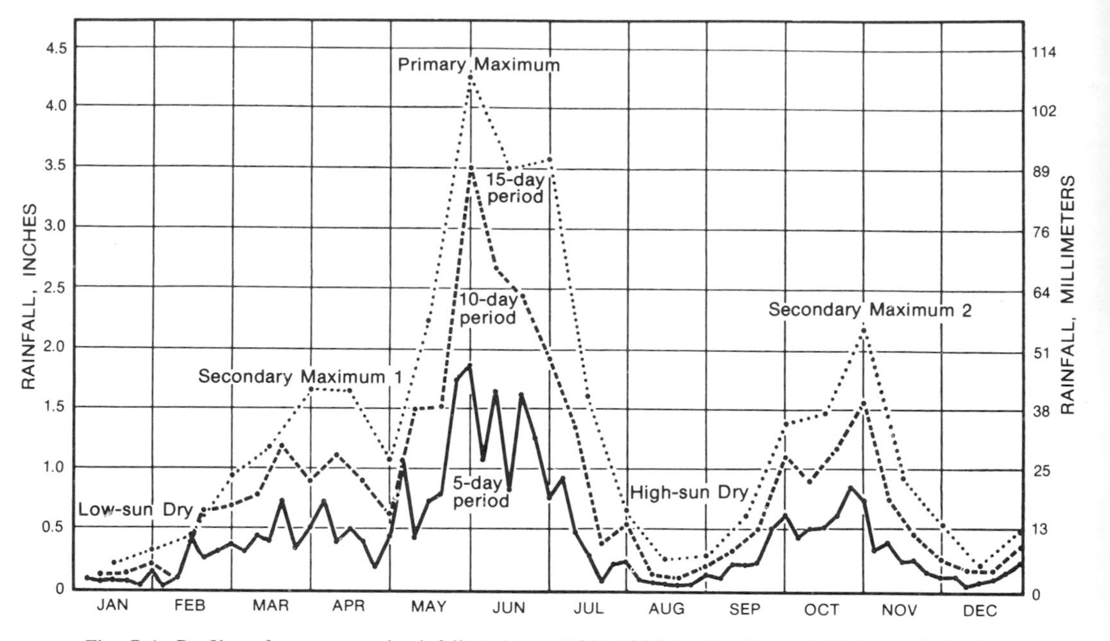

Fig. 7.4. Profiles of mean annual rainfall at Accra (1941–1956) by 5-, 10-, and 15-day periods. As a consequence of this more refined treatment of the rainfall data, a minor third rainfall maximum becomes evident in March–April. (After chart prepared at Ghana Meteorological Department, Accra.)

not to the perturbations arriving from the south in the southwesterly flow. By contrast, the somewhat later rains in June and July fall on occasions when winds are weaker (under 40 kph) and with surface winds having a westerly component, both features pointing to the southwesterly disturbance rather than to the disturbance line.

It would appear, therefore, that most of the rainfall in the Sudan and the Guinea coast lands is composed of two elements (omitting any possible effects of random convection), that associated with disturbance lines which seem to reach their maximum development in Zone *C*, and that falling in conjunction with southwesterly disturbances which are more a feature of Zone *D*. With the northward movement of the weather belts following the sun, the maximum rainfall effects of disturbance lines normally precede those of the southerly disturbances.

Climatic Peculiarities

The Secondary Rainfall Minimum in August along the Guinea Coast.—While over much of the greater part of the region being described the annual rainfall profile is a simple one, with a single high-sun maximum and low-sun minimum, this is less true of the Guinea coastal belt south of 7° or 8°N. Even on the southwest coast, south to about southern Sierra Leone, there prevails the simple rainfall profile described above. But along the littoral from Liberia eastward it becomes more complicated, there being two summer maxima, one in June and the other in October, separated by a secondary minimum in mid- and late summer (Fig. 7.6). The June maximum, normally the primary one, occurs with the northward advance of the *ITC* and the deepening of the southwesterly flow. But, oddly, at the time when the

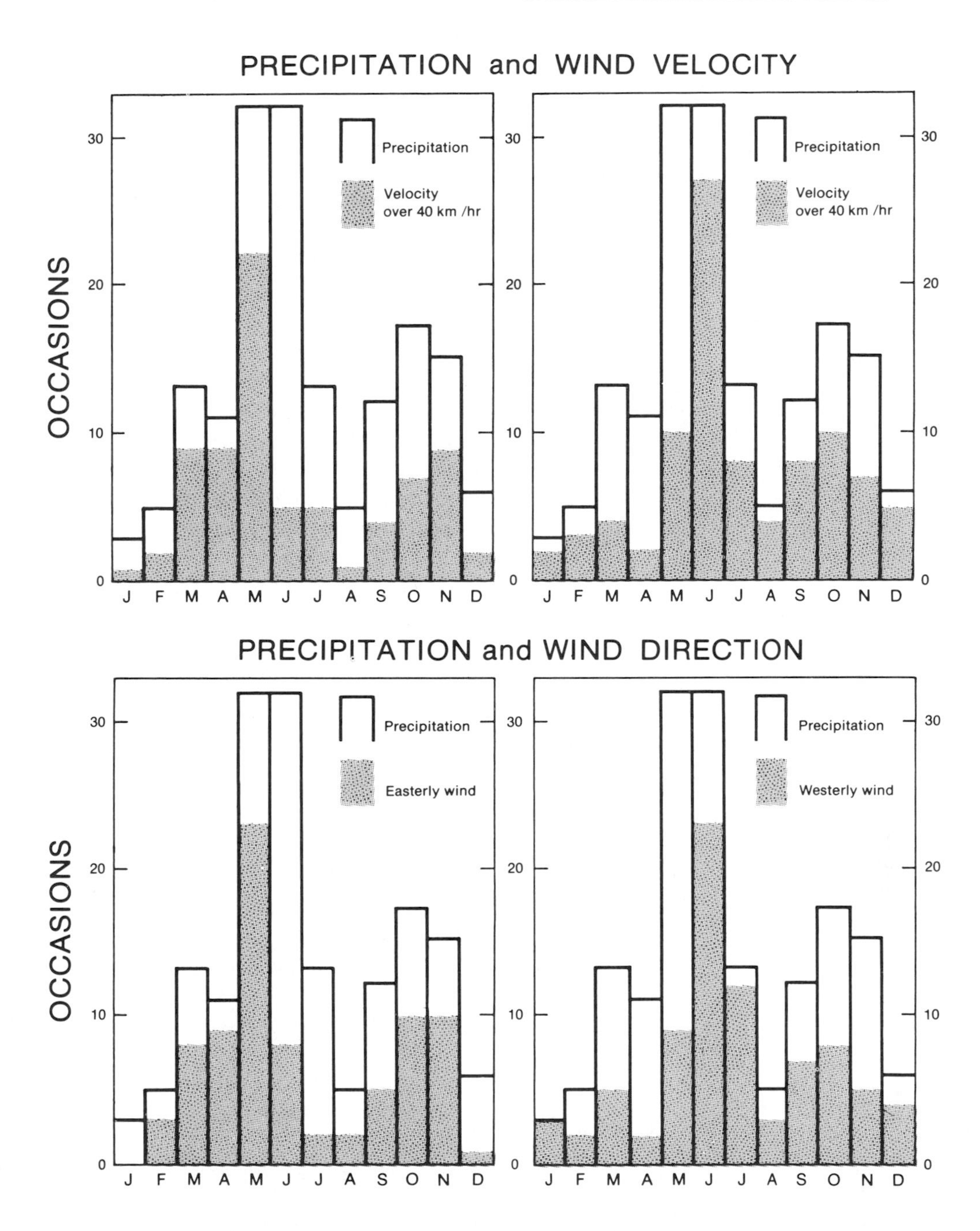

Fig. 7.5. Frequency of occurrence, by months, of rainfalls exceeding 5 mm in the Sudan-Guinea Coast during the period 1947–1951, separated into (a) those occasions when the rainfall was associated with winds above or below 40 km/hr and (b) those occasions when the rainfall occurred in conjunction with winds having either an easterly or a westerly wind component. It can be observed that the rains of March–April–May are associated with higher wind velocities than are those of June–July. Also the earlier rains of March–April–May occur more frequently with easterly winds, while those of June–July are to a greater extent associated with westerly winds. (After Swan, 1958.)

ITC should be at its maximum northern location, and hence the maritime current deepest over the coast, rainfall declines, and temperatures do as well.

The peculiar secondary minimum in August occurs at the time when weather Zone *E*, described previously, makes its appearance over the coastal lands and southward over the waters of the Gulf of Guinea. It does not extend far inland. Many of the weather features of Zone *E* are not well understood at present, and they obviously require much further observation and study, but certain tentative elements may be mentioned. Along the Guinea coast the circulation aloft in late July and August, when wind and rain belts are in their maximum northern positions, appears to become divergent and subsident, for at that time inversions and isothermals occur frequently at about 1,500 –2,100 m. It is possible that the seasonally strengthened South Atlantic anticyclone, which is responsible for the still lower July-August inversions at Kinshasa (Leopoldville), just south of the equator in Zaire (Belgian Congo), is able in these months to extend its influence a few degrees north of the equator so as to place weather Zone *E* of the Southern Hemisphere over the Guinea coast at this time. At Lagos, as shown by Table 7.1, August experiences the greatest frequency of inversions and isothermals, followed by September and July whose frequencies are less than half as great (Fig. 7.7). Unfortunately comparable data are not available for drier Accra and Lomé. Characteristically the rise in temperature in the inversion layer is accompanied by a fall in the wet-bulb temperature indicating the presence of warm dry air above the surface maritime current (see Table 7.2).

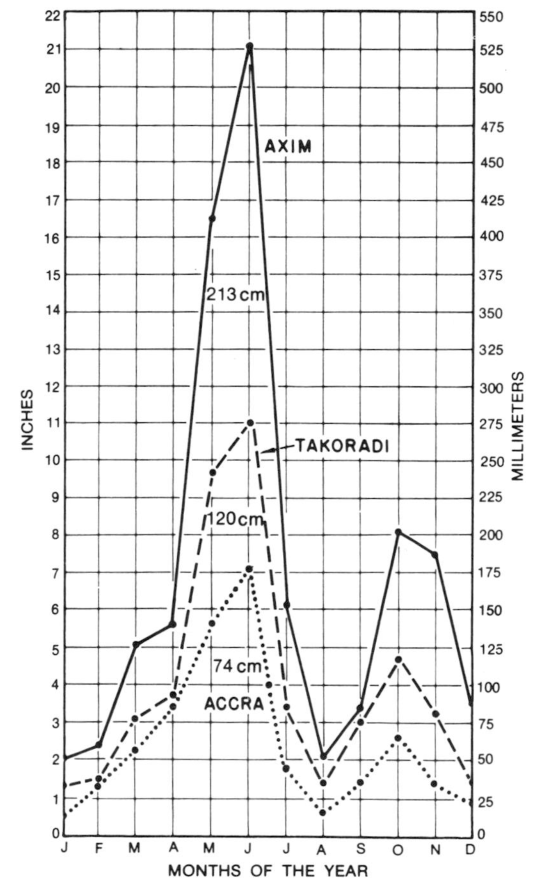

Fig. 7.6. Profiles of average annual rainfall at three stations along the Ghana coast. Although rainfall amounts are greatly in contrast at the three stations, the twinned maxima and minima are a common feature, with the abnormal August minimum equal in intensity to that of the low-sun period.

Not only is there a remarkable coincidence of inversions with the secondary dry season, but the frequency of inversions in different Augusts is likewise reflected in rainfall differentials. Thus, August, 1943, with a notably higher frequency of inversions than August, 1944, had only 42 percent as much rainfall (22 mm *vs.* 52 mm).

During the August secondary dry season the surface air is still very moist, but because of the frequency of inversions and isothermals, the height of cloud-top development is very limited. Characteristically the

Table 7.1. Percentage frequency of above-surface inversions and isothermals at Lagos, Nigeria, 1943 and 1944 (Great Britain Meteorological Office, 1949)

Millibars	J	F	M	A	M	J	J	A	S	O	N	D
700–750	0	0	0	0	0	0	2	2.5	0	0	0	1
750–800	1	1	0	0	0	1	7	18.5	4.5	0	0	0
800–850	2.5	1	0	0	1.5	0	5	9	4.5	1	1	1
850–900	1	2	0	0	0	1	1	2.5	2.5	2	0	1
900–950	1	3	2	0	2.5	0	2	4	0	2	0	1

cloud forms reveal one or more layers of stratus or strato-cumulus. By day solar heating may cause the cloud deck to break up to form small isolated cumuli, but the stratus reforms at night and when it is low and thick it may even yield some drizzle or light rain.

Winter Rainfall in the Guinea Lands.—An additional climatic feature of the Guinea coast lands which distinguishes this region from the Sudan is that the winter months are not rainless as they are farther north. Thus, Accra normally records nearly 75 mm in the period December to February inclusive and Takoradi over 100 mm. During the time of extreme southward positioning of the weather belts, dry easterlies prevail over most of tropical West Africa, but along the coast air movement continues to be from the southwest and from the sea. The maritime current has little depth, however, and its penetration of the land is shallow and fluctuating, normally not more than 150–300 km. It was noted earlier that the onshore air stream may be, at least in part, a sea-breeze effect, but the West African meteorologists are inclined to attribute it largely to the fact that the *ITC* remains slightly north of the coast even at the time of low sun. As the southwesterly air moves inland, rainfall may result from its elevation by terrain, or along a front where it converges with the northeasterlies which at this time are cooler than the maritime air. Density contrasts result in a frontal structure which differs from that of summer, with the cooler easterlies underrunning the westerlies in the form of a cold front. When the cold front is fairly smooth and without irregularities the weather is relatively undisturbed, but if the northerly flow strengthens so as to produce a sinuous *ITC*, showery weather with thunderstorms develops south of the cold thrusts. Normally the southwest current is too shallow to produce steady rain from a general overcast condition, but in the vicinity of an invasion of fresh northerly air the southwesterlies may be drawn into a convergence, so that the maritime air is strengthened and deepened and showery weather results.

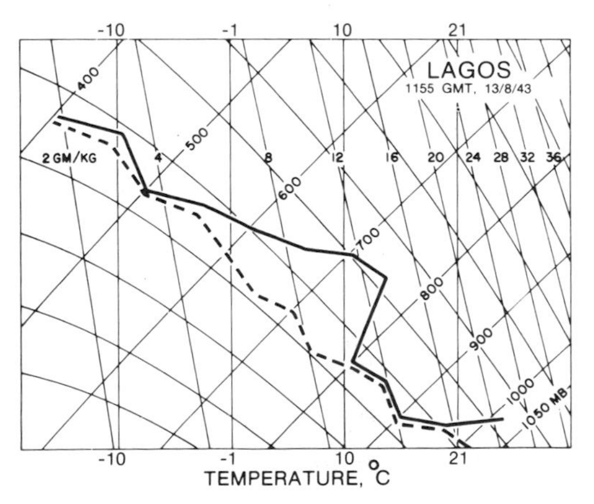

Fig. 7.7. Tephigram for Lagos, located on the coast in Nigeria, showing a temperature inversion. Dry-bulb temperatures are represented by a solid line, wet-bulb temperatures by a broken line. (From Great Britain Meteorological Office, 1949.)

Table 7.2. Percentage frequency of inversions and of dry air at Lagos, Nigeria in August (Great Britain Meteorological Office 1949)

Pressure level (mb)		850		800		750		700		650
Inversions						*Percent*				
	1943		12		22		2		3.5	
	1944		6.5		14.5		3		0.5	
Depression of wet bulb						*Percent*				
	1943	3.5		15.5		59		63		48
> 5°F	1944	1.5		9		26		31		23
	1943	0		0		33		16		3.5
>10°F	1944	1.5		3.5		4		6		0

The Dry Littoral of Ghana and Togo (Wise, 1944; Hubbard, 1954; Hubbard, 1956).—The most remarkable climatic anomaly of the Guinea coast lands is in that mid-section centering on central and eastern Ghana, Togo, and Benin (Dahomey). Here, wedged in between tropical humid climates (mostly *Ar*) both to the west and to the east, is a subhumid and semiarid littoral whose deficiency in precipitation is reflected in the vegetation cover which is a degraded wooded grassland representing a fire climax of grass and bush. From the air this enclave of savanna stands out conspicuously from the surrounding tropical rainforest. As a rule the vegetation belts in tropical West Africa are roughly parallel with the latitude lines. However, in the Ghana-Togo area this longitudinal pattern is deranged and the coastal savanna makes junction with that of the Sudan proper, thereby separating the forest area into eastern and western sections.

Scarcity of data does not permit drawing the boundaries of the dry-subhumid area with a high degree of accuracy. The 760 mm isohyet, marking the driest part, approximately follows the coast from Accra to Lomé. Accra records only about 720 mm on the average and it is not unlikely that parts of the surrounding plain receive even less. The 1,000 mm isohyet may be accepted as approximately delimiting the main region of rainfall deficiency. On the west it reaches to the coast at about Sekondi-Takoradi in western Ghana. Eastward it demarcates a fairly narrow dry coastal belt throughout Ghana, but as the highlands recede inland in Togo and Benin, the isohyet loops well inland some 200 km, before dropping southward to reach the coast again near the Benin-Nigeria boundary (Fig. 7.8). Coastal rainfall gradients are steep both to the east and west of the dry belt, the increase on the west between Takoradi and Axim being at the rate of about 15 mm per km. Even northward from the coastal dry belt annual rainfall remains relatively low, especially in the Volta River lowlands. As a consequence in the vicinity of the 0° meridian there is an almost continuous north-south corridor of semiarid or subhumid climates from the coast to the Sudan. Only the heavier rainfall on the hills

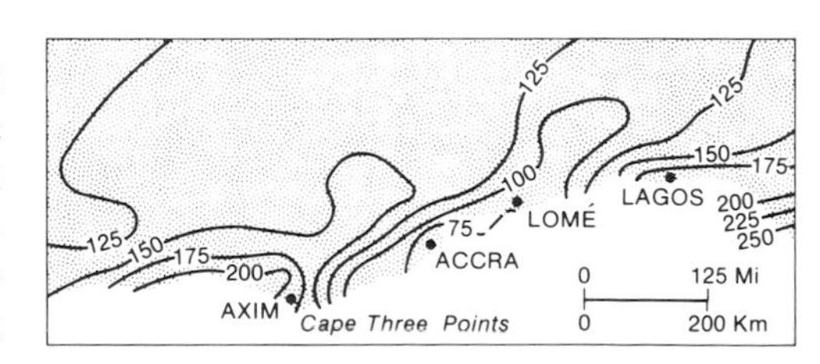

Fig. 7.8. Annual rainfall amounts along the Gulf of Guinea coast in Ghana, Togo, Benin, and western Nigeria. The 100 cm isohyet approximately defines the Accra dry belt. Compare with Fig. 7.1.

of southern Ghana and Togo interrupts the continuity.

Anomalous climatic features of the Accra-Lomé coastal zone include the following: (1) annual rainfall is abnormally low for an equatorial coastal lowland at about 5°–6°N where tropical maritime air is dominant throughout the year; (2) rain falls on fewer days and for briefer periods than in the areas surrounding it; (3) annual rainfall *increases* inland from the coast for 100 to 150 km; (4) the short dry season in July-August is unusually xeric, with August sometimes as dry as January; and (5) August in this area has the lowest average temperature anywhere in West Africa (Ojo, 1977).

In order to determine why the annual rainfall in the Accra-Lomé area is so much lower than that of the wetter parts of the Guinea coast to the east and west, it is necessary to compare the annual rainfall profiles of the dry and the wet stations to discover at what times of the year the dry stations have markedly less rain than the wet stations. To do this the differences in monthly rainfall means between dry Accra and wet Axim and Lagos were determined and subsequently annual profiles of the monthly differentials were drawn (Fig. 7.9). What becomes very clear is that Accra is much drier than Axim to the west and Lagos to the east chiefly because of the great reduction in rainfall at the times of the two maxima, one in May-June-July and the other in October-November, but chiefly at the time of the stronger first maximum. There is also slightly less rainfall at Accra at the times of the two minima, one in December to April and the other in August-September, but these differentials are not great enough to be of much consequence. The problem, then, is to explain why Accra and Lomé have so much less rainfall at the times of the two maxima, periods when the coast is feeling the effects of, first the northward advance, and later the southward retreat, of

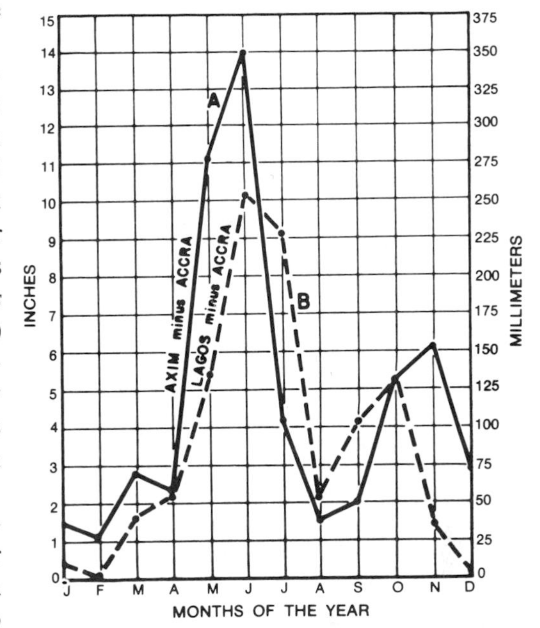

Fig. 7.9. Annual profiles of monthly rainfall differentials between Axim and Accra, and between Lagos and Accra. The great decline in rainfall in the Accra dry belt over that which characterizes the wetter areas both to the east (Lagos) and to the west (Axim) occurs during the time of the normal primary May–June maximum.

Walker's Zones *C* and *D* with their disturbance lines and southwesterly perturbations.

One clue is furnished by the observation that the drought reaches its greatest intensity along the coast, so that east-west rainfall gradients are much steeper along the littoral than they are at some distance inland. This seems to point to the coastline and the adjacent ocean as controls acting to depress the rain-making processes within the dry part of the littoral. It appears to be significant that one of the steepest rainfall gradients along the coast lies to the west of the dry belt in the vicinity of Cape Three Points where there is an abrupt change in the directional alignment of the coastline, while a second, but somewhat less steep gradient occurs between Lomé and Lagos where there is a more gradual change in the direction of the coastline. At Axim, just to the west of Cape Three Points, where the coastline is aligned in a northwest-southeast direction, the annual rainfall is about 2,100 mm; Takoradi just east of the Cape on a southwest-northeast coast records only 1,150 mm; while Cape Coast and Accra, a few score kilometers farther to the northeast, have totals of only 900 and 725 mm.

Wind direction at Accra has a westerly component throughout the year. At midnight surface winds show a highest frequency from the west, but with relatively high frequencies, especially at the two periods of rainfall maxima, also from WSW, WNW, and SW. At those times of day when the sea breeze is active there is a somewhat greater southerly component. It is evident that those parts of the coastline which are oriented in a WSW-ENE direction, as is the Accra-Lomé section, experience winds that are either parallel to the coast or very slightly offshore or onshore. Under such conditions, differential frictional effects of land and water should result in divergence in the surface air current in the vicinity of the coastline. Where the coastline is aligned more NW-SE, so that the winds have a larger onshore component, this effect is reduced or absent. The meteorologists at Accra are of the opinion that even within the dry coastal belt itself, minor bends in the coastline are reflected in rainfall variations of a small magnitude. If the previous reasoning concerning at least one cause of the Accra-Lomé dry littoral is valid, one might expect a somewhat similar reduction of rainfall east of Cape Palmas (easternmost Liberia) where the coastline also is aligned WSW-ENE. In other words, along the coast in the vicinity of Cape Palmas there ought to be a duplication of the steep rainfall gradients that obtain in the neighborhood of Cape Three Points. Unfortunately station rainfall data to the east and west of Cape Palmas are scarce. However, at Harper, close to the eastern boundary of Liberia at Cape Palmas, the annual rainfall is 3,400 mm and at Greenville about 170 km farther to the northwest, and also on a coast aligned NW-SE, it reaches 4,500 mm. But at Tabou on the Ivory Coast just a few miles east of Cape Palmas where the coastline has changed direction, the rainfall already has dropped to 2,300 mm and at coastal Sassandra and Grand-Lahou somewhat beyond, only 1,500 mm are recorded. It would appear, then, that coastal rainfalls east of Cape Palmas are only one-third to one-half what they are to the west of it in Liberia, although statistical evidence is lacking to show that any parts of the Ivory Coast are as dry as the Accra area.

Navarro (1950) has suggested a supplementary explanation for the Accra dry area when he points out that the disturbance lines which approach the Guinea coast from the northeast take several different routes, the terminal areas of these tracks ending in the vicinity of Mt. Cameroon, Contonou, Abidjan, and Tabou. According to this observation Accra appears to occupy a position between the terminal zones of these important rain-bringing disturbances. Data on frequency of occurrence of line squalls, however, show that drier Accra has considerably more of these disturbances than does much wetter Axim although, to be sure, this situation is reversed in terms of number of thunderstorms.

Another climatic uniqueness of the Guinea coast mentioned previously is the unusual dryness of the high-sun, or August, minimum in the vicinity of Accra and Lomé. Along the whole south-facing coast of West Africa from about the Sierra Leone-Liberia boundary eastward to beyond the Niger Delta, the annual rainfall profile shows a secondary minimum whose lowest point usually is in August. In an earlier section this widespread secondary summer minimum has been explained as being a consequence of Zone *E*, with its stable anticyclonic air and temperature inversion, invading the coastlands in late summer. But toward the midsection of the Guinea coast, the August minimum becomes increasingly strong, so much so that August is as dry as January and at Accra and Lomé the average August rainfall has declined to 15 and 8 mm. As the total amount of annual rainfall increases both to the east and west of the Accra dry belt, August rainfall increases not only in absolute amounts but also relatively (1 percent at Lomé and 2.2 percent at Accra, but 3.8 percent at Lagos and 4.2 percent at Tabou). This poses the question as to why there is an intensified August drought along the middle sections of the Guinea coast in the vicinity of Accra.

A number of writers have suggested the presence of locally cool water as either a primary or auxiliary cause of the Accra dry area. It is not to be doubted that cool water does make its appearance in the Accra area during the period July to September which precisely coincides with the second dry season, which here is abnormally dry. Data on ocean temperatures are very fragmentary, and sampling appears to be from one location in coastal Accra only. From these samplings one may judge that ocean surface temperatures here may drop as much as 8° to 11°C (15° to 20°F) in August over what they are during the more normal period January to May. How extensive the area of cool water may be is not known. During the short period of cool water the meager evidence indicates that the set of the ocean current at Accra is from the east. Among the local meteorologists the belief is held that the cool water represents a northward extension of the Benguela Current at the time when the South Atlantic anticyclone has pushed farthest north. It is admitted that there likewise may be an upwelling effect.

Since cool water usually is present along the middle Guinea coast only during the period July to September, its effect in stabilizing the air moving over it, and thereby damping the rain-making processes, must be limited to the late-summer period. Brooks and Mirrlees (1932) point out that in the vicinity of Ghana, "the winds as drawn from the resultants and pressure distribution show a curious divergence at this point, the stream lines turning to the left in the west and to the right in the east as if they were avoiding an obstacle. It seems probable that a local pool of cool air forms over this cool water, which acts as such a barrier and deflects the air streams to either side." But cool water can scarcely be invoked as an important control causing the greatly reduced annual rainfall in the Accra-Lomé area, for here the principal reduction in rainfall occurs during the time of the precipitation maxima, in May-June and again in October-November, when cool water is not present. On the other hand, the cool water probably is responsible for the intensification of the August drought period along the middle Guinea coast.

A third climatic peculiarity, noted earlier, for the Accra area is the abnormally low air temperature which prevails in August and is coincident, therefore, with one of the two insolation maxima (Fig. 7.10). The fact is, however, that July-September is the period of lowest average monthly temperatures along the entire coast from Liberia to Nigeria. This may be understood in terms of

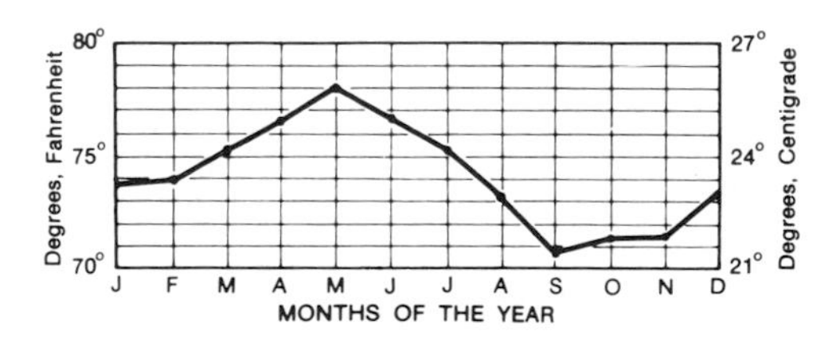

Fig. 7.10. Annual march of temperature at Accra. Note the unusual feature of the primary minimum occurring in August which is one of the periods of high sun.

the reduced insolational heating at the time the temperature inversion and its stratus clouds reach their maximum development. But along the Ghana section of the coast the average August temperatures (between 23°C [74°F] and 24°C [76°F]) at Takoradi, Cape Castle, and Accra are a few degrees lower than for most of the Guinea coast, suggesting that in this locality there is an auxiliary cooling agent, which is probably the cool current.

CHAPTER 8

Tropical Central Africa

The general region here under consideration is mainly the area of tropical wet (*Ar*) climate in interior equatorial Africa, most of it lying within the basin of the Congo River. Reliable rainfall data are too fragmentary to permit the precise defining of the boundaries or characteristics of climates in equatorial Africa, a fact borne out by the variety of isohyetal arrangements on maps prepared by different authors. Nevertheless, certain general facts of rainfall distribution and intensity are rather widely accepted. For one thing, the *Ar* climate appears to have a minimal sea frontage of only a few degrees and this faces west on the Gulf of Guinea. At some distance inland from the coast the area of tropical wet climate broadens, so that the latitudinal spread is nearly double what it is farther west, and the geographic equator roughly bisects the extensive area with over 1,500 mm of rainfall. Toward the coast, however, there is less latitudinal symmetry, for the isohyets are shifted northward, so that along the sea the heaviest rainfalls and the most extensive developments of *Ar* climate lie north of the equator (Fig. 8.1). On the east the tropical wet climate is terminated by the East African Highlands, and is largely lacking in easternmost equatorial Africa, even along the coast, only Madagascar exhibiting the common feature of a genuinely wet eastern littoral in low latitudes.

Circulation Features

Two contrasting circulations dominate the Congo Basin (Rainteau, 1955). At all seasons there is present a surface southwesterly flow of humid, maritime air, with an average thickness of 1,000–1,500 meters which originates over the South Atlantic Ocean and its cool Benguela current to the west. Entry of this southwesterly air into the Basin is moderately easy. The depth of penetration is variable, ordinarily to at least 16°E, but on occasions to 30°E and even beyond. Above the maritime southwest current, locally called the "monsoon," the tropical easterlies or trades prevail, mostly from the southeast, but also from the northeast, the latter more especially in those parts north of about 5°N. These easterly winds reach down to the surface where they are in directional opposition to the southwesterlies along a widely fluctuating zone of convergence, which changes position with the seasons and aperiodically as well. Seemingly, the stream of southwesterly maritime air varies considerably in thickness and also in vertical structure, and as a consequence its weather does as well.

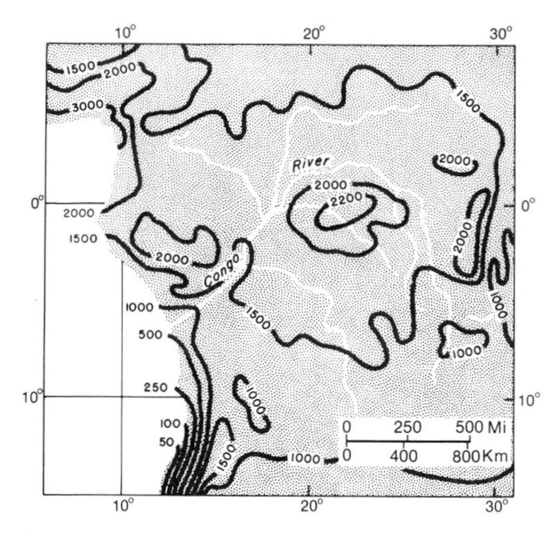

Fig. 8.1. Distribution of average annual rainfall (in mm) in the region of the Congo Basin. (After Pfalz, 1943, and Knoch and Schulze, 1956.)

Among meteorologists in central Africa there are widely divergent opinions regarding the roles played by the easterly and westerly circulations in Congo weather, as well as the relative importance of the Atlantic and Indian oceans as moisture sources for this region. In conversation, one at least contended not only that Congo rainfall depended in a large measure upon Indian Ocean sources, but likewise that the amount of rainfall is negatively correlated with the strength and the depth of penetration of the westerly flow. Another expressed his opinion that it was difficult to correlate wind direction and weather. On the other hand, Jeandidier (1956) and Rainteau (1955) of the meteorological service at Brazzaville, as indicated by their publications, look upon the southwesterlies as providing much the larger part of the precipitable moisture. According to this point of view, weather in the Congo Basin depends largely upon the vertical depth and the extent of penetration by the southwest current. Even as far east as Uganda on the East African Plateau the equatorial westerlies appear to play a significant role in the precipitation processes, for the convergence between westerlies and easterlies in this region probably has a rainfall effect resembling that of the *ITC*. When the westerly flow is weak or absent, rainfall on the western part of the East African Plateau in equatorial latitudes is below normal (Henderson, 1949). Rainteau (1955) contends that the East African Plateau and its mountain ranges block any important transfer of Indian Ocean moisture into the Congo Basin except for modest amounts entering along the corridor at about 4°N lying between the Ethiopian and Kenya highlands. Significantly, it is at the western end of this corridor that many of the squall lines affecting the Congo Basin seem to have their origin. Just what role the equatorial easterlies at higher elevations serve in transporting moisture westward beyond the eastern highlands remains undetermined.

Atmospheric Disturbances and Rainfall

While the origin of the rainfall in equatorial Africa remains a controversial topic among the local meteorologists, there is increasing evidence that much of it is derived from extensive mobile disturbances. To what extent random thermal convection plays a part is still debated. Bultot (1952) has pointed out, however, that at least some of the showery rainfall of the Congo Basin is not spread in random fashion over the region, but instead has a recognizable distribution pattern or organization in both time and space characteristics. Large areas of prevailingly showery weather in which more than 50 percent of the stations report rain, are designated by Bultot as "rain zones." Within them the rains do not necessarily follow a diurnal pattern, and are not confined to the warmest hours of the day. Bultot believes the organized rain areas of the Congo Basin are associated with extensive perturbations which, with rare exceptions, move slowly from west to east, not infrequently becoming stationary and following each other at frequent intervals. The duration of the period of perturbation rainfall is relatively brief, rarely exceeding three days, while one and two days are much more common.

While the meteorologists at Kinshasa (Leopoldville) and Brazzaville seem to agree with Bultot (1952) in his contention that some, or even much, of the Congo's showery rainfall is of an organized type associated with extensive disturbances, they doubt the validity of his conclusion that as a rule the disturbances characteristically move from west to east in the southwesterly current. Instead, they conclude that most of the disturbed weather moves in the opposite direction, or east to west, and is transported by the easterlies aloft. They point out that observation of the fact that rain areas on the synoptic charts appear to be located farther east on succeeding days is not sufficient evidence to warrant a conclusion that the perturbations producing the rains necessarily move in the same direction.

Jeandidier (1956) and Rainteau (1955) have provided one of the most complete analyses of disturbed weather in the Congo-Basin sector of equatorial Africa. They recognize three types of rain-bringing perturbations: (1) isolated random showers and thunderstorms presumed to be a consequence of insolational heating, (2) line squalls, similar to the disturbance lines of West Africa, and (3) surges in the southwesterlies. Whether the preceding classification includes the organized shower activity as described by Bultot (1952) is not certain. The conviction is strong with these two meteorologists that Congo-Basin rainfall is derived from Atlantic moisture advected by the cool southwesterly current, and that most of the disturbed weather occurs within, or along the margins of, the southwesterlies. Beyond the limits of this westerly maritime air, where easterlies prevail, the weather remains generally fair, with only scattered fair-weather cumuli. Obviously the accurate forecasting of the depth, vertical structure, and extensiveness of the southwesterly current is of high importance.

Vertical depth and extent of eastward penetration of the southwesterlies depend upon the seasonal location of the equatorial low and the position and strength of the South Atlantic anticyclone. Accordingly, a close correlation exists between the daily pressure at Pointe Noire on the Atlantic coast at about 5°S, and the strength and depth of penetration of the southwesterly flow the following day, so that a steep rise in pressure foreshadows an interior surge of the southwesterlies. That the relatively cool southwesterlies have certain qualities of above-surface stability is suggested by the fact that a deck of strato-cumulus usually prevails during the night and morning, which breaks up into isolated cumuli during late morning, with general instability and tall cumuli prevailing during the afternoon. Afternoon and evening thundershowers are the result. On days with widespread instability disturbances there is ordinarily a confused pattern of pressure and winds aloft, the latter being weak.

As a rule there is a tendency for these instability storms to be concentrated in greatest numbers in close proximity to the boundary marking the convergence between easterly and westerly circulations, and more especially where the convergence is strongest (Fig. 8.2). Here the thunderstorms may be so numerous that they are in the nature of organized storm systems. Moreover, within these areas of organized convection the thunderstorms may develop at night as well as during the heat of the day, suggesting the general lifting effects along the conver-

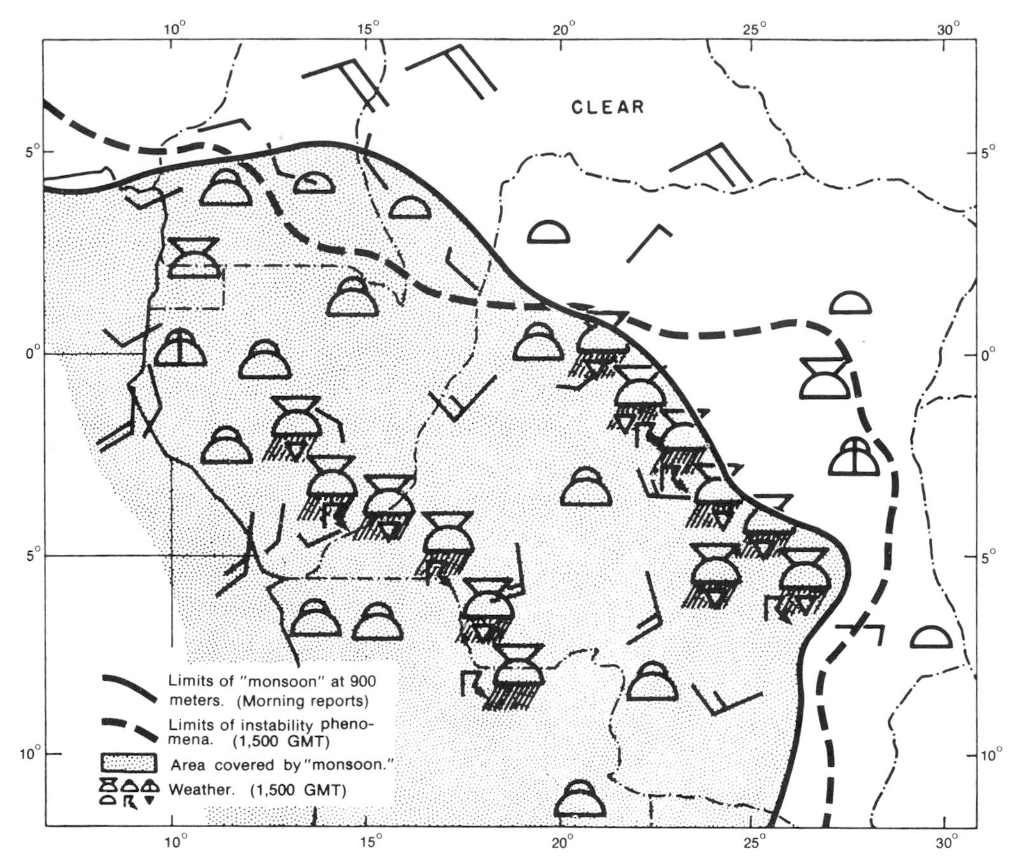

Fig. 8.2. Showing the depth of penetration of the southwest "monsoon" into the Congo Basin on the morning of December 6, 1955, and the associated weather. Note that the limits of the instability phenomena approximately coincide with the boundaries of the southwesterly invasion. Two major belts of disturbed weather are indicated, both of them associated with line squalls (see Fig. 8.3), the easternmost of which is positioned along the convergence between easterly and westerly air flow. (After Jeandidier, 1956.)

gence. It has been observed, also, that organized storm systems are more characteristic of some areas than of others on those occasions when the widely shifting boundary of the convergence between easterlies and westerlies happens to be positioned over them. Such favored areas are usually along slopes exposed to the southwesterly current.

As well as in proximity to the convergence along the margins of the southwesterlies, but not so frequently, areas of organized shower and thunderstorm activity also develop *within* the maritime southwesterly current. Two conditions favor their occurrence, (1) relatively high temperatures and dew points, and (2) a speed convergence produced by a surge in the southwesterlies. Such organized areas of rainfall developed within the maritime current ordinarily move in an easterly direction.

It is in close proximity to the convergence zone between easterlies and westerlies, also, that the more vigorous line-squall disturbances develop, and according to Jeandidier (1956), their origin is closely related to the storms of the convergence zone described above. The latter, especially when they are numerous and comprise an area of widespread vigorous convection, result in the development of an extensive cool pool over the area of occurrence, the drop in temperature amounting to 8°–10°C (14°–18°F) at an altitude of about 500 m, with the cooling effect probably extending up to 3,000 m. This cool air, carried upward by the equatorial easterly current, is displaced toward the west, and in the form of a vigorous cold front lifts the humid southwesterly air, thereby giving rise to the formation of an active line of *Cb*, the line squall of the Congo Basin (Jeandidier, 1956). The movement of the cold front corresponds approximately with that of the winds at elevations between 3,000 and 5,000 m (Fig. 8.3).

In exceptional cases the line squalls move from west to east instead of from east to west, especially in the southern part of the Congo Basin and in the Northern Hemisphere winter. At this time the Saharan anticyclone thrusts a ridge southward whose winds south of the equator have a westerly component. Thunderstorms and squall lines that develop along the convergence between the Saharan westerlies and the maritime

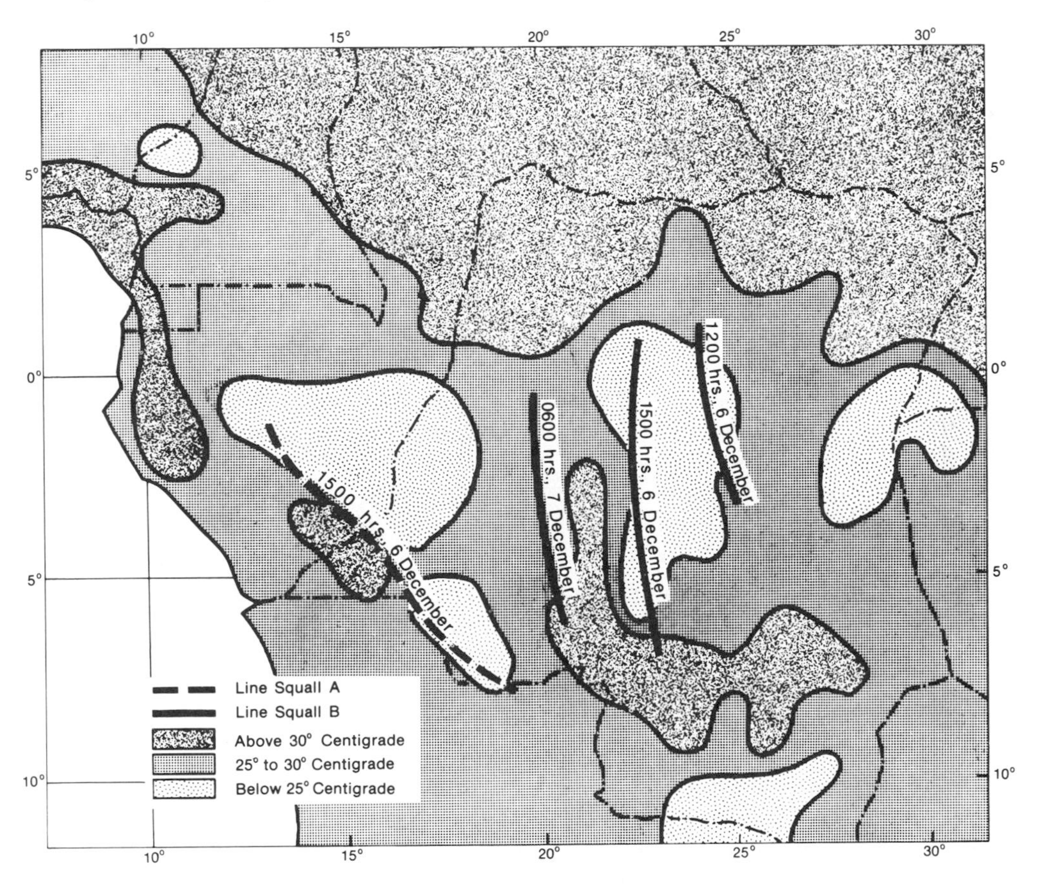

Fig. 8.3. Same area, date, and time as Fig. 8.2. The positions of the two line squalls are shown. At its position on the following day, December 7, at 0600 hours, line squall B was dissipating. Note that extensive areas of cooler air are located just to the east of the westward advancing line squalls in their positions at 1500 hours, December 6. (After Jeandidier, 1956.)

southwesterlies will move eastward in this combined westerly flow.

Rainteau (1955) offers a somewhat different explanation from Jeandidier's (1956) for the disturbance lines of the Congo Basin. He notes that most of these vigorous perturbations have their origin in the extreme northeastern part of the Congo at the western end of the corridor between the Ethiopian and Kenya highlands, which is the single natural route by which equatorial Indian Ocean air can enter the Basin. A favorable situation for the development of disturbance lines occurs when an anticyclone is located over Zimbabwe and Zambia and a relatively cool easterly flow derived from the eastern flanks of this high becomes well established at elevations of 3,000 to 8,000 meters. When this air converges with the equatorial trades line squalls appear to develop in the northeastern Congo near the western end of the corridor described above and move across the Basin to the west and southwest.

Climatic Peculiarities

As far as can be determined from the inadequate data, large-scale climatic anomalies are not present in the Congo Basin. A few of second-order significance are worthy of brief mention.

Over most of the Basin annual rainfall exceeds 1,500 mm but only in restricted areas does it surpass 2,000 mm (Fig. 8.1). Thus, equatorial Africa is less rainy than equatorial South America where the areas with over 2,000 mm are much more extensive. This reduced rainfall in equatorial Africa may derive from the fact that the central and eastern Congo Basin is affected by northerly air from the dry Saharan anticyclone in the Southern Hemisphere solstice season and by southeasterly currents from the South African anticyclone in the opposite season. No air from such dry continental sources floods the Amazon Basin. Hence there is considerably less *ITC* and post-*ITC* rainfall in equatorial Africa than in equatorial South America which is open to invasions of *mE* or *mT* air from the east at all times, and to unstable northwesterly *mE* air for well over half the year.

It is probable also that the southwesterly flow that enters the Congo Basin is less subject to convective overturning than is most of the air in the Amazon Basin. Rainteau (1955) characterizes this air at Brazzaville as stable in all seasons, which does not seem unusual considering that it is derived from the eastern end of the South Atlantic high and has crossed the cool Benguela current. The stratus deck so characteristic of this air during night and morning hours suggests a degree of stability. Especially in the Southern Hemisphere winter, and to a less degree in spring and fall, the air over Kinshasa (Leopoldville) is characterized by a temperature inversion at 850 to 800 mb and by a quasi-isothermal layer at 650 to 500 mb (Taljaard, 1955) (Fig. 8.4). In an earlier section it has been noted that an above-surface inversion is likewise present in the southwesterly flow along the Guinea coast in July-August-September.

One of the principal regions of heavier-than-average rainfall lies along the margins of the Bight of Biafra north of the equator, where the onshore equatorial westerlies are lifted abruptly by the Cameroon Mountains. Annual rainfall at Douala is 4,000 mm and at Debundja near sea level on the windward side of the Cameroon Mountains it averages 9,900 mm. In this area the isohyets turn sharply inland forming a cone, the apex of which lies along the elevated boundary zone separating Nigeria from Cameroon.

Inland from the west coast rainfall declines, but it increases again in the Cuvette Central where there is a moderately extensive area where rainfall exceeds 2,000 mm,

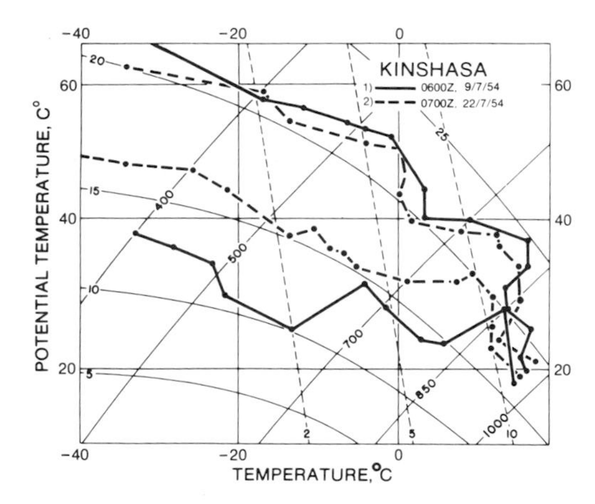

Fig. 8.4. Vertical structure of the atmosphere at what was Leopoldville, now Kinshasa, on two occasions in July, or winter. On most winter days there is a three-layered structure in which three distinct air masses are separated by two stable transition layers. Mild and nearly saturated Atlantic air extends up to about 850 mb. This is followed by a transitional layer with inverted lapse rate from 850 to 800 mb. A second transitional layer, with quasi-isothermal structure, extends from 650 to 550 mb. (After Taljaard, 1955.)

with one station, Boende, recording more than 2,200 mm. From this interior core area of heaviest precipitation, lying between 19° and 25°E and 2°N to 3°S, rainfall decreases in all directions. Annual amounts in excess of 2,000 mm are again characteristic of parts of the easternmost Congo Basin as the highlands are approached. Why the central and lowest part of the Congo Basin, the Cuvette Central, should also be the most extensive area of heavy annual precipitation is a problem requiring more investigation. The only explanation suggested by the meteorologists at Kinshasa (Leopoldville) was the increased local evaporation in this region of lower elevation where water surfaces are more extensive and the forest cover denser than elsewhere.

The Atlantic Littoral, 0°–12°S.—Also noteworthy is the fact that the littoral of equatorial and tropical western Africa south of the equator not only has a much lower annual rainfall than similar latitudes just to the north of the equator, but that it is also well below that of interior Africa in similar latitudes. Accordingly, the isohyets bend sharply northward along the coast in Gabon (French Equatorial Africa) and Angola. As the *ITC* moves south of the equator in the southern-solstice season, it is blocked in its southward advance along the west coast by the strongly subsident air of the well-developed South Atlantic high and the cool upwelled waters which the anticyclone's southerly winds generate (Fig. 8.5). In interior southern Africa the *ITC* is able to migrate 15° to 20° south of the equator, but along the west coast its regular southward migration into the Southern Hemisphere is limited to a few degrees. The average coastal location of the climatic equator is at about 3°N, for at Douala at 4°N there is a strong concentration of rainfall in June-July-August, while already at Libreville at 0°30′N, July is rainless and a Southern Hemisphere type of annual rainfall profile prevails. Total annual rainfall decreases very rapidly southward along the tropical west coast; 2,500 mm at Libreville at 0°30′N; Mayoumba (3°25′S), 1,550 mm;

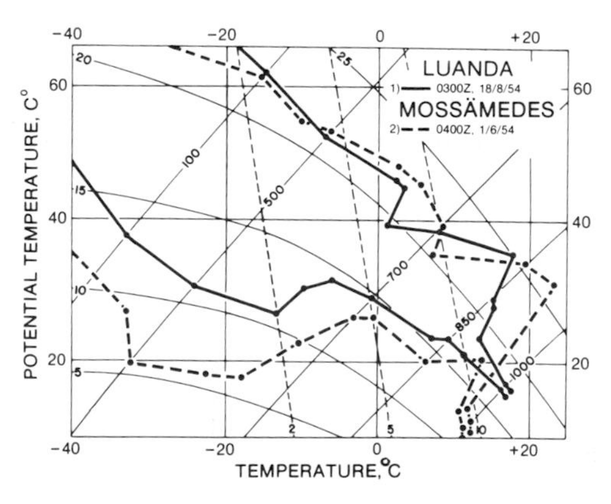

Fig. 8.5. Vertical structure of the winter atmosphere at Luanda and Mossämedes in Angola. Strong and low temperature inversions are present at both stations. At the latter station in winter the surface maritime air is less than 500 m thick. (After Taljaard, 1955.)

Loango (4°39′S), 1,220 mm; Banana (6°S), 690 mm; Luanda (8°49′S), 330 mm; and Mossämedes (15°12′S), 50 mm. This situation resembles that of the coast of South America in Ecuador where the rainfall gradient also is remarkably steep, both latitudinally and at right angles to the coast. Somewhere at about 5° or 6°S semiarid conditions are encountered.

Since the climatic uniqueness described above represents only the equatorward projection of a coastal rainfall deficiency which is much more intensively developed farther south, a detailed discussion of the phenomenon will be postponed to a later section dealing with the climates of southern Africa. Here, however, it is appropriate to mention that along the littoral in equatorial western Africa south of the equator the aridifying effects are not carried equatorward to the same degree and in the same intensity that they are in western South America, but more so than in western northern Africa. The climatic contrasts appear to be the result, in part at least, of coastal alignment. Where, with decreasing latitude, the coastline continues to bend westward into the major oceanic and atmospheric circulations, aridity is carried deep into equatorial latitudes. This situation is almost ideally represented in Peru. Western Africa south of the equator provides less perfect conditions, for north of Cape Frio the coastline recedes eastward, although north of about 12°S it again bends westward and coastal upwelling is maintained. It is north of Cape Frio that annual rainfall amounts increase more rapidly. In northwest Africa where the coast in lower latitudes recedes eastward and away from the oceanic circulation, the aridifying effects are not carried as far equatorward as they are in western South America and southwestern Africa (Lydolph, 1957).

A striking climatic feature of the equatorial west coast of Africa south of the equator is the remarkable variability of the annual rainfall, this in turn being a consequence of the variable average position of the *ITC* and its accompanying disturbances (Fraselle, 1947). The position of the convergence in any one summer depends upon the contrasting strength of the subtropical anticyclonic circulations in the North and South Atlantic oceans, and the accompanying oceanic circulations. This closely resembles the situation in coastal southern Ecuador. Variability increases southward along the coast as rainfall rapidly declines. At Santo Antonio do Zaire (6°7′S) in northernmost coastal Angola the coefficient of variability amounts to 40 percent, at Luanda (8°49′S) 50 percent, and at Lobito (12°33′S) 68 percent (Queiroz, 1955) (Figs. 8.6, 8.7).

However, this type of rainfall variability in coastal Angola is dubiously to be compared with the so-called El Niño spells which are so well documented for desert Peru in similar latitudes. Angola's variability is of a kind that is fairly normal for subhumid and dry climates in the tropics where annual rainfalls of individual years show strong oscillations either side of the mean. In Peru, by contrast, an intensely and consistently arid coast is on rare occasions subjected to revolutionary weather changes in which heavy equatorial rains may persist for as long as a month or more. As far as can be determined from the data and the literature, there is little evidence of similar catastrophic weather changes in coastal Angola, and conversations and correspondence with African meteorologists at Luanda, Pretoria, and Kinshasa (Leopoldville) corroborate this finding. The nearest approach to an El Niño effect is evident in the record for Mossämedes (average rainfall 50 mm), where over the 24-year period 1915–16 to 1954–55 there were three unusually wet years, 1932–33 and 1933–34 when rainfall reached 290 mm, and 1946–47, 210 mm. The difference between the situations in Angola and Peru lies in the fact that in the former, inland from the

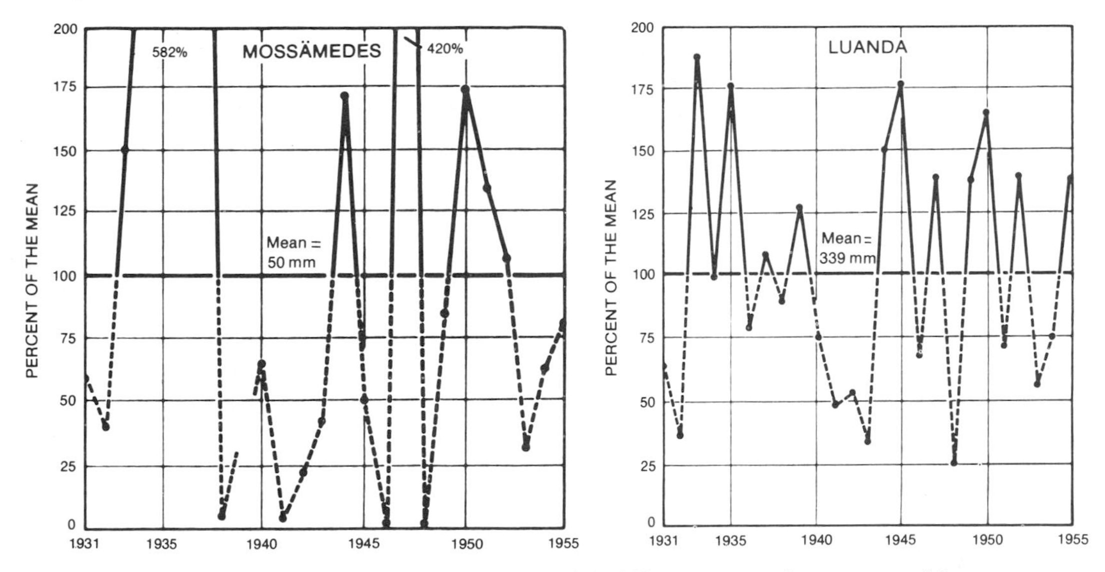

Fig. 8.6. Total annual rainfall of each year over a period of 25 years expressed as a percent of the mean annual rainfall for two coastal stations in Angola. (After Queiroz, 1955.)

coast, equatorial air of Atlantic origin, together with the *ITC* and its perturbations, shows a relatively normal southward migration in summer, so that moderately heavy rainfalls of 1,000 to 1,250 mm are common.

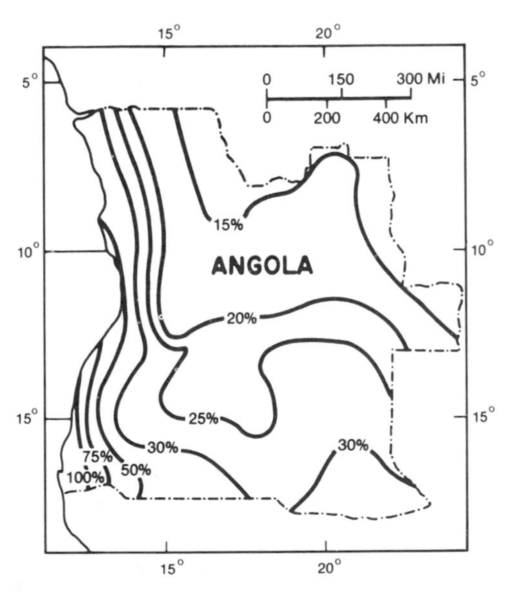

Fig. 8.7. Isolines of equal coefficient of precipitation variability in Angola. (After Queiroz, 1955.)

It is not unusual, therefore, for the *ITC*, here aligned north-south, to reach positions at variable distances from the coast in different years, resulting in highly variable annual rainfalls. In Peru, on the other hand, high mountains obstruct any invasion of *ITC* effects from the east, so that the superaridity can only be broken on those exceptional occasions when the equatorial trough and its disturbances are able to break through from the northwest in diametric opposition to the strong southerly flow of stable maritime air and cool water. Even in the intensely dry Namib of southwest Africa, the occasional thunderstorms, or widespread showers, occur when disturbances move in from the east, and not from the north along the coast. Such disturbances may originate along the equatorial trough when it lies unusually far south over the interior, or they may resemble the disturbance line or line squall of the Sudan and the Congo Basin. Thus, while the moisture source appears to be the equatorial Atlantic, the disturbances move from east to west as carried by the easterlies aloft (Boss, 1954; Flohn, 1956*c*).

CHAPTER 9

Tropical East Africa

Undoubtedly the most impressive climatic anomaly in all of Africa is the widespread deficiency of rainfall in tropical East Africa (Fig. 7.1). Yet, in the climatic and meteorological literature bearing on East Africa, this feature is seldom emphasized as atypical. On Carter's (1954) moisture regions map of Africa, based on Thornthwaite's 1948 system, it appears that approximately two-thirds of East Africa, as bounded by the 30°E meridian and the 10°N and 20°S parallels, has a dry climate, either arid, semiarid, or dry-subhumid (Figs. 9.1, 9.2, 9.3). Nowhere else in a similar latitudinal and geographical location does there exist such a widespread water deficit. The abnormality of this rainfall deficiency in an equatorial-tropical region is all the more remarkable when one notes that this is the eastern side of a continent, in the latitudes of the tropical easterlies, where an almost continuous north-south highland parallels the coast at no great distance inland.

To an unusual degree the restricted areas with genuinely humid climates in tropical East Africa coincide with the higher elevations. Only in limited areas do genuinely humid climates extend down to tidewater. One gets the feeling that if orographic effects were removed, almost the whole of tropical and subtropical eastern Africa would be dry-subhumid or dry. Along the whole of East Africa bordering the Indian Ocean, from Cape Guardafui, at about 12°N, south to 30°S, considerably less than half of the coastline has as much as 1,000 mm of annual rainfall, while only three relatively short stretches of the coast may be classed as humid following the 1948 Thornthwaite classification of moisture regions. A number of coastal stations south of the equator have annual rainfalls of only about 750 mm, while from the equator northward, much of the Somali coast is a desert. Back from the coast, along the valleys of both the Limpopo and Zambezi rivers, which are the most complete breaks in the highland wall south of the equator, and again along the depression in which Lake Rudolph is located, which separates the Abyssinian from the Kenya-Uganda highlands, semiarid, and in spots even arid, conditions prevail. Even extensive areas on the lower parts of the East African highlands, as for example in Tanzania south of Lake Victoria, where the elevation still is over 900 m, have such low and erratic rainfalls as to make crop cultivation virtually impossible (Glover, Robinson, and Henderson, 1954). It would appear that the dynamic and thermodynamic rain-generat-

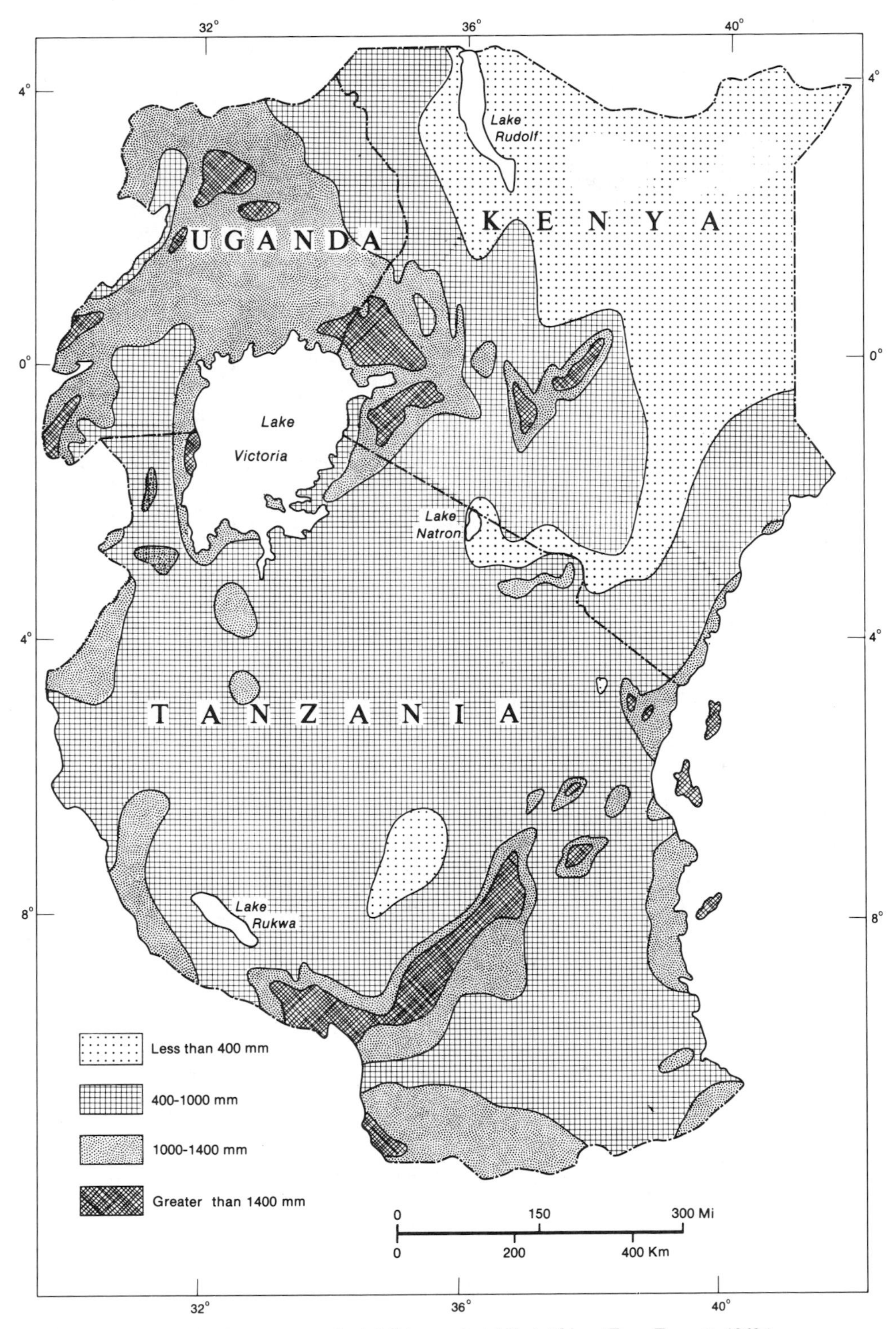

Fig. 9.1. Average annual rainfall in equatorial East Africa. (From Tomsett, 1969.)

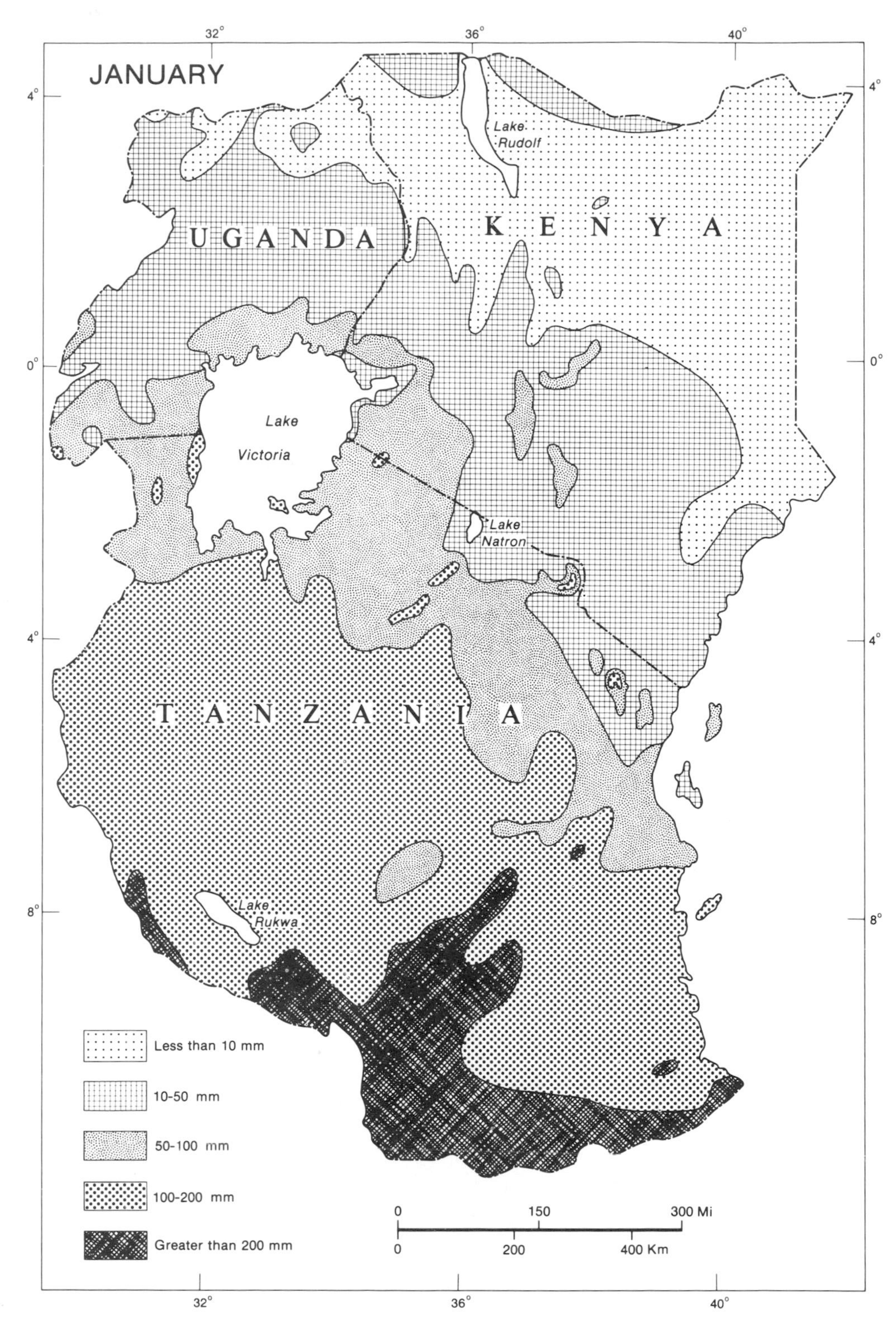

Fig. 9.2.A. Average January rainfall in equatorial East Africa.

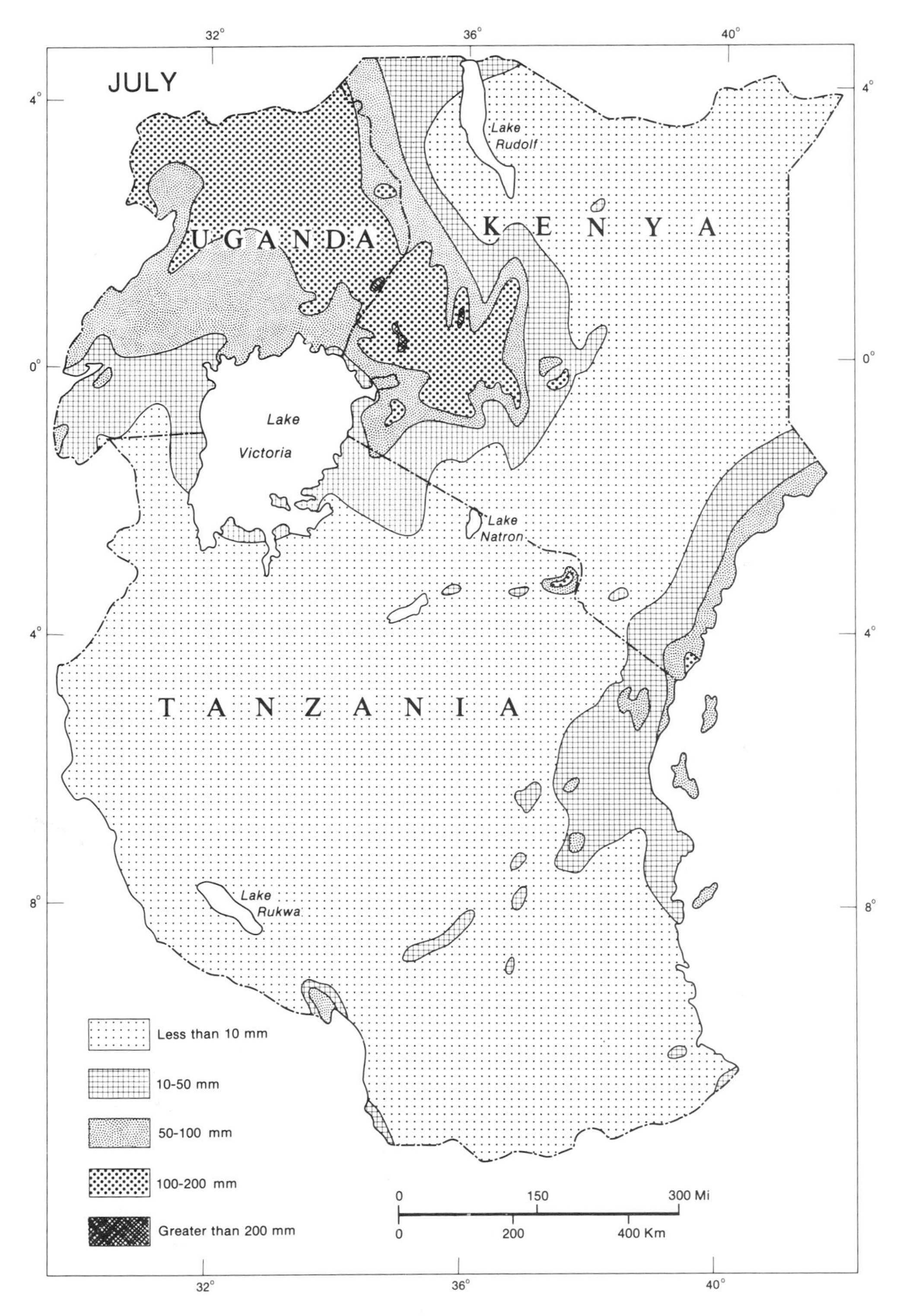

Fig. 9.2.B. Average July rainfall in equatorial East Africa. (From Tomsett, 1969.)

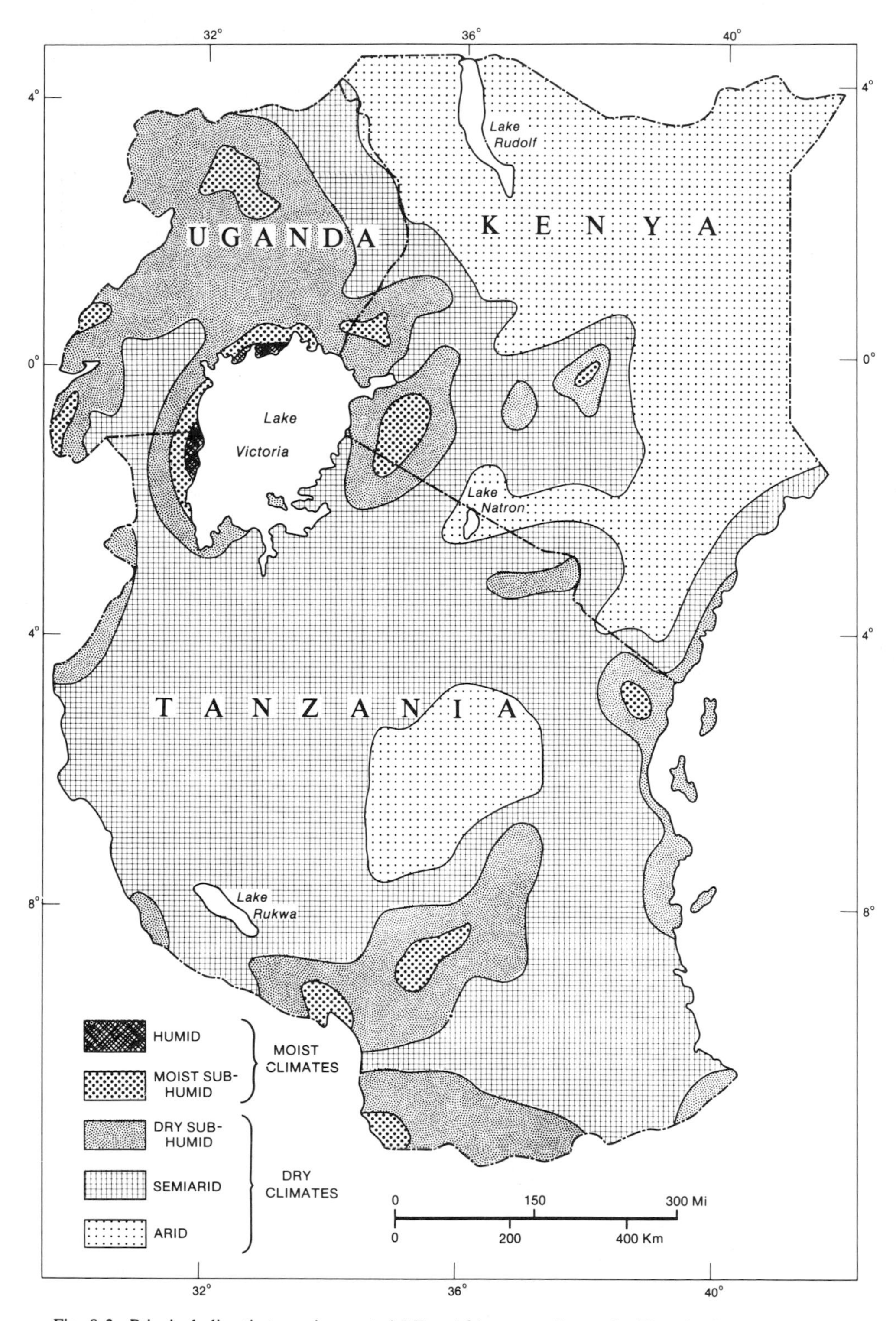

Fig. 9.3. Principal climatic types in equatorial East Africa, according to the Thornthwaite system of classification. (After Pant and Rwandsya, 1971.)

ing processes normally prevalent over equatorial-tropical lands of modest elevation are only weakly developed in the low latitudes of East Africa.

On an annual rainfall map (Fig. 9.1) of tropical Africa the isohyets, which in the western and central parts trend in an east-west direction, in the east turn abruptly and assume a general north-south alignment. Such a meridional arrangement is influenced most strongly by the rainfall patterns of the two transition seasons separating the northerly and southerly monsoons. Out of this seasonal isohyetal arrangement, with more humid conditions both toward the coast and also toward the interior, is derived the intervening "dry wedge," which is such a prominent feature in Tanzania and Zambia. Heaviest rainfalls are coincident with highlands.

Origin of the General Rainfall Deficiency

Circulation Patterns.—Preliminary to an analysis of the causes for the general meagerness of rainfall in East Africa it is necessary to have in mind the principal lineaments of the atmospheric circulation over this area (Fig. 9.4). A marked seasonal reversal of surface winds is conspicuous, a southerly flow prevailing when the sun is north of the equator (April to October) and a northerly flow during the Southern Hemisphere summer (October to March). By the local meteorologists these are designated as the SE monsoon and the NE monsoon. However, the seasonal wind reversal is scarcely the result of differential heating and cooling between the African landmass and the Indian

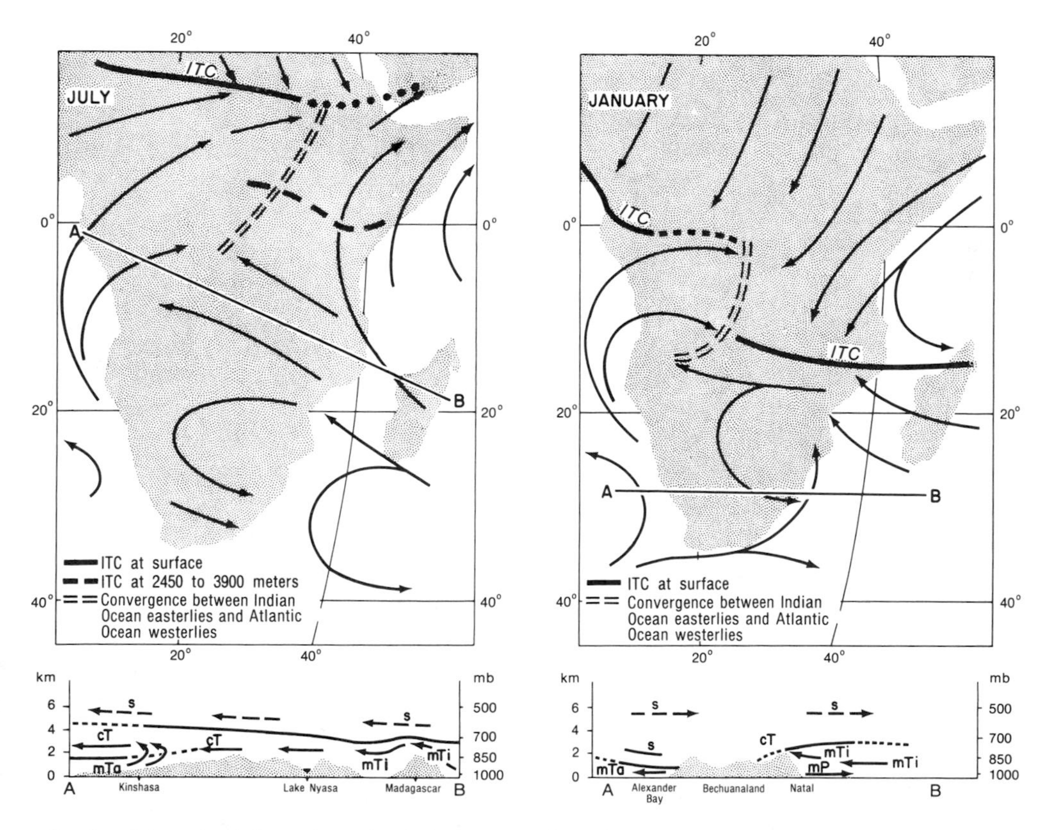

Fig. 9.4. Mean circulation at lower levels for eastern and southern Africa in July and January. Lower diagrams show vertical sections along line AB. (After Henderson, 1949; Taljaard, 1953; and Aspliden, personal communication.)

Ocean to the east. The so-called monsoons that influence East Africa's climate actually represent the western margins of the gigantic seasonal circulations of southern Asia and the Indian Ocean. Thus, eastern Africa's monsoonlike wind reversal and its climate are greatly influenced by the Asiatic landmass to the north and east, while the African continent itself has a more local and peripheral influence, especially in creating the divergent character of both monsoons which cause tropical East Africa to be so dry. So, although tropical East Africa lies under the influence of the western margins of the two Indian Ocean monsoons, it does not have so much to do with their origin and character. East Africa's monsoons are scarcely of the classical, seasonal on-shore-wet and offshore-dry variety. Rather, because of their strongly meridional trajectories, neither monsoon enters very deeply into the continent; instead they tend to roughly parallel the coastline.

The northerly flow is made up of two unlike air streams, a drier one that has traveled across Egypt and the Sudan, and a more humid one originating in much the same region but which in moving around the eastern side of the Arabian high has had a sea track of modest length. In the equinoctial transition seasons, between the retreat of one monsoon and the advance of the other, winds are fickle and more easterly. Above the surface monsoons the winds of higher altitudes are dominantly from the east.

The southerly monsoon of East Africa, with its sea-land parabolic trajectory, has been described by Findlater (1969) as a high-energy circulation resembling a low-level jet. In its root region in the Indian Ocean east of Madagascar, it is a part of the southeast trade-wind circulation. Curving northward through equatorial East Africa, after crossing the equator and experiencing the reversed effects of Coriolis force it takes a northeasterly course over Somalia and the adjacent sea, and thence across the Indian Ocean to the west coast of India and beyond. This is a high-speed circulation whose variations in cross-equatorial flow have important effects upon summer rainfall in the Indian subcontinent. Thus, the southerly monsoon of tropical East Africa and the adjacent Indian Ocean is the primary moisture source for the summer precipitation of southern Asia. It seems odd, therefore, that this southerly Indian Ocean monsoon should not result in more rainfall in tropical East Africa.

In addition to the northerly and southerly monsoon currents there are also occasional invasions of moist unstable westerlies (Henderson, 1949; Thompson, 1957*a*). A boundary between these equatorial westerlies on the one hand, and the monsoons or tropical easterlies on the other, not infrequently can be detected along the western margins of the East African highlands, and occasionally it extends farther eastward, even to the Indian Ocean (Fig. 9.5).

Origin of the rainfall deficiency over East Africa must be sought not in one but several causes, and admittedly there is much that still remains a puzzle. Among East African meteorologists there are differences of opinion about some features of the seasonal circulation patterns, but there is general agreement that both monsoons are divergent and subsident over extensive areas. "Somewhat surprisingly, British East Africa . . . is in the main a dry region although it is subject to two monsoons, which usually follow ocean courses and bring moisture-laden air inland from the sea. Their failure to deposit water seems to be in their being persistently diffluent or directionally divergent over large land areas and in this they markedly differ from the convergent monsoons of southern Asia" (Glover, Robinson, and Henderson, 1954). Forsdyke (1949) notes that on many occasions in East Africa the lapse rate is favorable for precipitation, being in excess of

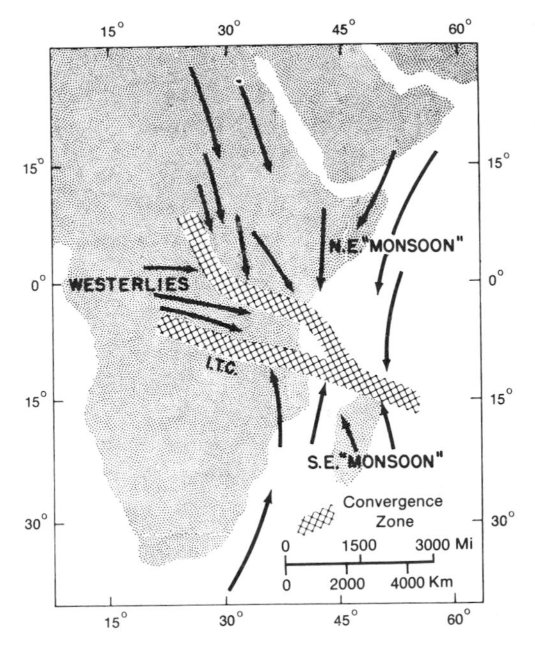

Fig. 9.5. On occasions, what appear to be equatorial westerlies invade the realm of the Indian Ocean easterlies in East Africa. Such episodes usually are occasions of above normal precipitation in East Africa. "Winter" conditions are shown here. (After Henderson, 1949.)

the wet adiabat, and the moisture content is likewise sufficient, but rain does not occur because the general atmospheric environment is divergent and subsident. Thompson (1957*a*) points out that the southeast monsoon which is moist and shallow as it approaches the coast of Kenya and Tanzania, becomes abruptly diffluent as it moves inland, one branch continuing westward across Tanzania and spreading north and south over the interior, while the other branch turns northward to parallel the Somali coast. Taljaard (1955) describes the air over Nairobi as usually stable as regards temperature distribution except during the short rainy seasons April to May and October to November, but potentially unstable if the wet-bulb potential temperature is considered (Fig. 9.6).

Both of the seasonal monsoon circulations, northerly and southerly, in the East African region tend to bifurcate in the vicinity of the coast. The main flows continue on a sea trajectory while branches turn obliquely inland. As it applies to the longer and stronger southerly monsoon, originally a southeasterly trade wind in its root region, it veers first into a southerly flow and then southwesterly as the equator is crossed; Coriolis force diminishes and eventually changes sign. Its original southeasterly direction would bring it directly onto the Tanzanian coast; but as it approaches the mainland it deviates, with the main flow veering to a southerly direction and hence paralleling the coast while a secondary branch continues northwestward in a weakened condition as it partially surmounts the plateau. From Table 9.1 it may be noted that at Dar es Salaam located on the coast at 6°50′S, wind direction which was mainly southeast in April is predominantly south in July. And

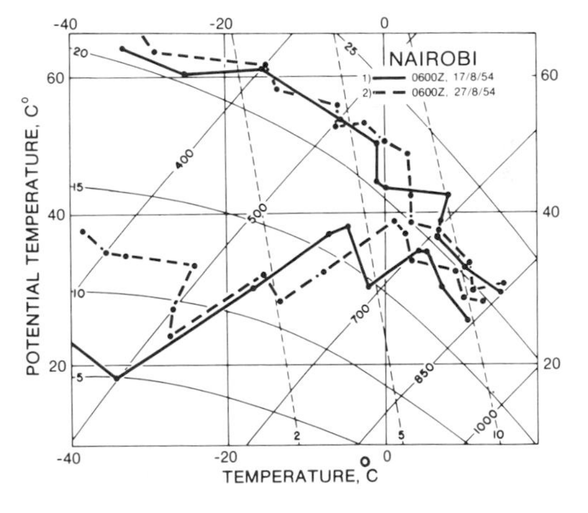

Fig 9.6. Vertical structure of the air at Nairobi in August. Except in the short transition seasons of maximum rainfall, the air over Nairobi has a stable temperature stratification. The surface layer of moist maritime air derived from Indian Ocean sources extends upward from the surface (840 mb) to between 700 and 650 mb, and therefore averages about 2 km deep. Inversions or quasi-isothermal conditions are characteristic of the transitional layer above the maritime air, with subsided air aloft. (After Taljaard, 1953.)

Table 9.1. Low-level wind direction and humidity at Dar es Salaam, Tanzania 1960 to 1963 (Kuepper, 1968)

Month	Winds at 2,000 m (percentage of occurrences)		Mean humidity mixing ratio at 850–800 mb (gms/kgms)
	Southeast	South	
April	48.9	24.8	11.2
July	19.7	71.0	8.4

between April and July the mean humidity mixing ratio had declined sharply. Reasons for this bifurcation and the resulting diffluent flow, which is a prime feature of the low-sun period south of the equator, are not entirely clear. It has been suggested that the barrier effect of the East African plateau may play a part or that the gradual weakening of Coriolis force as the equator is approached may be influential. But Ramage (1971) is of the opinion that the Great Rift heat low is mainly responsible for the diffluent character of both the southerly and northerly monsoons. In January the northeast monsoon divides, one branch continuing southward as attracted by the heat low over Mozambique, and the other veering westward toward the Great Rift's heat trough. In July, while the mainstream of the southerly monsoon continues northward off the coast, a lesser branch likewise diverges toward the heated Rift. The dominant synoptic patterns of the dry season of East Africa south of the equator during the southerly monsoon are all in the nature of slight variations of the cross-equatorial drift as described by Johnson and Mörth (1959).

Not only is the low-level circulation divergent but it likewise is not of great depth. In the coastal regions of South Africa the southeasterly flow is only 1–2 km deep and hence is too shallow to surmount the escarpment and reach the plateau. It deepens toward the equator, so that northward of about 25°S the easterlies are able to rise over the escarpment and extend inland over the highlands. In the vicinity of the equator the SE monsoon, although variable in depth, may extend up to 2,400 to 3,600 m above sea level, but this still means a relatively shallow current of humid maritime air over the East African highlands, where as a consequence the depth of convective overturning ordinarily is not great.

In addition, these shallow monsoon currents are capped by another current moving from a somewhat different direction, usually easterly, in which moisture content is low but variable and lapse rates weak or even inverted (Henderson, 1949). As a consequence clouds that do originate in the moister lower air are unable to develop and expand in the dry and stable easterly air aloft. Taljaard (1955) has described the vertical structure of the air at Nairobi and shown typical soundings. On the average the lower maritime air, transported either by the northerly or southerly seasonal flow, is about 2 km deep, and has a humidity mixing ratio of 8 to 12 gm/kg in the dry seasons and 11 to 15 gm/kg in the two rainy periods. Above this is stable air characterized by subsidence-type inversions or quasi-isothermal layers, and by a sharp drop in moisture content. The persistent stratus or strato-cumulus overcast, so common at Nairobi during the dry seasons when the monsoons are at their maximum development, is the visible evidence of the inversions. Such above-surface subsidence inversions are more typical of the east sides of tropical oceans than of the west sides.

A further element conducive to modest precipitation in East Africa is the strongly meridional flow characteristic of both monsoons over the land. The southeasterly current, the moister of the two, has had a long trajectory across the Indian Ocean before reaching the African coast, but the drier northerly monsoon has a much more meridional than zonal track both over the ocean and the adjoining land. But along the coast and over the land the southerly monsoon, likewise, becomes strongly meridional, so that at times it is nearly parallel to the coast or even offshore. The result is a much smaller transport of moisture from ocean to land than would be true if the air flow were more nearly normal to the coast, while the lifting effect of the eastward-facing plateau escarpment is greatly minimized. It is significant in this respect that it is during the transition seasons between the two monsoons, when the air movement is more zonal and from the east, that rainfall reaches its maximum

Accordingly, although mobile atmospheric disturbances are experienced in East Africa, the amount of rainfall which they generate is modest or even small. In addition to the disturbances it is essential for abundant rainfall that there shall be a layer of moist air which is at least moderately deep and not too stable and this requirement seems to be lacking too frequently over much of East Africa.

A minor feature of rainfall distribution which presents something of an unsolved problem is the markedly greater rainfalls recorded on the low islands of Zanzibar and Pemba (5° to 6°S), a few score miles offshore, than on the adjacent mainland. Thus, three insular stations report rainfalls of about 1,950, 1,625 and 1,400 mm, which amounts are to be compared with 1,125 and 1,200 mm at Dar es Salaam and Mombasa on the coast opposite, and 850 mm at Nairobi on the upland. One might surmise that a double sea breeze convergence could explain the increased insular rainfall, except for the fact that it reaches a peak in late morning, which is prior to the maximum development of the sea breeze (Thompson, 1957*b*).

Still another factor contributing to the stability of the southerly monsoon in East Africa is to be found in the fact that the Indian Ocean anticyclone, unlike others of this variety, shifts its meridional axis westward in the low-sun season, possibly as far as 60°E. Subsidence in the western parts of this oceanic cell doubtless accentuates the stability of the southerly circulation over East Africa. Such stability is evidenced by the fact that on occasions when there is a large scale penetration of the continent by southeast winds, the whole of East Africa is almost rainless. In order to obtain such a large scale penetration of the southeast flow there is usually a relaxation of the normal recurvature along the coast (Kuepper, n.d.).

A possible additional feature causing aridity in East Africa during the dominance of the southerly monsoon may be the likely downstream acceleration of the high velocity above-surface winds along the east African littoral (Findlater, 1969). If such a speed divergence does exist over the adjacent land, it may supplement the divergence produced by other factors previously described.

Large Scale Weather Situations and Mobile Rain-Generating Disturbances in East Africa

If the usual concept of synoptic-scale weather systems as those *traveling* atmospheric disturbance with a lifetime of a few days and lateral dimensions of several hundred kilometers is accepted, then the subject of synoptic models in the tropics may appropriately be described as one of disquietening confusion. In general, synoptic models represent the condensing of the results of numerous empirical investigations of a particular type of mobile weather system. All successful synoptic models are statistical concepts derived from a study of a large number of sample cases. As it relates specifically to *tropical weather*, one meteorologist (La Seur, 1964) has opined that, except for waves in the easterlies and the mature hurricane, the currently prevailing situation as regards tropical synoptic models "is one of incomplete description and understanding."

Still, D. H. Johnson and H. T. Mörth (1959, 1961), both formerly of the East African Meteorology Department, have described in some detail a classification of large-scale weather models (*Grosswetterlagen*) which have in their opinion proved to be of value in East African weather analysis and forecasting (Fig. 9.7). To be sure, their large-scale weather models do not fit the usual criteria for traveling weather disturbances, since their models are not advected by the winds, which appear to blow *through* them rather than to carry them along. The systems they describe seem to appear and disappear in somewhat random fashion. In many respects the Johnson and Mörth models resemble large-scale climatological features such as equatorial troughs, subtropical anticyclones, monsoons, and trade winds rather than less extensive advected weather systems. Regrettably, the authors give only meager attention to details of weather features associated with each of the synoptic models. Rather, the focus of their attention is on upper-air pressure and wind fields. Johnson and Mörth recognize three large-scale primary models and four or five secondary and composite ones.

Omitting orographic effects, daily rainfall distribution patterns in East Africa have a very patchy appearance. Still, the rainfall values are not randomly scattered but can be grouped into relatively wet and dry regions. Rainfall areas often remain *in situ* for several days; they do not appear to be transported by winds of the lower troposphere. This suggests that advected changes in moisture and stability are not fundamental to the areal distribution of precipitation; rainfall areas tend to be fairly stationary. So the weather forecaster is led to seek a solution to the weather

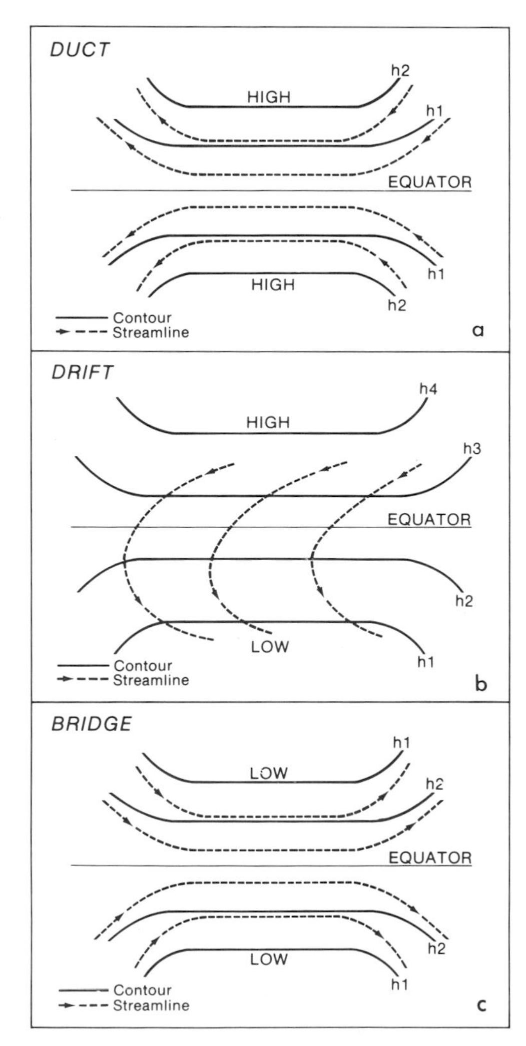

Fig. 9.7. Main large scale weather systems (duct, drift, and bridge) in equatorial East Africa. (After Johnson and Mörth, 1959.)

patterns and weather changes in the dynamics of the upper-level circulation as expressed in pressure contour charts. On a sequence of daily rainfall charts there appears to be a continuous evolution of the rain areas; they do shift about, but there is little of a progressive movement of precipitation patterns on time scales that might suggest a generic association with moving waves, troughs, or other types of ambulatory disturbances.

In East Africa, according to Johnson and Mörth (1959, 1961) the *equatorial duct* is probably the most frequently observed synoptic model (Fig. 9.7, part a). When it prevails the circulation is a consequence of two high pressure cells being positioned opposite each other across the equator and separated by a flattish pressure trough at or near the equator. The pressure profile is bowl shaped. The easterly air flow is approximately parallel to the contour lines, but with some frictional convergence. With the duct there is an entrance with wind acceleration on the east, and an exit with deceleration on the west. Acceleration of the air at the confluence requires a down-gradient component of flow leading to convergence near the equator. Similarly there is divergence at the western or diffluent part of the duct. When the confluence is of great depth and extends downward through the lower atmosphere, conditions are favorable for ascent and rain. The likelihood of rain is increased if either or both of the high pressure cells are intensifying and thereby increasing the convergence.

The *equatorial drift* (Fig. 9.7, part b) is characterized by a high pressure cell, or ridge, on one side of the equator and a low, or trough, on the opposite side, as a consequence of which a pressure gradient exists at the equator. Where the high is positioned north of the equator and a low to the south, the easterly flow is down gradient, with recurvature-to-westerly flow occurring somewhat south of the equator. The flow is remarkably cross-contour and resembles the Indian Ocean summer monsoon. There is divergence in the easterly flow to the north of the equator, and convergence, after recurvature, in the westerlies to the south. This situation is common during the months January to March when the lower troposphere drifts are associated with fine weather in the divergent easterlies over Kenya north of the equator, and seasonal rainfall in the convergent equatorial westerlies south of the equator. But duct and drift are not only seasonal features; they may also alternate over a period of several days, with corresponding changes in the weather.

The *equatorial bridge* model (Fig. 9.7, part c) is characterized by low pressure cells on opposite sides of the equator with a weak ridge of flat pressure gradient between them near the equator. This is an inverted form of the duct. Nothing is said by the authors about associated weather.

Four *secondary* synoptic models will be mentioned: the diamond, the displaced duct, the shear drift, and the zonal gradient drift (Mörth, 1963). Only the first two will be commented upon here.

The *diamond* occurs when a duct and a bridge occur simultaneously in longitudinal sequence. There is a high frequency of this model at lower medium levels near the East African coast. When the duct is positioned over the inland area and the bridge over the sea this situation may persist for several days. If the duct-forming high pressure cells build eastward, there is a prompt deterioration of the equatorial weather from fair conditions to severe storms.

The *displaced duct* occurs over land areas where solar heating creates extensive zonal heat lows in the near-surface atmosphere, and where there is a gradual displacement of the equatorial trough into the summer solstice latitude. Convergence and vertical updraft occur in the trough, with resulting rainfall; the maximum updraft occurs when the trough is positioned between about latitudes 7° and 12°.

Johnson and Mörth (1959, 1961) are of the opinion that humidity conditions in the upper atmosphere are of greater importance in forecasting local rainfall in equatorial East Africa than is the humidification of the lower atmosphere. Only very rarely, if ever, is there such a meagerness of moisture throughout the lower atmosphere as to stifle precipitation. And a sufficiency of upper level moisture is usually indicative of current or recent upward moisture transport, and similarly of an absence of divergence and subsidence in the lower troposphere.

Turning now to the *traveling* weather disturbances in tropical East Africa and their rainfall effects, any outsider who essays to identify the various rain-bringing disturbances of East Africa and to describe their climatic consequences should do it with full knowledge that he is attempting something which even the local meteorologists in the area find baffling and frustrating. It is not so much a situation in which there exist different schools of thought as it is a matter of numerous individual differences of opinion. Even among the staff members at the same weather station, as at Nairobi, for example, there is the greatest diversity of thinking as to disturbances and rainfall origin. Some, for example, remain convinced of the efficacy of random thermal convection, while Thompson (1957*a*), by contrast, states that "scattered afternoon showers in the tropics, except in specific rain areas, are a figment of the imagination." In the Rift Valley of Kenya, Oliver (1956) found that 90 percent of the annual rainfall occurred in association with large-scale disturbances and was of an organized type, while only 4 percent occurred in the form of random local showers.

Omitting the effects of orography and random thermal convection (if the latter exists), attention here will be focused on mobile disturbances and associated factors controlling rainfall production and its distribution, both areally and temporally. The precipitation control most commonly emphasized in the literature on East Africa is the zones of surface-wind discontinuity or convergence, while the seasonal march of precipitation characteristic of the region usually has been interpreted as the result of a latitudinal migration of the *ITC* following the course of the sun. Unfortunately for this interpretation synoptic practice can rarely detect this convergence, and likewise many rain areas develop, at least in the equatorial parts, which show a fine disregard for any probable position of the *ITC*. This is not to say that active convergence through its effects on atmospheric instability and the deeper humidification of the air does not create conditions favoring the rain-making processes. Along the more active parts of the *ITC*, and especially at those times when it is most distant from the equator, organized thunderstorms may develop. But commonly some large-scale disturbance is required to activate the convective processes, even in the presence of a diffuse convergence zone, and it is these disturbances which the synoptician has difficulty in locating, tracking, and understanding. Thompson points out that *ITC* theories in forecasting practice have failed dismally in East Africa.

One commonly observed type of disturbed weather generating rain, especially in coastal East Africa, exists in the form of organized belts of rainfall oriented NNE to SSW, or approximately normal to the direction of air flow (Thompson, 1957*a*). These disturbances originate over the ocean and advance toward the coast. Forsdyke (1949) has identified such features in the SE monsoon current of the western Indian Ocean. Henderson (1949) describes them as being in the nature of surges in the monsoon currents, and occurring most frequently during those periods when the monsoonal flow has become well established. They are not, therefore, associated with the *ITC*. What the structure of these easterly disturbances is, remains controversial. Thompson (1957*a*) suggests that they may be of that variety known as easterly waves, but so far it has not been verified that the organized belts of rain approaching the East African coast from the ocean have the regular pressure and wind variations characteristic of the easterly-wave model as described by Riehl for the Caribbean. Henderson (1949) and others have suggested the possibility that the easterly disturbances may have their origin over the ocean eastward from Africa as a consequence of surges of cold air from higher latitudes invading the trades, which subsequently carry the disturbed weather westward toward the continent (Forsdyke, 1944). It is noteworthy that they are largely confined to the SE monsoon which has had a long westerly track over the South Indian Ocean. They appear to be rare in the NE monsoon which is more meridional in flow and has had a shorter sea trajectory. This may help to explain the greater drought in East Africa to the north of the equator than to the south, and also why the Kenya coast has more rainfall in August-September than in January-February. Thompson (1957*a*) states that the easterly disturbances are principally phenomena of the littoral and uncommonly penetrate more than 80 km inland, the divergence in the monsoon current tending to dampen their convective effects so that rainfall diminishes rapidly westward from the coast. On the other hand, some consider them to be important rain-bringers as far inland as Uganda. Gichuiya (1970) points out that the frequency of easterly disturbances in East Africa is low. Also they are more frequent in September, a transition season when any monsoon circulation is weak, than in July, when the southerly monsoon is strong. Such disturbances show no thermal contrasts within different sectors. He is of the opinion that these easterly disturbances of East Africa are of a different character than the easterly waves of the Caribbean.

During the equinoctial seasons in equatorial East Africa, when wind discontinuities are most difficult to detect and rainfall is heaviest, the origin of the disturbed weather is hardest to understand. Air flow at such times is prevailingly easterly so that more oceanic moisture is advected landward, and the vertical structure of the air is less stable, both of these conditions resulting in an improved milieu for rainfall. But the expected equatorial perturbations from the Indian Ocean cannot be detected over tropical East Africa, although the reason for this may lie in the lack of a sufficiently dense network of wind-reporting stations, and the meagerness of weather reports from the western Indian Ocean. Likewise the high elevations inland from the coast make the interpretation of wind reports over the land extremely difficult. So at the very times when rainfall is at a maximum, the origin of the precipitation in equatorial East Africa remains very much a puzzle.

At Nairobi, where temperature and pressure elements vary little from day to day, the moisture content of the air is thought to be the best single indicator of weather. Relative humidity below the 500 or 600 mb levels does show wide temporal and areal variations, and the only really successful forecasting is based upon the rule that if the lower troposphere is moist it is likely to rain, and if it is dry, rain is unlikely to occur (Thompson, 1957*a*). Since radiosonde stations are few in East Africa this forecasting technique has not been widely tested. The origin of these upper-air humidity pools is not clear, but it would seem likely that they are the result of low-level convergence and upward movement associated with as yet undetected disturbances rather than with advection (Forsdyke, 1949). There are occasions when a moist atmosphere does not result in significant rainfall, which suggests the absence of a necessary synoptic feature. Precipitation associated with humidity "pools," although organized in character, since it is confined to specific areas, cannot at the present be related to a specific atmospheric perturbation of an extensive character.

Rainfall in East Africa, especially in the western interior parts, likewise occurs in association with irregular eastward invasions of unstable westerly air (Henderson, 1949). Uncertainty exists as to the cause for these eastward surges. Their occurrence is concentrated in the period December to April and they are largely confined to East Africa south of the equator. Coincident with the arrival of equatorial westerly air the humidity rises and showers and thunderstorms break out especially on western-facing slopes and probably in conjunction with the convergence zone between westerlies and easterlies. Whether there are distinct perturbations associated with these eastward pulsations of equatorial air is undecided.

In this connection it is worthy of note that Aspliden (personal correspondence and discussion) in Nairobi believes that some of the rainfall of interior East Africa originates in eastward-moving waves within the middle and high troposphere, imbedded in either westerly or easterly circulations, induced by low-pressure troughs in higher latitudes. These disturbances may occur in any season, although it seems reasonable to assume that their rainfall effects would be most marked when the monsoon currents are weakest (equinoctial periods), and the more nearly onshore easterlies bring added moisture into East Africa. This may explain the increased rainfall of the transition seasons. Significantly, it has been possible at times to correlate rainfall in Kenya and Tanzania with high pressure in southern Africa and a ridge of high pressure extending northward from it along the east coast. Aspliden's theory concerning westerly disturbances and rainfall in East Africa looks promising, but up to the present it has not been found to be of great use in current forecasting.

Occasionally tropical East Africa, even as far north as Somalia, experiences disturbed weather from surface cold fronts advancing up the coast from the south by way of the Mozambique Channel (Forsdyke, 1944). These appear to be infrequent phenomena in tropical latitudes, however, and probably make a relatively small contribution to the total annual rainfall. Most of the cold fronts moving northward to the east of Africa seem to take a route eastward of Madagascar so that their effects are missed along the mainland coast.

Annual March of Rainfall.—In contrast to the striking anomaly that exists in the form of an annual rainfall deficiency, the annual march of rainfall conforms somewhat better to the expected pattern for these latitudes. Thus, in close proximity to the equator most stations have annual rainfall profiles in which there are two wetter periods following a month or two behind the equinoxes. Farther to the north and south a single high-sun maximum and low sun minimum are more common, although this is by no means universal. Such patterns of annual rainfall variation are usually explained as being a consequence of the latitudinal migration of wind systems and the *ITC* following the sun. The drier period or periods are identified as coinciding with the times of relatively strong and steady northerly or southerly air flow, while increased rainfall coincides with the transition seasons of weaker winds and *ITC*. It must he admitted, however, that such a migration of windfield convergence is difficult to detect over much of East Africa. This is especially true in the transition seasons (the equinoxes) in equatorial East Africa which are the periods of maximum rainfall. Unlike the solstices, when either a northerly or southerly air flow prevails, the equinoctial seasons are occasions when one of these is retreating and the other advancing and both are weak. In the lower troposphere the flow is predominantly easterly. This might suggest that the two maxima of rainfall are the result of equatorial perturbations such as Palmer has described for the equatorial Pacific, but their detection in East Africa has thus far been impossible (Thompson, 1957*a*). Location of wind discontinuities in equatorial East Africa at the time of the equinoxes has proven to be so difficult in practice that their routine determination has been abandoned, but at the same time no satisfactory alternative tool has been discovered. Farther away from the equator the concept of low-level convergence zones may be more useful in explaining the annual rainfall variation, but even here surface wind discontinuities are often difficult to locate on the synoptic charts.

While a pattern involving the predominance of a biannual rainfall variation close to the equator, and a single maximum at slightly higher latitudes, does prevail in a general way, it is incorrect to assume that there are not frequent exceptions to this rule. Numerous localisms in annual rainfall profiles do exist, some of them, no doubt, consequent upon terrain peculiarities. But sufficiently strong and consistent regional patterns of profile character are difficult to detect, so that no attempt is made to treat this topic genetically.

Tropical East Africa North of the Equator

The rainfall deficiency characteristic of much of equatorial East Africa in general is intensified north of the equator and, more especially, on the Indian-Ocean side in Somalia where genuine desert conditions are extensive. There, semiarid climate actually extends across the equator and into Kenya and Tanzania. Except for the higher parts in the Abyssinian and Kenya-Uganda highlands, all of this extensive area, according to the 1948 system of Thornthwaite, is characterized by a dry climate—parts of it genuine desert. Yet it occupies latitudinal and geographical locations where normally tropical humid climates prevail. Such is the case in western and central Africa north of the equator in similar latitudes. And although East Africa has extensive frontage upon a tropical ocean, some of its driest parts are on the coast. Throughout this part of East Africa, annual rainfalls of under 500 mm are characteristic of both lowlands and extensive uplands. There are large areas with less than 250 mm and probably some with less than 120 mm. By contrast, the highlands are much wetter, annual rainfalls of 750–1,500 mm being common in the Ethiopian Highland.

In addition to the general rainfall deficiency in equatorial East Africa north of the equator, there is a further abnormality as it relates to the annual march of rainfall, for the high-sun period, or summer, is dry, when in these latitudes one would expect it to be the wettest season. To be sure, stations with reliable rainfall records are scarce and some station records are strongly local in character. The prevailing pattern, however, is one of bimodal annual rainfall profiles, with the maxima spaced 4–5 months apart. And while this is not unusual in equatorial latitudes, here it is the rule, even out to 10°N and beyond where a single July-August maximum would be more normal. Some drought-making control appears to prevail during the period June through September at the very time when the tropical rainfall should reach its peak. The dual rainfall maxima, with one peak in March-May and the other in October-November, together account for 80–95 percent of the meager annual total. The two peaks would seem to reflect the annual migration of the *ITC*, northward in spring and southward in fall, but oddly enough its effects fade out in summer.

The physical causes for the widespread *summer* dryness in the eastern Somali Peninsula have been summarized by Flohn (1965) as attributable to four divergence effects in the persistent and strong southwest monsoon: (1) a *directional* divergence between the southwest circulation over the Indian Ocean and a SSW flow in northern Somalia and eastern Ethiopia, which is deflected into the heat low over the Danakil Desert; (2) a strong *speed* divergence from the equator northward where just off the Somali coast the average speed of the southwesterlies increases from 3.9 Beaufort at 0°–5°N to 6.0 Beaufort at lat 10°–15°N, with a similar speed increase in the interior; (3) frictionally-induced divergence along the Somali east coast where winds roughly parallel the coastline, and the increase of surface stress over the land causes a slight deflection of the flow toward the interior; and (4) the effect of the anticyclonic deflection of the wind-induced offshore ocean currents toward the east, producing a coastal zone of upwelled cool water with increased stability, a temperature inversion, and low-level stratus with mist.

During the two rainy seasons the winds are predominantly easterly and at those times most of the rain falls in the more easterly parts of East Africa. When westerlies do occur during the February-May rainy period, widespread heavy rains are more likely. Statistically speaking, while the greater relative raininess of westerlies compared with easterlies is well recognized, it is scarcely explained by the eastward advection of "Congo air." Here there appears to be no significant difference between the average *low-level* moisture content of easterly and westerly flows. It is known, however, that upper westerlies are characterized by high relative humidities, chiefly explained by the upward transport of moisture. In equatorial latitudes there is a basic tendency toward ascent in the westerlies and descent in the easterlies.

Mörth (1970) has pointed out that major rainfall anomalies affect large parts of equatorial East Africa and that they alternate rythmically between deficits and excesses. This suggests that the rainfall variations over a period of years are not random, but are controlled by definite physical processes, within, or acting on, the equatorial atmosphere. He indicates that the time and space scales of the major anomalies rule out explanations relating to synoptic models, and that their cyclical character suggests extraterrestrial influences.

Low-Sun Period.—It does not appear to be unusual that most of tropical East Africa north of the equator should be relatively dry during the low-sun period. Such is the case in western and central Africa as well, for normal solar control shifts anticyclonic subsidence farthest south at this season. What is unusual, however, is that (1) the low-sun drought should reach even to the coast, as it does in Somalia, and also (2) the seasonal drought should extend so far south, even two or three degrees south of the equator. The winter isohyets which cross most of Africa north of the equator in a general E-W direction, but with a slight WNW-ESE alignment, at about 35°E turn abruptly southward and this trend continues to nearly 5° south of the equator (Fig. 9.8).

During the low-sun period two great dynamic anticyclones dominate eastern Africa north of the equator. One of these is centered over the northern Sahara with its long east-west axis at about 27° to 30°N, while a second cell, oriented N-S, overlies Arabia and the Middle East, and even projects as far south as Somalia in East Africa (Fig. 9.9). It is this displacement of the Arabian cell so far south that gives tropical-equatorial East Africa north of the equator its widespread drought during the low-sun period. Dry, subsident air with anticyclonic curvature streams southward over East Africa from the Sahara and Arabian highs. The northerly current from the Saharan cell has had a land trajectory across Egypt and the Sudan and consequently is very dry and stable. This air makes itself felt chiefly to the west of the highlands, although occasionally between October and March it arrives on the eastern slopes and lowlands. A second anticyclonic current, northeasterly in direction, originates in the Arabian cell and it is this which dominates most of East Africa in winter. The northeasterly current has followed an anticyclonic path around the eastern and southern side of the Arabian cell and as a result of a sea trajectory of only modest length, its lower levels are fairly moist, although the air mass is still strongly subsident and stable aloft.

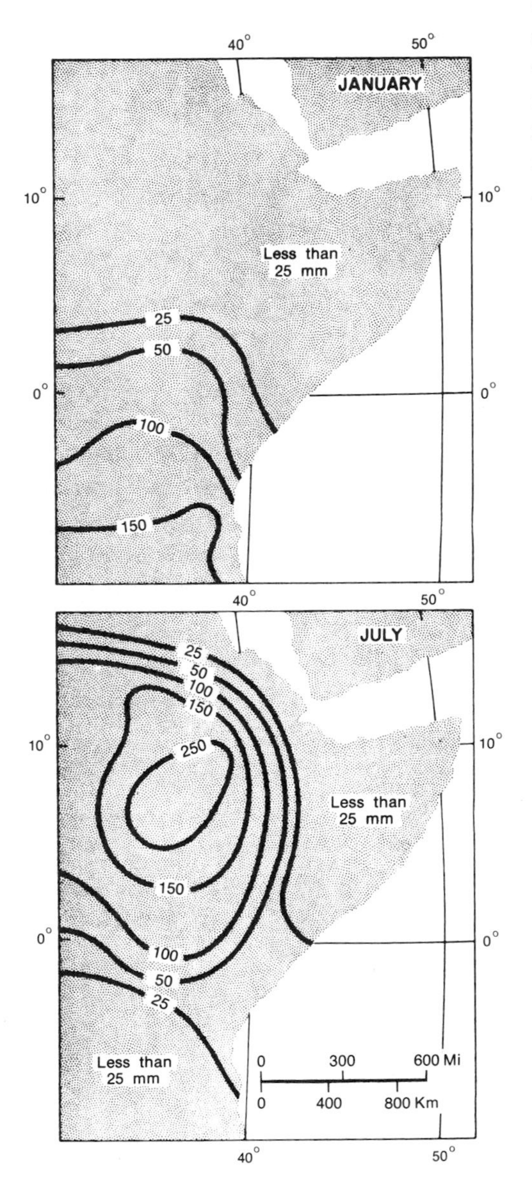

Fig. 9.8. January and July isohyets in tropical East Africa north of the equator.

It needs to be underscored that this northeasterly current is not composed of the neutral or unstable *mT* air which ordinarily prevails along eastern tropical littorals, derived from the western end of an oceanic subtropical high, for such a cell is lacking over the North Indian Ocean and Bay of Bengal. Supplementary to this is the fact that Africa north of about 12°N is bordered on its eastern side not by a tropical ocean, but by the Asiatic continent, a geographical situation having obvious climatic consequences. The northeasterly air stream, therefore, instead of stemming from the western side of an oceanic high, has originated on the eastern flanks of a land anticyclone centered remarkably far south, so that the air mass is very stable, even though moistened in its lower layers.

As this northerly current moves southward behind the shifting *ITC* it yields very little precipitation. Even the highlands remain relatively dry. In Somalia the northeasterly air flows approximately parallel with the coast, so that there is little inland movement of sea moisture and little orographic lifting. Its drying effects are felt even to the equator and beyond, for at coastal Malindi in Kenya, at about 3°S, where the annual rainfall is 1,000 mm, December, January, and February all show less than 25 mm of rainfall. This same station has over 325 mm in May, and 160 mm in June, so although it lies south of the geographic equator it is characterized by a seasonal distribution of rainfall which is strikingly Northern Hemisphere in character. But at Dar es Salaam, 7°S, January is wetter than July. In this subsident air stream disturbances of all kinds are remarkably few, including easterly waves and cold fronts, so that rainfall is meager.

High-Sun Period.—While it is not so unusual for tropical East Africa north of the equator to be relatively dry during the low-sun period, it is much more abnormal that the region's rainfall should be so meager at the time of high sun when the *ITC* and its disturbances have shifted to a position north of the equator. West and central Africa in similar latitudes are also dry at the period of low sun, but during the Northern Hemisphere summer the precipitation is so abundant that extensive areas north to 8° or 12°N are characterized by a tropical humid climate. This is not true of lowland East Africa north of the equator where actually June-July-August is a dry period, at certain stations even drier than January-February. Most stations in dry Somalia have two rainfall maxima, one in April-May and another in October-November.

The deficiency of summer precipitation on the lowlands of tropical East Africa north of the equator is chiefly a consequence of regional peculiarities in the circulation pattern. Somalia and Arabia appear to have almost the only tropical east coasts which nearly the year around are exposed to strongly subsident maritime air. A well-developed oceanic anticyclone is lacking in the North Indian Ocean–Arabian Sea in summer as well as in winter, so that neutral or unstable *mT* air such as normally develops in the western parts of such cells is absent.

As the sun advances into the Northern Hemisphere in April and May the arrangement of the controlling centers of action undergoes important change. Over northern Africa and Arabia the anticyclones of winter are replaced in summer at lower levels by thermal lows, an extensive Saharan cell with its long E-W axis centered at about 20°–22°N, and a deeper cell located over Arabia and southwestern Asia (Fig. 9.10). Above these surface thermal lows an anticyclonic circulation still prevails. The Saharan anticyclone which in winter was oriented E-W, in summer shifts eastward and its long axis is

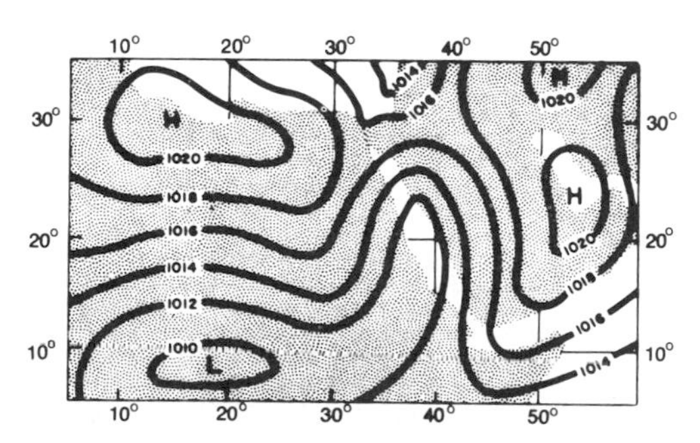

Fig. 9.9. Representative low-level pressure arrangement over the eastern Sahara and Sudan in winter. (After Solot, 1950.)

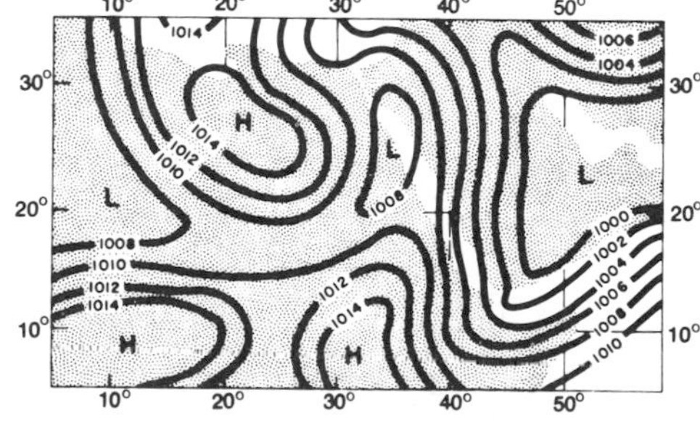

Fig. 9.10. Representative low-level pressure arrangement over the eastern Sahara and Sudan in summer. (After Solot, 1950.)

aligned N-S over the eastern Sahara, so that its influence extends unusually far south (Solot, 1950). As a consequence, the surface *ITC* which reaches to 20°–22°N in tropical West Africa, lies closer to 10°–15°N in East Africa. Significantly, the isohyets which assume a general east-west trend across most of tropical Africa north of the equator, just eastward of the Ethiopian Highland turn abruptly southward, precisely as was described earlier for winter.

As a result of the previously described pressure arrangement, the general atmospheric circulation in summer over East Africa north of the equator is as follows. North of the surface *ITC* there is a northerly anticyclonic current composed of hot, dry subsident air from the Sahara. South of the surface *ITC* are two moister currents whose temperatures are not so high. One of these, and far the more important one as it concerns East Africa north of the equator, is the southeast trade from the Southern Hemisphere, which here has developed a southerly or southwesterly flow. The second moist current, but one less important to northern East Africa, has a westerly component. It is more deeply humidified. Two major zones of wind discontinuity and convergence develop from the three air currents just described: (1) an east-west zone, the *ITC*, separating hot, dry, anticyclonic Saharan-Arabian air to the north from the moister and less hot southerly air of maritime origin to the south, and (2) a north-south zone separating relatively moist westerly air from southerly air derived from the South Indian Ocean (Fig. 9.4). Since equatorial westerly air rarely is able to penetrate east of the Abyssinian barrier, its influence on the weather of the Somali area is slight and is, no doubt, a partial explanation for that region's drought. The heavy rainfall in Ethiopia, on the other hand, is to a much greater extent derived from westerly air. But even if an invasion of such air should occur, it would be dry and subsident after crossing the Kenya and Ethiopian highlands. Most of the precipitation in northern East Africa no doubt is derived from the southerly monsoon air stream. Yet this source is far from good. Over the Kenya-Somalia area it is strongly diffluent, so that low-level subsidence must be prevalent. Moreover, because it has had a long meridional trajectory over the uplands of East Africa farther south, it has lost much of its moisture enroute. As a southwesterly current over Somalia it flows parallel with the coast, or even somewhat offshore, so that little moisture is imported from the adjacent ocean. Actually much of the time it may have a downslope component that adds to its stability.

Equally important in determining the inefficiency of the southerly monsoon as a rainmaker is the structure of the surface *ITC*, in which dry, stable northerly air is ascending over a surface of discontinuity marking the boundary with the maritime current, which slopes gently upward to the south (Solot, 1943). This frontal structure and its effects on weather have been described in an earlier section on tropical West Africa and it is reasonable to assume that in East Africa many of the same characteristics prevail. It only needs to be emphasized that such a zone of surface convergence is opposed to precipitation, and that even moderate amounts of rainfall are unlikely to occur for several hundred kilometers south of the surface front, or until the depth of the southerly current is sufficient to permit the development of clouds of considerable vertical dimensions. Thus, in summer, while at the surface the southwesterly flow prevails over East Africa north to about 10°–15°, the stable northeasterly air of Saharan origin predominates aloft. The upper convergence of the two air masses at about 2,400–4,000 m, Aspliden locates in the vicinity of the equator in East Africa in July, and since the surface convergence, as well as that aloft, between northerly and southerly currents is farther south in East Africa than in western and central Africa, it follows that the rain and weather belts are as well. The situation in northern East Africa and southern Arabia in summer closely resembles that of northwest India-Pakistan, where the surface *ITC* likewise coincides with a region of drought.

Supplementary causes for the meagerness of summer rainfall in northern East Africa should also be noted. As mentioned previously, the surface air flow in summer over Somalia and eastern Ethiopia is prevailingly from the southwest, so that those winds on the average probably have a downslope component, and at the same time are either parallel to the coast or slightly offshore, features which are opposed to precipitation (Fig. 9.11). One effect of the paralleling, or slightly offshore, winds is to produce cool upwelled water along the coast of Somalia (Fig. 9.12). The areas of exceptionally cool water appear to be localized along certain parts of the Somali coasts bordering both the Indian Ocean and the Gulf of Aden, although the exact positions are subject to change. The best known among the areas of cool coastal water lie in the neighborhood of Ras Hafun on the northeast coast of Somalia, in the vicinity of Bab el Mandeb (Great Britain Meteorological Office, 1944*a*). The average temperature of the sea surface near the coast in summer is as much as 6°C (10°F) cooler than the ocean surface farther to the east. This is especially true along the northern Somali coast where, significantly, the drought is most intense. On occasions the temperature differences are much greater than the averages indicate. One ship recorded a fall of 9°C (16°F) in sea temperature as it approached Ras Hafun from the south, followed by a rise of 14°C (25°F) as it rounded Cape Guardafui. Along the whole Somali coast there is a high frequency (over

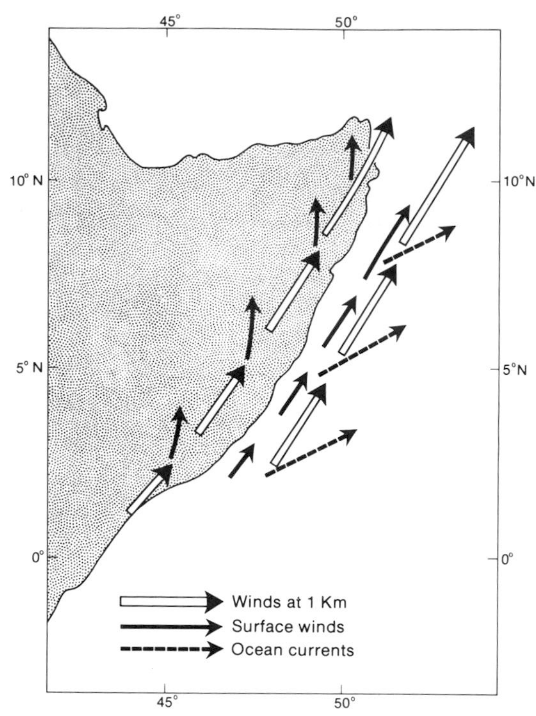

Fig. 9.11. Low-level winds and ocean currents along the Somali coast and over the adjacent sea in summer. (After Flohn, 1965.)

40 percent) of fog-mist-haze in the summer months, these phenomena being attributable chiefly to the cool upwelled water (Fig. 9.13). By contrast fog is rare in December

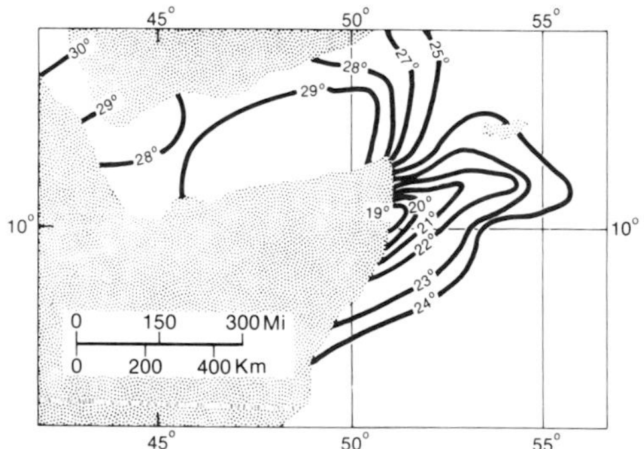

Fig. 9.12. Isotherms of surface sea temperature in July along the Somali coast. (From Great Britain Meteorological Office, 1944*a*.)

and January. The stabilizing effects of the cool water upon the surface air must damp any convective activity and hence reduce the likelihood of precipitation.

The same paralleling, or slightly offshore winds which produce the cool upwelled water may likewise act to produce coastal subsidence which combines with the cool water to intensify stability and hence aridity. Because of the differential frictional effects of land and water upon the paralleling or slightly offshore surface winds, there is a tendency for the land and sea elements of the southwesterly current to diverge, thereby creating subsidence in the vicinity of the coastline. Attesting to the general stability of the air along the coast of Somalia and over the adjacent waters is the great infrequency of atmospheric disturbances of any kind. Observations on the frequency of thunderstorms are very scarce, but the evidence indicates that they are infrequent over the whole area. Even in the equatorial parts where they should be most frequent, thunder is heard only 20 days of the year, and in the Gulf of Aden there are fewer than 10 such days. Along the Somali coast thunder is more frequent in winter than in summer. Tropical cyclones are lacking in the area south of 10°N, and even north of that latitude such storms are extremely rare. What little disturbed weather affects this region, most of it of unknown origin, appears to be concentrated in the transition seasons between the two monsoons.

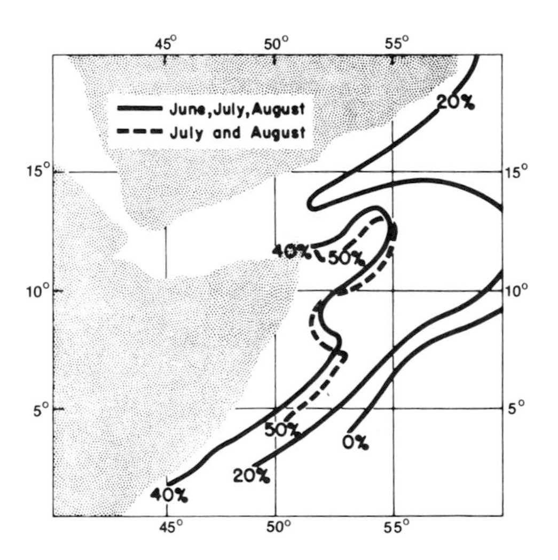

Fig. 9.13. Frequency of fog and mist along the Somali coast in summer.

Tropical East Africa South of the Equator

Rainfall distribution in tropical East Africa south of the equator is difficult to describe, for a clearly recognizable pattern is not readily discernible, and the isohyetal arrangement is complicated. As in East Africa north of the equator, the higher elevations are the sites of heaviest rainfall, but on the other hand, humid climates do exist in parts of the coastal lowlands, while dry climates also prevail over extensive sections of the Kenya-Tanzania uplands. The mean annual rainfall map of East Africa shows evidence of one somewhat moister zone along the eastern margins closest to the Indian Ocean, and another in those western parts closer to the sources of humid westerly air (Figs. 9.1, 9.14). In between these two is a drier central zone, where, except for spots with higher elevations, rainfall is generally under 750 mm, and there are extensive areas with even less than 500 mm. This arrangement may reflect the effects of the Indian Ocean moisture source and easterly disturbances on the maritime frontier, and of the greater frequency of incursions of humid westerly air along the inner or western margins. Significantly, rainfall variability increases from west to east, so that the highest values are in those parts bordering the Indian Ocean where easterlies dominate, while they are lowest in western East Africa where westerly air occurs more often (Griffiths, 1959).

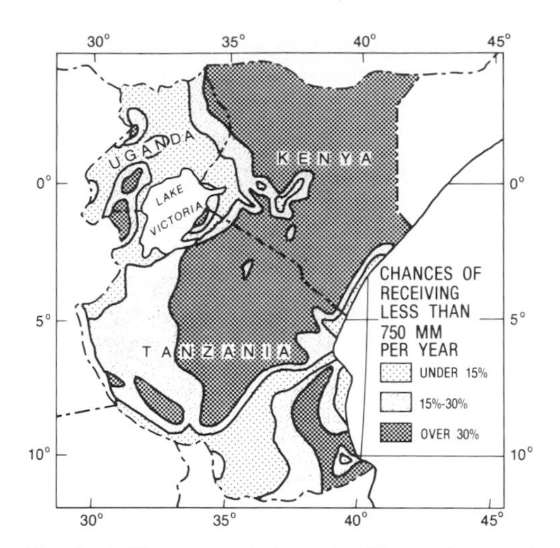

Fig. 9.14. Illustrating the low reliability of the annual rainfall in East Africa. (From East Africa Royal Commission 1953–1955 Report, 1955.)

On the whole, tropical East Africa south of the equator does not have as serious a water deficiency as its counterpart to the north. Genuinely arid climates are lacking except for a small area in eastern Kenya (Carter, 1954). But if deserts are limited in extent, by contrast dry-subhumid and semiarid climates are widespread, both along the coast and in the interior, according to Carter's map of Moisture Regions employing the Thornthwaite 1948 system. Such a condition is abnormal for the eastern side of a tropical continent, especially where there is ample trigger action provided by numerous highlands. Not only is rainfall modest in amount, but it is highly unreliable as well.

No satisfactory explanation either for the general rainfall deficiency or for the areal and temporal distribution of rainfall within this southern part of the larger area is at present available. It is not possible to supplement in any important way the reasons given earlier for the modest rainfalls characteristic of East Africa in general. These included (1) the divergent character of both monsoons over extensive land areas, (2) the modest depth of the southwest monsoon, especially over the highlands, (3) the strongly meridional flow in all but the transition seasons, a feature that limits the advection of sea moisture and reduces the orographic effects, and (4) the stable stratification of the air aloft, including a marked decline in moisture content. In an earlier section the stable vertical structure of the atmosphere at Nairobi has been described. At Salisbury at nearly 18°S in Rhodesia the characteristic winter structure shows a striking discontinuity between a lower air mass with steep lapse rate and modest moisture content, and a potentially warmer and excessively dry air mass above. This same type of discontinuity is observed likewise in spring and fall, but in a form less clearly marked and located at higher levels. In summer the upper-level dry air is still more infrequent and the discontinuity is evident on less than 20 percent of the soundings. The strongly meridional course of the winds is suggested by the data from Mombasa where at about 300m in May and June, 44 percent of the winds are from the south, 20 percent from the southwest, and 16 percent from the south-southeast. Accordingly, the southerly current appears to slide along the edge of the plateau, while only a residual quantity surmounts the barrier.

It seems likely also that the huge mountainous island of Madagascar, which parallels the African coast several hundred kilometers offshore from about 12°S to nearly 26°S, may have an appreciable effect upon reducing the mainland's rainfall. It should be noted that on Madagascar, with its 2,000–3,000 mm of annual rainfall on the windward eastern side and only 500 to 1,000 mm on the west and southwest margins, precipitation amounts and distribution are more in keeping with what is expected of tropical windward locations with highlands present, features which are conspicuously absent or only weakly developed in mainland East Africa. To be sure, elevations are usually greater in Madagascar than on much of the East African plateau, and the ascent from the east is also more abrupt. But probably equally important in explaining Madagascar's much heavier east-side rainfall is the fact that the SE trades in these longitudes are strongly zonal in flow so that after a long trajectory over the Indian Ocean they meet the Madagascar highlands at almost right angles, resulting in a maximum lifting of the air and consequent heavy rainfall. With such a strong zonal flow, the leeward side of the island and adjacent Mozambique Channel become a region of downslope winds and surface divergence. As a consequence disturbances from the east, and from the west also, are likely to deteriorate in this area of subsidence to the lee of Madagascar, so that they become difficult to track. Through this weakening of disturbances, especially those from the east, and likewise the deflection of what was a zonal current into a weaker and more meridional flow, Madagascar may appreciably and adversely affect the mainland's rainfall.

Among the important rain-bringers to eastern and northern Madagascar are the tropical storms, both violent and moderate, which infest the tropical western Indian Ocean in the Southern Hemisphere. As noted earlier, there are no such disturbances in a similar location along the Somali coast to the north of the equator. The majority of these storms recurve to the south and southeast before their centers reach Madagascar. Nevertheless, the eastern side of the island usually feels the rainfall effects of a number of tropical storms each year. However, only very occasionally do the storms enter the Mozambique Channel and reach the African mainland, so that the latter region largely misses their rainfall effects. The very fact that tropical storms are uncommon west of Madagascar suggests the shielding effects of that island and the prevalence of atmospheric stability to leeward, both of which discourage the propagation of such disturbances.

One is led to surmise also concerning the rainfall effects on East Africa of the anticyclonic circulation which prevails over southern Africa. In winter, local radiational cooling of the plateau surface plus advected cold air from more southerly latitudes strengthenes the land cell of the subtropical high over southern Africa. Such a land anticyclone when it extends its influence eastward and northward, over the coast and the Mozambique Channel, sometimes to 10°S or beyond, serves to weaken any onshore easterly wind and substitute for it a westerly flow which is both dry and stable, and which along the coast has a downslope component that causes it to be föhn-like in character. Such dry berg winds are strengthened by the passage of a cyclone to the south.

And finally, it may be pointed out that, while cold-front disturbances are not unknown along the east coast of tropical Africa south of the equator, they are by no means so important a feature of weather there as they are along the east coast of Brazil (Forsdyke, 1944) (Fig. 9.15). Southern Africa causes no such interruption in the belt of subtropical high pressure and the zonal winds as does South America, with the result that east of South Africa the tendency is for the cold fronts to be carried more eastward than northward, so that they do not ordinarily affect the littoral of tropical East Africa. It seems not unlikely, also, that the subsidence westward from Madagascar in the region of the Mozambique Channel may have a damping effect upon such northward-moving disturbances. Farther south along the south and southeast coast of South Africa, cold-front disturbances are numerous and in that region they are responsible for a

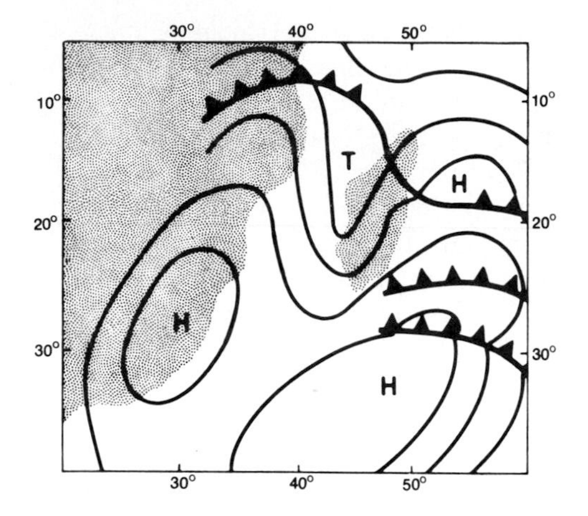

Fig. 9.15. A cold front from higher latitudes advancing equatorward over eastern Africa south of the equator. (After Forsdyke, 1944.)

large percentage of the rain in all seasons (Vowinckel, 1956). But their effects wane rapidly to the north. As a consequence low-sun rainfall is less in tropical East Africa south of the equator than it is in eastern Brazil in comparable latitudes. Along the whole east coast of East Africa south of about 5°S there prevails a strong winter minimum of rainfall, which in part may be attributable to the relatively few and weak cold fronts. This again is in contrast to the situation in coastal Brazil. Moreover, such invasions of cool southerly air as are experienced, more often than not have had a land trajectory, and as a result bring dry weather instead of wet.

CHAPTER 10

Southern Africa

Except for modifications imposed by terrain and altitude, the general climatic character and arrangement in much of southern Africa (south of about 20°S) are in large measure what one would expect from the operation of the great planetary controls (Wellington, 1955). From the equator to about 20°S the isohyets of annual precipitation across most of Africa have a general east-west trend, with amounts increasing equatorward, and rainfall strongly concentrated in the high-sun period. This expresses the operation of latitudinally migrating zonal controls such as the *ITC*, wind systems, and associated disturbance belts. South of about 20°S the isohyets make a right-angled bend and assume a north-south direction approximately parallel with the coastlines. The 400 mm isohyet, with a nearly meridional alignment, bisects southern Africa into almost equal parts, a drier western and a less dry eastern. This contrasting meridional alignment of the isohyets south of about 20°S reflects the waning effect on rainfall of zonal tropical controls, and the rapid taking over of subtropical anticyclones and westerly flow, with drought-producing controls in the form of a stable anticyclone and cool waters prevailing on the west side, and weaker subsidence, a warm current and more numerous disturbances, chiefly of westerly origin, on the east. An exception to the prevailing aridity on the west side is to be found in the extreme southwest, where, in the vicinity of the elevated Cape, cool-season cyclonic-orographic rains produce a limited area of Mediterranean (*Cs*) climate.

Like the distribution of total precipitation, the annual variation also conforms reasonably well to the expected pattern based upon the planetary controls. Only in the extreme southwest does over 60 percent of the annual rainfall occur in the winter half-year. This is the region of Mediterranean climate (*Cs*) and of Mediterranean steppe and desert *BS*(*W*). The extreme north (15°–20°S) with its E-W isohyets has a high-sun maximum, the low-sun period being almost, if not quite, without rain (*Aw*). Eastern South Africa has over 60 percent of its annual rainfall in the summer half-year, but winters are not without rain, a feature which distinguishes it from the more tropical latitudes along its northern margins where winters are nearly rainless. Along the eastern littoral especially, the winter months are moderately wet, but with increasing distance from the coast the accent on summer becomes greater. A restricted transition zone, becoming a broad belt only along the south coast where rainfall is well distributed throughout the year, separates the winter-rainfall area of the southwest from the summer-rainfall areas of the interior and east. Thus, much the larger part of southern Africa has a relatively strong summer maximum of rainfall.

Seasonal Circulation Patterns

Most of Africa south of about 15° or 20°S is dominated by subtropical anticyclones. Conspicuous oceanic cells of high pressure exist in the South Atlantic and South Indian oceans both to the west and to the east of South Africa. The strong eastern limb of the South Atlantic cell remains relatively fixed in position along the west coast, so there subsidence is strong and persistent, the inversion low, and atmospheric stability remarkably well developed. By contrast the east-west position of the Indian Ocean high fluctuates more widely, that cell withdrawing eastward in summer and advancing westward toward the African coast in winter (Vowinckel, 1955*b*). This anticyclone is unique in that it is probably the only one of the maritime subtropical highs whose center of gravity seasonally moves into the western part of an ocean. Here the overall stability is less than on the west side of the continent, but there is a seasonal variation, with increased stability and lower and more frequent inversions in winter than in summer. In addition the Indian Ocean cell, at least in its eastern and western parts, is more in the nature of a composite of eastward-moving individual cells of high pressure than it is a permanent element (Fig. 9.4).

Over the plateau of South Africa the pressure and circulation patterns are not so simple. It has been customary to explain the contrasting seasonal climates of most of South Africa, with their generally wetter summers and drier winters, as a consequence of a monsoonal reversal of pressure and winds. Thus, the sea-level pressure charts show a trough of low pressure over the interior plateau in summer and an anticyclone in winter. But the winds at the plateau level do not conform to this pressure reversal, for there is no evidence of a cyclonic wind system in summer. The error arises from reducing mean surface pressures of the plateau to sea level using the pressure-height equation, which is not applicable in the case of South Africa. S. P. Jackson (1952) has prepared seasonal pressure charts of South Africa at a height of 2,000 meters, which is somewhat above the plateau level, and he finds the January and July maps to be essentially similar, both showing isobaric arrangements that are anticyclonic. This pattern is in reasonable agreement also with the observed wind systems. Thus, the persistent pressure distribution over the South African plateau is anticyclonic, but somewhat weaker in summer than in winter when the cell moves northward bringing the surface westerlies over the southernmost parts of the continent.

Taljaard's (1953) maps representing pressure distribution at 1,250 m, which is closer to plateau level, show at least one significant difference from those prepared by Jackson (1952) representing the 2,000 m level. On Taljaard's summer map there is a southward extension of a shallow thermal trough over the interior, while in winter there is a weak anticyclone over the Transvaal. That the summer low at plateau level is shallow is indicated by the fact that at 2,000 m an anticyclone prevails. Moreover, the surface wind system over the plateau does not conform to the pressure distribution at 1,250 m, so that there is no evidence of a monsoonal indraft of air. Besides, any such inflow from the Indian Ocean at this level would be effectively blocked by the high escarpment to the east. It appears likely, therefore, that it is an error to explain the seasonal rainfall concentration of South Africa in terms of a monsoonal system of winds. An anticyclonic circulation persists at the 2,000 m level in summer as well as in winter.

The stronger high in winter over the plateau is partly a consequence of the cool land surface but also of the invading cold highs from higher latitudes. Many of the latter skirt the south coast and move northeastward to replenish the Indian Ocean cell, while a smaller number invade the continent and make juncture with the land cell there, giving rise to the coldest spells of winter weather. Thus, the mean pressure map of the winter season well represents the day-to-day synoptic situation, in which anticyclones dominate the land area, with few interruptions by rain-bringing disturbances. By contrast, the average weaker high of summer is less persistent and the daily synoptic chart reveals a greater number of disturbed weather situations. Nevertheless, it is the persistence of an anticyclonic circulation over South Africa in both summer and winter which accounts for three-quarters of the continent south of about 20°S being dry, according to the 1948 Thornthwaite system.

As a consequence of the year-round anticyclonic development, resultant winds over the plateau are mainly from the west, both summer and winter. Hence the widely held concepts of South Africa being dominated by the southeast trades, and of the rainier east side being a consequence of windward location, obviously are false. In reality, even along the east coast, northward to about the latitude of Durban, winds are more westerly than easterly, so that this section of coast is more leeward than windward. Farther north the southeast trades do prevail along the eastern littoral, but they are shallow and do not enter the continent deeply because of the escarpment. Normally the change from easterly to westerly flow occurs at between one and two km above sea level (Taljaard, 1955).

Atmospheric Disturbances and Weather Types

It is not unexpected that disagreement still prevails among the meteorologists in South Africa concerning an identification and classification of the disturbances which are responsible for the climatic pattern of that region. It seems to be farily well agreed, however, that the earlier studies of weather types based upon sea-level isobars leave much to be desired. At the present time the synoptic charts published by the Meteorological Office at Pretoria employ the 850 mb surface over the upland and sea-level isobars for the lowlands. It is recognized, also, that while a representation of synoptic conditions near the surface is of great value, such charts need to be supplemented by above-surface data, and here the information is inadequate. As a consequence, the origin and distribution of rainfall and its relation to particular synoptic situations continue to be imperfectly understood.

The most useful publication attempting to identify and classify the weather types of southern Africa, while its author has served on the Meteorological Office staff at Pretoria, is not an official publication of the Meteorological Office and should not be considered as necessarily representing the majority opinion of that staff (Vowinckel, 1955*a*). Using Pretoria as the characteristic station, Vowinckel's identification and classification of weather types is particularly applicable to the high veld, but it does not neglect other parts, including the coastal lowlands. Supplementing this more general study is another which specifically focuses on the eastern lowlands, using Lourenço Marques as the central station (Vowinckel, 1956). Seven principal synoptic types are recognized (Fig. 10.1 and Table 10.1).

Types 1 and 1a. Fair-weather types. Anticyclonically controlled fair weather is a common feature of upland South Africa in

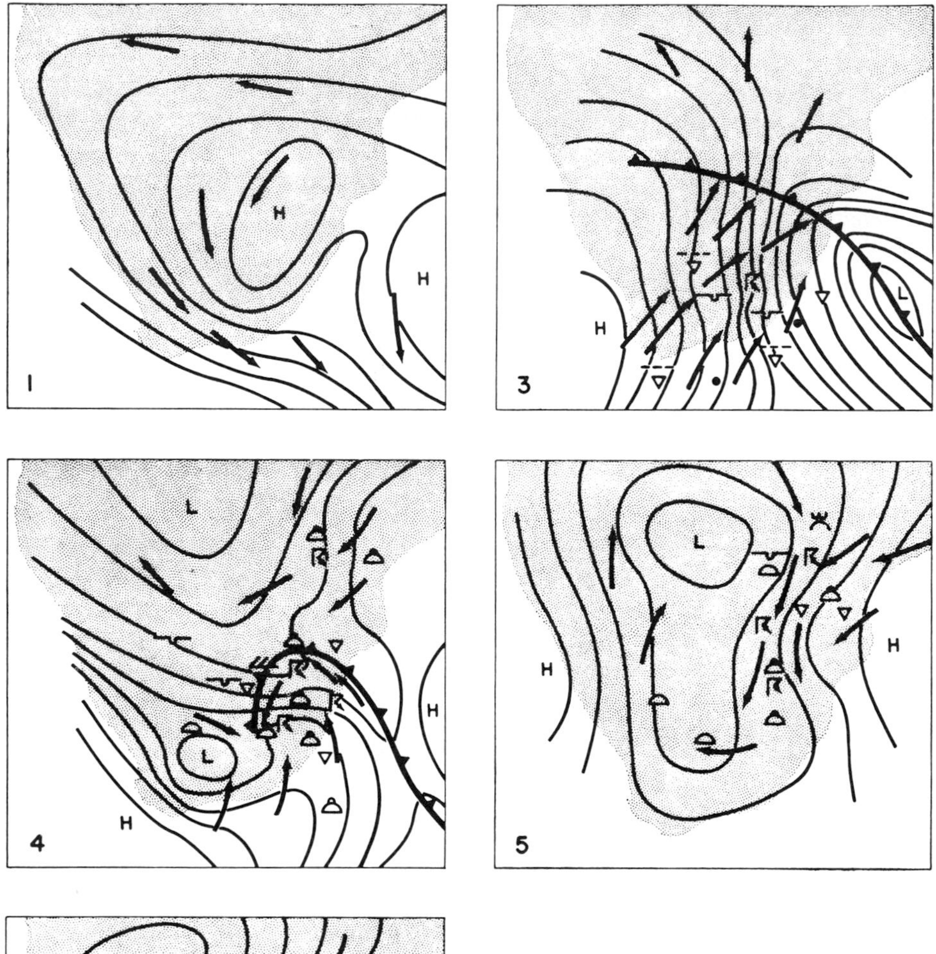

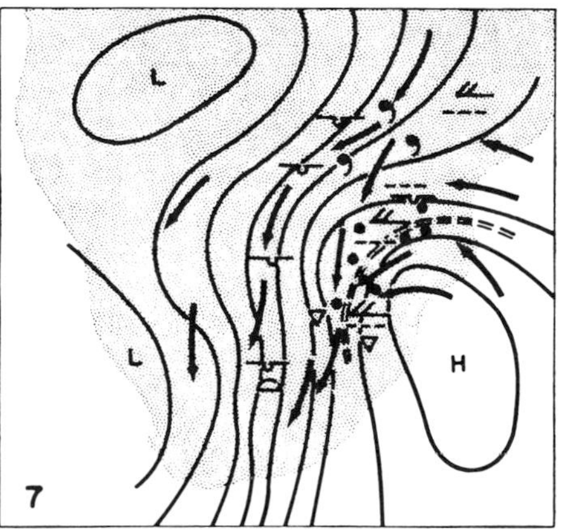

Fig. 10.1. Five important weather types of South Africa. 1. Fair-weather type; 3. Cold-air type; 4. Monsoon type; 5. Equatorial type; 7. Bad-weather type. (After Vowinckel, 1955*b*.)

all seasons, but it strongly predominates in the winter months (see Table 10.1), at which time temperature inversions are very persistent. The combined frequency of these two anticyclonic types reaches 79 percent in June and July, but drops to 12 percent in December, 18 in January, and 17 in February. Type 1a is only a weaker form of Type 1, with slightly more morning cloud.

Type 2. Warm-air type. Its occurrence is conditional upon a strong trough development to the south and west and an increased east-west pressure gradient, which result in cloudless skies with strong solar radiation, and an advection of warm land air from the north and northwest. In this situation, windy, hot weather, with a dust haze, prevails at Pretoria. The warm-air type is most common in late winter and spring, with the maximum in October (19 percent).

Type 3. Cold-air type. The cold air invades from the southwest in the form of an anticyclone from higher latitudes. Since before reaching Pretoria this air has had a long continental trajectory, it is relatively dry in the region of the high veld. Further radiational cooling of this cool air may result in killing frosts on the uplands. Unusually strong thrusts of cold air may be accompanied by cumulo-nimbus clouds with thunderstorms and hail along the eastward-advancing front, although the amount of rainfall usually is small. This type normally passes over into the fair-weather type. Usually absent in summer, it reaches a modest maximum frequency in winter and early spring.

Type 4. Monsoon type. Like Type 3, this one similarly is initiated by a thrust of cool anticyclonic air, but from the south and southeast instead of from the southwest, so that it has been modified over warmer waters and has had a shorter land trajectory before reaching Pretoria. As the trough and its cold front move northward across the southeastern parts, warm, humid northeasterly air from the Indian Ocean is drawn into the depression, and along the discontinuity between tropical and polar maritime air, as well as within the two air masses, showery rainfall occurs which is much more plentiful and widespread than that accompanying Type 3. The shower activity is strongly diurnal, with less cloud in the mornings and more in the afternoons and evenings. Such

Table 10.1. Frequency of different weather types at Pretoria, South Africa (in percent) (Vowinckel, 1956)

	TYPE 1 Fair-weather type	TYPE 1a Weaker fair-weather type	TYPE 2 Warm-air type	TYPE 3 Cold-air type	TYPE 4 Monsoon type	TYPE 5 Equatorial type	TYPE 6 Shower type	TYPE 7 Bad-weather type
January	8	10	1	0	19	27	28	7
February	8	9	0	0	16	35	23	9
March	14	5	1	0	17	37	20	6
April	11	14	3	3	18	26	17	8
May	24	14	3	4	14	19	14	8
June	68	11	2	4	6	7	0	2
July	70	9	4	5	6	3	1	2
August	60	12	8	8	10	1	1	0
September	55	10	14	6	7	5	1	2
October	27	11	19	3	9	15	5	11
November	14	5	5	3	12	25	27	9
December	3	9	1	1	14	35	28	9

showers may go on for several days and only die out as the flow of humid air from the south wanes and anticyclonic control is re-established. Orographically induced rains are plentiful on the eastward-facing escarpments. While the name, monsoon type, is dubiously valid, it is nevertheless true that an inflow of tropical air plays an important role in this synoptic situation, which is one of the frequent ones in eastern South Africa. Occurring in all seasons, it reaches a maximum frequency in the warmer months November through May when the subtropical high is weakest and displaced farthest south.

Type 5. Equatorial type. This type of disturbed weather occurs with a large-scale advection of tropical Congo-Basin and Indian Ocean air from the north and northeast initiated by a low-pressure system over the interior of South Africa. No air-mass discontinuities are present, but instead there is widespread shower activity within the humid tropical air induced by insolational heating, terrain, air-mass convergence, and probably disturbances in the form of weak surges. Of very frequent occurrence in the warmer months, November through May, it is infrequent during June through September.

Type 6. Shower-weather type. Included in this type are all those days on which, while mornings are cloudless, showers or thunderstorms break out in the afternoon. Thus, cloudiness is usually much less than in either the monsoon or equatorial types. Commonly the shower type is transitional and occurs when the monsoon or equatorial type is waning, and weak anticyclonic control is strengthening. From Table 10.1 it becomes evident that this is a more frequent type of warm-season weather, reaching a minimum during the cooler months June through September.

Type 7. Bad-weather type. Such persistently cloudy and rainy weather occurs when simultaneously with an outbreak of moist, cool air from the east, occasioned by an anticyclone offshore, equatorial air from the north glides up over the cool air. It does not occur so frequently as the three previous types, but like them it is least characteristic of the cooler months.

From the previous sketch of the principal weather types affecting South African weather, certain climatic generalizations can be made. Winter is the period of least disturbed weather except along the southwest and extreme south coasts. At this time strong and persistent anticyclonic control over the interior makes for settled air-mass weather there. But since the subtropical anticyclones are farthest north in winter, middle-latitude disturbances skirting the southwest and the south coasts cause cool-season rainfall in those areas, while the cold fronts moving equatorward along the southeastern margins add modest amounts of precipitation there as well. Elsewhere it remains dry.

The analysis of weather types and their monthly frequency of occurrence likewise makes clear why the warmer part of the year shows a strong maximum of precipitation except in the southwestern and extreme southern parts. It is not a consequence of a summer monsoon, but rather is associated with a greater prevalence of rain-bringing disturbances in the warmer months, and of a deeper penetration of maritime air masses, both tropical and middle latitude, induced by these disturbances. Likewise, the subtropical anticyclonic belt is weaker over the continent in the warm season and is also farthest south, and the *ITC* has advanced to the northern margins of the area, so that humid tropical air is in a position to enter the region from the north on occasions when synoptic situations are favorable. Accordingly, the summer rains are of the shower type and the summer weather is relatively tropical, although cool, middle-latitude air continues to be an active element in most of the weather types. The heated land surface is an additional factor making for low-level summer instability. Moreover, because the Indian-Ocean cell recedes eastward in summer, the southeast trades are not only deeper at that time, but the inversion is likewise higher and less prevalent, and the air less stable.

It also becomes clear from the survey of the weather types why the eastern side of South Africa is the wetter side. It is not so much that this is the windward side in the realm of the southeast trades, for ordinarily these shallow winds do not move inland beyond the escarpment. Moreover, southward of about 30°S the southeast coast has a westerly rather than an easterly circulation, so that a windward situation does not prevail. Rather, the greater rainfall along the east side is a consequence of anticyclonic control in this part being so frequently interrupted by perturbations and their maritime air masses from both middle and tropical latitudes. Most of the disturbances affecting South Africa appear to originate in the westerly air stream and do not arrive from the low latitudes. But on the other hand, the tropical air is an important source of moisture for the rainfall. Jackson (1952) is of the opinion that more vertical ascent leading to rainfall within the tropical air stream is a consequence of convergence within the air mass itself than of vertical displacement along frontal surfaces, although this view is not held by some others.

Climatic Uniquenesses

The Extensiveness of Dry Climates.— Probably the most conspicuous climatic peculiarity in southern Africa is the fact that dry and subhumid climates occupy such a large proportion of the total area, even extending across the continent and in places reaching almost to tidewater on a subtropical east coast which is paralleled by a warm ocean current (Schulze, 1958). Between latitudes 20° and 33°S only parts of a very narrow coastal strip along the Indian Ocean, and the higher elevations to the rear, show annual rainfalls in excess of 1,000 mm. In the lowlands of the Limpopo drainage basin which separates the highlands of Rhodesia from those of South Africa, the isohyets bulge far to the east, so that semiarid climates as defined by Thornthwaite (1948) reach almost to the sea. In parts of this basin in northernmost Transvaal, annual rainfall is only 300–400 mm, representing desert conditions. A somewhat similar belt of low rainfall follows inland along the Zambezi River valley farther north. Again, to the south and west of the highlands, semiarid climate reaches the coast in the vicinity of Port Elizabeth. One gets the impression that almost the whole of eastern South Africa is saved from being dry in some degree only by the orographic effects of the highlands, for except for a very narrow coastal strip, humid climates are coincident with elevated lands. It seems probable that the Limpopo Basin represents in a general way what the rainfall situation over much of eastern South Africa would be if lowlands prevailed.

A satisfactory explanation for the dry climates being so extensive, and extending so far eastward, as in the Limpopo Valley, is not yet available. No doubt the dryness has its origin in the general prevalence of anticyclonic circulations centered over the eastern plateau at and above the 2,000 m level and over the adjacent oceans to the east and west. Rubin (1956) has pointed out the close correlation between seasonal anomalies of precipitation and of pressure departures in southern Africa. Thus, an abnormally wet summer over the eastern parts is associated with a general negative departure of pressure, stronger cyclonic activity in the westerlies, and a deeper continental trough. Dry summers, by contrast, coincide with positive pressure anomalies over, and to the east of, the continent, with the Indian Ocean high located well to the west and north of its usual position. Similarly, winter precipitation in the southwestern and southern parts is below normal when the Indian Ocean high is far to the west of its normal position, resulting in the blocking of cyclones from the Atlantic. Except in winter, the prevalent anticyclonic pattern is interrupted sufficiently often in the eastern parts by disturbances and their accompanying invasions of maritime air so that even the lowlands are saved from being as dry as are the more persistently anticyclonic western and central parts.

Other factors making for reduced rainfall over tropical East Africa have been analyzed in an earlier chapter and a number of these causes are applicable to eastern subtropical Africa as well. The maritime air masses along the east coast of South Africa are prevailingly shallow, although somewhat deeper in summer than in winter, and above the humid surface air and separated from it by an inversion, is continental subsided air. The inversion is lower and more intense in winter, the average height at Durban and Lourenço Marques in that season being about 1 km or below (Figs. 10.2, 10.3). In summer the discontinuity has risen to 2 to 2.5 km (Taljaard, 1955). Such a stratification is opposed to abundant precipitation. On most days the South African plateau rises above the shallow layers of maritime air around it, important penetrations of the continent occurring chiefly on those occasions when disturbances deepen the maritime currents and attract them inland.

It bears reemphasizing, also, that along the east coast southward from 25° or 30°S easterlies do not prevail and winds ordinarily are onshore only when disturbances induce a strengthened southerly or southeasterly flow. North of 25° or 30°S the eastern littoral benefits from a more persistent windward location in the easterlies. Most of the cold-front disturbances instead of progressing northward along the coast are carried eastward by the westerly flow and therefore away from the continent, so that their rainfall effects are minimized. This reflects a less active exchange of air between high and low latitudes east of Africa than east of South America where cold fronts follow the coast well into the tropics. As in tropical East Africa, a considerable part of the movement of maritime air, both from the north and the south, is parallel with the coast or only obliquely onshore, so that the advection of moisture is lessened and the stowing effect reduced. In such meridional currents, the low rainfall in the Limpopo and Zambezi valleys might be viewed as rainshadow effects. It has been suggested, also, that the Limpopo dry area is situated in a kind of no-man's-land between the rain-bringing pertur-

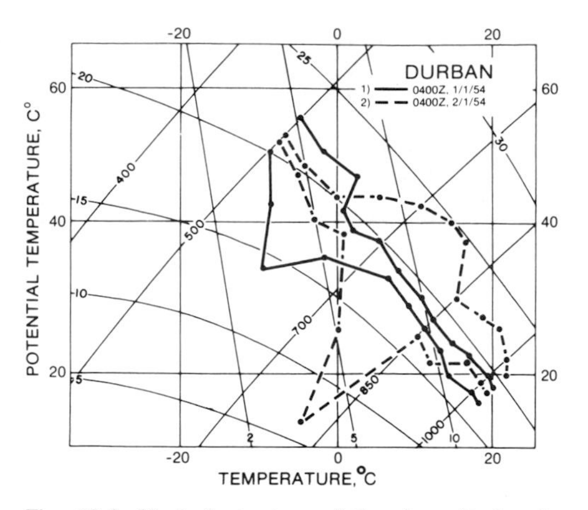

Fig. 10.2. Vertical structure of the air at Durban in summer. At this season stable stratifications are less frequent, and they occur at higher levels than in winter, so they decrease upward from the surface in rate of occurrence. The maritime layer is 2–3 km thick at this time and inversions and quasi-isothermal layers occur most frequently between the 800 and 900 mb levels. (After Taljaard, 1955.)

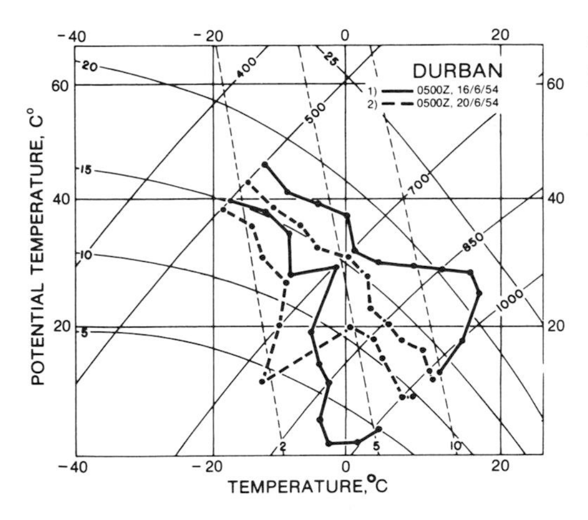

Fig. 10.3. Vertical structure of the air at Durban in winter. At this season the maritime layer normally is less than 1 km in thickness and inversions occur most frequently near the surface and decrease upward. Dry subsident air prevails above the maritime layer. (After Taljaard, 1955.)

bations of the westerlies, which ordinarily terminate in southern Mozambique, while tropical precipitation associated with the *ITC* normally does not reach this far south. Significantly, the Limpopo zone of reduced rainfall is conspicuous in all seasons except winter, when the whole eastern region is dry.

A minor puzzling feature on the annual rainfall map of eastern South Africa is the occurrence of a somewhat irregular and discontinuous belt of decreased rainfall inland from the coast, in spite of an increase in elevation. Along a narrow belt of littoral, 15–30 km broad, rainfall is mostly over 1,000 mm. Further inland it drops below 1,000 mm and there are a number of closed isohyets delineating irregular areas, most of them with a long axis oriented NNE-SSW and hence at right angles to the grain of the relief, where rainfall is as low as 800, 700, and even 600 mm. In a very few instances the spots of lower rainfall can be correlated with terrain features, but this is not true of most of them. Farther seaward, along the coast, the belt of somewhat heavier rainfall may reflect convergence associated with the sea breeze and the frictional effects of the land surface.

The Namib.—The Namib coastal desert of southwestern Africa belongs to that unusual class of deserts (*Bn*) typically located along the tropical and subtropical western littorals bordered by cool ocean currents and cooler upwelled water. Like *Bn* deserts in general, the Namib is characterized by intense aridity, negative temperature anomalies, small annual and diurnal ranges of temperature, and a high frequency of fog or low stratus with drizzle. Within the *Bn* class of deserts, that of southern Africa is distinctive in that, after the Chilean-Peruvian, it is latitudinally the most extensive and also the most arid. An annual rainfall of under about 120 mm is maintained along a narrow coastal strip some 30–60 km in width over a latitudinal stretch of about 15°, and while this may be 10° short of the dimension of a similar arid belt in Chile-Peru it is, on the other hand, much more than the latitudinal spread of those in western North America and northwestern Africa. Three stations, Swakopmund, Luderitz Bay, and Walvis Bay, record less than 25 mm of annual rainfall, so that the most intense aridity seems to be concentrated between about parallels 22° or 23° and 27° or 28°S. Lowest precipitation amounts appear to coincide with the coastline and there is a slow increase inland, with the isohyet of 25 mm paralleling the coast about 60 km inland and that of 100 mm about 110 km inland.

Most of the same weather controls described earlier for the desert of Chile-Peru are operating in the Namib but at somewhat different intensities and with slight locational contrasts. The locationally stable South Atlantic high is centered at about 32°S making for a maximum subsidence along the coast at about 25°S. With the cell's eastern end terminated abruptly at the coastline, smoothly curved atmospheric and oceanic circulations develop along the coast with accompanying upwelling. Most likely the upwelling is localized, for the circulation along the coast appears to be in the form of local whirls with interlocking tongues of warm and cool water (Currie, 1953) To what extent the localisms in oceanic circulation and temperature are paralleled by localisms of climate is not known. The location of the coolest water along the coast oscillates over a distance of about 110 km between Hondeklip, 200 km north of the Orange River mouth, and Port Nolloth. Where the land projects westward so that it maintains contact with the southerly circulations of air and cool water around the eastern margin of the anticyclone, pronounced drought prevails. Where the land recedes eastward so that contact with these air and water circulations is weakened, rainfall increases rapidly, as it does northward from Mossämedes (Lydolph, 1957). Supplementing the stabilizing effects of anticyclonic subsidence and cool water is the coastline subsidence resulting from the differential frictional effects of land and water upon the southerly air stream roughly paralleling the coast. Bryson and Kuhn (1961), who have computed this coastal divergence in southwest Africa, find that it is present both in winter and in summer, and is particularly strong in winter.

Soundings at Walvis Bay and Mossämedes indicate that the maritime air commonly is not over 500 m deep and is capped by a strong inversion with dry subsided air aloft, a situation reflecting extreme stability (Taljaard, 1955). Rain rarely, if ever, falls from the overlying deck of low stratus, but it does generate a drizzle called the *moltreen* or moth-rain. It cannot be measured by a rain gauge, but filter-paper measurements indicate that it may be equivalent to about 47 mm a year (Wellington, 1955).

North of about 16° or 17°S the coast, which farther south bends westward into the major atmospheric and oceanic circulations, changes its direction and recedes eastward and away from these circulations. But while the main Benguela moves away from the continent in a broad arc to the northwest, a weakened branch of the main current continues northward closer to the coast. As a consequence inshore upwelling wanes. North of about 16°S, therefore, the curve of precipitation plotted against latitude begins to rise very rapidly, so that while Mossämedes at about 15°S has only 50 mm of annual rainfall, Luanda at about 9°S has 339 mm and Banana at 6°S, 857 mm. Nevertheless the weakened aridifying effects of the minor cool current along the coast are carried deep into equatorial latitudes, perhaps even north of the equator to the Accra area. In the section north of about 16°S the east-west rainfall gradients become excessively steep, for a short distance inland from the coast heavy tropical rainfall prevails. Thus, while Mossämedes on the coast at 15°S records only 50 mm, Sa' da Bandeira, inland about 160 km, experiences 887 mm.

That the cool waters of the main Benguela are an important control making for aridity along the western coast of southern Africa, is suggested by the fact that Ascension Island (7°57′S, 14°22′W), which is in direct line with the main cool-water oceanic circulation as it diverges from the coast at Cape Frio, has a meager 140 mm of precipitation. This is only one-fifth of the rainfall characteristic of the mainland coast in the same latitude. Accordingly, Ascension Island in the South Atlantic forms a counterpart to the Galapagos Islands in the Pacific west of Ecuador. Both of them represent a continuation of the dry climate of the littoral in a direct line with the major cool-water circulations which have moved away from the coast.

The fact that intense aridity is continued nearly 10° farther equatorward in western

South America than in southwest Africa, reflects the fact that the westward-bending Peruvian coast continues to maintain strong contact with the principal subtropical atmospheric and oceanic circulations to within about 5° of the equator. This is not true to the same degree in Africa where the coast has a more variable alignment.

PART IV
Southern and Eastern Asia

CHAPTER 11

Monsoon Circulation and the Indian Subcontinent

Seasonal Atmospheric Circulation Patterns in Monsoon Asia

So much of the literature on the climates of both southern and eastern Asia has been written in terms of a simple thermally induced monsoon-type system of surface winds, that it seems appropriate to preface the analysis of Indian climates with a brief summary of the recent concepts concerning these seasonal atmospheric circulations as background for the climatic discussion to follow. And since the elements of the circulation for the extensive area including both southern and eastern Asia are closely interrelated, it is believed that efficiency will be served by viewing the broadscale lineaments of air flow for this region as a whole, and subsequently adding such details as are necessary when each regional subdivision is considered separately.

Although the last few decades have witnessed important additions to our knowledge of individual features of the so-called monsoon circulations of southern and eastern Asia, there is still much that is not well understood. The known fragments do not always fit together to form a coherent whole.

The original concept of monsoon, as first described by Edmund Halley in 1686, visualized a thermally generated, closed, and stationary atmospheric circulation system, where the direction of flow reversed between the extreme seasons of winter and summer. This seasonal reversal in wind direction was believed to have its origin in the contrasting thermal properties of continents and oceans, both in winter and in summer. Thus, the seasonal rhythm of the monsoon winds was thought to be similar in origin to the diurnal rhythm of land-sea breezes. So, in summer, as in daytime, the surface wind flow is from cooler sea to warmer land (aloft the reverse is true), while in winter, or at night, the opposite prevails and the surface air flow is from cooler land to warmer sea (again, reversed aloft). In each case, it was in the nature of a convective system. Consequently, onshore summer monsoon and daytime sea breeze were both considered to be moist, potentially unstable, and rain prone, while the seaward-directed winter monsoon of land origin, like the nighttime land breeze, was considered to be dry, stable, and usually rainless. Thus, in time monsoons came to be associated more strongly with air masses and seasonal rainfall than with a particular type of atmospheric circulation. In India, for example, the term monsoon now usually refers to the main rainy period of summer, but including as well that of late fall on the southeast coast of the Indian peninsula. Yet the seasonal shift in prevailing wind direction does not necessarily coincide with the beginning and end of the rainy season.

In reality, the seasonal reversal of wind direction, which is conspicuous in many tropical regions including South Asia, northern Australia, Brazil, and the Guinea Coast–Sudan region of West Africa, is essentially a consequence of a latitudinal migration of the planetary wind belts following the annual course of the sun. Thus easterlies, or trades, prevail at the time of low sun, and equatorial westerlies at the period of high sun. Mainly it is *middle-latitude* East Asia, where land-water seasonal temperature contrasts are strong, that has a monsoon circulation resembling the classical model.

Moreover, the classical or thermal school of thought regarding monsoons fails to involve important synoptic problems. It scarcely explains the fact that while the lands of southern Asia heat regularly each summer, the onset of the monsoon is irregular in time of occurrence and strength and is regionally variable. Also in the southern and central parts of the subcontinent May is the month of highest temperatures, yet the southwest summer monsoon normally does not come to dominate that region until mid-June. It warrants noting, too, that in South Asia genuine density fronts are only occasional features in rainfall formation. They are not permanent or year-round features as in middle latitudes. In India, fronts are not significant features in predicting daily weather.

By now it is well recognized that the Asiatic monsoon is neither a single, local, or regional phenomenon, nor mainly one of convectional origin. It is, rather, a gigantic three-dimensional aspect of the general atmospheric circulation and is intimately related to features of the middle and upper troposphere, such as jet streams and quasi-stationary troughs and ridges. Moreover, it is greatly influenced also by terrain features, especially high mountains and plateaus such as the Himalayas and Tibet.

Winter Circulation.—Anticyclones dominate the weather over most of eastern and southern Asia during the cooler months. Over southern Asia these are of the deep, dynamic, subtropical type, while a shallower thermal high, with its center in the Baikal (USSR) area, prevails over eastern Asia. As a consequence, the low-level flow is mostly from a northerly direction, much of it having had a land trajectory (Fig. 11.1). Chiefly the southernmost parts escape the effects of continental northerly air, and by contrast are dominated by the zonal trades. Over eastern Asia, and including the northern parts of the Indochina pennisula, the vigorous northerly flow is derived from the Siberian anticyclone and is relatively cold, although its temperature may vary considerably depending on the length of its land or sea trajectory. Farther west in India and Pakistan, which are protected from large-scale invasions of Siberian air by highlands of great altitude, the weak northeasterly flow is subsident air from the subtropical anticyclone and from the locationally fixed subtropical jet stream just south of the Himalayas, and is essentially a continental flow, although farther south it becomes more maritime in character. Two principal surface

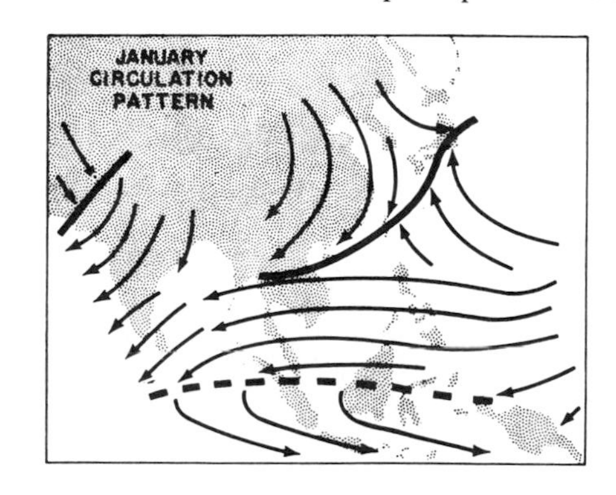

Fig. 11.1. Main features of the low-level circulation over eastern and southern Asia in the cool season.

convergence zones are present in winter. Along the southern and eastern margins of the Siberian air, where it makes contact with the maritime tropical easterlies of the North Pacific, is the fluctuating polar front. South and west of Japan, this convergence is one of the world's most active regions of cyclogenesis, but it becomes progressively less effective in its westernmost parts. A second fragment of the polar front is to be found in northwestern India-Pakistan where northwesterly continental air invades the realm of the Indian trades. Not infrequently these westerlies and their front extend well down the Ganges Valley. Weak depressions develop along the front in northern India, providing a modest amount of winter rainfall.

At higher levels the winter circulation pattern is different in important respects (Fig. 11.2). Because the Siberian anticyclone is relatively shallow, at an elevation of some 3,000 m the zonal westerlies prevail across eastern Asia and even extend their influence over southern Asia southward of Tibet, reaching down to the surface in northern India. As the zonal westerlies shift southward

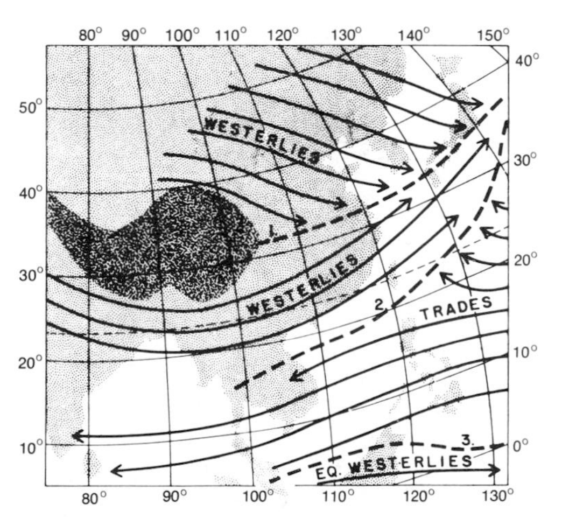

Fig. 11.2. Characteristic features of air flow over eastern and southern Asia at about 3,000 m, November to March. 1. Tibet lee-convergence zone; 2. polar front; 3. *ITC*. (After Thompson, 1951.)

in winter following the sun, they are increasingly obstructed by the Tibetan Highland and its higher flanking mountains, which act in such a way as to bifurcate the westerly air stream in winter, causing a southern branch of the westerlies and its jet stream to wrap themselves around the southern flanks of the Himalayas, while the main westerly stream and its jet continue to flow eastward on the northern side of the terrain barrier. The two branches of the upper westerlies, one to the south and the other to the north of Tibet, converge again to the east of the obstructive highlands, to form the Tibetan Lee Convergence Zone, this confluence feature aloft signifying subsidence at lower levels. Two other significant discontinuities are present at about the 3,000 m level, one of these, an upper polar front, between the southern branch of the zonal westerlies and the North Pacific trades, and the other, in equatorial latitudes, separating these same trades and the equatorial westerlies. The upper polar front is remarkably steady in its position, while its low-level counterpart fluctuates widely as determined by the seaward surges of Siberian air. Weatherwise, the equatorial discontinuity separating trades from equatorial westerlies is not very active at this season, but there is a marked contrast in weather on either side, with stable conditions to the north and unstable to the south.

Summer Circulation.—In summer the low-level circulation is greatly in contrast to that of winter, but at the same time there is more similarity between that at the surface and the one aloft (Fig. 11.3). The summer circulation represents the combined functioning of three of the earth's great planetary winds: the middle latitude westerlies, to a modest degree; the tropical easterlies or trades; and the equatorial westerlies (SW monsoon), most of all. At this season the middle latitude westerlies and associated subtropical jet stream are for the most part

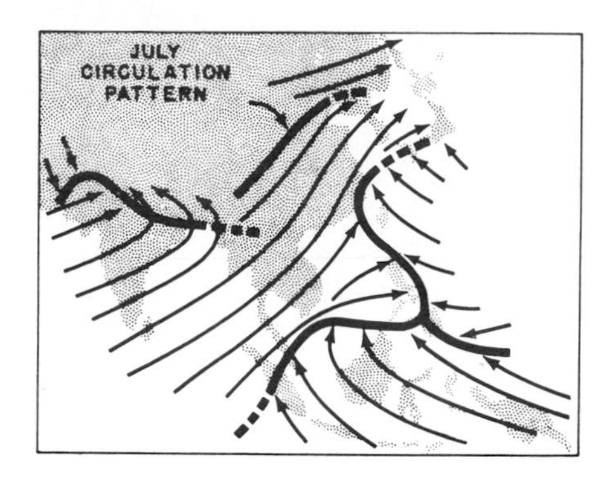

Fig. 11.3. Principal elements of the low-level circulation pattern of eastern and southern Asia in the warm season.

north of Tibet, but surface westerlies are present regularly in the extreme northwestern part of the subcontinent, and on occasion westerlies and jets reappear briefly south of the Himalayas in the Ganges Lowland. Tropical easterlies are especially conspicuous to the north of the monsoon pressure trough in northernmost India, and equatorial westerlies prevail over the rest of the subcontinent. This latter circulation is commonly designated the southwest summer monsoon (Figs. 11.4, 11.5, 11.6, 11.7). At this season the zonal westerlies and the southern branch of the subtropical jet have disappeared, except on occasions, from south of Tibet. A deep and extended surface pressure trough, at least partly thermal in origin, extends across northern India-Pakistan into Indochina China and even China. This is a part of the intertropical convergence zone which here reaches its maximum poleward displacement for any part of the earth. The heat low over north-

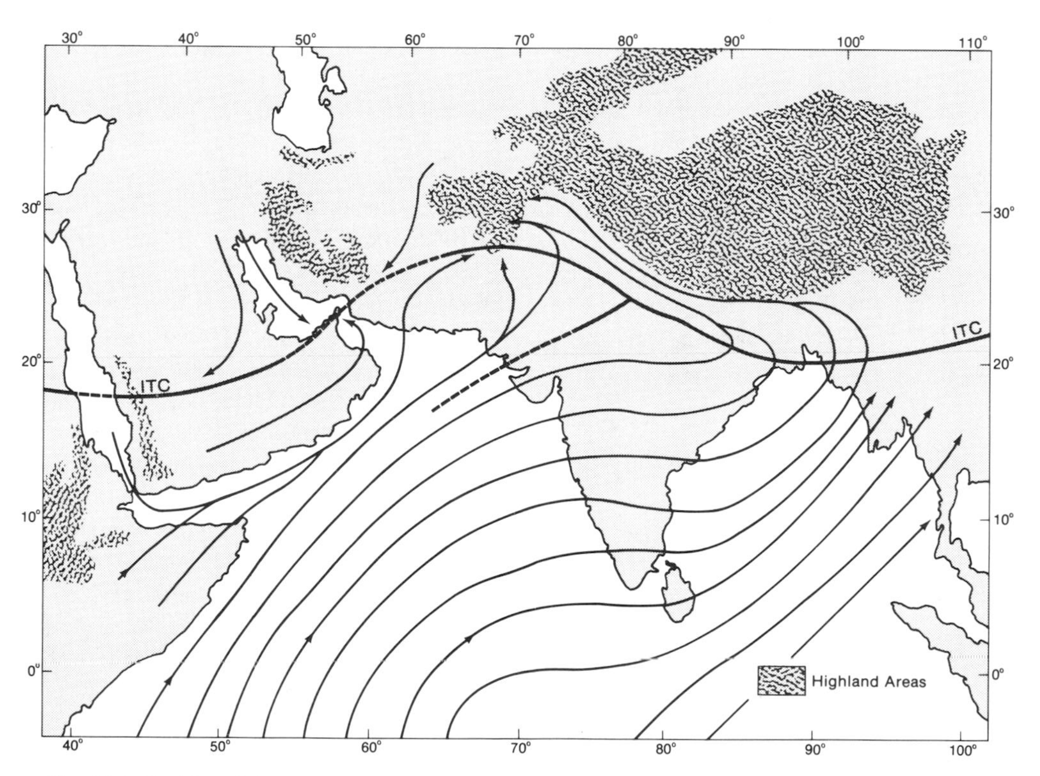

Fig. 11.4. Position of the monsoon pressure trough (northern *ITC*), and streamlines of the resultant winds at 0.9 kms, over southern Asia in summer. (After Flohn, 1970*b*.)

west India-Pakistan is restricted to the layer below 700 mb, for aloft the anticyclone continues to persist. South of the trough is a deep and relatively unstable current of southwesterly maritime air, usually called the southwest summer monsoon, but actually only the northward-displaced equatorial westerlies. To the north of the India trough is an easterly current of maritime air which forms a zone of discontinuity with the southwesterly flow to the south. In the far northwest, dry continental air from western Asia converges with the southwesterly monsoon current.

The same potentially unstable southwesterly current of maritime air that crosses India continues eastward over the continental parts of Southeast Asia and then northeastward over much of eastern China and also Japan. It is a principal humidity source for the summer rainfall of southern and most of eastern Asia as well. Over northern China the southwesterly current makes contact with Siberian air along a widely fluctuating polar front. On its eastern flanks it converges with the deflected North Pacific trades which may be somewhat more stable than the southwesterlies. This latter convergence is not a very active weathermaker. It should be emphasized that the net water vapor transport over eastern Asia in summer is not mainly from the east or the Pacific (except for Japan and perhaps eastern China), but instead is from the southwest and hence from the Indian Ocean, which is contrary to the generally held concept of an easterly summer monsoon in these middle latitudes.

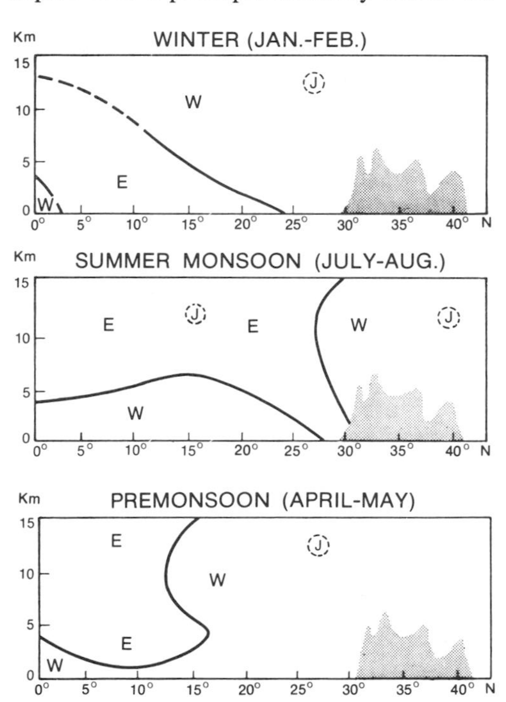

Fig. 11.5. Schematic profile of the planetary wind belts along meridian 78°E over India in the different seasons. J, jet stream. (After Flohn, 1970*b*.)

Seasonal Circulation and Weather

Cool-Season Weather

Winter is the season when, because of the prevalence of anticyclonic subsident air over the subcontinent, the weather element is feeblest. A cross section of the atmosphere along about the 78°E meridian (Flohn, 1956*a*) shows the southern branch of the subtropical jet stream to be located over northern India just south of the Himalayas, with the middle latitude westerlies reaching down to the surface, or nearly so, north of about 25° (Fig. 11.5). Within the upper levels of this westerly circulation, over the northern parts of the subcontinent, is to be observed a succession of eastward-moving troughs and ridges. The low-level convergences associated with these mobile troughs not infrequently lead to strong convective activity with associated cumulo-nimbus clouds and some precipitation. Over lowlands, the dry air evaporates some of the falling rain before it reaches the ground, but in the highlands to the north a durable snow cover is formed, resulting in the creation of a cold source. South of about

25°–29° a northeasterly flow prevails, the so-called winter monsoon, but in reality it is locally subsident air and hence, a land trade wind. To this normal anticyclonic subsidence, there is added a component derived from the southern flanks of the orographically located and positionally stable jet at high altitudes. Thus, the northerly flow of dry continental air over central India is not a monsoon in the usual meaning of that term, for it is not a consequence of the differential heating of land and water. In no sense is it Siberian anticyclonic air, for such an invasion is successfully blocked by the mountain ramparts of central Asia. South of 15° to 18° the easterly current becomes progressively less continental and increasingly derived from the maritime trades, so that humidity increases. In the extreme south in winter, equatorial westerlies are felt on occasions when the fluctuating discontinuity separating maritime trades and equatorial westerlies shifts northward.

From what has been just said it becomes understandable why in winter over the subcontinent as a whole clear skies and fair weather prevail and rainfall is meager or even absent. Only at the northern and southern extremities where weak disturbances associated with wind discontinuities are present is the weather element significant and cloud and rainfall appreciable (Fig. 11.6, January).

WESTERN DISTURBANCES IN THE NORTH.—The modest winter rainfall over northern India-Pakistan is associated with disturbances which enter the area in the extreme northwest, or originate there, and subsequently take a course which approximates that of the jet stream across northern India, Southeast Asia and into south China. These disturbances reach their maximum development in winter when the jet lies south of the highlands, but they also occur, though less frequently, in fall and spring, and as above-surface waves their southern extremities may not be completely absent even in summer. On the surface synoptic charts the western disturbances usually first appear in the northwest in the vicinity of the surface polar front. There relatively cold continental air from eastern Europe and western Asia breaks through the lower highlands and spills out on to the Indus lowlands. The front formed between the modified *cP* air and the dry but warmer Indian air may not in the beginning be very active weatherwise. But if the resulting incipient depression acts to pull in a vigorous inflow of humid *mT* air from the Arabian Sea and the Bay of Bengal, the convergence is likely to produce extensive light rains. Some of the western disturbances appear to be weakened counterparts of storms that originated farther to the west over the Mediterranean Sea or even western Europe.

Not all of these perturbations are evident on the surface synoptic charts, but exist rather in the form of waves in the high westerly current over northern India. Although they are middle-latitude disturbances, a goodly proportion do not have well-developed cold or warm fronts, either at the surface or aloft. Significantly, since about 1920 when upper-air soundings made their detection more possible, a larger number of western disturbances have been reported (See Table 11.1).

Although the primary disturbances characteristically keep to a track which approximately follows the Indus-Ganges Lowland, not uncommonly they are associated on their southern margins with secondary perturbations which may extend the rain area of the winter storms well southward of the main track of the primary disturbance, so that the northern parts of the Deccan (north of 20°N) receive some winter precipitation as well as the Indus-Ganges Lowland. The areas most commonly receiving widespread rains from western disturbances are Kashmir, Punjab,

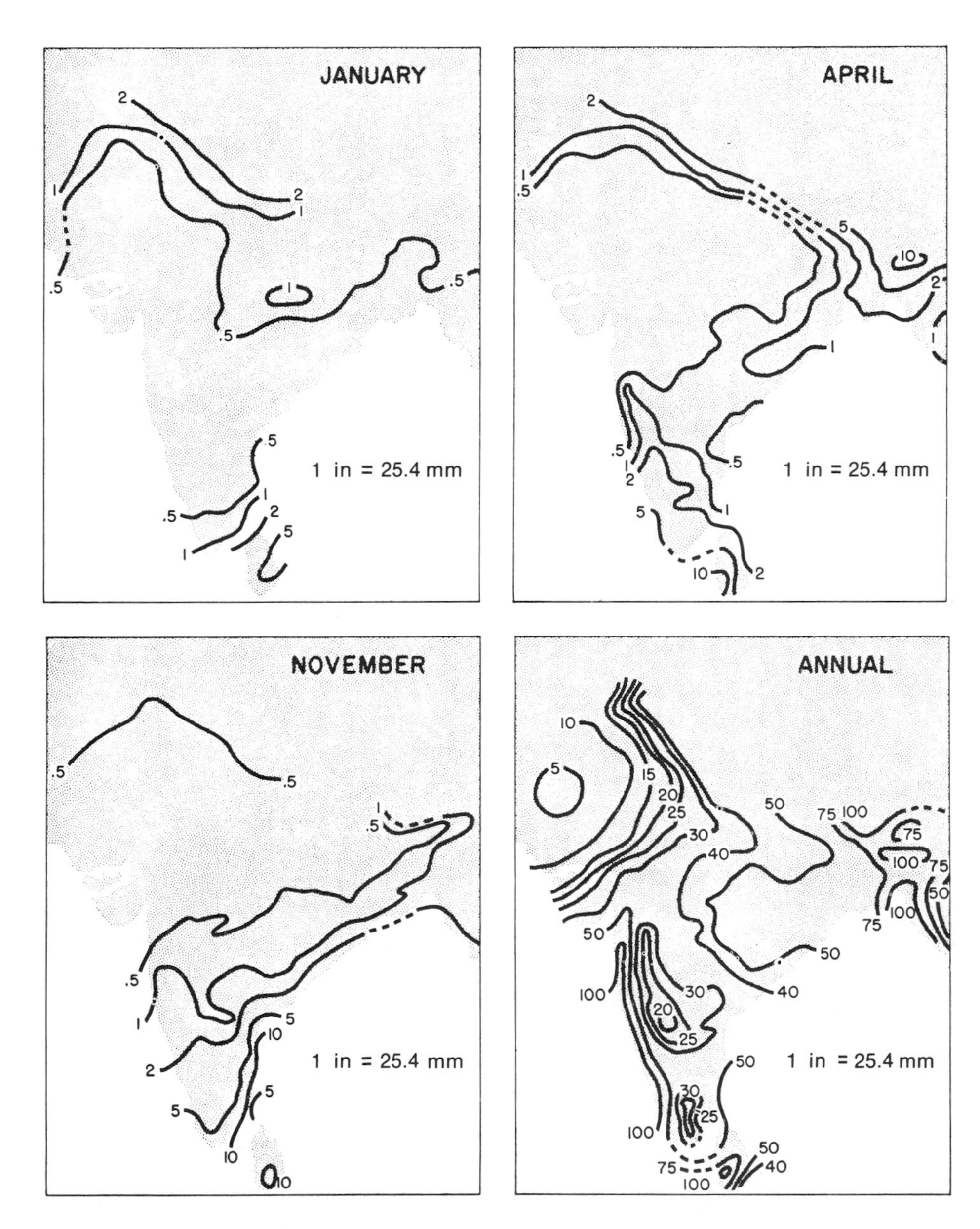

Fig. 11.6. Average rainfall (in inches) in the Indian subcontinent for a winter, spring, and fall month and for the year as a whole. Summer rainfall distribution resembles that for the year.

and northwestern Uttar Pradesh. But although the winter rainfall of the western disturbance is largely confined to northern India-Pakistan, individual storms, during the passage from the Punjab to Bengal, involving three to five days, may produce very different rainfall patterns. The belt of winter rainfall produced by the western disturbances can be traced along the Himalayas from Kashmir through Nepal and Assam

Table 11.1. Average number of western disturbances per year affecting northern India

Nov.	Dec.	Jan.	Feb.	Mar.	Apr.	May
2	4	5	5	5	5	2

into upper Burma. In this belt the monsoon rains of summer overlap the winter rains of the western disturbances. No dry season separates the two rainfall regimes (Flohn, 1968).

The precipitation generated by these western disturbances of the cooler seasons is usually fairly widespread and light to moderate in amount. It is locally heavy principally where thunderstorms are associated with the disturbance. Since it falls in the cool season when losses from evaporation are relatively low, this precipitation is highly effective for the growth of winter crops. On the plains the total fall for the three winter months usually is only 25 to 75 mm, and yet this is of vital importance. Such storms also provide a much larger amount of winter precipitation in the form of snow in the highlands, whose melt-water in spring and summer furnishes the indispensable irrigation water for the Indus-Ganges Plain. Over much of the northern part of the subcontinent annual rainfall curves of individual stations show a very modest secondary maximum of precipitation in winter, the origin of which is the western disturbances and their secondaries. Compared with the primary summer maximum, the winter secondary may appear rather feeble over most of the northern subcontinent. However, in the extreme northwest the winter rains may equal or even exceed those of summer. Thus, Peshawar has almost as much rain in winter as in summer and in Kashmir winter is rainier than summer, although March and April are the wettest months (Fig. 11.15).

Western disturbances have temperature as well as rainfall effects. As such a perturbation moves across northern India, warmer and moister southerly air precedes it, while cooler, drier air from northerly latitudes follows. Relatively deep depressions are often followed by genuinely cold northwesterly air from considerably higher latitudes which spreads east and south in the form of a cold wave. Such invasions of cold air following a western depression are essential to the maintenance of a cool season, while the associated cloud cover of this type of disturbance likewise acts to decrease the daily maximum temperatures (Mooley, 1957).

COOL-SEASON WEATHER IN THE EXTREME SOUTH.—Even in winter southernmost India and Ceylon Island lie in sufficiently close proximity to the equatorial westerlies and one or more surface and above-surface discontinuities so that they are subject to weak tropical disturbances, very occasional tropical cyclones, and active convectional systems. Certainly some of these weaker disturbances resemble the monsoon depressions which in summer are conspicuous in the northern part of the Bay of Bengal and over northern India. As the equatorial westerlies retreat southward with the sun, the tracks of the monsoon depressions do likewise, and in winter they are present only in the extreme south. At this season the eastern sides of Ceylon Island and southern India have more rainfall than the west, suggesting the prevalence of easterly winds and the concentration of monsoon depressions in the southwestern part of the Bay of Bengal.

Spring Weather—The Pre-Monsoon Period (March-May)

This is sometimes described as the hot-dry period to distinguish it from the cool-dry season of winter and the hot-wet season of summer and early fall. Main circulation features resemble those of winter. In March,

April, and early May the anticyclonic subsidence and clear skies characteristic of the winter months still prevail and this in combination with a much stronger sun sets the weather pattern for the season. Temperatures are excessively high and a heavy, dry haze envelops the interior, but drought still grips most of the country.

Still, there is some premonition of change in the basic circulation. With the northward advance of the sun and an increase in solar radiation, shallow lows begin to appear over southern peninsular India. These features gradually expand northward as the season advances until by June an intense heat low is positioned over the northwestern part of the subcontinent. On the southern flanks of the thermal lows weak and shallow westerlies develop, but they are lacking in much weather. Over the Bay of Bengal and the lands bordering it on the north, moist southerly winds begin to replace the dry northeast monsoon.

Actually the areal rainfall distribution in March or April is not fundamentally different from that of winter (Fig. 11.6). The central latitudes where maximum subsidence prevails are still the driest part of the subcontinent and the two extremities, the south and the northeast, are the wettest. The most perceptible areas of rainfall increase are (1) the extreme south, (2) the northeast, and (3) the seaward margins of peninsular India. In the far south the increased rainfall reflects the slow northward creeping of the *ITC* and the equatorial westerlies.

The increased premonsoon rainfall in the northeast is mainly provided by the western disturbances. Such perturbations are unable to produce much precipitation in the hot dry air of the northwest. But in the more humid air of the northeast, at the head of the Bay of Bengal where a shallow southerly flow provides a supply of moisture, the western disturbances produce a goodly amount of showery precipitation. Low-level convergence, importantly aided by orographic lifting, are contributors. As a result the region including West Bengal, Bangladesh, Assam, and northern Burma shows a greatly augmented rainfall during the spring months. Much of the Ganges Delta region records 50 to 120 mm in April and 120 to 250 mm in May, and the amounts are much greater in the hills and mountains.

These rains of the premonsoon season precede any marked reversal of winds aloft. But the greatly increased heat of condensation associated with this increased rainfall in the northeastern part of the subcontinent is a preparatory feature for the advent of the summer monsoon.

The increased rainfall in spring along the seaward margins of peninsular India has its origins in the onshore movement of well-humidified sea air attracted toward the southern flank of the shallow heat low over the land. Surface heating and terrain obstacles produce upward movement of this air, resulting in showery rains along a littoral belt and over the hills for some distance inland. It warrants reemphasis that this increased rainfall in April-May, within the three regions mentioned, precedes the development of the Tibetan high-level anticyclone of summer.

Spring especially is characterized by violent types of weather in the form of thundersqualls in the northern part of the subcontinent (Table 11.2). In the drier sections of northwest India-Pakistan the rainfall accompanying these vigorous convective systems is slight, but they do produce well-developed cumulo-nimbus clouds, strong squall winds, and violent dust storms. Here they are known as "andhis." In the more humid lower Ganges valley they are designated as "nor'westers" and such storms are accompanied by heavy shower rainfall. Nor'westers are most numerous in the spring months and there is a marked decrease with the onset of the summer monsoon. A large majority (70 to 80 percent) occur simultaneously over extensive areas and they are highly concentrated between the hours of 1700 and 2200. Most of them travel from northwest to southeast.

In spite of the fact that there is a considerable literature dealing with the origin of andhis and nor'westers, the topic is still a controversial one. Probably more than one cause is operative. A few of the nor'westers appear to be of the cold-front variety occurring in zones or belts along lines of discontinuity. An occasional one may be associated with a monsoon depression as it moves toward the coast from the Bay of Bengal. But according to one group, three-quarters of the nor'westers are linked with the approach of western disturbances or with the accentuation of east-west pressure gradients over Bengal (Desai, 1950; Newton, 1951). With the approach of a western disturbance in the Lower Ganges Valley a situation develops in which a moist southerly or southeasterly current from the Bay of Bengal is overrun by dry northwesterly air. During the hours of maximum heat the lapse rate is much greater in the dry than the moist air, so that at the separation layer of the two air masses the dry air is cooler than the moister air beneath. This temperature contrast is not so marked at other times of day. The great instability at the separation layer during the afternoon hours acts as a trigger for releasing the latent energy. When the head of a cumulus cloud developing in the moist lower air reaches the surface of separation there is an almost explosive growth into cumulo-nimbus, the severe downdrafts from this cloud producing the violent squalls of the nor'wester. Such developments are absent in winter for at that season moist air is lacking at lower levels. They are fewer during July and August at the height of the monsoon for then the clouded skies reduce insolational heating and the upper current is moist instead of dry.

Table 11.2. Average number of thundersqualls per month in Delhi and Calcutta (Ramaswamy, 1956)

	Mar.	Apr.	May	June	July	Aug.
Delhi	0.6	2.8	7.0	4.2	2.0	0.4
Calcutta	2.6	3.6	5.2	2.4	0.2	—

Ramaswamy (1956) has expressed doubt that explanations for nor'westers and andhis can be found exclusively in lower troposphere phenomena. He is convinced that they are linked with vorticity patterns at the 500 and 300 mb levels. The actual convective developments, he finds, take place in advance of a trough or at the rear of a ridge at the 300 mb level. Other things being equal, thunderstorms are found to be more numerous and extensive, and the squalls more severe, when the jet stream is most intense. Fair weather prevails when winds are unusually light in the upper troposphere.

Bose (1957), likewise, is convinced that factors other than low-level convergence are equally, if not more, decisive in nor'wester genesis. Koteswaram and Srinivasan (1958) have sought to discover the relative importance of high-level and low-level conditions for the formation of nor'wester weather and they arrive at the conclusion that both are essential. Low-level convergence alone is not sufficient to trigger off the strong convective processes if simultaneously there is no high-level divergence to provide for mass removal of the ascending air. According to these authors, only when a condition of high-level divergence is superposed over a low-level convergence zone does nor'wester weather result.

Tropical storms, some of them hurricanes, provide another severe weather type of the spring season which contributes to the rainfall of parts of the subcontinent. Most of the storms which affect the land come from the Bay of Bengal and hence their effects are chiefly felt on that side of the subcontinent, and toward the head of the Bay. The hurricanes of the Bay of Bengal show two seasonal maxima, a primary one in October-November (46 percent) and a secondary in April-May (26 percent).

Weather of the Hot-Wet Period (June-August)—The Summer Monsoon

In India the transition from the relatively dry, and in most parts weatherless, spring period to the more cloudy, rainy season of summer, with its numerous perturbations, is abrupt and usually is associated with strongly disturbed and turbulent weather. The so-called monsoon rains begin over Burma in May or even late April, but in India they are delayed for several weeks, and when they do arrive, beginning in the extreme south, they tend to spread rapidly northward over the country, so that usually by the end of June the southwesterly summer monsoon is established over the whole subcontinent. And while its day-to-day northward advance is bewilderingly complex, its major weather changes follow pretty much the same sequence. Usually the advance northward of the monsoon current over India is accompanied by turbulent weather in the form of thunderstorms and squall winds, but these decrease in number after the summer circulation has been established. The onset of the summer monsoon rains does not coincide with the seasonal reversal of the wind field in all parts. In northeastern India the rains precede the monsoon wind shift; in the far south low-level westerly winds precede the onset of monsoon rains. The monsoon begins to retreat from northern India in late August, and the withdrawal southward continues throughout September and October.

ONSET OF THE MONSOON.—The turbulent advent of the summer monsoon over India, which occurs between late May and early July, depending upon latitudinal location, is recognized as a noteworthy climatic singularity. The origin of the summer monsoon appears to be associated with certain basic changes in the general high-altitude circulation over southern Asia. Maung Tun Yin (1949) has suggested that the later arrival of the monsoon in India than in Burma may result from the fact that during the winter and spring months there is an orographically determined upper trough, oriented north-south at about 85°E, over the western Bay of Bengal, which acts to accelerate the southwesterly monsoon flow over Burma, located east of this trough, while at the same time retarding it over India to the west.

Dey (1977) briefly summarizes the synoptic conditions associated with the onset of the summer monsoon of southern Asia as follows: "(a) the initial formation of the monsoon trough near 95°E at 7000 millibar level and subsequent shift of this trough westward, (b) the displacement of the subtropical westerly jet stream to the north of the Himalayas and the establishment of an easterly jet well south of the Himalayas, (c) the northward displacement of the North Pacific High from 13°N in the first half of May to 23°N by the end of June, and (d) the retreat of the Arabian Sea High from central India to over western Arabian Sea and Arabia." As the subcontinent heats intensively in April and May the zonal westerlies over northern India begin to move northward but are resisted by the mountains. As a result the jet stream, which has been south of the highlands at about 30°N during winter and spring, tends to alternately disappear and then reappear south of the mountains. Disappearances become more frequent as the season advances and each disappearance is associated with a northward surge of the summer monsoon. Finally in late May or early June, the jet largely disappears over northern India and takes up an average position at about 40°N, to the north of the Himalayas and Tibet. Simultaneously there occurs a shift of the low latitude trough and ridge positions, and the upper trough which previously was located at about 85°E quickly moves westward some 10° and takes up a position over western India at approximately 75°E. With the more complete disappearance of the jet over northern India and a westward shift of the upper trough the equatorial westerlies, or summer monsoon, surge northward over India accompanied by unsettled weather.

This is scarcely a monsoon in the original meaning of that term, however, for it is not a result of the differential heating of land and water. Such a northward advance of southwesterly equatorial air would occur in summer in these latitudes even if the tropical-equatorial area south of Asia were entirely land. What is here called a monsoon is mainly the normal seasonal migration of the planetary winds following the sun. Apparently, the heating of the subcontinent and the development of a surface pressure trough are unable to produce a northward advance of the *ITC* until large-scale dynamic features of the circulation aloft become favorable. When the jet reappears south of the Himalayas again in fall, the summer monsoon again retreats southward.

Flohn (1957) relates the disappearance of the westerly jet over northern India and the onset of the monsoon to still another factor. He sees the heating of the surface of the Tibetan Highland in summer as resulting in the mid-troposphere (500 mb) temperatures being higher there than in surrounding areas at this level. Large-scale upward movement of air in this cul-de-sac region of heated uplands, with the consequent release of vast amounts of heat of condensation, also plays an important role. As a consequence, a thermal anticyclone develops over the Highland which in turn creates an easterly flow of air on its southern side that reaches down to low levels over northernmost India. Thus, the summer warming of the air at the 500 mb level over Tibet gradually weakens the western subtropical jet south of the highlands and eventually reverses the gradient and wind flow in the subtropical latitudes over India.

Koteswaram (1958) finds that the burst of the monsoon and the development of an anticyclone over the Tibetan Highlands is synchronized with the appearance of an accelerating easterly jet over India whose mean summer position is approximately 15°N. This probably is not a fortuitous event, but a part of the readjustment in circulation patterns. Like the monsoon, the easterly jet appears to develop first in longitudes east of India and subsequently to extend its influence westward across India and the Arabian Sea to eastern Africa.

Thus, a conjunction of closely interrelated events accompanies the onset of the Indian summer monsoon. But how these changes are connected, and which are cause and which are effect are still not clear. Perhaps of highest importance are the thermal and orographic effects of Tibet and its bordering mountains. From satellite photographs it has been observed that central and southeastern Tibet are almost snow-free throughout even the coolest seasons of fall, winter, and spring. At this high altitude, where already by spring solar radiation is intense, the bare plateau surface becomes superheated, and thereby creates a high-level radiation heat source. Pressure surfaces are thereby raised and a midtroposphere anticyclone is created. Combined with the direct heating of the atmosphere from the elevated plateau surface, there is an additional prime source of energy, namely the release of latent heat in the orographically forced ascent of humid air, as well as in the giant convective cells in the region of southeastern Tibet and its associated mountains. This applies especially to summer but also to spring. The result is a warming of the high troposphere and a consequent weakening of the upper westerlies. Finally, in early June the wind pattern over latitudes about 18° to 20°N reverses direction and the southwest monsoon sets in. Thus, the Tibetan Highland becomes a great dynamic pivot having both mechanical and thermal effects of unusual magnitude. Accordingly, Tibet, together with associated mountains, must, in summer, act as a huge heat engine, with a giant chimney in the southeastern part, where vast amounts of heat are carried upward (Flohn, 1968; 1970*b*).

SUMMER CIRCULATION AND WEATHER TYPES.—To recapitulate, the generalized average summer circulation is cyclonic and is developed around the low-level pressure trough situated over northern India-Pakistan (Figs. 11.4, 11.7, 11.8, 11.9). At higher levels, above the surface trough, is the Tibet anticyclone, which links the circulations of the middle latitudes and the tropics. To the south of the low-level trough is a southwesterly flow of humid and potentially unstable air (southwest monsoon or equatorial westerlies), having a depth of six to seven km in southern India and four to six km in northern India in the vicinity of the Ganges Lowland. A return easterly flow prevails at higher levels. It warrants emphasizing that such a monsoonlike circulation, with westerly flow at low levels and easterlies aloft, is opposite to the normal trade-wind circulation of these tropical latitudes. To the north of the trough, an easterly maritime current moves up the

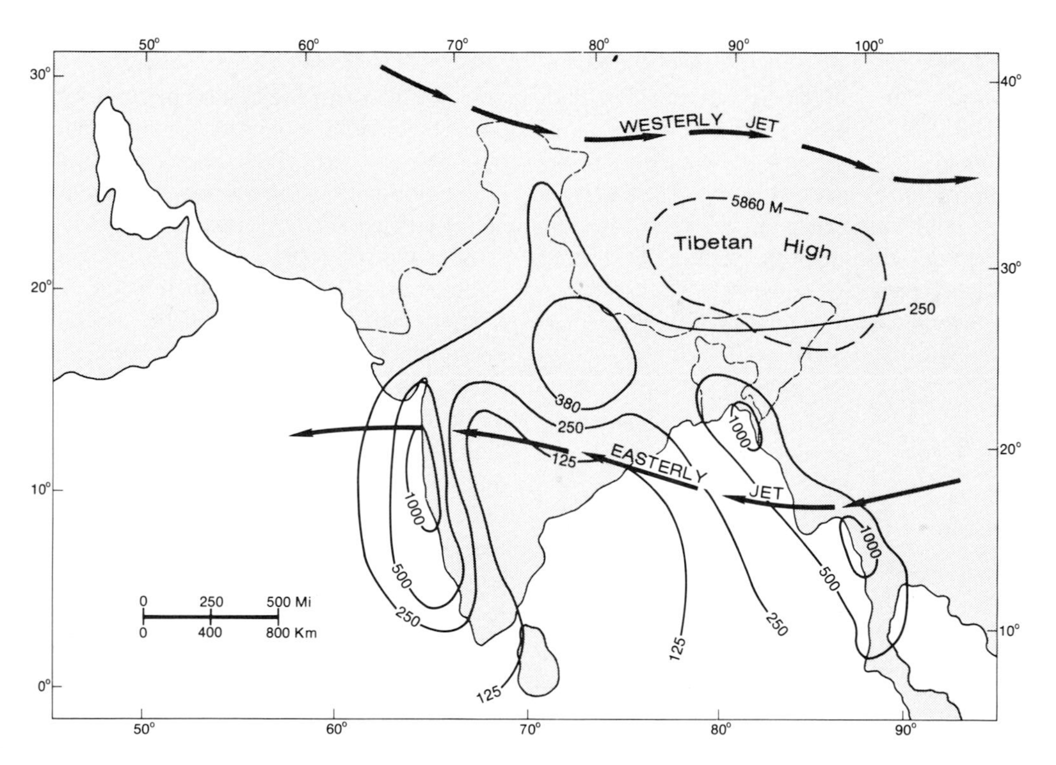

Fig. 11.7. Mean positions of the two jet streams in July–August and mean July rainfall (in mm) over South Asia. (After Flohn and Koteswaram, as observed in Chakravarti, 1968.)

Ganges Lowland. The convergence between the easterly and westerly currents is the *ITC*. In the extreme northwest, dry continental air meets the monsoon current along the *ITC*. The deep southwesterly current of equatorial air exists as two branches. One of these, from the Arabian Sea, meets the elevated west coast of India at almost right angles, with resulting heavy orographic-perturbation rainfall, and subsequently continues eastward across the peninsula in somewhat modified form. A second, or Bay of Bengal, branch has exclusively a sea trajectory. Toward the head of the Bay of Bengal the two branches merge, and some of this air appears to recurve westward around the eastern end of the pressure trough and contributes to the easterly current on the northern side of the trough. To what extent this easterly current is maritime trade wind air from the Pacific is controversial.

These currents are not steady but on the contrary are subject to variations in both speed and direction. Moreover, the variations in speed and wind-flow pattern are accompanied by marked changes in the position and alignment of the *ITC* and its associated disturbances, resulting in shifting weather patterns. Thus, the *ITC*, which on the average is aligned in a WNW-ESE direction over northern India, not only undergoes wide fluctuations in a north-south direction, but on occasions even assumes a north-south orientation, and at other times appears to be an interrupted zone of convergence.

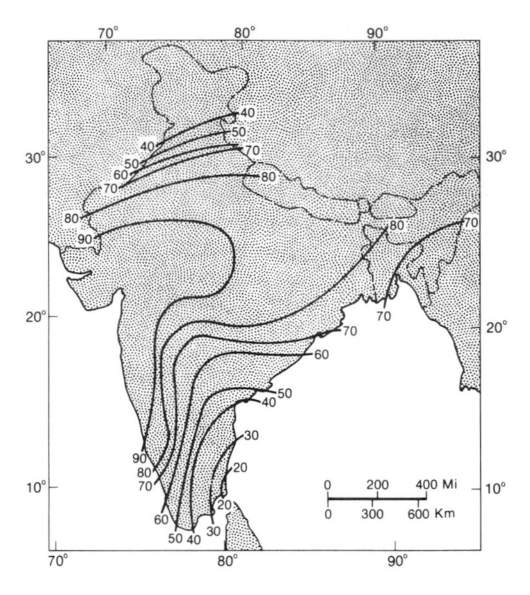

Fig. 11.9. Percentage of the annual precipitation that falls over India during the southwest summer monsoon, June–September inclusive. (From Dey, 1977.)

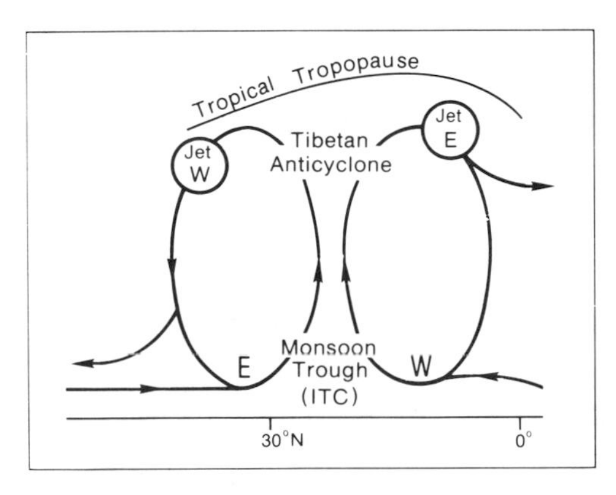

Fig. 11.8. Schematic model of the Indian summer monsoon circulation. (After Koteswaram, 1958.)

With the establishment of the southwest summer monsoon in India, there takes place a rapid increase in atmospheric moisture. Precipitable water vapor nearly doubles between March and June. Whether the prime moisture source is the Arabian Sea and Indian Ocean north of the equator, or the Indian Ocean south of the equator, is a topic that continues to be discussed. Throughout the summer monsoon period the mean monthly precipitable water is very high, and it increases from southwest to northeast, reaching 7.0 gms/cm^2 over Assam in northeast India. It is this high water vapor content in the equatorial southwesterlies that makes possible the relatively abundant summer rainfall over large parts of the subcontinent. Some 80–90 percent of India's annual rainfall occurs during the summer monsoon months, June-September.

To what extent the several currents involved in the summer circulation over India represent genuine density contrasts, and the weather-active convergence zone over the north a bona fide front, is a highly controversial question. Some Indian meteorologists represent the *ITC* as a warm-front type of discontinuity and the numerous disturbances associated with it as having frontal origin. But increasingly weather scientists in India have become skeptical regarding the existence of authentic density fronts and believe it is sufficient to speak of the monsoon trough as merely a convergence zone and the depressions which develop in conjunction with this trough, and produce much of the summer precipitation over India, as nonfrontal in origin.

The southwesterly current of equatorial air which prevails over much of the subcontinent in summer, while it has a high rainfall potential, is not the direct cause of the rain. Its chief role is to provide a favorable environment as regards moisture and temperature stratification in which the rain-making processes readily develop. A steady and uninterrupted southwesterly flow is usually associated with fair weather. In fact, when the southwest monsoon is strong, in a homogeneous field, rainfall is characteristically light over lowlands, although it may be heavy along the windward side of obstructing highlands (Malurkar, 1956).

Two phases of the southwest summer monsoon are usually recognized (1) the active monsoon when winds are more or less steady and continuous, and (2) the so-called "break" in the monsoon when the average circulation is weak (Fig. 11.10). The latter is in the nature of a deviant, with the interruption in the circulation lasting some three to ten days (Chang, 1972). During breaks, when surface pressure gradients slacken, the southwest monsoon weakens and decreases in depth over peninsular India, but it is more active in the far north in the Sub-Himalayan

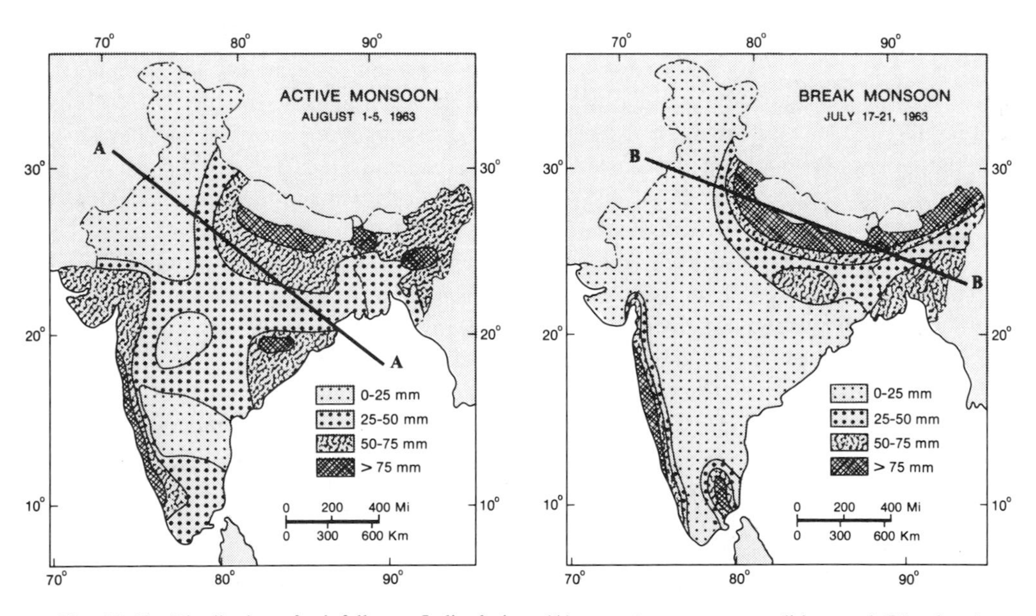

Fig. 11.10. Distribution of rainfall over India during (A) an *active monsoon* condition, and (B) a *break monsoon* condition. The heavy lines show the position of the monsoon trough. (From Dey, 1977.)

Region. Breaks appear to be more frequent in late summer than in July; they also show a tendency to occur most often between the occurrence of monsoon depressions entering from the Bay of Bengal. On the occasion of breaks, there is a northward shift of the general monsoon circulation, with the pressure trough (*ITC*) over northern India being displaced some 250 km northward to the foothills of the Himalayas, while the low-level easterlies in the northern Ganges plain are replaced temporarily by midlatitude westerlies, and there is a reappearance of the subtropical jet south of the highlands. Simultaneously, the tropical easterly jet over peninsular India is displaced somewhat farther poleward. Rainfall totals for the subcontinent may not be much different during weak and active monsoon periods; the main difference seems to be in the regional distribution of the rains. During breaks there is usually more rain farther north, especially in the Sub-Himalaya belt, while farther south in peninsular India it is drier than usual, and the region of heavy rainfall associated with the Western Ghats shrinks in size (Fig. 11.10).

The typical weather sequence during the summer months is one of spells of disturbed weather with organized cloud and rain areas, alternating with intervals of relatively fair weather. Rain, other than possibly that from random convection, or associated with orographic lifting, is almost invariably caused by some form of convection-activating agent in the guise of extensive atmospheric disturbances. These appear to be of several kinds, although they cannot always be clearly differentiated.

Surges in the Southwesterlies.—One such disturbance exists in the form of a surge in the above-surface southwesterly flow. The southwest monsoon is not a steady current, but instead it is not infrequently interrupted by pulsations which move downstream with the main current. During these surges wind velocities may increase to twice what they are normally. Disturbed weather characterized by organized cumulus activity and associated thundershowers accompanies the surges. Both branches of the southwesterly flow experience these pulsations, those from the Arabian Sea traveling northward along the west coast of India and then advancing eastward into the interior of the country. In a similar manner surges in the Bay of Bengal branch bring spells of showery rain to the Burma coast and interior and probably some to southeastern India as well.

Little seems to be known about the origin and structure of the surges in the southwesterly current. They probably are in the nature of speed convergences. There are indications that the surges and their associated bad weather over India and Burma are preceded by the occurrence of more southerly components in surface winds and squally weather near the equator four or five days before the arrival of pulsations of fresh equatorial air in India. This suggests an influx of Southern Hemisphere air (Malurkar, 1950). It likewise has been observed that the strengthened southerly components of the surface winds and accompanying showery weather are associated with shallow low-pressure areas moving westward in the equatorial latitudes just south of the equator, which cross over into the Bay of Bengal or the Arabian Sea. As the surge marches northward in the Bay of Bengal, it appears to be one element favorable for the generation of another form of disturbance known as the monsoon depression to be described subsequently. It seems probable that a large part of the summer rainfall over India south of about 20°N is associated with the passage of these surges in the monsoon. When the supply of fresh *Em* air weakens and the current again becomes steady, rainfall declines and fair weather sets in. Such spells of fair weather in India are associated with, and preceded by, the occurrence of northerly components in the surface winds in equatorial latitudes and an unusual amount of shower activity there, which suggests a retrogression of Southern Hemisphere air from the Indian area.

Ratcliffe (1950) points out that since most bad weather in the tropics is of the instability type, it is not surprising that areas of disturbed weather in India can be correlated with cool lows, and fair weather with warm highs. The bad weather occurs when offshoots of the cool pools over the adjacent seas invade the warmer land areas where heating at the base is usually sufficient to produce large-scale instability. Ratcliffe believes that the cool pools over the waters east and west of India are fed and replenished by surges of cool air from the Southern Hemisphere high latitudes, which accompany the northward advance of cold fronts in the Indian Ocean. He also suggests that the surges in the southwest monsoon may be the result of these Southern Hemisphere cold fronts.

Monsoon Depressions.—Still another type of rain-generating atmospheric disturbance of the summer season is the so-called monsoon depression. It is a medium scale synoptic type which moves into peninsular India from off the Bay of Bengal. Most of these lows are extensive but weak disturbances whose weather effects are largely restricted to the cloud and rainfall which they produce. A few develop into more vigorous storms, and a still more restricted number into the violent hurricane type. Apparently these disturbances of the Bay of Bengal cannot be correctly classified into weak and violent types only. Rather, there are disturbances of all gradations of intensity, though the milder ones are much the more prevalent. In the summer months the monsoon depressions originating over the Bay of Bengal generally take a course toward the northwest which tends to concentrate their rainfall effects over northern, and especially northeastern, India. The abundant summer rainfall of this area is a result. In the transition seasons the direction of their movement is more uncertain, but usually they follow a course which lies between west and northeast. In *late* summer when the *ITC* trough is positioned farther south, the depressions form in the central part of the Bay of Bengal and take a westerly course that routes them inland over peninsular India. It is estimated (Ramage, 1971) that on the average about eight well-developed monsoon depressions from the Bay of Bengal enter India in the periods June–September.

Their origin is not completely clear but they appear to form at the eastern end of the trough, along the *ITC*, especially on those occasions when the trough extends farther eastward than normal and over the Bay of Bengal, and when in addition certain other favoring conditions prevail. One such is the presence of an above-surface surge in the southwesterly current, which carries fresh monsoon air to the head of the Bay. Additional favoring elements are (1) the entrance into the Bay from the Burma area of an easterly wave disturbance, and (2) a strengthening and northward shifting of the easterly jet stream over the northern part of the Bay of Bengal (Koteswaram and George, 1958). The disturbances arriving from the east, with an average frequency of about one every six days, exist as waves in the deep North Pacific trades which overlie the equatorial southwesterly monsoon current in summer. They can be traced westward from the China Sea across the Indochina Peninsula into the Bay of Bengal (Ramanathan, 1955). When the above-surface disturbances, either from the east or from the west, are superimposed upon a pre-existing monsoon trough at sea level the situation is ripe for the development of a monsoon depression. The latter then moves northwestward along the axis of the trough separating the easterly and westerly air streams. Since the position of the *ITC* and of its wind convergence fluctuates, the routes of the depressions do so as well. As the cool pool of maritime air accompanying the monsoon depression invades the warm land, heating at the base supplements the effects of general convergence in such a vortex disturbance, so that widespread instability and organized convective activity result (Ratcliffe, 1950).

Cyclogenesis is first observed at elevations of two to three kilometers, and subsequently it extends downward to the surface. Widespread and locally heavy rains of the shower variety accompany the passage of the disturbance, with the southwest qradrant extending for as much as 400 ± km from the center, receiving the heaviest precipitation.

While the hurricane is not absent from the Indian space in summer, this weather type is less frequent in that season than in the transition seasons.

High-Level Westerly Troughs.—Still another type of summer weather, and one largely confined to the extreme north, is associated with a few weak western disturbances still to be detected at high altitudes. It is true that only when frequent and well-developed western depressions cease to move across northern India is a strong flow of humid easterly air likely to extend up the Ganges Valley from the Bay of Bengal. This westward extension of the easterly current is considered to be synchronized with the disappearance of the southerly branch of the jet stream from its winter and spring position south of the Himalayas. But even in summer after the arrival of the monsoon, an occasional westerly upper trough passing over the highlands to the north of India may affect weather conditions in the northernmost parts of the country. These upper troughs do not develop into surface depressions, so that the Indian Weather Service does not refer to them as western disturbances. Nevertheless, the high-level troughs are associated with changes in the pattern of the general circulation. In June before the onset of the general monsoon, such a trough may cause a temporary invasion of the Punjab and Kashmir by the monsoon current, resulting in widespread rains in that area. The same effects may occur occasionally even during the height of the monsoon in July and August. If several such westerly troughs pass in rather quick succession, the surface monsoon trough shifts northward to the margins of the Himalayas, and moderate to heavy rains occur in the highlands and foothills. Simultaneously there is a cessation of the movement of monsoon depressions from the east over northern and northeastern India in general, which causes a break in the so-called monsoon rains over most of the country (Mooley, 1957).

An additional summer type of disturbance, localized along the west coast of India, is described by George (1956). He characterizes those disturbances as vortices of small diameter, normally 50–160 km, which vary in depth from 300 to 1,500 m. Heavy showery rainfall accompanies these small vortex disturbances as they move northward along the coast. George attributes their origin to the sharp wind shear along boundaries separating local air streams of different velocity and moisture content within the general southwesterly flow. He likewise indicates that the Ghats themselves play an important role in developing the local air streams along whose boundaries the vortices originate, for while no part of the west coast escapes their effects, there are, nevertheless, favored local areas. Although much of the heavy westside rainfall undeniably is orographic in origin, it is significant that even here the rainfall is intermittent and occurs in spells of disturbed weather, suggesting the influence of nonperiodic disturbances.

Fall Weather (September-December)—The Season of the Retreating Monsoon

As the thermal trough over northern India weakens, paralleling a decline in insolation, the flow of southwesterly equatorial air across India and up into the Bay of Bengal likewise weakens and the *ITC*, as well as the paths of the monsoon depressions, retreat slowly southward and rainfall declines in the north (Fig. 11.11). With this retreat of the equatorial westerlies there is a greater intrusion of the North Pacific trades into the Bay of Bengal and over India, so that by late fall the trough of low pressure separating the easterly and westerly air currents becomes established over the southern part of the Bay of Bengal and adjacent southern India. As a consequence the Tamilnad or Coromandel coast of southeast India experiences an onshore easterly flow of air. Along the discontinuity between the equatorial westerlies and the zonal easterlies various kinds of perturbations develop ranging all the way from weak monsoon depressions to hurricanes. The depressions follow less well defined tracks than in summer, but in general their progress is westward, so that their rainfall effects are concentrated in coastal southeastern India, which lies in close juxtaposition to the earth's most active region of cool-season tropical cyclogenesis, located in the southwestern part of the Bay of Bengal (Ramage, 1952*b*). Thus, the November rainfall map shows a distribution in which the heavier rainfalls are oriented toward the southeast and east, where 250 mm are recorded for the southeast coast south of about 15°N. Rainfall declines away from the southeast coast and toward the northwest, with the minimum of about 3 mm prevailing in an extensive area bounded by Karachi, Lahore, Benares, and Ahmedabad (Fig. 11.6).

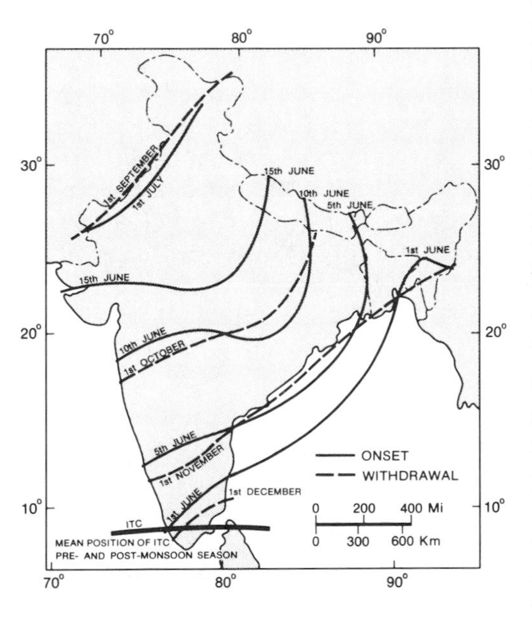

Fig. 11.11. Normal dates of the onset and withdrawal of the summer monsoon over the Indian subcontinent. (After Amanthakrishnan and Rajagopolochari, as observed in Dey, 1977.)

Within the Indian sector severe hurricane storms are most numerous at this season, 46 percent of those in the Bay of Bengal occurring in October-November and 62 percent in the three months September through November. Some of these storms, in weakened form, appear to have come from the western North Pacific, regenerating over the Bay of Bengal. More, seemingly, develop over the Bay itself and are genetically associated with the southward-moving discontinuity between equatorial westerlies and zonal trades. The trajectories of these storms lead them onshore somewhere between Madras and Calcutta where they add appreciably to the total annual rainfall and occasionally work fearful havoc.

During the fall months the dynamic features of the circulation aloft, including jet stream and the orographically imposed troughs and ridges, begin to approach their cool-season positions. With the reappearance south of the highlands in October or November of the middle-latitude westerlies and the westerly jet, and the reestablishment of the polar front in the extreme northwest, the western disturbances once more become an important control of weather in northernmost India-Pakistan. Thus the November rainfall map shows an appreciable increase in precipitation along the extreme northwestern frontier of the subcontinent over what it is farther south at Multan or Lahore (Fig. 11.6).

Unusual Climatic Features

Temperature

A notable feature of India's climate is the abnormally high temperatures of the hot season, coupled with the fact that the maximum usually is reached in May, and therefore precedes the period of maximum insolation (Köppen's *g*, or Ganges Type). In May over one-half of the subcontinent experiences average temperatures above 32°C (90°F), and a considerable area in the interior averages above 35°C (95°F). No other regions in southern Asia in similar latitudes (the Mandalay Basin in Burma excepted) show either such high warm-season temperatures or the pre-solstice temperature maximum.

These temperature peculiarities stem from the fact that in the interior skies remain clear and rainfall meager until very late in the spring, as a consequence of the retardation of the onset of the monsoon which does not occur until late May or early June. With the increase in cloud and rain in July and August, temperatures drop markedly (Delhi from 32°C [90°F] in May and June, to 29° or 30°C [84° or 86°F] in July and August). Over South China and most of Southeast Asia spring is cloudier and rainier than in India, so that temperatures are lower.

On the other hand, winter temperatures in India are by no means so low as they are in subtropical South China in similar latitudes. Thus, Delhi and Chungking, interior stations at about the same latitude, have January average temperatures of 15° and 9°C (59° and 48°F) respectively. The explanation for this contrast is to be found in the unlike origins of the northerly air which the two stations experience in winter, China's being of Siberian origin, while that in India is from local subsidence, the highlands to the north effectively blocking any advection from the cold source over Siberia.

Precipitation

One of the most conspicuous features of both the summer and the annual rainfall maps of the subcontinent is the fact that the rainfall gradient in the north is the reverse of that in the south (Fig. 11.6). Over northern India precipitation decreases toward the northwest; in the south it is highest on the west side and decreases eastward, although, to be sure, on the map of annual rainfall it picks up again toward the southeast coast. Most of peninsular India is dominated in summer by the southwesterly air stream from the Arabian Sea which unloads much of its moisture in surmounting the Western Ghats, so that it is less humid as it undergoes subsidence east of the Ghats. Here perturbations are few or weak, for surges in the westerly flow are greatly weakened east of the Ghats, and the monsoon depressions in summer ordinarily do not reach this far south.

In the north, by constrast, it is the southerly and easterly currents which supply a part, if not most, of the moisture. Moreover, it is the easterly troughs and monsoon depressions from the Bay of Bengal, traveling toward the north and west, which generate much of the warm season's, and also the year's, total precipitation in these northern parts. Such disturbances weaken as they move toward the west, few of them ever reaching the northwestern parts. To be sure, in the cooler months it is the northwest which is the region of origin or entrance of the western disturbances, but the total rainfall from these perturbations is insufficient to have an important effect upon the annual rainfall distribution.

Regions of Annual Rainfall Deficiency. —What appears to be a still more striking peculiarity of the subcontinent's rainfall, considering the fact that almost the entire region is dominated by a surface flow of equatorial-tropical maritime air in summer, and is crossed twice by the *ITC*, is the widespread deficiency of rainfall. According to the Köppen classification, arid and semiarid climates occupy nearly two-fifths of the total area, while the figure is very close to one-half if the Thornthwaite (1948) system is employed. When dry-subhumid climate is included, well over 50 percent of the subcontinent's area is indicated as having a rainfall deficiency. The most extensive area of dry climates is located in the western half of northern India-Pakistan, while a smaller area of semiarid climate lies east of the Western Ghats in interior peninsular India (Fig. 11.12).

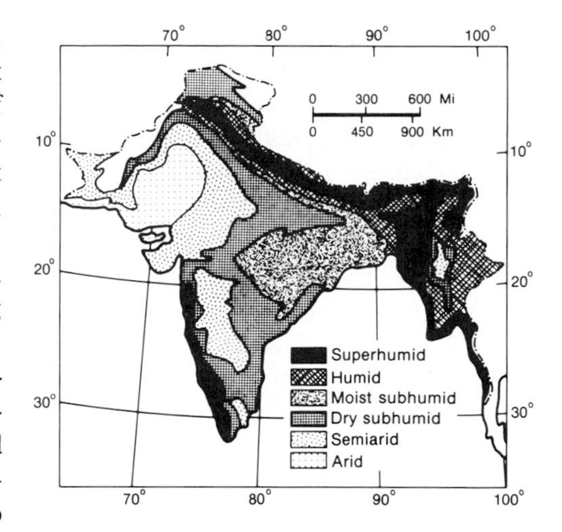

Fig. 11.12. Moisture regions of the Indian subcontinent. The west coast is actually a combination of humid and superhumid. (Simplified from Shanbhag, 1956.)

The fact that the northeastern and northwestern extremities of the subcontinent exhibit such impressive contrasts in rainfall amounts, and this in spite of the fact that the two regions have so many similar climatic controls, presents a problem. Accordingly, while the northeast is one of the wettest parts, much of the northwest is a desert. Yet both areas have similar positions at the northern ends of tropical seas and in summer are invaded by surface equatorial air from these seas. In terrain arrangements the two regions likewise show strong resemblance, for in each case air from the south would appear to be boxed in by highlands so as to produce directional convergence and ascent. In summer both are regions of surface low pressure, cyclonic circulation, and *ITC* positioning, but the dry northwest is the center of lowest pressure. Moreover, in winter the northwest is the principal region of depression origin, while it is the locus of the highest temperatures in spring and summer which, other things being equal, should produce a maximum of convectional overturning. For a number of reasons, therefore, the rainfall contrasts between the northeast and the northwest, and more especially the drought in the northwest, appear to present a climatic problem.

As a first step in explaining the existence of the Thar Desert it should be emphasized that this is the normal latitude of subtropical anticyclones and the deserts which they generate. The strong summer low, centered in this area, is thermally induced and is not of great depth, for at an elevation somewhere between 600 and 1,200 m the circulation becomes anticyclonic, and hence opposed to precipitation.

In addition, it needs to be noted that while in summer the surface air flow from the Arabian Sea does enter this area, the monsoon current is relatively shallow, so that the total mass of humid air carried across the coastline is small but variable. Stable anticyclonic air not only prevails aloft above the shallow maritime current, but dry continental air is also drawn into the surface low from the more elevated west and northwest, and suffers downslope subsidence and further stabilization in the process.

THE THAR DESERT OF THE NORTHWESTERN INDIAN SUBCONTINENT.—One of the most intriguing of the earth's problem climates is found in the Indus lowlands at the head of the Arabian Sea. Here intense aridity extends down to tidewater and out over the sea (Fig. 11.13). In its driest part the annual rainfall of the Thar declines to 100 mm. Making this intensely arid desert appear all the more remarkable is the fact that in the northeastern part of the subcontinent in similar latitudes and similarly located at the head of a tropical bay (Bay of Bengal) is one of the earth's most rainy climates.

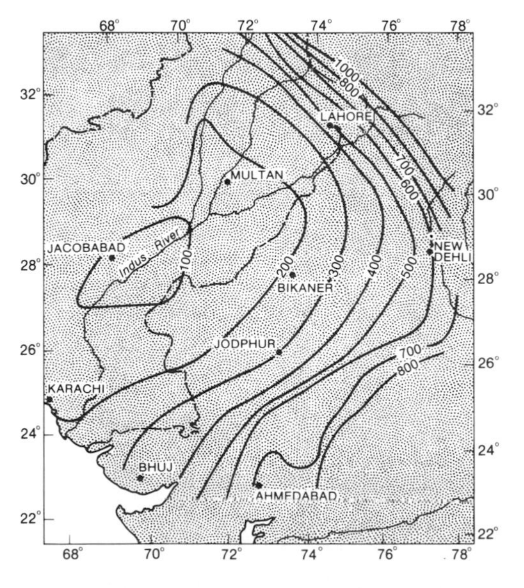

Fig. 11.13. Mean annual precipitation (in mm) in the northwestern part of the Indian subcontinent. (From Rupprecht, 1970.)

Until recently Sawyer's (1947) air-mass explanation of arid Thar, which was used in the 1961 edition of this book, was looked upon as reasonably convincing. But within the last 10–15 years as air-mass principles have gradually waned in repute, climatologists have turned increasingly to the atmosphere's thermodynamic processes for an explanation; in other words, they have attempted to discover the nature of the vertical wind components to determine whether a net ascent or net subsidence of the atmosphere prevails over the region.

Das (1962) and Ramage (1966) computed the vertical motion of the atmosphere in the Thar region in summer, and found that subsidence prevailed at nearly all levels. The intense solar radiation under clear skies creates a very shallow thermal low over the Indus plain in the warm months. Humid southwesterly air flow from the Indian Ocean has easy entrance to the lowland and creates a superficial surface layer of humid air, but the prevailing subsidence with its accompanying temperature inversion stifles vertical ascent so that little rainfall results. The July streamline maps for 0.5 km, 0.9 km, and 1.5 km show a remarkable diffluent flow of air over the Thar, but there is also a speed convergence present, probably due to surface friction (Fig. 11.14). The first of these two features fosters aridity; the second favors upward movement and precipitation, but the former dominates the situation. Rupprecht (1970) has calculated the vertical motion over the Thar up to 2.1 km; he feels that calculations at still higher levels would scarcely lead to different results. His analysis verified that of other meteorologists: the Thar is dominated by large-scale subsidence, and subsidence increases with altitude. Directional divergence in the southwesterly circulation predominates over the speed convergence, and this diffluence extends up to higher levels. The divergence appears to be caused mainly

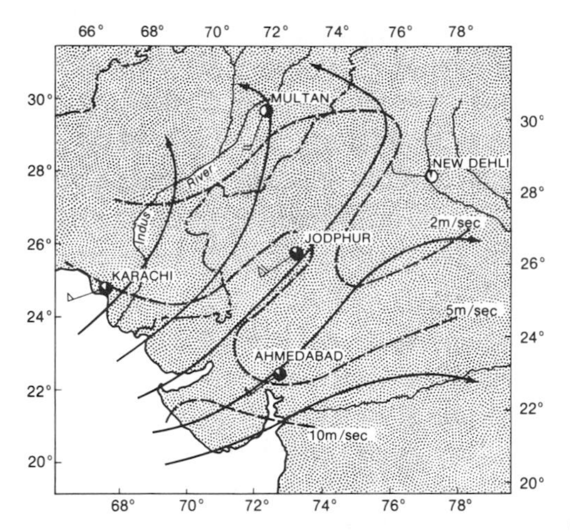

Fig. 11.14. Mean resultant surface winds over the northwestern part of the Indian subcontinent. (From Rupprecht, 1970.)

by the heat low over Baluchistan, which attracts the westerly branch of the general southerly air stream.

The mean subsidence over the Indus lowland is in agreement with the vertical circulation of the tropical easterly jet, for the Thar desert is situated in the region of subsidence on the warm flanks of the exit region of the jet. So it appears that the average vertical motion of the atmosphere over the Thar is one segment of the much larger scale circulation over southern Asia (Rupprecht, 1970). Because of diffluence in the lower air, the overall subsidence extends down to ground level, so that the shallow humidified surface air is unable to release its moisture in the form of rain. A sharp temperature inversion prevails at about 1.5 km.

There is archaeological evidence to show that the Thar Desert in times past may have supported more flourishing civilizations than prevail there today (Das, 1968). Bryson and Barreis (1967) indicate there is evidence that three different major cultures may have existed in this region. Such evidence suggests that there have been periods when the climate was less repellent than it is at present. However, there is little evidence of a flourishing period here during the last thousand years. It has been suggested that the more recent dessication may be related in part to an increase of dust in the atmosphere, as a consequence of a decline in soil texture resulting from past and present occupancy. The dense pall of local dust over the Thar is believed to be a significant factor in the development of subsidence over the desert. Thus, man may have had a role in making the present desert.

An additional uniqueness in the Thar desert-steppe area is the fact that, although its latitudinal location at about 25° to 33°N, as well as its geographical location, suggest it might normally show a wet winter and a dry summer (Köppen *BShs*), the reverse is the case. Such a winter-maximum rainfall variation does prevail over the dry Iranian highlands just westward of Thar, but not on the Indus lowlands, although there, significantly, a slight secondary winter maximum is to be observed. It would appear that the Thar Desert with its summer rainfall maximum is displaced 5° to 10° too far poleward. In truth, this is the fact, for nowhere else on the earth are the equatorial trough and its *ITC* displaced so far poleward in summer, with the result that the annual rainfall profile of the tropics is carried northward into subtropical latitudes where a winter maximum would be normal, as it is in Iran, for example. Such being the case, the shallow indraft of maritime air drawn into the thermal low is able to deliver a small, but well-defined, July–August maximum.

The semiarid climate in interior peninsular India eastward from the Western Ghats is partly the result, no doubt, of a lee location with respect to the escarpment in summer, plus modest downslope subsidence as the westerly current continues eastward toward the Bay of Bengal. Actually, however, the

Western Ghats are not so high; mostly they are between 900 and 1,200 m above sea level. Moreover, the downslope of the terrain surface to the east is not striking in most parts, and semiarid conditions exist on the upland at elevations of 450–900 m. Altogether, the height of the Western Ghats and the amount of downslope subsidence to the east scarcely seem sufficient to account for the rainfall deficiency on the upland. One might conjecture, however, that a supplementary effect of the Western Ghats is to induce anticyclonic-flow curvature in the broadscale southwest monsoon current over the obstructing highlands and just beyond. A similar situation and its climatic effects in Patagonia to the lee of the Andes, as described by Boffi (1949) have been noted earlier in Chapter 3.

Significantly, this dry region of peninsular India lies outside the main routes of the rain-bringing disturbances during the summer-fall wet season. The monsoon depressions from the Bay of Bengal normally follow a route to the north of the dry area, while the surges in the southwesterly current from the Arabian Sea are made less effective rain-bringers after crossing the Ghats. Moreover, in fall and early winter, the easterly disturbances which bring abundant rainfall to the southeastern coastal area ordinarily do not penetrate this far inland.

Koteswaram (1958) suggests that the modest summer rainfall over much of peninsular India may be related to an easterly jet stream whose average position is at about 15°N over the subcontinent in July. Since in the longitudes of India it shows an acceleration of wind speed, there is more ascent of air to the north of the jet axis and subsidence to the south, which is in accord with the distribution pattern of summer rainfall over India.

The Annual March of Rainfall.—On the whole, the annual march, or seasonal concentration, of precipitation over most of the subcontinent is in general what one would expect (Ludwig, 1953). Preponderantly India is a region of strong summer concentration of rainfall, but parts of the north and northwest show a slight secondary maximum in the cooler months associated with the western disturbances (Fig. 11.15). The three major departures from the prevailing summer maximum of rainfall are (1) the winter-spring maximum in the highlands in the extreme northwest, (2) the late fall-early winter maximum south of about 15° along the southern Coromandel coast and eastern Ceylon Island, and (3) the double warm-season maximum characteristic of much of interior peninsular India south of 18°–20° (Fig. 11.15). The first of these departures (Fig. 11.15, Type 5) is characteristic of a dry region dominated by anticyclonic continental air in summer and feeling the effects of western disturbances in winter and early spring. Here the profile of annual rainfall is relatively flat, but there is a primary maxi-

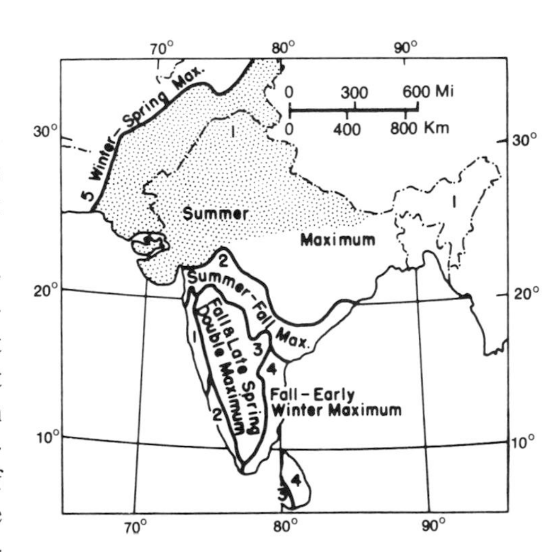

Fig. 11.15. Precipitation subdivisions of the Indian subcontinent based on the time of rainfall maximum. Stippled area indicates extent of region showing some evidence of a modest secondary rainfall maximum in winter. (Modified after Ludwig, 1953.)

mum in winter-early spring, and a slight secondary in summer. It is definitely transitional between the region of strong summer maximum to the east and the winter-rainfall area of the Middle East.

Southern India shows greater complications in the annual march of rainfall than exist elsewhere, as fall waxes in importance at the expense of summer. Ludwig (1953) recognizes a transition belt characterized by summer and fall rain, south of which fall becomes the rainiest season (Fig. 11.15, Type 2). The Malabar coast likewise falls within this Type 2. The entire eastern littoral south of about 20° exhibits a fall maximum of rainfall with the profile's peak occurring progressively later as latitude decreases (Type 4). Thus, between about 15° and 20°N, October records the largest monthly total. South of about 15° in coastal India and northern and northeastern Ceylon Island the maximum is shifted to November, and in east-central Ceylon Island the peak is reached in December. Here, then, is that relatively rare feature of a cool-season maximum of rainfall in tropical latitudes (Köppen's *Afs*, *Afs'*). Over extensive areas along the southeast coast 50 to 60 percent of the year's rainfall is concentrated in the three months, October through December.

While the fact that the flow of air is onshore in the cool season (the so-called northeast monsoon) it cannot be dismissed as unimportant, this is probably not the primary cause of the fall-winter rainfall maximum in southeastern India, especially on the lowlands. Of greater significance is the fact that Tamilnad in fall and early winter feels the effects of numerous easterly disturbances developing along the southward-displaced *ITC*, located at this time over the southwestern Bay of Bengal. A goodly number of these perturbations move west and northwest across the Tamilnad coast causing widespread and locally heavy rains. Those taking routes to the north and northeast provide little rain. The variability of fall rainfall is high, actually higher in the coastal areas where rainfall is abundant than farther inland where it is less. This is related to the fact that rainfall in the coastal districts is largely dependent on depressions and cyclones from the southern Bay of Bengal and the number and paths of these disturbances are highly variable. Large excesses of rainfall in fall occur when depressions and storms are numerous and take a route which brings them either onshore or close to the land. Large deficiencies in rain result when the disturbances are few or when they take a northward course so that they do not greatly affect the coast (Rao and Jagannathan, 1953). The rain-bringing disturbances are not limited to depressions and cyclones which are evident on surface weather charts, for some fall seasons with few such surface disturbances are still wet. Upper troughs also move from east to west across this area and several of these can result in a rainy fall even in the absence of surface depressions.

If it were only the effects of onshore northeast winds that were responsible for the fall-winter maximum in the southeast, the rainfall would be concentrated in midwinter when the easterly winds are best developed. But both in southeastern India and in eastern Ceylon Island February is relatively dry, and in Madras January as well. Furthermore, the fact that the maximum comes a month or two later at Trincomalee than at Madras suggests that the principal rain-producing control progresses from north to south, so that by February it is so far south that its influence is weak anywhere north of the equator. Such a situation points more to a migrating wind discontinuity and storm belt than to the effects of onshore easterlies.

Much of interior peninsular India south of about 20° has an annual march of rainfall which is unlike that of either the eastern or the western littoral. Here low sun is still the dry season, but there are two maxima, a primary one in September–October and a secondary one in May–June (Fig. 11.15, Type 3). Midsummer has somewhat less than the months just preceding and following. Ludwig (1953) recognizes this area of double maxima as one of five principal rainfall regions and labels it "Fall and Late Spring Rain." The twinned maximum of the warm season appears to be a consequence of the northward advance and southward retreat of the disturbances accompanying the *ITC*.

CHAPTER 12

Tropical Southeast Asia

Seasonal Circulation Patterns and Associated Weather Types

As a result of the complicated character of the air streams and their boundary surfaces, the fragmented insular and peninsular nature of the land areas, and the presence of numerous highlands having a variety of directional alignments, it is to be expected that the weather and climate of this region lying between China and Australia will be unusually complex. But much of the surmised climatic complexity must remain unverified because of the meagerness of well-distributed weather observations. Consequently, in the description to follow, what appears as a relatively simple overall pattern of climate with few large-scale abnormalities, is probably a simplicity stemming from ignorance. A few isolated spots such as the Philippines, Java, and the Malay Peninsula must provide the key data from which extensive extrapolations are made. As might be expected, there is still lack of agreement among the local meteorologists on how the available fragmentary evidence should be interpreted in defining the seasonal patterns of air streams and their boundary surfaces in Southeast Asia, and in evaluating their climatic implications.

The area between southern China and Australia, including the Burma-Indochina peninsula and the archipelagos to the south and east, is dominated by three main air streams whose average locations and boundaries have been broadly generalized by Watts (1955) as follows. To the north and northeast are the North Pacific trades, supplemented by the northeast monsoon in the cooler months (Fig. 12.1). To the south and southeast are the South Pacific trades. In the east the two trades meet along a convergence which is the *ITC*. Farther west the two trade streams are separated by a wedge of westerly air, the equatorial westerlies or Indian southwest monsoon, and instead of one there are here two zones of convergence, a northern one separating northern trades and equatorial westerlies and a southern one marking the boundary between the southern trades and the equatorial westerlies. In modified form this system moves north and south with the sun, resulting in seasonal weather changes as determined by the different air streams, their boundary surfaces, and the atmospheric disturbances that develop in them and along them (Figs. 11.1, 11.2, 11.3).

According to Watts (1955), the seasonal shifts of air stream boundaries at upper levels resemble those near the ground. The slopes of the boundary surfaces may be either side of the vertical and they are not

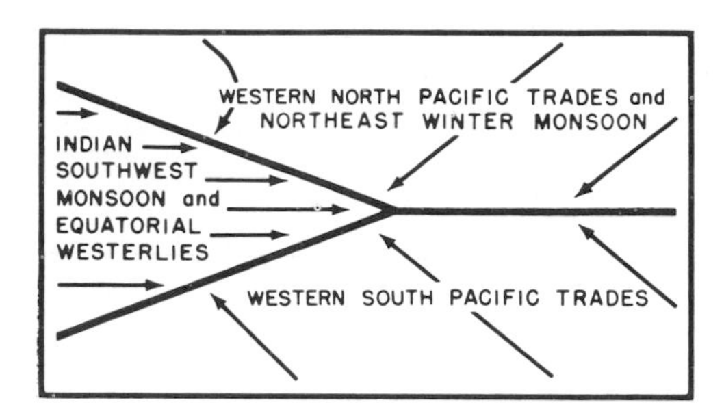

Fig. 12.1. Diagrammatic representation of the three principal air streams and their zones of confluence in Southeast Asia. (After Watts, 1955.)

uniform with height or along different parts of the same boundary within any one layer of the atmosphere. Changes in the slopes of the boundary surfaces follow small fluctuations in density and wind speed. Thompson's (1951) portrayal of the seasonal air streams and their boundaries at the 600 and 3,000 m levels show significant departures from the surface positions as described by Watts (1955).

Weather Disturbances

Equatorial Disturbances.—Watts (1955) clearly differentiates between rain-producing disturbances of genuinely equatorial locations (between about 10°N and 10°S) where, because Coriolis force is weak or absent, revolving storms and wave depressions are of less consequence, and those of tropical areas farther poleward where the expression, cyclonic vorticity, is meaningful. That author is inclined, therefore, to attribute most rainfall in the equatorial parts, other than that due to the effects of orography and possible thermal convection, to convergences, but without attempting to differentiate and classify the perturbations. Two type locations of disturbed weather associated with convergence are recognized, (1) convergence zones and centers along air-stream boundaries, and (2) convergence in minor perturbations embedded within an air stream.

Air-stream boundaries are by no means alike in their effects on weather and rainfall. When they separate air streams that are emphatically convergent, the upward escape of the converging air, even in the absence of significant density contrasts, will produce abundant cumulo-nimbus clouds and heavy rainfall. But even where the air streams are decidedly convergent the weather activity is variable depending on the slope of the boundary surface. In addition, horizontal shear along such a boundary may produce alto-stratus cloud and some rainfall far back into one or both streams. Where the air streams are parallel and nonconvergent, usually the towering cumuli are absent but alto-stratus and some rain may result from shear if the velocities of the two air streams are decidedly different. Where strong outflow downstream causes subsidence, the air-stream boundary is comparatively cloud-free.

Without doubt some of the most active weather, with maximum cloudiness and rainfall, is associated with the proximity of air-stream boundaries and their disturbances. For example, the heaviest seasonal precipitation is characteristically associated with the northward and southward movement of major air-stream boundaries following the movement of the sun. Such is the early-summer maximum concurrent with the northward passage of the boundary between equatorial westerlies and North Pacific trades as the southwest monsoon invades the tropics and subtropics of southern Asia north of the equator.

But air-stream boundaries in Southeast Asia exist not only between the great planetary winds, for there are likewise more numerous local ones generated by islands, peninsulas, and terrain barriers which act to bifurcate an air stream and are therefore the cause of the lee convergences. Moreover, while an air-stream boundary with convergence present is conducive to cumulus development, this process is greatly affected by diurnal variations in temperature. Thus, the daytime heating over the land greatly intensifies cloud development along an air-stream boundary, but it may be much less conspicuous over the adjacent sea and along the immediate coast. More than 30 km inland where the sea-breeze effect is small, cumuli may again mark the air-stream boundary. Diurnal temperature variations also may act to displace an air-stream boundary. Accordingly where a boundary lies close to a sizable land mass, the sea breeze may shift the low-level convergence well inland by day, while the land breeze may shift it seaward at night.

Watts (1955) makes little or no mention of extensive perturbations which develop along these boundaries. In this respect his treatment of weather in equatorial Southeast Asia differs strikingly from that provided for India by the meteorologists of that region. Watts does not so much deny the existence of such boundary perturbations, but indicates that the lack of observational material makes it impossible to apply this type of weather analysis to equatorial Southeast Asia.

Some rain also falls from minor disturbances of a convergent character that develop within an air stream and, therefore, are not associated with boundary surfaces. These disturbances appear to be diverse in character, difficult to detect, and are not well understood. Watts's (1955) treatment of these intra-air-stream disturbances is very general and vague, suggesting that observations in the region are too few to permit of their detection and analysis. He states, "During much of the year weather analysis is complicated by pressure troughs too weak to be determined from the observations at Southeast Asia's scattered stations." Schmidt (1949) notes that the irregularity of the weather in Java is a consequence of weak disturbances which originate in the southern Indian Ocean and move northward across Java as east-west troughs.

Some of the minor disturbances noted within an air stream appear to be in the nature of speed convergences or surges in the southwesterly flow. Others arriving from the east bear some resemblance to the easterly wave. One of the most common is that associated with land and sea breezes, which in an insular-peninsular region such as this must be widespread in its effects. The sea breeze may have a boundary structure resembling a cold front of middle latitudes with cooler air undercutting warmer, resulting in convective clouds and a line squall.

Line squalls of various origins appear to be common phenomena in the several parts of Southeast Asia. Some of them certainly are associated with the passage of air-stream boundaries, others lie wholly within an air stream. Few line squalls seem to occur during the northeast monsoon, although they are not unknown. The origin of many is not clearly understood.

Tropical Disturbances.—That part of tropical Southeast Asia lying north of about latitude 10°N is affected by additional types of disturbances, particularly those having cyclonic vorticity and appearing on the synoptic chart as revolving storms and pressure waves.

In the cooler seasons two great air streams dominate the upper troposphere over this region (Fig. 11.2). One of these is the low-latitude branch of the zonal westerlies and its jet which skirt the southern margins of the Himalayas and then proceed eastward across Southeast Asia, southern China, and Japan. The other high-troposphere air stream, probably of equatorial origin, flows from the southwest across southern India and the Bay of Bengal and converges with the zonal westerlies over the Indochina peninsula. The general cool-season rainfall deficiency of this region results from the compensating subsidence and divergence in the lower troposphere underneath the high-level convergence. Occasionally, however, the prevailing subsidence is temporarily halted by the advance into the region of an atmospheric disturbance and on such occasions considerable rainfall may result. One such type of perturbation is a weak western disturbance which has moved eastward from northern India. These disturbances affect chiefly the mountainous and sparsely settled northern parts of the area, so that little is known about them. Southeastward-moving cold fronts accompanying surges of the Chinese winter monsoon provide still another winter weather type for the Vietnam region which brings cloud and light precipitation. The average position of the oscillating front in January is about 12° to 14°N.

Ramage (1955) describes other varieties of cool-season disturbances, among them the tropical storm (including typhoons), the easterly wave, and the tropical trough. Because the subtropical high-pressure ridge in early winter remains unusually far north with its axis at 15°–17°N over the China Sea, disturbances typical of the deep easterlies in the warm season, such as easterly waves and typhoons, continue to be felt well into the winter season. Maung (1955) has been able to trace the courses of individual disturbances across the Burma region from their centers of origin in the western Pacific and the China Sea. Both easterly waves and tropical storms are capable of producing winter rainfall as far north as southern China.

Ramage's (1955) third cool-season disturbance, the tropical trough, is first detected over southern India above the 500 mb level. Here it is too weak to affect the weather significantly, but as the trough moves eastward in the high-level southwesterly air stream it intensifies, resulting in low-level convergence and rainfall in Thailand and Vietnam. This situation ordinarily occurs from three to six times during the period from January to March or April.

In the warmer seasons the area north of 10° or 15°N in Southeast Asia experiences weather associated with tropical storms and easterly waves, and it is the latter perturbations which, when they arrive at the head of the Bay of Bengal, combine with surges in the southwest monsoon to form the monsoon depressions of India. Monsoon surges, which can be traced across the southern Bay of Bengal, likewise play an important role in producing the summer rainfall of Burma and probably other parts of Southeast Asia.

CLIMATIC PECULIARITIES.—In spite of the complications imposed upon air streams and their boundaries by the very complex physical nature of the area, and the numerous local climates resulting from these complications, the broad lineaments of the regional climates, as expressed largely through the annual march of rainfall, are surprisingly similar to what might be expected. Thus between about 5°–7°N and 5°S there is no dry season and the normal *Ar* symbol prevails. To the north and to the south of these latitudes the *Aw* or *Am* (Köppen) prevails, indicating drought in the low-sun period (Braak, 1931). In the equatorial latitudes where a double rainfall maximum is common, the spring maximum is usually the weaker (Otani, 1954).

Regions of Reduced Rainfall.—One of the climatic peculiarities is the small amount of annual rainfall (1,000 –1,500 mm) on the southernmost islands of Indonesia, including eastern Java, southernmost New Guinea, and the islands lying between them, together with the fact that they have such a marked and long dry season (Fig. 12.2). There are 10 to 15 stations in this area with annual rainfalls of less than 1,000 mm and at least

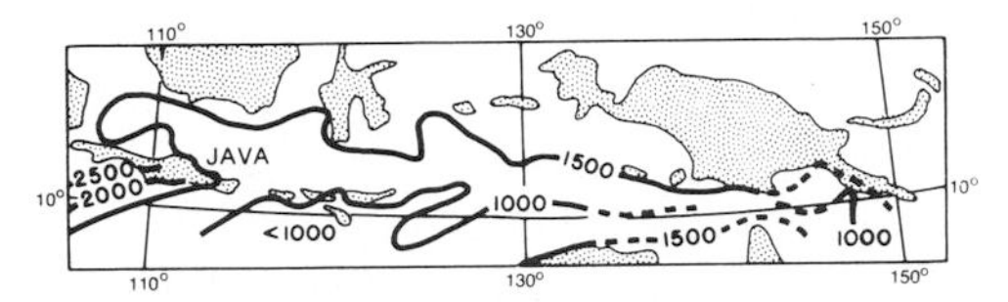

Fig. 12.2. Isohyetal arrangement showing the general location of the subhumid zone to the east and south of Java. Rainfall in mm. (After Wyrtki, 1956.)

4 where it is so meager as to qualify as semiarid. Palu on Celebes records only 547 mm and on the average eight months have less than 60 mm of rainfall each. Over the Savu and Timor seas in the vicinity of the islands, annual rainfall is only about 1,000 mm, but there is a rapid increase westward over the waters south of Java, from 1,000 to 1,500 mm in the east, to 2,500 to 3,000 mm in the west. Seventy-eight percent of the Savu Sea has less than 1,000 mm of annual rainfall (Wyrkti, 1956). Throughout this equatorial region of reduced rainfall the long dry season coincides with the prevalence of southeast trades, while westerly flow prevails during the wet season.

The cause of this decreased rainfall in the equatorial area between eastern Java and southern New Guinea, a condition which probably continues southeastward in the direction of New Caledonia, is by no means clear. At least a partial explanation appears to lie in the greater prevalence here of a drier and relatively stable easterly air current and the more modest development of unstable equatorial westerlies (Deppermann, 1941; Schmidt and Ferguson, 1951). These easterlies in the New Guinea area do not appear to be a westward extension of those dry trades of the equatorial eastern and central Pacific which are responsible for the dry tongue along, and just south of, the equator. In the western Pacific the dry tongue close to the equator is separated from this second area of diminished rainfall south and east of New Guinea and extending westward to Java, by a conspicuous wet tongue projecting ESE from the Solomon Islands toward Samoa and Tahiti. The latter approximately coincides in location with the *ITC* in summer and with shearing zones between two branches of the easterlies during the remainder of the year (Fig. 12.3). The dry easterlies in the southern New Guinea region appear to have their origin, instead, in the eastern end of a well-developed high-pressure cell in the western South Pacific centered at about 180°E and 20°S (Gentilli, 1952).

In the vicinity of eastern Java and the Lesser Sunda Islands the contrast between the easterly and westerly air streams is remarkable, the westerlies being deeply humidified and unstable, while the easterlies are drier and more stable (Deppermann, 1941)

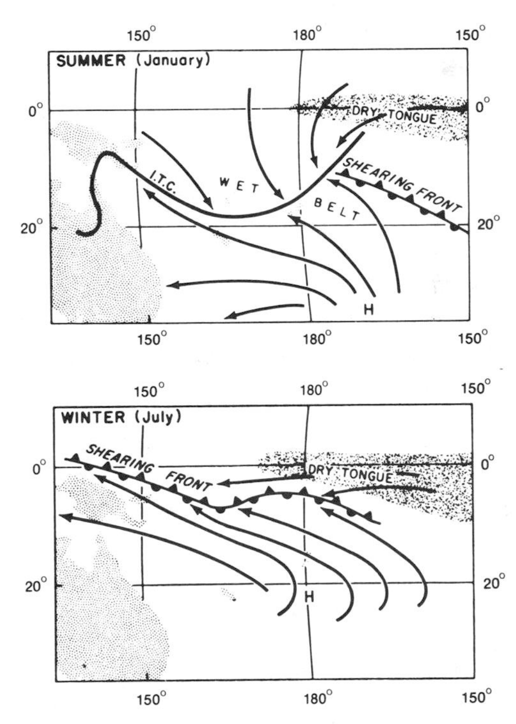

Fig. 12.3. Seasonal circulation patterns for January and July in the western South Pacific, showing location of the anticyclonic source region, and of the convergence zones. (After Gentilli, 1952.)

(see Table 12.1). Thus eastern Java, the Sunda Islands, and southernmost New Guinea, where unstable westerlies prevail at high sun, are dominated by the easterly air stream for 7 months or more during the period of low sun, with a resulting lengthy period of drought.

Other relatively extensive subhumid, and perhaps even semiarid, areas are to be found in interior upper Burma and central Thailand. In both areas there are stations with less than 1,000 mm of rainfall. But these situations are not so exceptional since during summer, when the westerly monsoon air stream prevails, both basin-like regions lie in the rainshadow of north-south mountain ranges to the west. Thus, in July, when as much as 1,000 mm of rain is falling on the western, or Arakan, coast of Burma, Mandalay in interior Burma receives only 80 mm.

Since the equatorial westerlies are convergent and also humid, while the easterlies are subsident and less deeply humidified, it may be expected that western sides of elevated islands and peninsulas will generally be wetter than eastern sides. This is a conspicuous feature in most of the area. It is not true in the Malay Peninsula, however, whose drier west side lies in the lee of Sumatra, and it fails also in the Philippines which are too far east to feel the effects of the westerly air stream for a lengthy period.

Peculiarities of Annual Rainfall Variation.—In much of the tropics, a high-sun rainfall maximum is so universally the rule that departures from this condition are viewed as exceptional. To be sure, within close proximity to the equator where the wetter and less wet periods are less synchronized with the elevation of the sun, local areas with a drier season at the time of high sun are not so unusual. Several such local areas are to be observed in the equatorial regions of Southeast Asia (Fig. 12.4). Braak (1931) notes three such to the south of the equator, one in southern Ceram just west of New Guinea, and two in eastern New Guinea. Otani (1954) likewise observes that there are a number of stations on or near the eastern and western margins of New Guinea which show either an annual rainfall variation with the maximum in, or close to, July, or a semiannual variation with the primary maximum in the period June to September. This peculiarity may have its origin, at least partly, in local orographic effects.

In equatorial latitudes north of the equator

Table 12.1. Surface and upper-air mean monthly relative humidity (in percent), Djakarta, Java (Deppermann, 1941)

Month	Surface	0.5–3 km	3.5–5.5 km	6.0–8.0 km	Wind
January	87	83	71	65	West
February	87	77	76	72	West
March	86	75	84	70	West
April	85	72	73	63	West
May	84	72	53	36	East
June	84	71	50	45	East
July	82	72	53	46	East
August	79	70	53	40	East
September	78	74	41	33	East
October	80	65	60	46	
November	82	73	68	54	
December	85	91	63	60	West
Mean	83	74	63	52	

Otani points out a similar area with a low-sun rainfall maximum located to the north-east of Celebes, while Wyrtki (1956) notes that the general area of the Celebes Sea has its rainfall maximum in the period November to January. Otani (1954) believes there is some evidence of a cool-season convergence between Asiatic winter monsoon air and the North Pacific trades in this locality.

In the tropics poleward of 5° or 10°, however, where an annual rainfall variation characterized by low-sun drought is so much the rule, a winter maximum is more truly exceptional. At least three such fairly extensive areas in tropical Southeast Asia north of the equator may be noted, all of them located along eastern littorals with highlands to the rear. They are (1) the northeastern Philippines, (2) the Vietnam coast, and (3) the eastern Malay Peninsula. The fact that in winter they are all on the windward side of elevated areas suggests that the onshore easterlies in this season are orographically lifted by the highlands causing the cool-season rainfall maximum. Doubtless this is part of the explanation, but not the whole of it. Because the rainfall maximum is in fall and early winter rather than midwinter when the easterly flow consisting of combined trades and winter monsoon is strongest, there is good reason to believe that the westward-moving perturbations of various kinds which develop along the air-stream boundary between easterlies and equatorial westerlies as it retreats southward are an additional factor accounting for this cool-season rainfall maximum along the east coasts. In the Philippines the vigorous disturbances are most numerous in September and October and the weaker ones in November, while combinations of the two types are about equally numerous in September, October, and November. Along the Vietnam coast the disturbances are most numerous in September to November. Ramage (1955) has pointed out that in the western North Pacific and the China Sea, the subtropical high remains unusually far north in fall and early winter, retreating southward only very slowly. As a result disturbances typical of the deep easterlies and the *ITC* in summer remain active in this area well into late fall and early winter. It is these depressions and storms which greatly aid in establishing the fall-winter maximum of pre-

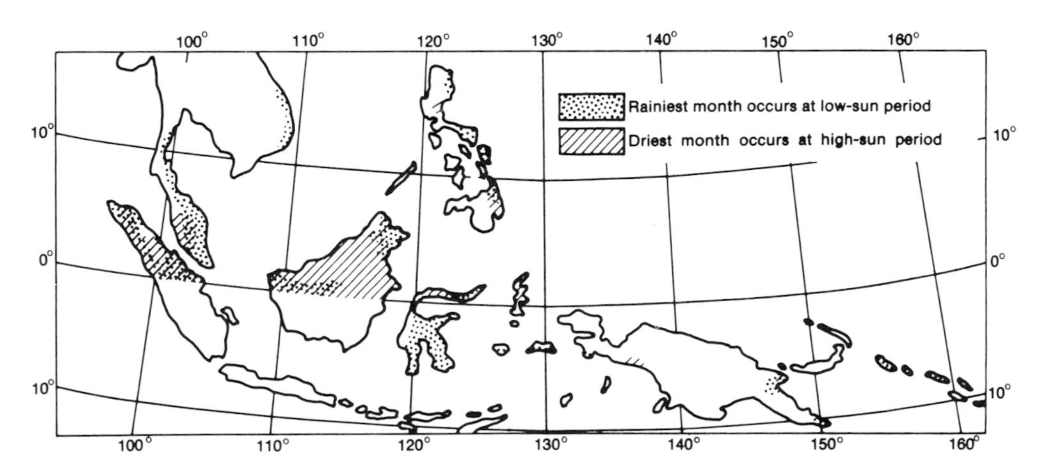

Fig. 12.4. Showing distribution of those areas in Southeast Asia where the rainiest month occurs in winter, or the period of low sun, and the driest month occurs in summer, or the period of high sun. (After Braak, 1931.)

cipitation in the eastern Philippines, the Vietnam coast, and perhaps in the Malay Peninsula as well. This situation resembles in many respects the circumstances in southeastern India where the southward retreating *ITC* and its numerous perturbations produce a similar late fall and winter rainfall maximum.

Noteworthy also are two additional climatic features of eastern Vietnam, viz., (1) the unseasonable winter cold of the Tonkin Delta area in the northeast, and (2) the weather type known as the "crachin." The cool winters of the Tonkin area are simply a projection into tropical latitudes of the abnormally cool winters of subtropical China just to the north, both areas feeling the effects of advected polar air with the vigorous winter monsoon. In the Tonkin Delta minimum temperatures of 10°C (50°F) are common in winter and occasionally the minimum drops below 7°C (45°F). In modified form the winter cold is carried farther southward along the Vietnam coast, as evidenced by the southward bending there of the January isotherms.

Crachin is derived from the French verb, *cracher*, meaning "to spit," and is applied to the spells of drizzle and light rain that occur along the coasts of Vietnam and South China in winter, accompanied by low stratus and sometimes fog, producing a condition of low visibility. Such weather is not continuous, but instead it occurs in spells of 3 to 5 days for a total of 10 to 15 days per month throughout the winter period (Sanderson, 1954). The crachin may occur anywhere between the mouth of the Yangtze and southernmost Vietnam, but its chief concentration is along the coast from Hongkong to about 15°N. It is at its worst in late February. Crachins appear to be of both frontal and non-frontal origin. Ramage (1954) states that the best developed crachins are non-frontal and are the result of the advection of warm humid air from the east over the cool water which prevails along the coasts of South China and northern Vietnam in winter and early spring. He associates the westward advection of warm air with two synoptic types: (1) an anticyclone moving offshore from the continent, or (2) a westward extension of the subtropical anticyclone. Hare (1944) describes a frontal type of crachin which develops in the cool air behind a cold front moving southward into the tropics. Such a crachin exhibits less diurnal variation than one which develops in tropical air. It is most prolonged and severe in the Tonkin Area. While this weather type is typically a feature of the coastal districts, it may move inland through the valleys and envelop the hill lands as well.

CHAPTER 13

Eastern Asia

Seasonal Circulation Patterns

Since the overall features of the seasonal atmospheric circulations for the whole of southern and eastern Asia have been broached in Chapter 11, here a brief recapitulation is followed by an amplification of remarks as they apply specifically to eastern Asia, mainly China and Japan.

In spite of a relatively abundant literature on climatic aspects of this extensive region, the task of updating the discussion is made difficult by the contrasting emphasis given by different authors to such fundamentals as circulation patterns, water-vapor sources and fluxes, and rain-generating disturbances. Then too, many of the most useful published papers are based on a limited period of observation, not uncommonly of a single year, and also of different periods of observation. And in a region where interannual variations in atmospheric conditions are notorious this may lead to different conclusions arrived at by the various writers. Some of the disharmonies in the literature are especially noticeable in the diverse points of view expressed by Japanese and Chinese aerologists, due in part no doubt, to authentic atmospheric differences that actually do exist between continental China and insular Japan, the latter some 800 km removed from the mainland. A few of the difficulties may stem also from textual ambiguities resulting from the imprecise use of the English language by some Oriental scholars. However, at least one generalization is valid for Monsoon Asia in general, and East Asia in particular, *viz.*, that precipitation there is characterized by seasonality, interannual variability, heavy downpours, and sharp regionality.

Winter Season.—In winter the upper-level zonal westerlies have migrated sufficiently far south so that they feel the obstructive effects of the Tibetan Highlands, are bifurcated by them, and flow around them on the north and on the south (Figs. 11.1, 11.2). There exist two westerly jets, therefore, the one to the south of the highland being positionally stable, while that to the north fluctuates more widely. The southern branch of the zonal westerlies and its jet, flowing eastward south of the Himalayas, crosses southern China in latitudes 25°–30°N and passes over southern Japan. To the east of Tibet at 3,000 m and above there is a confluence of the northern and southern branches of the zonal westerlies, each with markedly different temperature characteristics, to form the upper Tibetan lee convergence zone. The high-level trough continues to be positioned along the eastern margins of the continent.

In the lower layers of the troposphere the winter flow pattern is essentially different. Here the cold anticyclone over eastern Siberia results in a strong northwesterly flow, which floods eastern Asia with successive outpourings of cold *cP* air. The leading edge of each surge of anticyclonic air forms a new polar front moving to the south and east.

According to Lu (1945) the winter fronts over eastern Asia are somewhat more complicated than the previous analysis would suggest (Fig. 13.1). The Siberian air that reaches northern China has had a land trajectory exclusively, and hence is dry and cold even though its surface layers have been considerably modified by terrain. When it moves eastward off the continent and crosses the Sea of Japan, it arrives in Japan retaining many of its temperature characteristics but well humidified in its lower layers. The Siberian air that reaches central and southern China, however, may have had either a land or sea trajectory. If it arrives by land it is not greatly in contrast to what it was in northern China, especially at the upper levels. But when a bud of the winter anticyclone surges eastward into Manchuria the *cP* air may reach central China from the northeast after a trajectory over the cool waters of the Gulf of Pohai and the Yellow Sea. This sea-modified *cP* air is warmer and moister than land *cP* and hence convectively unstable in its lower layers. On other occasions the anticyclone surges southward into the Yangtze Valley where it may stagnate for a time before moving eastward toward Japan or the regions south of it. An easterly flow from the western end of this cold anticyclone, when its center lies well to the east of the mainland, may enter southern China warmed and humidified as a consequence of having traveled over subtropical waters. There are, therefore, at least three polar air masses

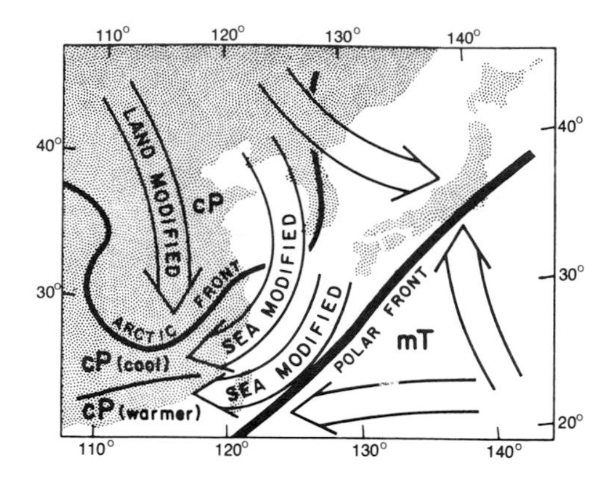

Fig. 13.1. Representation of air masses and fronts over eastern Asia in winter. South China's weather is affected not only by dry polar air which has had a land trajectory, but likewise by (1) polar air modified over relatively cold waters and (2) other more strongly modified polar air that has had a longer ocean stretch over warmer waters. (After Lu, 1945.)

which may be in conflict in central and southern China: (1) dry, cold *cP* air with a land trajectory; (2) *cP* air, modified over cold northern waters, which is moister and less cold than land *cP*; and (3) *cP* air transformed over warmer waters farther south which is warmer and moister than either of the other two. Sea-modified *cP* air of both varieties predominates in South China in the winter half-year. Weather associated with this air is generally fair except in hilly regions where clouds, light rain, and drizzle may occur as a result of moderate lifting. But where the two sea-modified *cP* air masses come into conflict with each other along lines of convergence, gloomy weather with light rain, and occasionally even heavier showers, results. Lu (1945), therefore, shows two principal surface fronts in winter in East Asia, the East Asia arctic front which marks the convergence between land *cP* and sea-modified *cP*, and the polar front delimiting the boundary between *cP* of any and all kinds and *mT* air from over the tropical sea. Both fronts shift greatly both longitudinally and latitudinally, but the arctic front commonly is located inland and north of the polar front, which in its mean position usually lies offshore aligned somewhat parallel with the coast.

In winter in southern China where the polar air comes into conflict with tropical or subtropical air of marine origin, the latter is forced to rise over the colder, denser air. The upper westerlies are dry and warmed, as a consequence of subsidence to the south of the axis of the subtropical jet stream positioned just south of Tibet. Between the dry upper westerlies and the *mT* air below is a dynamically stable layer characterized by a temperature inversion of up to 10°C (18°F) and a dew-point increase of up to 20°C (36°F). During those periods when the temperature inversion prevails the weather is characterized by a cover of stratus and strato-cumulus clouds over the lowlands in southern China. Such a situation is likely to prevail for 20 days or more in January in southernmost China, but inversion and cloud decline in frequency northward, reaching about half that figure in the Yangtze Valley. It is, therefore, a common winter weather type in subtropical China.

SUMMER SEASON.—East Asia in summer is dominated by three basic currents, the middle-latitude westerlies, the equatorial southwesterlies, and the tropical easterlies (Fig. 11.3). North of about 40°N are the deep zonal westerlies, with the subtropical jet stream aloft concentrated at about 40° to 45°N, but fluctuating widely in latitudinal position. South and east of the zonal westerlies is a southwesterly current which has its origin in the equatorial latitudes of the Indian Ocean. This is commonly spoken of as the southwest monsoon and it is an eastward and poleward extension of the same equatorial westerlies which dominate most of tropical South Asia in summer. Extending up 3,000 m and beyond, this deep and humid southwesterly current from tropical latitudes covers most of continental East Asia south of about 40°N, and it is this same current which supplies a goodly amount of the moisture for the abundant summer rainfall of that region. Flohn and Oeckel (1956) have shown that in summer the mean water-vapor flux over Korea, Japan, and much of China is directed from west to east. This is opposed to the widely held concept that the abundant summer rains are mainly derived from moisture evaporated from the Pacific Ocean and transported to the continent by monsoonal easterly winds. Clearly the Indian Ocean is a primary moisture source in summer for most of eastern Asia, and especially Japan, although the role of land evapotranspiration cannot be overlooked. Pacific moisture no doubt enters the continent on the northern flank of eastward-moving cyclonic storms and also on those occasions when the tropical southeasterlies invade the mainland.

The onset of the summer monsoon in eastern Asia has some resemblance to its counterpart in India. In both regions the upper air circulation makes the shift from a winter to a summer regime in late May and early June, in conjunction with a series of interrelated atmospheric events. Above the 300 mb level an anticyclone forms over the Tibetan Plateau; the southerly branch of the westerly jet stream anchored south of Tibet during the cool season progressively disintegrates and shifts to a position north of the highlands; and the N–S oriented upper trough that has been positioned over the Bay of Bengal moves westward to central India at about 80°E. These mutually related events, not entirely understood in terms of their effects, are roughly coincident with the advent of the monsoon in India, and somewhat concurrently, with the invasion of southwestern China by the equatorial southwesterlies from the Indian Ocean and Bay of Bengal. At about the same time southeastern China, and especially Japan, begin to experience the southeast monsoon from off the subtropical southwest Pacific. The onset of the summer monsoon occurs on the average about May 20 in southernmost China, between June 10 and 20 in the Yangtze Basin, and between July 10 and 20 in North China. More recently it has been noted that the onset of the early summer rains (called Maiu in China and Baiu in Japan) is closely related to the positioning of the subtropical ridgeline in the upper troposphere, which in turn is associated with the location of the polar front at the surface, and is the key factor in determining the duration of the Maiu rainy season. In China the northward advance in early summer of both the subtropical anticyclone and the polar front, as indicated by the average dates of the onset of the summer monsoon at increasingly higher latitudes (Chang, 1972, fig. 146), is in the nature of two or three surges. The Maiu appears to continue as long as the high-level subtropical ridgeline is located between 20° and 25°N, and the surface polar front is situated some 5° to 8° north of the ridgeline, or over the Yangtze country.

The two branches of the summer monsoon in East Asia, the equatorial southwesterlies and the tropical-subtropical southeasterlies, both rich in water vapor, become confluent in eastern southern China and southern Japan, resulting in an E–W belt of concentrated humidity with associated rainfall (Figs. 13.2, 13.3, 13.4). The southeasterly circulation is essentially a deflected trade wind of the Northern Hemisphere as it circles clockwise around the western end of the North Pacific subtropical anticyclone. In the early Maiu season the confluent southwesterly and southeasterly circulations over southern China continue across the East China Sea as a southwesterly flow in a position just southward of Japan. In this humid west-east circulation, where upward movement is widespread, the environment is favorable for

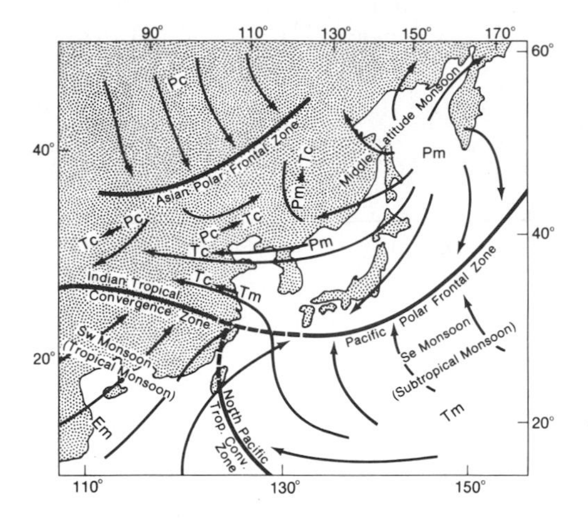

Fig. 13.2. Mean circulation and frontal zones over East Asia during the Baiu season. (After Kurashima and Hiranuma, as observed in Yoshino, 1971.)

precipitation and for the advent of the Maiu or Baiu rainy season. With the beginning of Maiu pressure falls, wind subsides, humidity increases, the air becomes sultry, and rain falls frequently from local clouds of the cumulus type. The Maiu phenomenon appears to be best developed and most prolonged in the Yangtze region, but even there it is a fickle weather type and shows variety in intensity and duration. In some years it is so weak as to appear almost absent.

In this early part of the summer monsoon rainy period, precipitation is mainly generated by synoptic disturbances associated with what by Orientals is called the Maiu or Baiu front (Fig. 13.2). According to some Japanese meteorologists, however, this so-called front differs in structure in its eastern and its western parts. In the former, or the Japanese sector, where the discontinuity separates polar maritime (*mP*) air on the north from tropical maritime (*mT*) air on the south, it is more in the nature of a density polar front. But farther to the west in China the discontinuity has more attributes of a convergence zone between tropical-equatorial maritime (*mT*) air to the south and a va-

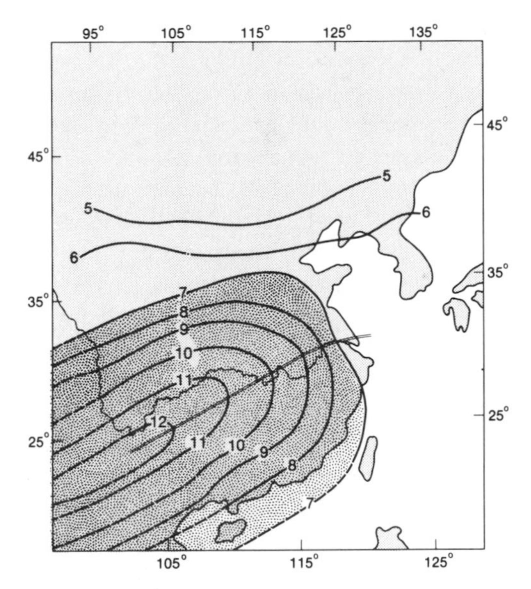

Fig. 13.3. Mean specific humidity (g/kg) at the 700 mb level over China in July 1954. (After Shen as observed in Yoshino, 1971.)

riety of air masses (*mT*, *cT*, and modified *mT* and *mP*) to the north (Kurashima and Hiranuma, 1971, fig. 1). Toward the east the Baiu front is characterized by a sharp *temperature* gradient; by contrast its western section, mainly in China, the zone of air-mass separation, is defined mainly by a marked *water vapor* gradient.

Although there is a fairly abundant recent literature bearing on the summer monsoon circulation over eastern Asia and the resulting horizontal and vertical transfers of water vapor, the conclusions relative to effects on rainfall distribution, spatial and temporal, are not always harmonious. In part this comes about because individual published papers are likely to focus on different parts of eastern Asia, some emphasizing conditions as they exist in insular Japan and others reaching conclusions based on observations made in mainland China.

In one of the early studies Chu (1939) concluded that the southwest monsoon, originating as it does in maritime-equatorial-tropical environments, has a higher water vapor content, at least in the lower and middle troposphere levels, than does the southeast monsoon coming from the southwestern periphery of an oceanic subtropical anticyclone. This conclusion is in harmony with the concept that in low latitudes westerly circulations in general have a total environment more conducive to rainfall than do those from an easterly direction.

Hsü's (1958) analysis of monsoon circulations and resulting transfer of water vapor over eastern China is based on data for July and January, 1956. He points out that in both winter and summer, but deeply only in the latter season, the inflow of water vapor into eastern China is largely from southerly directions. In that region in summer the *land* surface itself is an important source of water vapor, for evapotranspiration exceeds precipitation. In the warm season there are two pronounced streams of water vapor transfer, one from the southwest (equatorial or Indian

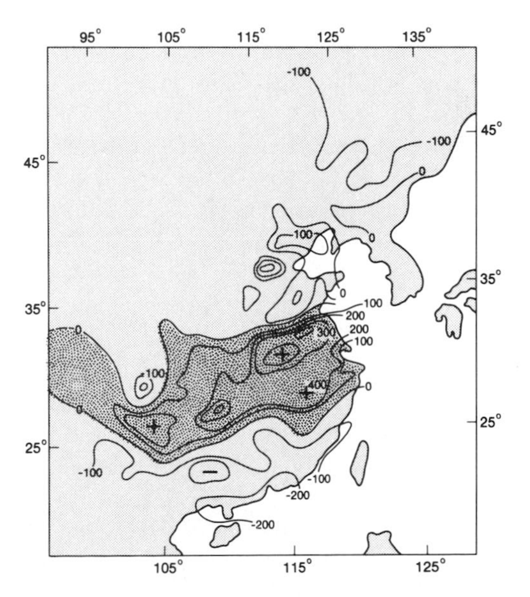

Fig. 13.4. Distribution of the precipitation over China, in July 1954 (in mm). Compare with Fig. 13.3. (After Chen as observed in Yoshino, 1970.)

westerlies) and the other from the southeast in the form of a deflected trade wind. Hainan Island at 110°E approximately demarcates the zone separating the southwest from the southeast flow. The former enters the continent in that sector of Southwest China situated between Hainan Island and Tengcheng, or through Yunnan, Kwangsi, and Kweichow provinces (Fig. 13.2). The southeasterly flow of water vapor enters the coastal sector between about Hainan Island and Foochow. In reality, the inflow of water vapor east of Hainan Island probably includes moisture from various southerly directions, (including SW, S, and SE). Transfer of water vapor takes place mainly in the lower half of the troposphere, but since land areas in southwestern China (Kwangsi and Kweichow), which the southwest flow of water vapor must traverse, are some two kilometers above sea level the humidified air column is thereby shortened. Thus, Hsü (1958) reasons that although the *intensity per unit area* of water-vapor influx from the southwest in July 1956, was larger, the amount of transfer throughout the entire shortened air column was less than the inflow from the southeast. The moisture-carrying southerly air currents which move directly toward the north are able to reach to the lower Hwang River region east of the Ordos Bend to the North China plain and even to the plains of Manchuria, where in July 1956, they may have penetrated to the northernmost limits of the polar front. But at this season, the upper-atmosphere (above 750 mb) trough, formerly positioned over the coast, has already shifted westward over the mainland and the dry northerly current along the rear of the upper air trough in northwestern China thrusts southward into the region south of Szechwan, thereby blocking the entrance of the moisture-bearing southwesterly current into northern interior China. In the atmospheric layer below 750 mb Hsü judges that the influx of water vapor via the southeast coast is greater than that from the southwest. Above the 750 mb level the amount of inflow of water vapor from southeast and southwest are similar. But because the northward transfer of moisture is blocked in western China by the northerly current west of the trough, transfer of moisture by the southwesterly current is limited mainly to subtropical latitudes. However, the moisture coming from the southeast penetrates far into northern China via the eastern flanks of the upper trough (Hsü, 1958, fig. 3).

There is some merit in having available a second major paper dealing with the atmospheric circulation and water vapor transport over eastern Asia in summer, in this instance authored by a Japanese climatologist, Murakami (1959). His analysis also is based on data for one year only, in this case the summer of 1957.

Murakami (1959) found that an outstanding feature of East Asia during the period June 11 to July 10, 1957, was the existence of a well-defined, moist tongue of atmosphere conspicuous both at the surface and up to the 500 mb level; it was most marked at 700 mb. The location of the moist tongue closely coincided with the axis of the Indian or equatorial westerlies over China and of the confluence between Indian westerlies and the southeast monsoon farther east over the East China Sea and the Pacific Ocean, just to the south of Japan. A well-defined zone of upward motion coincided with the moist tongue at all levels. Thus, the moist tongue is at least partly maintained by vertical water-vapor transfer from the lower atmosphere, which is in contact with the land-water surface. Both upward and downward motions are characteristically weaker in the easterly winds from out of the western flanks of the subtropical high, so in this case the moisture is mainly contained at lower levels, or below 850 mb. So Murakami (1959) reasons that the Indian southwesterlies, with their stronger vertical motion, are responsible for transporting vast quantities of water vapor into East Asia, and thereby greatly influencing the rainfall of that region. He is also of the opinion that the chief moisture source for China and Japan in the early stage of the summer rainy season is the Indian southwesterlies, but in the later stages the southeasterlies may be a larger contributor. In addition, he suggests that while the transfer of water vapor over China is governed mainly by the Indian southwesterlies, in Japan both the southwest and the southeast currents are important, the latter probably more so.

Fukui (1964) is of the opinion that in the Japan region, at the beginning of the summer rainy season, Baiu, the required water vapor is provided mainly by the convectively unstable Indian southwesterly circulation. The Baiu season terminates when the North Pacific subtropical anticyclone expands westward, thereby bringing in more subtropical easterly air, which is less humid and probably less unstable than the Indian southwesterlies, the latter being progressively squeezed out. This ushers in the midsummer less wet season. At this time of slackening rains there is a striking negative correlation of rainfall and temperature—the less rain and cloud, the higher the midsummer temperature.

Saito (1966), noting that in lower latitudes the source regions of very humid air are characteristically regions of air-flow convergences, finds that the prevailing air masses in southern China in June and early July (the Baiu season) are mainly equatorial Indian southwesterlies; however, to a greater degree Japan is dominated by an easterly flow from the subtropical-tropical western Pacific, which has a somewhat lower water vapor content at both the 850 and 700 mb levels. He is of the opinion that the core region of strongly humidified air in South China in early summer is derived from a convergence between southerly currents (both SW and SE) and subtropical-tropical continental air over the interior parts of Central China.

Chang (1972) holds that the main moisture source for the early summer rains, or Maiu, in South China is the southwesterly circulation from the Bay of Bengal, but this may not be the case in the late summer rainy season. The key factor determining the beginning and ending of the early summer Maiu rains appears to be the northward shifting of the subtropical midtroposphere crestline of the North Pacific subtropical anticyclone, to which the location of the surface polar front is closely related. At the 300 mb level the average position of the zonal axis of the North Pacific subtropical anticyclone in East Asia shifts from about 18°N in May to 25°N in June and 29°N in July. In subtropical China the Maiu rains accompanying the front slacken as the subtropical ridgeline advances northward and the less wet phase of the summer rain takes over. After the termination of Maiu the rainfall belt shifts northward by almost 10°, moving from Central China to North China. The latter as a consequence, shows no midsummer dip in its annual rainfall profile.

Yoshino (1971), basing his conclusions on Murakami's 1959 study, believes that in the early Baiu period the abundant water vapor over southern China is provided mainly by the Indian southwesterlies, which generate a pronounced moist tongue, oriented WSW by ENE over the Yangtze lowlands. In the last stage of Baiu the atmospheric moisture over southern China is chiefly provided by easterly winds from off the subtropical Pacific. Yoshino (1971) agrees with Flohn and Oeckel

(1956) that the land surface of humid southern China is also an important source of water vapor.

Principal Weather Systems in Eastern Asia

It is a misnomer to speak of the rainfall of eastern Asia, even that of the summer season, as monsoonal in origin and resulting from an inflow of humid air from seas to the east and south. Admittedly the circulation from equatorial and tropical source regions provides an atmospheric environment which has a high rainfall potential, but to an overwhelming degree the precipitation originates in extensive traveling disturbances. This is the case in all seasons.

Winter and the Transition Seasons.—Since in the cooler seasons the zonal westerlies prevail over nearly all of eastern Asia, the synoptic patterns are chiefly those of that wind system. (Staff members, Peking, 1957, 1958; Thompson, 1951). Foremost among these are (1) the cold wave and (2) the extratropical cyclone, the latter including disturbances of several types and origins. The succession of cold waves invading from Siberia results in eastern Asia having the lowest average winter temperatures for these latitudes in any part of the earth. Each fresh outpouring of cold air as it streams southward and eastward develops either directional or speed convergences with air of different character. Over northern China the fresh *cP* makes contact chiefly with older modified *cP* so that the resulting cloud and precipitation are meager. Here the cold wave is a weather type characterized chiefly by strong northerly winds and sharply falling temperatures. But farther south and east where the fresh *cP* converges with sea-modified *cP* or even *mT*, cloud and precipitation are more abundant and wave cyclones may develop along the fronts thus formed. Also where sea-modified *cP* is forced to surmount terrain barriers, rain or snow showers may result.

Chang (1972) identifies three low-pressure synoptic systems which are responsible for most of the cool season precipitation in China: (1) traveling upper-air troughs, (2) wave cyclones, and (3) shearlines. The first of these systems, the upper troughs, usually are not conspicuous on surface weather charts; they generally are detected by the accompanying wind shift and cloudiness. Such upper-air troughs moving into eastern Asia from the west may have quite different life histories. In the northern latitudes they often deepen in the vicinity of Lake Baikal and then move eastward, but in the dry, cold air prevailing in northern China they yield little precipitation. Southern sections of some deep upper troughs are able to surmount Tibet and reach southwestern China. There, as they move eastward, they may rejuvenate and produce precipitation, usually concentrated in an east-west belt in China south of the Yangtze River.

Wave cyclones show two general regions of concentration over the mainland, in northeastern China and in central and southern China, with the two tracks tending to converge in the vicinity of Japan and adjacent seas (Sung, 1931; Biel, 1945). Cyclones in eastern Asia, more especially the southern group, are characteristically smaller, less intense, and move more slowly than their counterparts in central and eastern North America. This feature is attributed to the subsidence underneath the upper-level lee convergence zone to the east of Tibet which tends to stifle cyclone development (Staff members, Peking, 1958). For this reason the vortices which develop in interior China tend to dissipate as they move downstream and only regenerate as they reach the longitudes of the surface polar front and the upper trough along the eastern margins of the continent (George and Wolff, 1953). Over northeastern China, which largely escapes the effects of the upper convergence zone, cyclone development is more marked.

The northern disturbances principally originate as secondary cyclones from primary disturbances arriving from the west in a weakened occluded form over western Siberia. As the latter reach the Baikal region a secondary cyclone characteristically forms near the occlusion point and the new disturbance deepens rapidly as it moves eastward, reaching its maximum development over northeastern China or the adjacent seas. Since the air at lower levels is dry and stable such storms do not yield much precipitation, but somewhat more in spring and fall than in winter, the latter showing less cyclonic activity than the transition seasons.

Cyclogenesis in subtropical latitudes is of different types. In the first instance there are those cyclones which develop along cold fronts as they begin to decelerate in their southward and eastward progress. This type of origin is common in eastern North America. Other disturbances develop in association with upper troughs which move into the China area from the west. Such troughs commonly dissipate in the lee subsidence zone as they move eastward, but if there is a strong advection of warm humid air, cyclogenesis occurs. Some of the westerly troughs invade China from the northwest, others from the west by way of Tibet, while still others are imbedded in the zonal westerly current which encircles Tibet on the south, and arrive in subtropical East Asia via India and Indochina (Staff members, Peking, 1958). The latter bear close resemblance to the cool-season western disturbances of northern India. Characteristically these southern disturbances are weak and are not conspicuous on the surface synoptic chart, but they are accompanied by bad weather, nevertheless, and are responsible for a large part of the cool-season rainfall of subtropical China.

In the immediate lee of eastern Tibet, in the region of the upper-level convergence, wind speeds are low, corresponding to the stagnation point in the flow around a cylinder. Lee eddies of small diameter, some of them resembling shearlines, probably develop in this region of stagnation and then move southeastward into subtropical China south of the Yangtze; but because of the low-level divergence, they remain weak with only occasional showers until they feel the regenerative effects of the jet and the upper trough in the vicinity of the coast or over the adjacent seas (Yeh, 1950). Ramage (1952*c*) shows a strong maximum of continental cyclogenesis in interior China just to the lee of Tibet (Fig. 13.5). In addition, there are occasional winter disturbances of tropical origin which affect chiefly southernmost China. Strong subsidence south of the axis of the Indian jet results in prevailingly dry weather over most of southern Asia in the cooler months, but occasionally the subsidence is interrupted by an atmospheric disturbance. These are of two kinds, (1) easterly waves and tropical storms coming from the China Sea and (2) troughs moving from the southwest in the high-altitude southwesterly current (Ramage, 1955). Both of these affect the weather of genuinely tropical latitudes more than they do China, but their effects are not absent in southernmost China where they are responsible for a modest amount of winter rain (Cheng, 1949; Yao, 1940).

It is significant that the two principal concentrations of cool-season disturbances in eastern Asia, one in the north and the other in the south, coincide with the two jet streams which traverse the region. The southern and more positionally stable jet encircles Tibet on the south while the northern and more widely oscillating one keeps poleward of the highlands. Each of the jets is

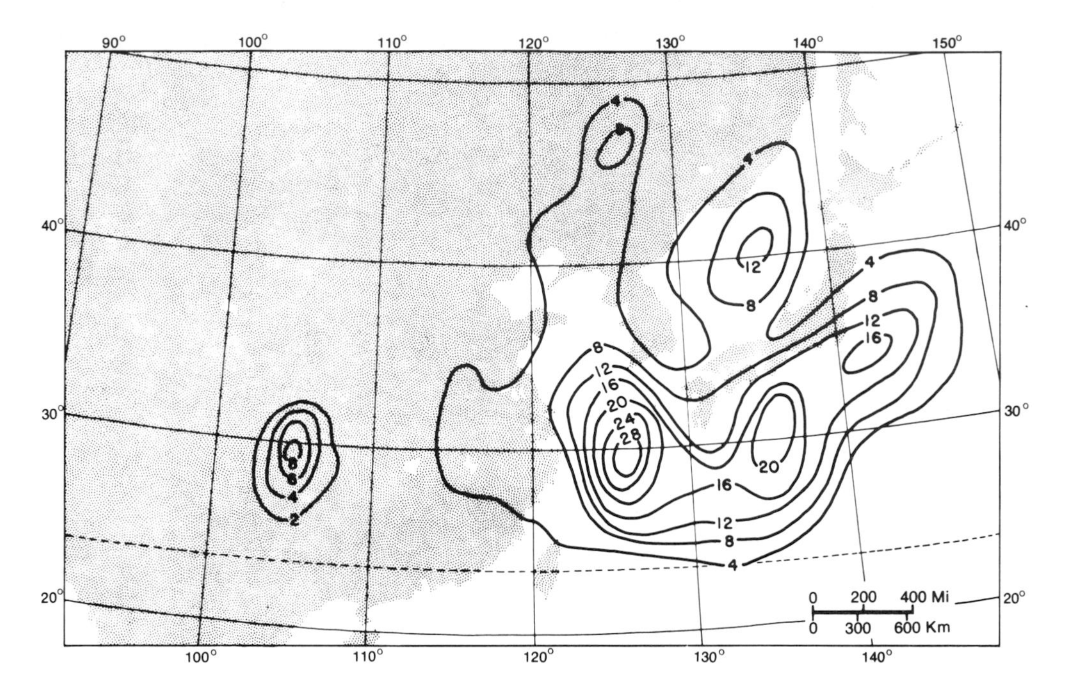

Fig. 13.5. Showing frequency of cyclone formation. The principal region of cyclogenesis in the Far East is over the waters just to the east of the continent, with major centers located over the East China Sea, the waters south and east of Japan, and the Sea of Japan. The principal continental cyclogenesis center is in the vicinity of the lee convergence zone east of Tibet, and in Manchuria. The value of any isoline at any point represents the number of cyclones that formed within a radius of 2.5 latitude degrees from that point in the months October through April 1932–1937. (After Miller and Mantis, 1947, and Yeh, 1950.)

positionally related to a distinct frontal system at the surface (Mohri, 1953). Jets probably do not originate cyclones, but through their removal of upward-moving air in the high-level convergence, they create an environment which favors cyclonic propagation and intensification. For this reason the jet appears to have the effect of steering the courses of cyclones and of causing them to deepen when superposition occurs. Thus, the northern disturbances intensify rapidly as they move across northeastern China and northern Japan and the same is true in southern China after they are far enough east to escape the low-level subsidence leeward of Tibet. In these latitudes a great deal of the reactivation of the disturbances occurs over the seas just east of China and surrounding Japan. For this reason Japan's winter weather is much more dominated by cyclonic activity than is that of subtropical China, since the southern storms tend to take a northeastward course which carries them along the whole length of the Japanese archipelago. Miller and Mantis (1947) have shown that the principal areas of cyclogenesis in the period October through April in the coastal region and adjacent waters are (1) the East China Sea between latitudes 25° and 35°N; (2) the ocean area south and east of Honshu, Japan; and (3) the Japan Sea. There is a strong frequency maximum in March (Fig. 13.5).

In *spring*, a transition period, weather systems become more numerous and more active so that in most of China interdiurnal pressure and temperature fluctuations are at a maximum. A large proportion of the spring rainfall continues to be generated by wave cyclones, upper-air troughs, and shearlines, which now are more vigorous than in winter, with some of them developing into major disturbances in southern China. Still, at ground level, well-developed fronts with accompanying windshifts are not commonly observed. An additional synoptic system of the spring season is the so-called heat low which originates in interior southern China and often spreads eastward to cover much of the territory south of the Yangtze. The increase in temperature accompanying the heat low is attributed to the advection of tropical air masses, intensified by radiational heating under clear skies. Such thermal lows normally persist until displaced by a polar air mass or weakened by a cloud cover.

SUMMER DISTURBANCES.—In summer zonal westerlies have retreated to the north of the Himalayan-Tibetan Highland complex and a single westerly jet prevails. China, south of about 40° ± N, is dominated then by equatorial and tropical air streams. Although most disturbances of the warmer months generally have weak circulations, the relatively favorable temperature-moisture environment provided by the southwest and southeast monsoon circulations causes them to be relatively effective rain-bringers. Chang (1972) refers to several significant rain-generating disturbances of the summer season in China: (1) cold fronts; (2) warm fronts of low pressure systems that develop along the oscillating polar front in northern China; (3) the monsoon low; (4) the blocking high with cold vortex; and (5) tropical storms, including typhoons. The weak polar front disturbances bring a short season of heavy summer rainfall to northern China, including Manchuria, which season accounts for 60–70 percent of the total annual rainfall in those parts. In these disturbances rainfall is mainly of the cold-front type. When a cold front is displaced southward into central and southern China, the several air streams are fragmented by terrain irregularities so that the rainfall produced is likely to be of a scattered nature. Warm-front rains are fewer and of less importance than those associated with cold fronts and are characteristically dispersed spatially, brief in duration, and light in rate of fall.

Monsoon heat lows in summer are mainly characteristic of South China; in them rainfall is particularly heavy and widespread on the southern side of the disturbance. The summer cold vortex is best developed when the general circulation is sluggish, and intermittent thrusts of cold air are likely to develop into isolated circulations. It is most persistent in southwest China where it is maintained by the continuous supply of cool air flowing southward along the eastern slopes of the Tibetan Plateau.

Tropical storms of varying degrees of intensity are more frequent in the western North Pacific Ocean and the South China Sea than anywhere else in the world. Of the land areas, they affect mainly subtropical southeastern China and the Pacific side of southern Japan. Most of them originate in the region of Yap or Guam, although a few develop in the South China Sea. In the cool season they only infrequently affect the Asiatic coast, being more frequent at that time in Japan than elsewhere. Beginning in May and June they are felt along the China coast as well as in Japan, but they reach their greatest frequency in eastern China during July–August–September. The China coast escapes the rainfall associated with those numerous tropical storms which recurve to the north over the sea, but Japan is likely to be affected by most of these recurved disturbances. Consequently there is no time of year when Japan is completely free of tropical storms, although they are more frequent in the warmer seasons.

Ramage (1951) comments on two other weak rain-bringing tropical-type disturbances affecting subtropical East Asia in summer. They develop in tropical air that is essentially homogeneous in character, and in which surface isobars bear no close relationship to rainfall occurrence and distribution. In them flow patterns at 3,000 meters are better weather indicators than surface circulations and speed convergence is more important than directional convergence. One summer disturbance is in the nature of a pressure wave, which resembles the well-known easterly wave of the Caribbean trade-wind area. In the warmer months the waves do not ordinarily get far west of the Philippines, but like tropical storms, some recurve northward and pass around the western end of the subtropical high, and become waves in the subtropical flow moving northward. An appreciable surface circulation is lacking in these wave disturbances, but the rainfall over southeastern China associated with the recurved pressure waves at times is extensive and heavy. Six such perturbations were recorded at Hong Kong during the summers of 1948 and 1949 (Ramage, 1951).

The confluence between the tropical southwesterly and southeasterly currents over easternmost China yields little weather except on occasions when an easterly wave recurves northward around the western end of the subtropical high.

A second tropical-type summer disturbance, and one that is characteristic of the southwesterly equatorial flow over subtropical China, is in the nature of an eastward-moving speed convergence. It resembles the surge in the summer monsoon circulation over India and Southeast Asia noted in Chapter 11. The origin of the speed convergence is not clear. As such surgelike disturbances move downstream in the equatorial westerlies current, they are accompanied by cloud and showery precipitation.

Because of the extensiveness of China, the distribution of rainfall there in any one season or year will depend largely on the nature of the general circulation at the time, and it may vary greatly from one year and from one region to another. During periods of high zonal index, when north-south pressure gradients are steep, the Pacific subtropical anticyclone pushes northwest toward the Yangtze Valley; fair weather is then the rule in Central China, but rainfall is much heavier in North China and is also above average in South China. By contrast, during periods of low zonal index (weak pressure gradient in a N–S direction), when the subtropical anticyclone withdraws to a position off the coast of China, rainfall is likely to be concentrated in Central China where polar and tropical air masses clash, while the north and south experience drier weather.

THUNDERSTORMS.—A notable feature of the climate of eastern Asia is the relative infrequency of thunderstorms, particularly as compared with the situation in central and eastern North America in similar latitudes (Biel, 1945). What makes the contrast even more striking is that over much of North America east of the Rockies lowlands prevail, while a large part of East Asia is rugged hill land and mountain, where terrain obstacles should favor the development of strong convective overturning. Northward from about the Yangtze River at 30°N, the number of thunderstorms is fewer than 20 a year, and over considerable areas it is fewer than 10. By contrast, in the United States, northern Florida and southern Louisiana along latitude 30°N record 70 days with thunderstorms, and even at latitude 40°N there are 40 or more. Subtropical hilly Japan and Korea experience only 10 to 20 days with thunderstorms. It appears, then, that lowland central and eastern United States records more than three times as many thunderstorms as does hilly East Asia in similar latitudes. No adequate explanation for East Asia's relative deficiency in thunderstorms has been given.

Certain other aspects of thunderstorm distribution are worthy of comment. A reasonably steep gradient of thunderstorm frequency prevails at about 30°N, which is approximately the latitude where a steep annual rainfall gradient is likewise located. It suggests the dominance of dry continental air masses in northern China for most of the year, while south of the Yangtze humid *mT* air is much more prevalent. Noteworthy also is the fact that the region of maximum thunderstorm frequency is inland from the east coast and concentrated in the interior of South China, with the isolines of 30 and 40 thunderstorms a year showing broad loops that extend in a northeast direction over interior southern China (Fig. 13.6). Along the whole southeast littoral south of about 23°N there are fewer than 20 thunderstorms a year. This pattern of distribution south of the Yangtze may be closely related to the areal dominance of either the southwesterly or the southeasterly air stream. Accordingly, the greatest number of thunderstorms is in those interior parts over which the unstable southwesterlies prevail, and there is a diminishing number along the eastern littoral where the more stable southeasterlies, derived from the Pacific anticyclone, reach their highest frequency of occurrence.

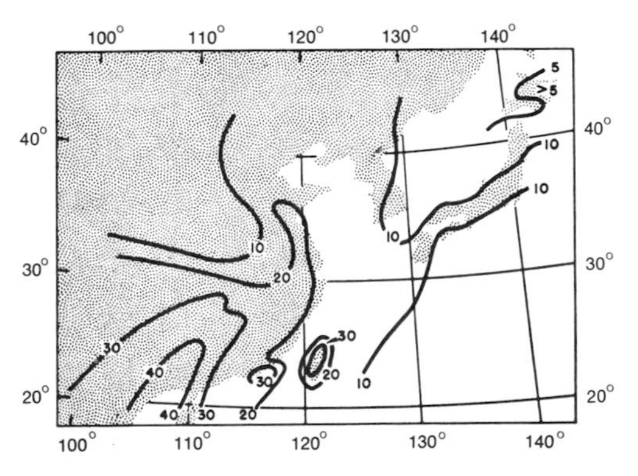

Fig. 13.6. Annual number of days with thunderstorms in eastern Asia. The isolines of highest frequency in southern China approximately coincide with the course followed by the equatorial southwesterlies in the warmer months.

Distinctive and Unusual Climatic Features

Temperature

Extreme winter continentality as expressed in such features as striking negative anomalies of mean January temperature, abnormally steep winter temperature gradients, and similar steep gradients of growing-season isolines, are not unexpected in eastern Asia. Somewhat abnormal, however, is the fact that in China south of about 35° the sea-level isotherms bend equatorward over the eastern parts, indicating a colder winter nearer the sea. In spite of a considerable rise in altitude toward the interior, Shanghai at sea level near the coast is 10° colder in January than Chungking deep in the interior and 230 m above sea level. This is due in part to the protective influence of the Tsingling and other highlands which shunt the northerly invasions of cold *cP* air eastward and at the same time impose a barrier against their entering the western basins and lowlands. It is possible, also, that föhns have a moderating effect in the interior basins.

The excessively steep winter temperature gradient characteristic of eastern China may result in part from an unusual feature of upper-air flow. In an earlier section it was pointed out that the high-troposphere westerlies flowing around the central Asian highlands on the north and south acquire different temperatures because of their contrasting latitudinal trajectories. When these two streams converge again in the lee of the highlands, the thermal contrasts are accentuated in the general region of the confluence. Moreover, the central Asian highlands which block the entrance of Siberian air into India and Burma, force the cold air to concentrate its southward movement farther east between longitudes 110° and 130°. But when the cold anticyclones come under the influence of the jet stream at about 30°N they may be steered abruptly eastward, which operates to emphasize a strong thermal contrast throughout the whole troposphere in the vicinity of latitude 30°N (Mohri, 1953).

Although insular, and for this reason characterized by winters somewhat less severe than those on the mainland of eastern Asia, Japan, nevertheless, has winter and spring temperatures which are 4° to 8° lower than those of corresponding latitudes along the Atlantic side of the United States and similar to those of the interior. Summer and fall temperatures in the two areas show less contrast. Emphatically Japan's is a continental climate, the cold winters being a consequence of the East Asia monsoon. Mohri (1953) suggests that Japan's steep temperature gradients in winter, like those in China, are in part a consequence of its location in a region of jet confluence.

Likewise somewhat unexpected is the fact that in Japan there are no important air-temperature contrasts in winter between the side facing Asia and the side facing the Pacific. On first thought one might anticipate that western Japan, which is exposed to strong invasions of fresh *cP* air from the continent, would be markedly colder in winter than the east side. To be sure, in terms of sensible temperatures, the windy, cloudy west side does feel colder in winter, but there is no significant difference in average January air temperatures. This is explainable in terms of the much greater prevalence of low cloud on the west wide (about 9/10 as compared with 5/10 on the east) which results in reduced night cooling.

In the warmer months the Pacific side of Japan north of about 37° or 38° has below-normal summer temperatures. This anomaly is caused by the cool Okhotsk current which washes the shores of northeastern Japan, the counterpart of which is the Labrador Current in eastern North America. Its prime effects are felt along the eastern, including northeastern and southeastern, littoral of Hokkaido where the cool foggy summers do not permit the successful maturing of a rice crop. While the chilling effects of the cold current are less pronounced in northeastern Honshu, even there the east-coast stations have July and August mean temperatures 1° to 3°C lower than those on the west side, and rice growing is still precarious.

Precipitation: China

Annual Precipitation.—A conspicuous feature on the map of annual rainfall is the contrast between the subhumid-semiarid north and the abundantly watered south. In the vicinity of latitudes 30° to 34°N the east-west isohyets are closely spaced, producing a steep gradient, so that north of about 34° precipitation is only half or less what it is to the south of parallel 30°. In part this is a result of the jet steering the cold anticyclones abruptly eastward after the northerly air arrives at about 30°N. Hence the rainfall contrast between north and south reflects the general prevalence of dry, cold anticyclonic air over North China during much of the year, the invasion of humid southerly currents in that region being felt for only a restricted period during the warmer months. By contrast, the south is either dominated by humid equatorial and tropical air, as is true in the warmer months, or is a region of air-mass conflict as is the case in the cooler seasons (Fig. 13.1). Typhoon rains are almost exclusively confined to the south. The Tapei Hills approximately mark the line of demarcation between the subhumid-semiarid north and the humid south.

Over most of China south of about 30°N annual rainfall exceeds 1,250 mm and in most parts it exceeds 1,500 mm. Something of an anomaly exists along the southeast coast, however, where in spite of its marine location, with highlands close to the coast, the frequent occurrence of maritime easterly winds in summer, and its exposure to tropical storms, the annual rainfall is no higher, and in places it is actually lower, than in the interior. Most of the coastal area has somewhat more than 1,250 mm, but parts in the vicinity of Amoy have only 1,000 to 1,250 mm and on some of the offshore islands surprisingly low rainfalls of only 750 to 1,000 mm prevail. As a rule the islands are drier than the coast, particularly in the warmer months. They also have a more uniform annual distribution of rainfall than does the mainland. A satisfactory explanation of this situation is not at present available. In part it may reflect a rainshadow effect in the lee of highland Taiwan. It is also the part of South China which in summer is most frequently under the influence of the more stable southeasterly air stream and is less subject to thunderstorm activity. Lee (1937) associates the decreased rainfall on the islands with the lower air temperatures and stronger winds there as compared with the mainland, which reduce warm-season convection. He also points out that owing to a cool current, water temperatures along the China coast are noticeably cooler than those along the west side of Taiwan, which is an additional feature tending to suppress summer convection. It is likewise possible that the lower rainfall along the littoral may be associated with the recurving northward over the sea of many of the tropical storms which approach this coast. A spell of fine weather invariably precedes these storms, and if the disturbance recurves so that the center is well offshore, the fine pre-typhoon weather may prevail for some time, or until the storm reaches the Korea Strait. Some ascribe this spell of fine weather to a föhn effect, but Ramage (1951) thinks it more likely is due to divergence in the lower troposphere associated with the forward semicircle of the storm.

Seasonal Rainfall Distribution.—Salient features of the winter rainfall map of China are (1) the meager precipitation north of about latitude 32° or 33°, (2) the fairly abundant rainfall throughout southern China except in the deep interior, and (3) the existence of a striking zone of concentration extending in a WSW–ENE direction across southern China.

Winter is the season of minimum precipitation throughout all of China, but it is emphatically the case north of about 32° or 33°N where anticyclonic air-mass weather is dominant in the colder months, and likewise in the deep interior farther south. At the surface or aloft maritime air in winter only rarely invades the region north of the Tapei Hills. Thus, in Manchuria only 3 to 4 percent of the annual rainfall occurs in the three winter months and this is reduced to 2 percent at Peking and Tatung in North China. It is these winter-dry areas that bear the Köppen symbols *Dw* and *Cw*; elsewhere in South China it is *Cf*. On first thought it may seem strange that northeastern China should be so dry in winter when it is also one of the two principal concentrations of winter depressions, and likewise the location of a winter jet stream. However, this northern jet fluctuates widely in location, and the depressions which are steered by it are operating in cold and stable anticyclonic air, so that the resulting precipitation is very modest in amount. Since the latitudinal position of the jet is highly variable, there is little tendency toward storm and precipitation concentration.

Winter rainfall increases rapidly to the south and southeast of latitudes 33° or 34°N, for this is a region of air-mass conflict and active density fronts, numerous weak disturbances, and a strong and locationally fixed jet stream. It is more expecially those weak perturbations entering China from the southwest, some of them no doubt having originated much farther west over India, which regenerate and intensity underneath the jet and are steered by it in an easterly direction across subtropical China. Significantly, the zone of greatest concentration of winter precipitation extends diagonally in a WSW-ENE direction across southern China, closely coinciding with the positionally stable axis of the southern jet (Yeh, 1950; Mohri, 1953) (Fig. 13.7). Rainfall decreases both to the north and to the south of the jet axis, but more rapidly to the south where the subsidence on the jet margins is most marked. The strong rainfall concentration in the belt noted above is confirmed by the fact that for some distance to the north and south of the jet axis, the gradient of mean winter precipitation is in excess of 200 mm per 160 km (Riehl et al., 1954). Yeh (1950) points out that there is also a belt of minimum winter rainfall variability which coincides with the belt of maximum winter precipitation and the position of the winter jet over southern China. Such remarkable coincidence between precipitation and the jet axis is possible only because the latitudinal fluctuations of the jet are slight, both from day to day and from year to year. As noted earlier, a small amount of the winter precipitation in extreme southeastern China is derived from tropical disturbances.

The *crachin* winter weather type described earlier for coastal Vietnam is likewise characteristic of the China coast southward from Shanghai. It is most common, however, in the extreme south.

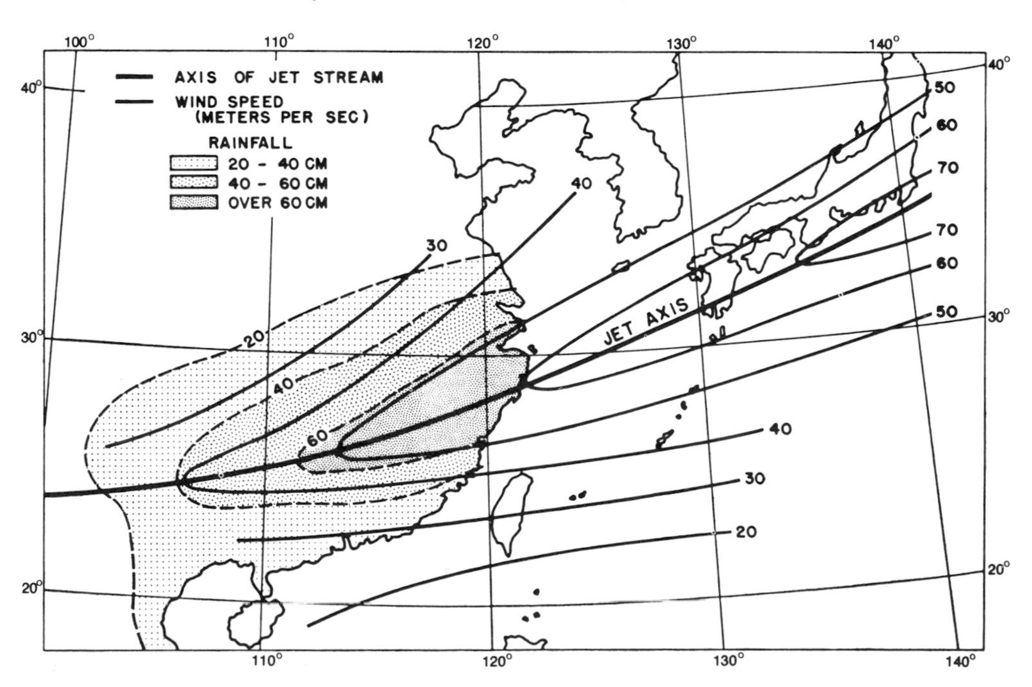

Fig. 13.7. Distribution of mean west wind speed (in m/sec) at 12 km over the Far East in winter (solid lines), and mean winter precipitation (broken lines) over China. The heavy line marks the mean position of the southern branch of the jet stream in winter (After Yeh, 1950, and Mohri, 1953.)

Spring shows an increase in rainfall over the whole of China although it is less in the north, where cold anticyclonic air continues to predominate, than in the south. Spring is the season of maximum cyclonic activity in South China and this is reflected in the greatly increased rainfall over that of winter, so much so, in fact, that May–June represent one of the two peaks of warm-season precipitation.

While spring receives less rain than fall in North China and Manchuria, in most of central and southern China it receives much more. Thus, Amoy on the southeast coast has only 12 percent of its rainfall in winter and 15 percent in fall, but 33 percent in spring (Table 13.1).

The origin of the spring rains is predominantly cyclonic. At this time air-mass contrasts are strong, southern maritime air is beginning to penetrate farther northward, the southern jet is still present over South China, and disturbances from northern India may still reach southern China. At Nanking, 84 percent of the spring rainfall is cyclonic in origin and 15 percent thunderstorm. At Canton the proportions are 82 percent cyclonic, 11 percent thunderstorm, and 7 percent typhoon (Biel, 1945).

Summer is the rainiest period throughout almost all of China, but the degree of summer concentration varies regionally, the north and the west having a much stronger summer maximum than the south and southeast. Thus, while summer accounts for 76 percent of the annual total at Peking and 65 percent at Haerpin, it shows only 38 percent at Hangchow, 40 percent at Amoy, and 46 percent at Canton. The most uniform annual march of rainfall is in the Yangtze Valley which is a main route of weak cyclonic disturbances. Advective and dynamic processes combine to produce a summer rainfall maximum. To be sure, summer is the season of maximum inflow of maritime equatorial air, but this in itself is an insufficient cause. Overwhelmingly the summer rains are associated with weak but extensive disturbances. Co-Ching Chu (1939; 1954) has pointed out that the stronger the summer monsoon in the lower Yangtze Valley in July, the less the rainfall in that area. But a stronger-than-average southerly flow in the Yangtze Valley, on the other hand, usually is associated with an abnormally wet summer in North China. This suggests that under such conditions *mT* air and the polar front and its associated disturbances are displaced farther poleward than usual. Chu also noted that frequency and amounts of summer rainfall were both less during periods of southerly winds than with all other wind directions, which clearly indicates that the rainfall is associated with perturbations and not with an undisturbed southerly flow. Most of the summer rainfall originates in weak disturbances, although typhoons provide an increasing amount in coastal locations. Thus while at Nanking 82 percent of the summer rainfall is of cyclonic origin, 10 percent thunderstorms, and 8 percent typhoon, at Canton the comparable figures are 54, 15, and 31.

Table 13.1. Seasonal distribution of precipitation in China in percent of the annual total (Yeh, 1950)

	Winter	Spring	Summer	Autumn
Haerpin	3	14	65	17
Peking	2	9	76	13
Hanchow	15	25	38	22
Hankow	10	33	41	16
Amoy	12	33	40	15
Canton	10	30	46	14

The Bimodal Summer Maximum.— An important feature of the summer wet season over extreme southern and southeastern China, and probably characteristic of a more extensive area that reaches as far northward as the Yangtze, is the two warm-season maxima separated by a secondary minimum occurring during the first half of July (Fig. 13.8). It is not known at present just how extensive this rainfall feature is, for monthly rainfall averages many times provide too coarse a screen for detecting it. Ramage (1952*c*), using 5-day overlapping means, has confirmed it as a feature of the annual rainfall distribution at 9 stations in southeastern China and Taiwan, and at 5 stations in southern Japan and the islands to the south. Even from the monthly rainfall averages it can be detected in scattered stations elsewhere in South China. The composite rainfall curve of the 9 Chinese stations used by Ramage shows a striking early-summer maximum with its peak between June 10 and 15. In Japan the maximum is about 10 days later. This May–June maximum is the so-called Maiu (China) or Baiu (Japan) season and its beginning coincides with the establishment of the genuine summer circulation pattern. When the southern branch of the zonal westerlies and the jet stream, which during winter and spring have held a course to the south of Tibet and across southern China, abruptly shift to a position north of the highlands, the equatorial southwesterlies, or the Indian monsoon current, push rapidly northward over India and likewise over China, Korea, and Japan. In eastern Asia this is sometimes referred to as the Maiu or Baiu current. Its establishment over China coincides with the first, or early-summer, maximum of precipitation, with the rainfall originating in a variety of weak disturbances within the southwesterly flow. It is largely non-frontal. Tropical storms begin to make an appearance in early summer also, but at that time they are not an important

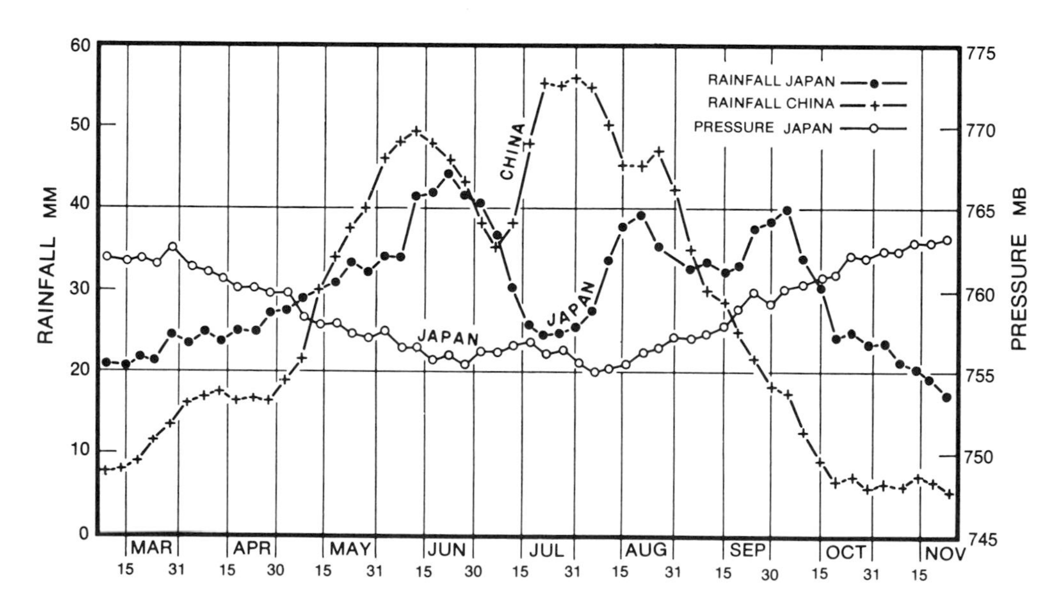

Fig. 13.8. Composite profiles of 5-day means of rainfall and pressure (overlapping method) for 5 stations in southern Japan, and composite of 5-day means of rainfall for 9 stations in southern China and Taiwan. (After Ramage, 1952*c*.)

weather element and probably add little to the first of the two rainfall maxima.

The midsummer secondary minimum in the annual rainfall curve, which occurs during the last half of July in South China and in early August in Japan, is associated with the intensification and northward and westward advance of the North Pacific subtropical anticyclone following the sun, and a decline in vapor pressure at the mid-troposphere level. Pressure rises over southern Japan and China, the Indian Ocean southwesterly current is interrupted more often or weakened as easterly trade winds from the Pacific increase in frequency, and rainfall decreases (Ramage, 1952*c*). Disturbances are forced to follow more northerly tracks and the upper-level surges in the southwesterlies are weakened or deflected. To be sure, there is an increase in the number of tropical storms, but their addition to the rainfall is insufficient to compensate for the other losses.

The second summer maximum (late July and August in South China, and mid-August to early October in Japan) is usually the primary maximum in South China, but this is not necessarily the case in Japan. According to Ramage this second maximum in Japan is itself bimodal, but the Chinese stations give only faint evidence of such a feature. In late summer the subtropical high weakens in the western Pacific so that atmospheric pressure in Japan and the China Sea reaches its lowest point in August which approximately coincides with the beginning of the late summer and fall rainfall maximum. The increased rainfall in late July and August in southern China Ramage (1952*c*) attributes to tropical storms, including typhoons, which not only are most numerous and most intense at that period, but in addition more of them travel in a westerly course toward the China coast instead of recurving before reaching the China Sea as they do earlier in the season.

Ramage (1952*a*) has found that over China south of Manchuria there is a striking diurnal periodicity of summer rainfall, although that is not true of the other seasons. In most Chinese stations there is a significant morning maximum, and at most inland stations and a few coastal ones an afternoon maximum as well. Thus, most Chinese stations appear to possess a twinned diurnal maximum, one in the morning and the other in the afternoon. The explanation of this climatic phenomenon is not so clear from Ramage's report. He states that the summer monsoon results in a maximum advection of warm air below 3000 m in the morning. This combined with radiational cooling of cloud tops, he believes, is responsible for the morning maximum. Along the coast the morning maximum may be accentuated by the movement onshore of the sea-breeze convergence. The afternoon second maximum may be a consequence of surface heating.

Fall witnesses a rapid decline in amount and frequency of rainfall as the winter circulation strengthens and cold anticyclonic air dominates East Asia farther and farther south. South of about 32 or 33° where frontal activity continues throughout the cool months, rainfall is moderately abundant, but over North China it is light, although somewhat more than in spring.

Precipitation: Japan

Japan lies in latitudes similar to those of subhumid and semiarid North China and Manchuria, but it is climatically unlike these Chinese counterparts in a number of respects: (1) the annual amounts of rainfall are two to three times as great and the annual reliability is likewise greater; (2) the summer concentration of rainfall is less marked; and (3) the nonperiodic weather element is more striking, for disturbances are more numerous and better developed over Japan than in China in similar latitudes.

These differences as they apply to Japan have their origins in a number of factors, most of which are locational. Japan's insular location assures a predominance of either genuinely maritime air or of continental air which has had important moisture and temperature modifications through a sea trajectory of several hundred kilometers. Moreover, the general highland character of the islands assures a marked orographic lifting of invading air masses. The country, likewise, is centrally located with respect to fronts, jet streams, areas of cyclogenesis, and storm tracks, so that the perturbation element is both strong and persistent. In the cooler months it lies within the domain of the oscillating polar front, both at the surface and aloft, which maintains an average southwest by northeast alignment seaward from the mainland, and hence roughly parallel with the archipelago. Japan's northern parts continue to be within the domain of the polar front even in summer, while its more subtropical parts feel the effects of weak disturbances within the southwesterly Baiu current. Asiatic jet streams are positioned over some parts of Japan at all seasons and act to regenerate disturbances in their vicinities and steer them through this area. In the cooler seasons that branch from south of the Himalayas has a position near to southernmost Japan, while at the same time the northerly branch of the jet oscillates over the northern parts (Mohri, 1953). And in the warmer months when the Himalayan jet has disappeared, the Siberian jet still continues to direct the flow of disturbances across northern Japan. In the general longitude of Japan there is likewise a relatively persistent upper-level trough which has a regenerative effect upon disturbances arriving from the continent as well as favoring actual cyclogenesis. Maps of cyclogenesis indicate that Japan lies very much in the midst of one of the earth's most active centers of origin for middle-latitude perturbations (Miller and Mantis, 1947). In addition, it is situated almost directly in the path of tropical storms which recurve northward around the western end of the North Pacific anticyclone, a route which is active throughout the year. Thus, more tropical storms affect Japan than China, which chiefly is influenced by that smaller number of cyclones which take a more westerly course.

Annual Precipitation.—All of Japan is humid, no section suffering from either an annual or seasonal deficiency of rainfall. Such a relative abundance of precipitation in all seasons is unusual for a continental climate in middle latitudes, a situation which is in contrast to that prevailing over most of China in comparable latitudes. It is difficult to generalize concerning the areal distribution of annual amounts of precipitation, for the variable relief and the limited size of the lowlands result in a confused and patchy rainfall map on which larger patterns are obscured by numerous closed isohyets and very circuitous courses of others. In general three areas of heavier-than-normal precipitation may be recognized: (1) the Pacific side south of about 34°N which faces the southerly air currents of the warmer months, where altitude is relatively high, and where storm tracks are concentrated; (2) the Japan-Sea side north of about 35° or 36°N where highlands face the humidified and moderated *cP* air masses of winter; and (3) the highlands of central Honshu. Rainfall of over 2,000 mm is characteristic of these three areas, 2,500 mm are common, and in isolated areas over 3,000 mm occur.

There are at least four areas which have distinctly less than the country average of precipitation, and where the annual total is under 1,200 mm: (1) most of Hokkaido; (2) large areas on the eastern side of northern Honshu; (3) the central part of the Inland Sea area; and (4) some of the interior basins of central Honshu. Over almost all of eastern Hokkaido rainfall is under 1,000 mm. Hokkaido's below-average rainfall reflects its peripheral location with respect to major regions of cyclogenesis. Those cool-season depressions which develop under the southern jet and move northeastward do not greatly affect this northern island. Genuine tropical storms likewise reach this area less frequently, and the same is true of northeastern Honshu. Moreover, the width of the Japan Sea in the latitudes of Hokkaido is not so great, so that the *cP* air masses entering Hokkaido in winter are less humidified and warmed than are those farther south where the Japan Sea is wider. Northeastern Honshu is definitely on the lee side during the cool season of strong northwest winds and experiences downcast winds. In all seasons it escapes the full effects of those cyclonic disturbances which move in a northeasterly direction along the south coast. Cool water along the east coasts of Hokkaido and northern Honshu likewise may inhibit the rainmaking processes in summer. The other two areas of below-average rainfall are definitely topographic rainshadow locations.

Seasonal Rainfall.—In contrast to the mainland, there are no winter-dry climates in Japan. Thus while Köppen's *w* symbol is widespread over North China, it is absent in Japan. This reflects the prevalence of more humid winter air masses in Japan and the much greater activity of disturbances, especially in the cooler months.

Over much the larger part of the archipelago precipitation is heaviest in the warmer months of the year and lowest in the cool months, a reflection of its continental climate. But, as noted previously, the proportion of the annual rainfall which falls in the warmer months is by no means as great as in North China.

Like subtropical South China, subtropical Japan, and even some stations as far north as northern Honshu and Hokkaido, exhibit two general peaks of warm-season rainfall, one in June and the other in September and late August, separated by a secondary July–August minimum (Fig. 13.9). As a rule, both the maxima and the secondary summer minimum are two to three weeks later than in southern China, so that the major dip in the top of the annual rainfall profile occurs in the latter half of July and early in August instead of the first half of July. Even when monthly averages of precipitation are employed, July and August usually show lower totals than June on the one hand and September on the other. But where overlapping 5-day means are used the early and late maxima, as well as the intervening minimum, are more strikingly revealed and more precisely located in time (Ramage, 1952*c*).

As noted previously, and illustrated in Figure 13.8, Ramage has described the second, or early fall, rainfall maximum in Japan, as itself comprising two peaks separated by a modest sag. However, his composite rainfall profile was constructed from data for five stations, only two of which were located in subtropical Japan proper. Annual rainfall profiles constructed from 5-day overlapping means for more numerous stations in subtropical Japan, cause the author to be unconvinced concerning the validity of a well-marked and widely distributed bimodal profile in the late summer-fall rainfall maximum (Fig. 13.9).

The causes for these temporal variations in summer rainfall in Japan were discussed briefly in conjunction with the analysis of comparable features in China's annual march

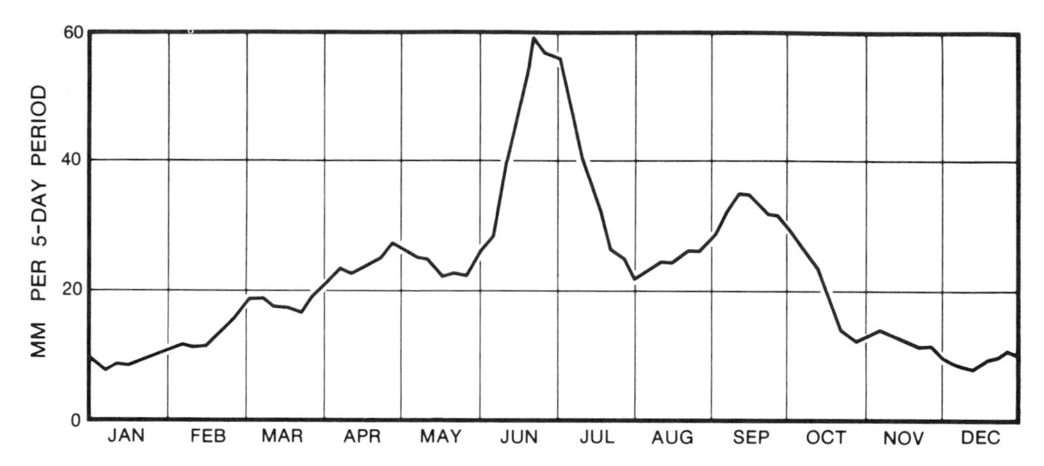

Fig. 13.9. Composite annual profile of 5-day means of rainfall (overlapping method) for 5 stations (Kumamoto, Nagoya, Okayama, Saga, Shimonoseki) in subtropical Japan.

of rainfall. It remains at this point only to amplify such aspects of the topic as are peculiar to Japan.

The Baiu Maximum.—Causes for the early-summer or Baiu maximum involve factors whose importance is weighed differently by various writers. Indisputably one element involved is the northeastward advance over China and Japan in early summer of a southwesterly air stream arriving from equatorial latitudes. In this southwesterly air stream high temperature and humidity and an unstable vertical structure create an environment in which above-surface surges and other weak disturbances are able to produce much cloud and precipitation.

The normal pressure for June shows a cell of high pressure over the cool Okhotsk Sea to the north of Japan, the subtropical high to the south, and a shallow low over the continent (Suda and Asakura, 1955). A sign of the approaching Baiu season is the intensification of the Okhotsk high and a deepening of the low over China. Some of the links in the chain of events are as follows. When the Indian jet suddenly shifts to the north of Tibet, and the Indian-Ocean southwesterlies set in with full force, conditions become ripe for the Baiu in Japan. The fact that the two events are so concurrent suggests a cause and effect relationship, which seems reasonable since the Baiu airstream of Japan originates over the equatorial waters south of Asia (Fig. 13.10). As the continental jet moves to a position north of Tibet in late spring, blocking action by a high occurs at about 110°E, which results in the splitting of the zonal westerlies into two branches downstream from this point. Japan is situated in the confluence of the two jets, the southerly branch usually being located south of 35°N, while the northern branch is north of Sakhalin (Yasui et al., 1955). This branching of the summer jet is fairly concur-

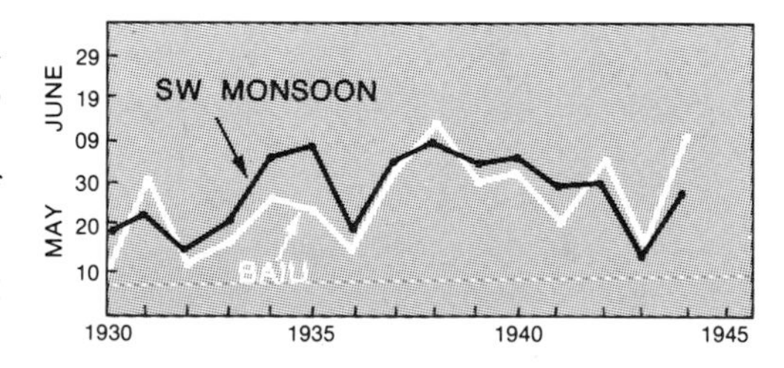

Fig. 13.10. Showing the degree of synchronization of the first appearance of the southwest monsoon in India and the beginning of the Baiu rains in East Asia. (After Suda and Asakura, 1955.)

rent with the development of the Okhotsk high and the intensification of the southwest monsoon. Upper-level pressure troughs from the continent travel slowly eastward over Japan. Air originating in the Okhotsk high moves far southward and forms fronts with tropical air along the coast of southern Japan, and along those fronts, positioned underneath the southerly jet, developing cyclones pass at intervals of two or three days which move toward the northeast, bringing rain to southern Japan (Katow, 1957) (Figs. 13.11, 13.12). Because of the frequency of Okhotsk air and the abundance of cloud, temperatures are moderate during the Baiu period, although sensible temperatures are relatively high. The southern branch of the jet is always to be observed over Japan during the Baiu season and it is when this jet suddenly disappears and Japan no longer lies within the field of confluence between the two branches of the jet stream that the Baiu season terminates (Murakami, 1951). Significantly, the ending of the Baiu rains likewise coincides with a greatly increased prevalence of easterly winds, indicating that the Pacific trades have increased at the expense of the Indian-Ocean southwesterlies (Nakamura and Arai, 1957).

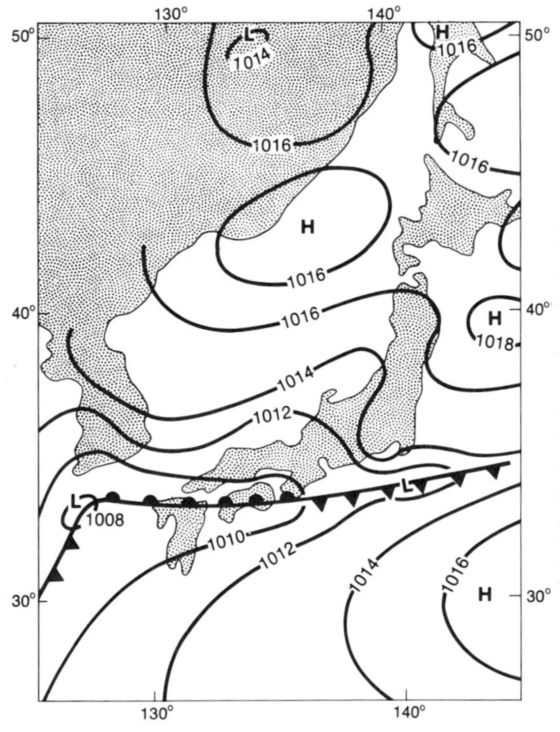

Fig. 13.11. Showing a typical location of the Baiu front over southern Japan. (After Katow, 1957.)

From a study of the pressure record of 34 years, Oosawa (1951) finds that even in the Baiu period there are a number of singular points in the station pressure curves. The first one, a relatively weak pressure trough, occurs about June 5, with others of greater intensity following on about the 18th and the 27th. The two maximum rainy spells within the general Baiu season are found to coincide with the later troughs and their accompanying depressions. Around the 21st of June when a pressure ridge appears the Baiu rains slacken. During the first Baiu rainy spell which begins about the 18th of June, it rains almost every day, and because there is much cloud the weather, while not so hot, is overcast and oppressive. During the rainy period coinciding with the second major pressure trough the rainfall is usually heavier and the temperatures are higher because the lows are traveling farther north. It has been suggested that the Baiu rains are heavier over Japan when a single jet prevails and less when there is a double jet (Matsumoto, Itoo, and Arakawa, 1954). That is, the rains are heavier over Japan when the confluence of the jets occurs over the coastal regions of East Asia than when the juncture occurs farther east.

The rains slacken and the Baiu rainy season is brought to a close as the subtropical high advances to its northernmost position in midsummer, and both pressure and temperature increase over Japan, while water-vapor pressure at the 700 mb level declines. Con-

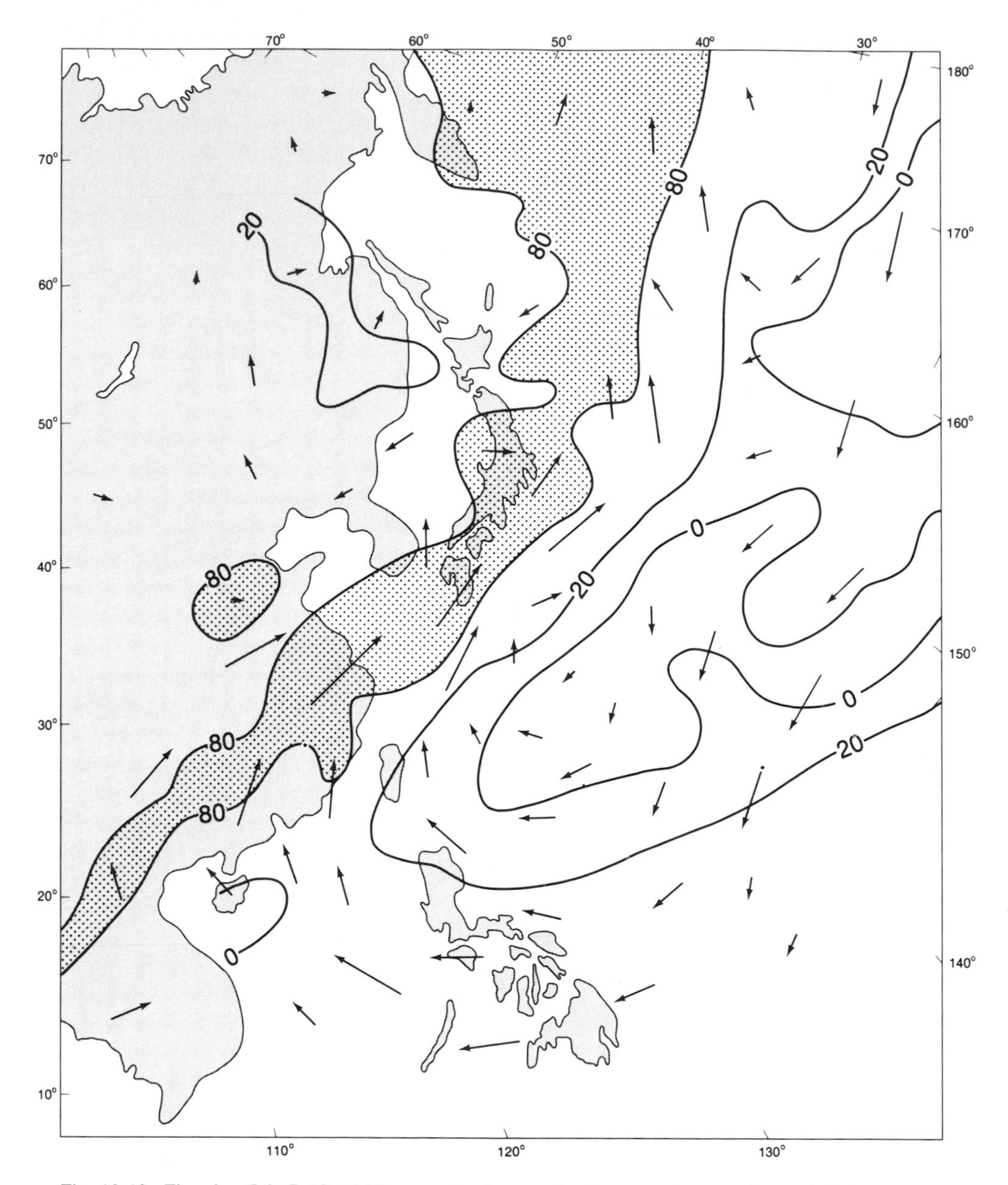

Fig. 13.12. Five-day (July 7–12, 1969) mean cloudiness and water-vapor transport (arrows) during a recent Baiu season in East Asia. (After Asakura, 1971*b*.)

currently the Okhotsk high weakens, the southern jet retreats northward, more interruptions occur in the southwesterly flow as the tropical easterlies intrude, and atmospheric disturbances become fewer. Thus begins the midsummer secondary minimum.

The Shurin Maximum.—As the subtropical high begins its retreat in late summer and pressure falls, the equatorial southwesterly flow again temporarily strengthens, and rainfall increases, producing the second peak in the rainfall profile, that of late summer and early fall. A large part of this fall rainfall maximum occurs in conjunction with tropical storms forming in the general area of the Carolines, which recurve northward around the western end of the subtropical high (Ramage, 1952*c*) (Fig. 13.13).

This second of the two warm-season peaks of rainfall beginning in late August and lasting into October, which is also the typhoon period, is known as the Shurin rains. In many of its features Shurin resembles Baiu, for there is a high-pressure cell over the Okhotsk Sea and a stationary front over southern Japan along which moves a succession of depressions. Cool northeasterly air from the Okhotsk high, much cloudiness, and frequent rains cause air temperatures to be appreciably lower than they are in drier July.

Asakura (1957) interprets the Shurin rainy season to be a consequence of modifications in the zonal westerlies which are of hemispheric proportions. In midsummer, preceding the Shurin, he found there existed four pronounced hemispheric waves, one ridge of which was positioned in the vicinity of Japan. With the onset of Shurin the wave number changed to three and the previous ridge over Japan was replaced by a trough. Concurrently what had been a negative height anomaly of the 500 mb surface over the North Pole changed to a positive height anomaly. According to Asakura it was the positioning of the pressure ridge over Japan in midsummer that was responsible for the high temperatures and relative dryness of July. When a trough replaced the ridge in late August there resulted an influx of cooler northeasterly air, followed by increased frontal activity, more depressions, and a greater frequency of rainfall, all acting to initiate the Shurin season.

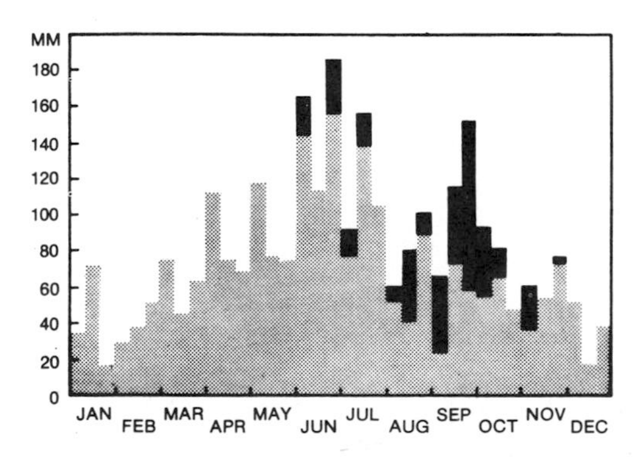

Fig. 13.13. Average 10-day precipitation amounts (1951–1955) for Shionomisaki in Japan. Precipitation resulting from typhoons is shown in black. Precipitation of other origins is shaded. Typhoon rainfall is heaviest during the September–October maximum. (After Saito, 1959.)

An unusual feature of Japan's summer rainfall is the fact that a majority of the stations show a relatively strong maximum frequency of occurrence in the early-morning and late-night hours, 0400 to 0600 representing the peak period in most instances. No adequate explanation of this peculiarity has been presented. A secondary diurnal rainfall maximum is evident between the hours of 1200 and 1800 (Landsberg, 1944).

Winter is the least rainy season throughout most of Japan, but unlike China in similar latitudes the cool season is by no means dry. Thus the ratio between the wettest month of summer and the driest of winter is 29 to 1 at Haerpin and 100 to 1 at Peking, while it is only 5 to 1 at both Tokyo and Osaka. The prevailingly wetter winters in Japan than in similar latitudes in North China, or even in most of South China, are related to the much more numerous and vigorous cyclones in the Japan area, the reasons for which have been noted previously.

Maps of Seasonal Rainfall.—The maps of January and July precipitation show important distributional contrasts. The July map is noteworthy in that it exhibits no very strong windward-leeward rainfall effects. In this season the flow of maritime air from the south, including southeast and southwest, is weak, which tends to make for only modest orographic effects. To be sure, the highlands of Kyushu, Shikoku, and Kii peninsula in Honshu facing most directly the weak on-shore drift do show somewhat higher totals in July than the northern side, but farther north this is not the case. From the latitude of Tokyo northward, the Japan-Sea littoral has more summer rain than does the cooler Pacific side, which is bordered by a cool ocean current.

The January rainfall map shows quite a different pattern. Most striking is the strong concentration of precipitation along the windward or Japan-Sea side of the country, so that, unlike the condition in summer, there is a marked windward and leeward rainfall effect. This seasonal contrast derives chiefly from the contrasting strength of the seasonal winds, those of winter being much stronger. The cold, dry winds from Asia are significantly warmed and humidified over the Sea of Japan, so that they yield abundant precipitation when forced to ascend the western highlands. Stations along this western side show a winter cloudiness of 8/10 to 9/10, and in January there are around 25 days with precipitation and 15 to 27 days with snowfall. Since much of the winter precipitation falls as snow, the accumulation on the ground becomes very deep. Snow sheds are conspicuous features along the rail lines. Accordingly, along most of the Japan-Sea side of Japan north and east from Matsue, winter is the season of maximum precipitation amounts and also of days with precipitation. But summer is likewise wet, only less so than winter.

Like China, Japan is a region with relatively few thunderstorms. This indeed is strange considering the subtropical latitudes of much of the country, the prevalence of tropical maritime air in summer, and the rugged terrain. North of about latitude 38°, days with thunderstorms are fewer than 10, and by far the largest part of subtropical Japan has fewer than 20 such days.

PART V
Europe and the Mediterranean Borderlands

CHAPTER 14

Western and Central Europe

Europe has the distinction of being climatically the most maritime of all the continents. Nowhere else are the *Do* climates so extensively developed. This uniqueness has its origin in the strong and persistent westerly circulation derived from the combined effects of two centers of action, Icelandic low and Azores high, only modest development of meridionally oriented highlands along its Atlantic windward margins, and the presence of an extensive area of unusually warm water in the eastern North Atlantic. In an eastward direction, the western limb of the Siberian anticyclone makes its influence progressively more frequently and more strongly felt during the colder months. On occasions when the zonal westerly flow is weakened, an easterly circulation derived from this thermal high may carry severe cold weather westward to the Atlantic margins of the continent.

Europe's climates are likewise greatly influenced by the fact that along much of its subtropical southern margins it is bordered by extensive bodies of water—the warm Mediterranean and the smaller and less warm Black and Caspian seas. Most certainly the southern peninsulas have their temperatures moderated and their rainfall accentuated by the direct effects of the warm Mediterranean surface. Indirectly, also, the effects of the Mediterranean are felt through the fact that it is a major region of cyclogenesis and cyclone propagation in all but the summer season. The southern branch of the jet stream not infrequently overlies the Mediterranean route in the cooler seasons, acting to steer the atmospheric disturbances along this subtropical course. Nowhere else on the earth is there a subtropical concentration of winter cyclonic activity with such a longitudinal spread, the consequence being that dry-summer subtropical, or Mediterranean, climate (*Cs*) is carried 3,000 km or more inland from the Atlantic, while the effects of the depressions following the Mediterranean tracks are felt for another 3,000 km farther eastward in the steppes and deserts of southwestern Asia where a winter maximum of rainfall and summer drought are the rule.

Most of the distinctive climatic features of Europe are best treated in the regional analyses to follow. At this point it may be useful to outline briefly the general features of the annual march of rainfall for the continent as a whole, which will provide the necessary background for the more detailed regional treatment to follow.

A map showing the regional distribution

of the season of maximum rainfall reveals two general regions where a winter maximum is prevalent: (1) over the eastern Atlantic and adjacent parts of the continent, including westernmost Scotland, Ireland, and extreme southwestern England, and (2) the southern and eastern Mediterranean. The core area of summer maximum is the continent inland from the Atlantic and north of the Mediterranean. These two contrasting types of seasonal rainfall maximum are separated by belts of variable width in which neither summer nor winter, but rather the intermediate seasons are richest in precipitation.

The regional distribution of the seasonal minimum of rainfall is somewhat simpler. Throughout the Mediterranean basin, and including Iberia and the Biscay coast of France, a summer minimum predominates. The borderlands of the rest of the Atlantic and also the Baltic are characterized by a spring minimum. Over the remainder of the continent, and including the interior of Scandinavia, winter is the season of minimum precipitation.

This introductory section on Europe may appropriately be terminated by a brief description of the climatic arrangement in Europe following the Köppen scheme of classification (Fig. 14.1). Since most of Europe is freely open to invasions of the maritime zonal westerlies, the change eastward from marine to continental climates is very gradual. As a result, only in Europe are humid continental climates (*Da-Db*) found on the windward or western side of the dry interior. In both North America and South America, the *Cb* climates along the west coast change abruptly to dry climates (*B*) east of the mountains. But in Europe the Atlantic moisture enters so deeply that going eastward continental temperatures, including severe winters, are reached before rainfall drops low enough to produce a dry condition.

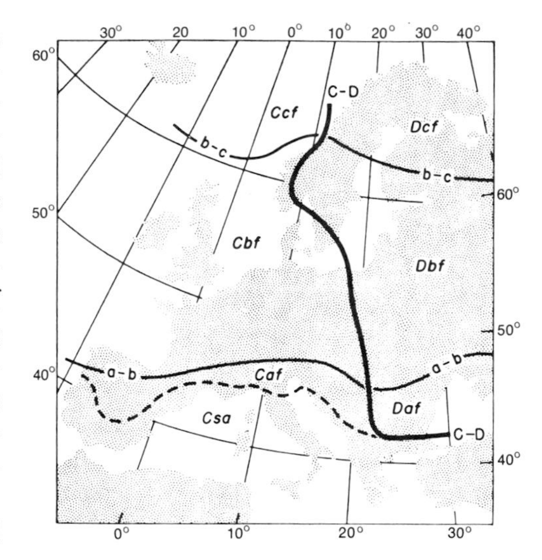

Fig. 14.1. Major climatic boundaries in Europe according to the Köppen classification. Note that the *C/D* boundary, based upon a winter-month isotherm, has a general north-south alignment. By contrast, the *a/b* and *b/c* boundaries, defined by summer-month isotherms, trend in a general east-west direction. (After Gregory, 1954.)

Nevertheless, in the continental climates of east-central and eastern Europe, separated by hundreds of miles from the Atlantic, the marine effects are still moderately obvious, so that for any latitude temperatures are not as severe as in their climatic counterparts in eastern Asia and central and eastern North America. Consequently, it is difficult to find homoclimes for the continental climates of central and eastern Europe in Asia and North America. (See also *Do–Dc* climates in Europe as shown on endpaper map.)

Gregory's maps show the *C/D* boundary as having a strong meridional trend with a tendency toward a SSE by NNW alignment, but becoming E–W in the extreme south in Bulgaria where the influence of the Mediterranean Sea becomes effective (Gregory, 1954). This general N-S alignment of the *C/D* boundary is to be expected since it is based on a cold-month isotherm. On the other hand, the *a*/*b* and *b*/*c* boundaries, which are defined in terms of warm-month temperatures, have a marked east-west alignment. Thus on the oceanic side of the *C*/*D* boundary the succession of climates from north to south is *Cc, Cb, Ca* and this is paralleled by *Dc, Db, Da* on the continental side. But the *Ca* subdivision is divided into two parts, the much more extensive *Csa*, or Mediterranean, type which includes much of the Mediterranean's borderlands, and the *Caf* subdivision which comprises parts of central Iberia, southern France, the Po Basin, and a considerable part of lowland Yugoslavia and Hungary. This is a unique location for a humid subtropical climate which normally is situated on the subtropical eastern side of a continent. In Europe, however, *Caf* lies on the poleward side of a *Cs* climate which is most unusual, but this becomes possible because *Cs* extends so far inland from the ocean. Accordingly, the warm summers and mild winters of the *Caf* are a consequence of the protective influence of highlands to the north and of the warm Mediterranean to the south.

Chiefly as a matter of convenience the discussion of European climates is organized under three regional headings: (1) western-central Europe, where the Atlantic influence is strong, but gradually wanes eastward; (2) the Mediterranean borderlands, including southwestern Asia and northern Africa; and (3) the Eurasian northlands, chiefly included within the USSR, where the continental influence is paramount.*

Within the subdivision, western-central Europe, is included the general area north of the Mediterranean lands and west of the USSR. It is not a single climatic type. In general, it includes the regions of strong Atlantic influence and also the transitional area of central Europe where, proceeding eastward, there is decreasing marine, and intensifying continental, sway. Because of the lowland nature of most of this region the transition from marine to continental character is slow and gradual, and sharp climatic divides are usually lacking.

* In this book no attempt is made to analyze the climates of the USSR. For a very recent and comprehensive treatment of this region see Paul E. Lydolph, ed., *Climates of the Soviet Union*, "World Survey of Climatology," vol. 7 (New York: Elsevier, 1977).

Temperature

At first glance western-central Europe appears to have a temperature structure which is approximately what might be expected in an area largely lacking in meridional highland barriers, and lying on the windward side of a continent in a latitude mostly poleward of 40°N. Still there are certain modest regional peculiarities.

Atlantic air consistently enters more deeply into western-central Europe in the warmer months than in winter. This is reflected in the fact that the sea-level isotherms have a zonal trend in summer and a strongly meridional one in winter. Flohn (1954) has pointed out that this contrasting arrangement is the result of the unlike seasonal circulations. In summer the general circulation is zonal and from the west, and any weak monsoonal effect of the continent operates to produce a flow in a similar direction. Consequently the two controls are cooperative in their effects and the resulting isothermal pattern is zonal. In winter, on the other hand, the planetary westerly flow is opposed by the easterly flow generated by the cold anticyclone over interior Eurasia, resulting in a conflict of air masses, with the meridionally oriented frontal zones shifting aperiodically eastward and westward as now one, and then the other, air mass is dominant. The result is a meridional alignment of the sealevel isotherms.

The effects of these controls are likewise expressed in the seasonal temperature anomalies. The unusual warmth of the waters in the eastern North Atlantic in winter, combined with the strong cyclonic circulation over the ocean, is opposed by the cold anticyclone over the continent, with the result that there is a strong plus anomaly of 28°C (50° F) located some 300 km off the coast of Norway at about 65° to 70°N, and a negative anomaly over eastern European USSR. The isanomalous lines thus have a SW by NE trend and the isanomalous gradient is relatively steep.

In July the eastern North Atlantic between Iceland and Norway is too cool, rather than too warm, for its latitude, but the degree of departure is only about 4°C (7°F). Thus, while the coast of Norway is 17° to 22°C (30° to 40°F) too warm for its latitude in January, it shows little or no departure from the latitude temperature normal in July. Since in summer the sea has only a modest negative anomaly and there is a zonal air flow from the west, it is to be expected that isanomolous temperature gradients in that season will be weakly developed over the continent. As a result, while the zero isanomal is on, or close to, the coast in July, it is 800 km or more inland in January.

The zone of contact between unusually mild maritime air and cold dry continental air is especially sharp in January along the coast of Norway. Accordingly, the coastal station of Bodö (67°13′N) is prevailingly under the influences of continental east winds, while Röst, on the southern extremity of the Lofoten Islands, is dominated by south and southwest winds (Sandström, 1926). This frontal location not only accounts in part for the very heavy winter precipitation in western Norway, but also develops strong isanomalous temperature gradients in this same region. In winter the Scandinavian Highlands are coincident with one of the sharpest climatic divides in Europe. To be sure, the Baltic Sea and Gulf of Bothnia carry the Atlantic influence northwards and inland behind the highlands, somewhat ameliorating the continentality of their borderlands.

Abnormally Cool Uplands.—A distinguishing temperature feature of the maritime sections of northwestern Europe is the very rapid decrease with altitude in the length of the growing season. For this reason a remarkably low tree line is characteristic of central and northern Britain, landscapes resembling tundra being conspicuous in flattish areas of the Pennines at elevations as low as 500 meters (Manley, 1952). In cool temperate climates the growing period for crops and vegetation is approximately determined by that part of the year during which the mean daily temperature is above 6°C (42°F). Since sea-level temperatures in the warmer months are relatively low, it requires only a modest increase in altitude to drop the mean daily temperature below 6°C (42°F). The rapid shortening of the growing season with altitude ultimately depends on the degree of dominance of polar maritime air in Atlantic Europe (Manley, 1945). In such air masses strong winds, frequent low cloud, and a steep lapse rate are characteristic. In northern Britain, for example, the rate of fall of the mean temperature with height is about 1°C for 146 m (1°F for 270 ft), and it is considerably greater on occasions when fresh *mP* air prevails. The altitudinal rate of change in the effective growing season is nearly twice as rapid in Britain as in New England (Manley, 1945). Consequently, even low highlands in maritime northwestern Europe are likely to have unusually low temperatures, a short growing season, and a native vegetation which resembles tundra. It is worthy of mention, also, that in those areas where cool maritime air predominates, the vertical rainfall gradient, like that of temperature, is relatively steep.

Effects of Anticyclonic Blocking.—Across the North Atlantic and most of Europe the westerly flow at the 500 mb level is strongly zonal. Not infrequently, however, this zonal flow is terminated, and a cellular pattern substituted, when a quasi-stationary high in the eastern North Atlantic in the vicinity of 5°W to 15°E splits the westerlies into two branches, so that a double jet stream results (Rex, 1950; 1951). One branch flows northward (about 75°–80°N) around the blocking high and the other southward (about 36°N). There is a distinct seasonal periodicity in blocking activity, with a general low incidence from June to November and a primary minimum in July-August-September (Fig. 14.2). The maximum is in April, while December to June is the general period of most frequent blocking (Rex, 1950). During the period of maximum 30 to 40 percent of the days are affected by anticyclonic blocking, while during the minimum period it is under 20 percent.

As one might expect, the blocking action of the warm high and the resulting bifurcation of the jet stream cause a dislocation of the routes of traveling disturbances. The general effect is to reduce the number of active centers moving eastward in the vicinity of 50°–55°N, and to concentrate them to the north and to the south of these more central latitudes of Europe, in the vicinity of the northern and southern branches of the jet.

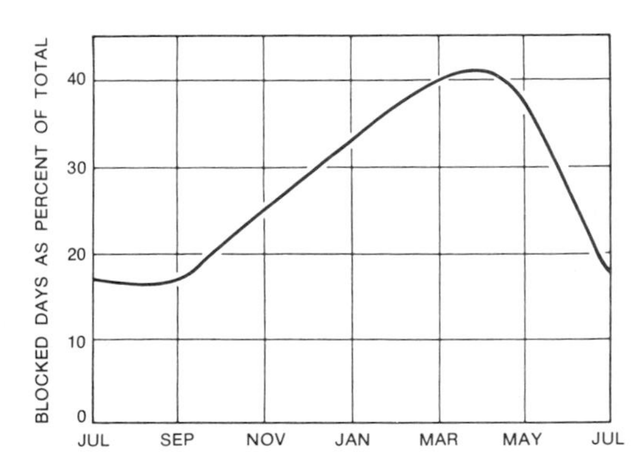

Fig. 14.2. Frequency of blocking action by a quasi-stationary high in the eastern North Atlantic, a feature which importantly affects the weather of western Europe. (After Rex, 1951.)

Thus, one main route of disturbances lies well to the north over Iceland, Spitzbergen, and North Cape and is associated with that branch of the jet following a course around the northern side of the blocking high. A second track follows the southern jet into Mediterranean latitudes. Two other disturbance tracks are associated with the converging branches of the jet to the lee of the blocking high. One of these marks the path of perturbations moving south and southwest from northwestern European USSR toward the Low Countries. The other follows the southerly jet as it bends northward over southern USSR, with the storms traveling northward into European USSR from the Black Sea area.

The effects of blocking action are felt sufficiently often so that they appear to leave their imprint upon the general pattern of disturbance routes in Europe as shown by van Bebber's (1882) map of storm tracks and A. Schedler's (1924) maps representing the yearly frequency of cyclones by means of isolines (Figs. 14.3, 14.4). Considering all disturbances with central pressures of less than 1013 mb, it can be noted that the two areas of concentration are (1) across Scotland and Scandinavia and (2) along the axis of the Mediterranean Sea, the two areas of jet positioning at the time of blocking activity. By contrast, the area with a minimum of cyclonic activity is in the general vicinity of parallel 50°N where the blocking high tends to obstruct the advance eastward of the Atlantic cyclones. Much the same regional pattern of storm activity is shown on the map which represents the deeper cyclones (central pressure <1000 mb) only. If maps of seasonal, instead of annual, cyclone frequency are employed, there is still further corroboration of the effects of blocking action (Figs. 14.5, 14.6). Deep cyclones in winter are concentrated to the north of 55°N and also in the central Mediterranean, while there is a minimum of storm activity in the vicinity of parallel 50°N. In summer deep cyclones have nearly disappeared entirely from the Mediterranean Basin and there is a distinct concentration north of latitude 55°N.

Important effects of blocking action upon weather and climate are to be expected. Rex (1950; 1951) has shown that during a period of winter blocking action a tongue of posi-

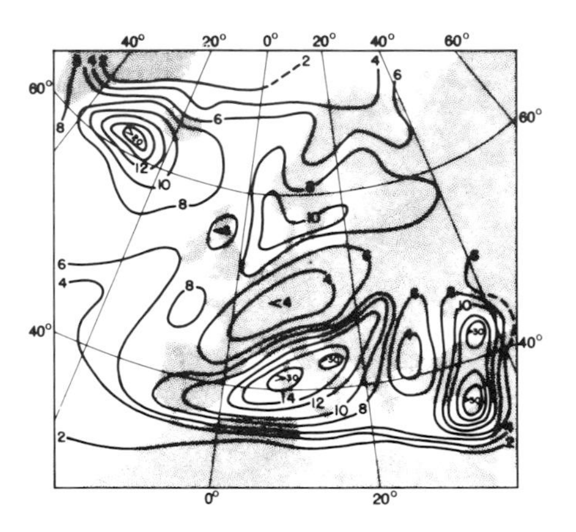

Fig. 14.3. Annual frequency of cyclones (with central pressure less than 1,013 mb) in Europe. High frequency is characteristic of the Mediterranean-Black Sea area in the south, and likewise of the higher latitudes in the vicinity of 60°N over the ocean south of Greenland. Low frequency is characteristic of western and central Europe in the general proximity of latitude 50°N. (After Schedler, 1924.)

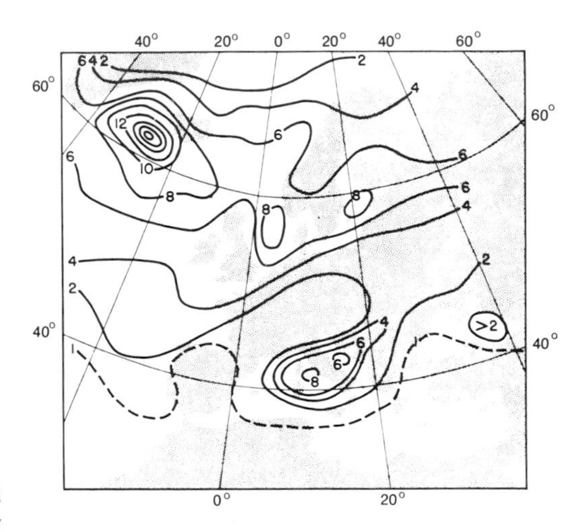

Fig. 14.4. Annual frequency of deep cyclones with central pressure less than 1,000 mb. The distribution is similar to that described for Fig. 14.3. (After Schedler, 1924.)

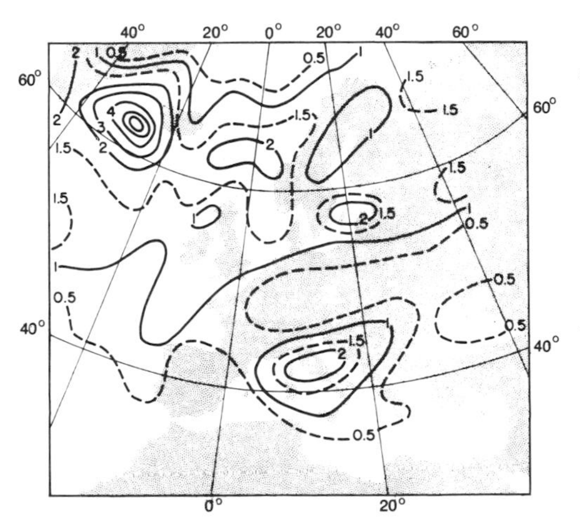

Fig. 14.5. Frequency of deep cyclones with central pressure less than 1,000 mb in winter. Minimum frequency is conspicuous in the general vicinity of latitude 50°N. (After Schedler, 1924.)

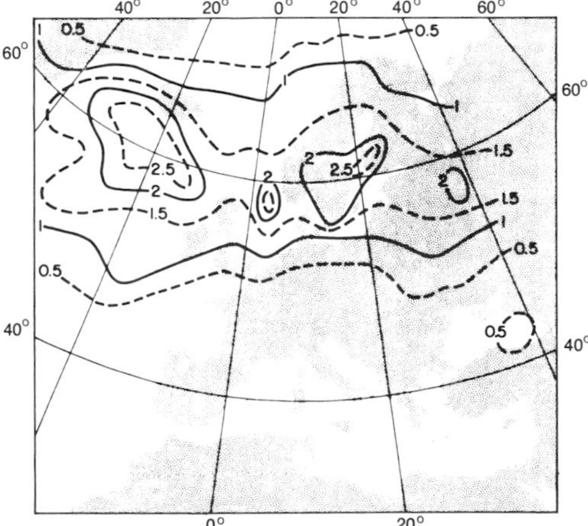

Fig. 14.6. Frequency of deep cyclones with central pressure less than 1,000 mb in summer. Deep cyclones are concentrated along the 60°N parallel, and the Mediterranean center is lacking. (After Schedler, 1924.)

tive temperature anomalies coincides with the western and northern margins of the blocking high where a strengthened southwesterly flow prevails (Fig. 14.7). Most of Norway, Sweden, Finland, northwestern European USSR, and the whole North Atlantic between Greenland and Scandinavia, on such occasions, may be warmer than normal. Negative temperature anomalies, by contrast, are concentrated to the rear of the blocking high where cold *cP* surface air, following the path of the northern jet, moves southward into a well-developed pressure trough which is very conspicuous at the 500 mb level (Rex, 1950; 1951). The trough and its polar air are very prominent in the vicinity of parallel 50°N, the region of least cyclonic activity. Thus, on occasions of winter blocking the strongest negative anomalies of temperature are to be found in an east-west belt along about latitude 50°N that extends from southern England, France, and the Low Countries, across Germany, the Danube countries, and into southwestern USSR. Negative anomalies are likewise characteristic of the Mediterranean, which feels the invasions of cold air from the north along the upper-air trough, located downstream from the blocking high.

Summer blocking appears to produce effects upon temperature distribution in Europe similar to those of winter (Rex, 1950). While extreme western and northern Europe show weak positive anomalies during blocking episodes, much of central and eastern Europe, together with the western Mediterranean, is characterized by a negative temperature anomaly (Fig. 14.8). The same mid-troposphere trough of low pressure lying eastward of the blocking warm high tends to channel cool northerly air southward over much of the continent.

In summary, the absence of a blocking high and the existence of a strong zonal flow of westerly air with a single jet stream result in stronger positive anomalies of temperature over most of western-central Europe, especially in winter, and precipitation amounts that are also above normal. With blocking action, on the other hand, temperatures are below normal over most of the continent except in the extreme north and northwest, and rainfall is below normal except in local areas. It is to be expected, therefore, that

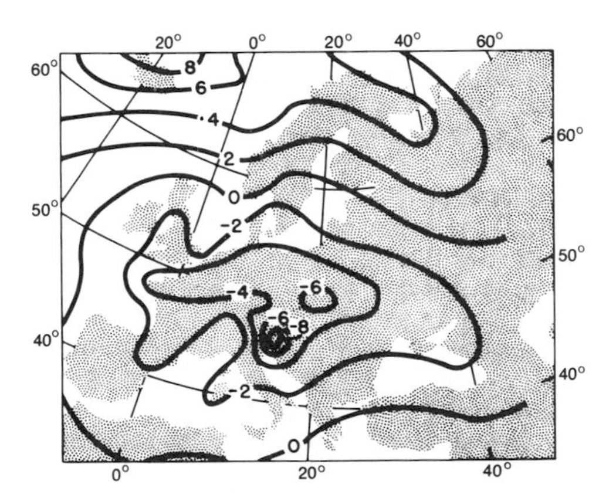

Fig. 14.7. Mean European surface air temperature anomalies in *winter* during occasions of blocking action by an anticyclone in the easternmost Atlantic and adjacent western Europe (in °C). (After Rex, 1951.)

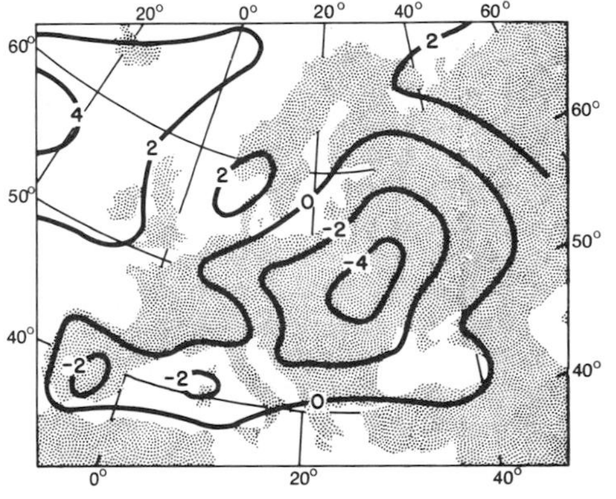

Fig. 14.8. Mean European surface air temperature anomalies in *summer* during occasions of blocking action by an anticyclone in the easternmost Atlantic and adjacent western Europe (in °C). (After Rex, 1951.)

variations in the frequency of blocking activity will be reflected in the variable seasonal weather from one year to another, and such is the case (Rex, 1950).

In view of what has just been said, it appears that the frequent blocking action may contribute to an understanding of some of the temperature peculiarities of Europe as described earlier. In winter the strong plus anomaly of temperature over the northeastern Atlantic and adjacent coasts of northwestern Europe is related to the positive action of the strong southwesterly flow of air and its cyclones which follow the northerly branch of the bifurcated jet. The sharp decline in the positive temperature anomaly to the east and south reflects not only increasing continentality, but likewise the effects of the southward surges of cold air associated with the northerly jet and the cold trough aloft over the continent. The fact that isolines of the winter temperature anomaly have a distinct SW-NE trend south of about 55°N suggests the effect of these northern invasions of cold air east of the blocking high. Likewise, the north-south alignment of the January isotherms and the relatively steep January temperature gradient across France and Germany both point to a moderately sharp boundary between maritime air from the southwest over the sea, and the not infrequent thrusts of cold northeasterly air over the adjacent lands.

In summer the near absence of significant temperature anomalies over most of Europe suggests that the continent does not warm up in summer to the degree that one might expect. This may be interpreted as due in part to the cool air from the north entering Europe along the high-level trough east of the blocking warm high. To be sure, summer and fall are the seasons of only minimum blocking action, and of maximum zonal flow, so that the effects of blocking action upon summer temperature characteristics are subordinate to those of the stronger zonal flow. The effects of blocking action also appear to be reflected in the isolines of thermal continentality as described by Holm (1953).

It is a temperature feature worthy of emphasis that western-central Europe has a minimum amount of warm-summer climate. If one excludes southern France and the Po Valley, which really belong to the Mediterranean Basin, the only sizable area of warm-summer climate northward from the highlands that separate the Mediterranean lands from western-central Europe, is to be found in the Balkan Peninsula, where in the plains of the lower Danube (Romania, Bulgaria), and possibly the middle Danube (Hungary, Yugoslavia) as well, the warm-month average temperatures in most years may exceed 22°C (71.6°F). That the summer warmth is greater here than in most of western-central Europe is partly a matter of lower latitude, but in part, also, it reflects an increasing distance from Atlantic influences.

Precipitation

Annual Amount

Two noteworthy features of the total annual rainfall in western-central Europe are (1) that rainfall declines but very slowly inland from the Atlantic and (2) that only modest amounts of rainfall characterize the European lowlands in general.

Slow Decline of Rainfall Inland.—Although the annual rainfall on the lowlands of Atlantic Europe is modest (commonly 500–750 mm), the decline inland is so slow that the annual isohyet of 500 mm is not encountered until east of the Urals, or 3000 km from the Atlantic. If a line is drawn from about Hamburg through Vienna to Belgrade, the lowlands to the west and south of the line usually have over 600 mm of annual rainfall, while to the east and north, with a few significant exceptions, the amounts are less. Actually, Budapest and Belgrade have slightly more rainfall than Paris, and Cambridge and London in southeastern England. Still, the axis of the continental zone of heavier precipitation extends in a SW-NE direction across central-eastern Europe, suggesting that it is in the southwesterly air streams and their cyclones, derived from the Atlantic source region, that much of the interior precipitation originates.

The very slow decline in annual rainfall inland attests to the easy entrance of Atlantic moisture and of cyclonic disturbances from the west, except on occasions when blocking highs interfere. Noteworthy is the fact that cool-season rainfall declines somewhat more rapidly toward the interior than does that of the warm season, suggesting that the increased moisture capacity and greater thermal turbulence of the warmer summer air inland operate to prevent an accelerated decline in summer rainfall toward the interior. The greater frequency of blocking activity in the eastern Atlantic in winter than in summer, and the increased prevalence of a cold anticyclone over central-eastern Europe in winter, likewise contribute to the more rapid decline of precipitation eastward in the cooler months.

Significantly, the number of rainy days declines more rapidly toward the interior than does the total amount of precipitation, suggesting a greater prevalence of showery convective rainfall in the interior. Thus, while the lowland Atlantic margins of the continent experience 180–200 days with rain, central Europe is more likely to experience under 180 days, and 150 and fewer are more common.

Modest Lowland Rainfall.—In spite of location on the windward side of a continent, much of lowland western Europe experiences annual rainfalls of only 500–750 mm, and some areas have less than 500 mm (Fig. 14.9). Hellmann (1928) points out that a number of the Danish islands have less than 450 mm, while one station records only 417 mm. In the Paris Basin there is an extensive area with less than 600 mm and a smaller area in the valley of the Loire with similar small amounts. A few stations in France record less than 500 mm and the same is true for a limited area bordering the River Thames estuary in southeastern England. In Germany along the middle Elbe, and also along the Oder, rainfall is less than 500 mm and there are extensive areas in Poland where this is the case. To be sure, even though the amount of rainfall is modest, it is, nevertheless, very effective for plant growth since it is distributed over many days and is accompanied by temperatures and relative humidities that make for small losses through evaporation-transpiration. Hellmann (1928) suggests that the spots of unusually low precipitation in western-central Europe have one of three type locations: (1) in the lee of higher land, (2) along flat coasts, or on low islands in the midst of water bodies, where winds are strong, and (3) in the higher latitudes where specific humidity is low.

A part of the explanation for the prevailingly modest amounts of precipitation lies in the nature of the air masses which prevail in western-central Europe. The maritime tropical air is chiefly derived from the cooler and more stable northern and eastern margins of the Azores high, so that its surface temperatures are not unusually warm and its vertical structure is such that convectional overturning is not favored. Moreover, the northward trajectory of *mT* air as it is drawn into passing disturbances tends to further stabilize it at the base. As it enters a vortex system over the eastern Atlantic, the modest temperature and density contrasts between *mT* and *mP* air favor the development of fronts and depressions of only moderate intensity. In addition, most of the disturbances are al-

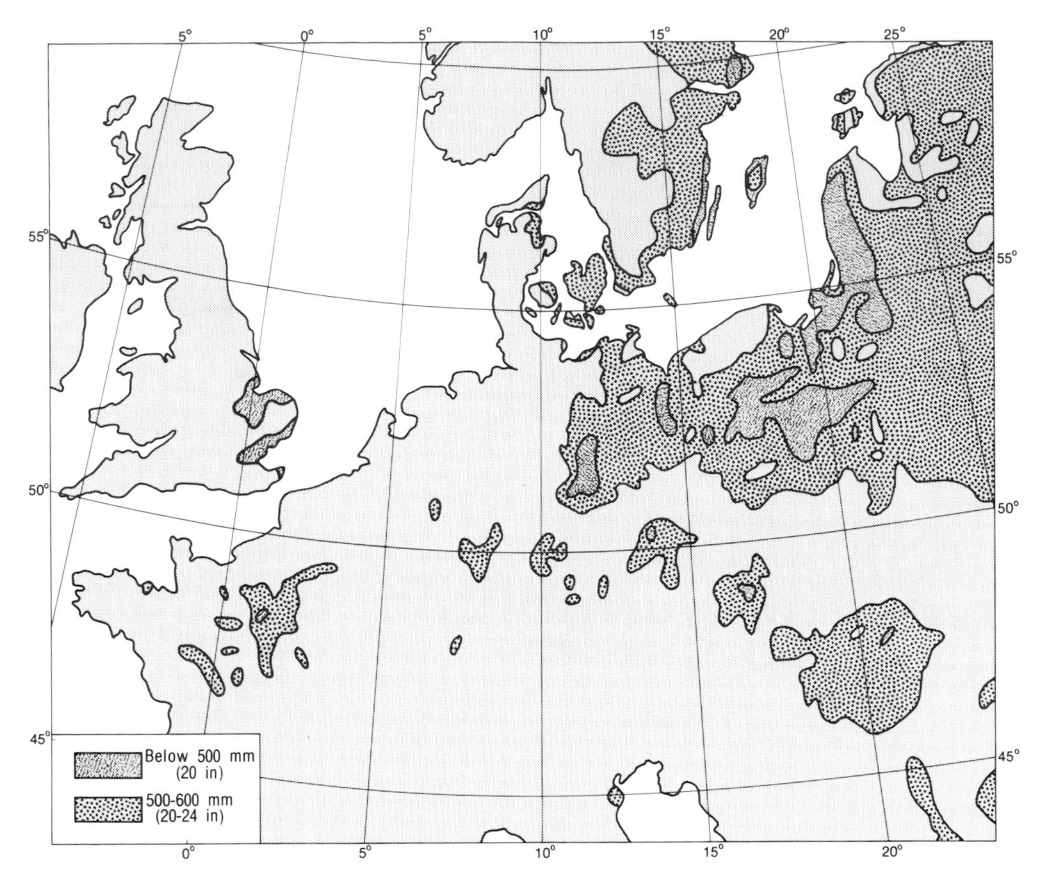

Fig. 14.9. Showing the distribution of areas with very modest amounts of annual precipitation. Extensive areas in lowland western and central Europe experience less than 750 mm of precipitation.

ready in an advanced stage of occlusion when they arrive along the Atlantic coast of Europe, so that while they produce much cloud and numerous rainy days, the intensity of rainfall is light. So although, in the absence of blocking activity, cyclonic disturbances enter the continent freely and frequently, they do not yield an abundance of precipitation.

Further tending to reduce the total annual amount of precipitation over the lowlands of western Europe is the fact that these lowlands are concentrated between about 45° and 55°N which is the latitude of fewest cyclones, both weak and strong, the zones of concentrations lying both to the north and south of these latitudes. Conversely, latitude 50°N is approximately the location of the high-pressure axis of the continent, which tends to reduce the number of rain-bringing disturbances as well as their effectiveness as precipitators.

The minimal frequency of cyclones in and around the 50°N parallel is, in turn, at least partly controlled by the frequent blocking action instituted by stagnant warm highs centered at about latitudes 45° to 60°N. During such periods of blocking activity not only are depressions prevented from passing inland in these latitudes, but in addition there are the effects of subsidence in the high itself. During periods of blocking, precipi-

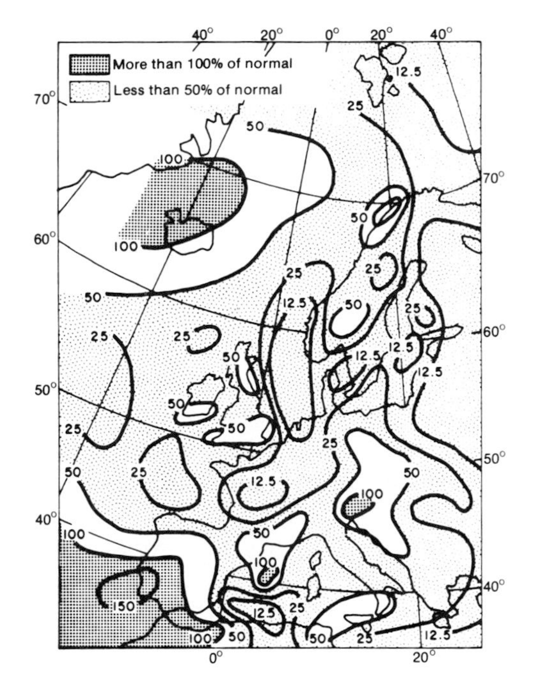

Fig. 14.10. Mean European precipitation anomalies in *winter* during occasions of blocking action, expressed in percent of normal. Rainfall is in excess of normal in the North Atlantic around Iceland and also in the subtropical eastern Atlantic to the north and to the south; it is well below normal over much of western and Mediterranean Europe in intermediate latitudes. (After Rex, 1951.)

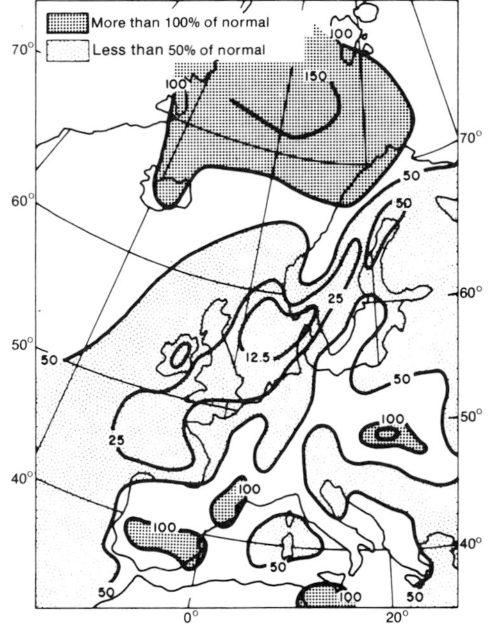

Fig. 14.11. Mean European precipitation anomalies in *summer* during occasions of blocking action, expressed in percent of normal. Rainfall in excess of normal is likewise concentrated in the far north and in scattered areas in the Mediterranean region. Deficiencies are most conspicuous in the intermediate latitudes. (After Rex, 1951.)

tation is well below normal, not only on the plains of western Europe, but on those farther east as well (Rex, 1950). Positive anomalies of rainfall are characteristic chiefly of the Scandinavian area on occasions of winter blocking, and of certain highlands and the Mediterranean Basin in summer (Figs. 14.10, 14.11). During anticyclonic blocking the reduced rainfall over the medial latitudes is less of general cyclonic origin, and more the result of orographic effects and convective processes occurring in the polar air masses moving southward in the high-level trough to the east of the blocking high, with resulting rain and snow showers.

Unusual Features of Coastal Rainfall. —An appreciable maximum of total precipitation appears to exist at the coastline along certain stretches of the flat coasts of western Europe, for example in Holland and Germany, where winds are prevailingly onshore. Bergeron (1949) points out that along the Bothnian coast of Sweden during the period of August to April there is a sharp maximum of precipitation coincident with the coastline and a decline toward the interior, in spite of an increase in elevation. He finds the same situation to hold for winter synoptic situations along the flat coast of the Netherlands. Erwin Prager (1952) reports similar findings for the Netherlands' littoral (Table 14.1). For that region he compares the precipitation records for (1) the light ships at some distance seaward from the Frisian Is-

Table 14.1. Rainfall distribution in the Netherlands (Prager, 1952)

	Average in mm			
Location	Year	Summer	Winter	Average number of rainy days per month
At the light ships	39.3	34.3	44.3	11.7
Coastal strip	65.0	57.3	72.7	15.2
Land stations	58.9	61.8	55.2	16.0

lands, (2) the Frisian Islands themselves, and (3) stations inland from the coast. For the year as a whole, as well as for the winter and the summer months separately, his data show the coastal strip to have heavier precipitation than that recorded by the light ships offshore. The average precipitation for the year as a whole, as well as for the winter months, shows a maximum at the coast and a decline inland. In summer there is a slight increase inland. The increased precipitation at approximately the coastline is ascribed by both Bergeron (1949) and Prager (1952) to the piling up of the onshore sea air as it is frictionally retarded by the land surface. These effects of the coastline upon condensation and precipitation processes are most marked in winter when the sea air is closer to the saturation point and is more unstable than that over the land, so that a slight amount of lifting is all that is necessary to bring it to the condensation level. Inland from the coastal convergence there may be a zone of weak subsidence, resulting in a slight decline of precipitation, except in summer when the warmer land favors convectional overturning.

The shores of the Baltic Sea present a number of regional variations in rainfall which are difficult to explain. Certainly the evidence is not strong that the sea itself is an important local source of precipitation moisture, for there is no continuous belt of heavier precipitation coincident with its margins. Thus, along parts of the Baltic coast of Sweden and adjacent islands annual precipitation is less than 500 mm. Also, the west-facing coasts on the eastern side of the Baltic have extensive areas where rainfall exceeds 600 mm but at the same time there are a number of localities where it is below 500 mm, and at least one where it drops below 400 mm. This localization of areas of heavier and lighter rainfall along the Baltic coast may be explained partly in terms of the trend of the coastline and its relation to the direction of the prevailing winds. Hans Maede (1951; 1954) has interpreted the rainfall deficiency along the southern Baltic coast east of Lübeck as a consequence of a coastline trend such that the prevailing southwest winds are obliquely offshore. Under such a condition Maede reasons that there is an acceleration of the lower frictional layer as it passes from the land to the water surface, so that subsidence and divergence occur at the coastline. The result is a narrow coastal band of lower winter rainfall with somewhat heavier precipitation inland. This may also account for the lower coastal rainfall along parts of the Baltic coast of Sweden where winds are offshore. It may explain likewise some of the drier spots along the west-facing coast of the same body of water. Most parts of the coast whose trend is such that the southerly and westerly air flow is onshore have more than 600 mm of rainfall, while those sections which, due to their directional alignment, have lee positions are drier. This relationship is most clearly exhibited along the shores of the Gulf of Finland where the windward south-facing coast has over 600 mm and the leeward north-facing coast less than 500 mm. Dry spots are likewise to be observed on those coasts of the Gulf of Riga and the Gulf of Danzig which face the northeast.

Seasonal Rainfall Characteristics

Small Annual Range of Precipitation. —Of foremost importance in any discussion of seasonal rainfall in western-central Europe is the fact that there is no dry season. Everywhere the Köppen *f* symbol prevails. What is more, except in highlands, there is usually no very marked seasonal accent, so that the annual range of precipitation is small. This fact may be shown by representing the difference between the average amounts of precipitation in the rainiest and driest months expressed as a percentage of the annual amount. Over most of western and northwestern Europe this is less than 8 percent and over large areas it is less than 6 percent. It increases somewhat inland but there are only a few areas where it exceeds 10 percent and there are extensive areas in southern central Europe where it is under 8 percent (Alt, 1932). Other methods of representing the amplitude of the annual rainfall variation reveal the same general absence of a strong seasonal variation.

That the range in the annual rainfall profile is small does not mean that the weather types producing rainfall are similar throughout most of western-central Europe. On the other hand, it does suggest that weather types producing rainfall are common to all seasons throughout the region, and that in no season are rainy weather types so scarce as to result in an average condition of rainfall deficiency.

Lack of Latitudinal Progression of the Month of Maximum.—A second noteworthy feature of the seasonal rainfall is that stations along the west coast of Europe do not show a simple latitudinal progression of the month of maximum precipitation as is the case in coastal western North America. In the latter area the orderly latitudinal movement of the westerlies, the jet stream, and the belt of cyclones following the sun is clearly reflected in the annual rainfall profiles. Thus, at San Diego in the extreme south the maximum occurs in February while it is progressively earlier with increasing latitude: at San Francisco in January, Portland in December, Vancouver in November, Ketchikan in October, and Anchorage in September-August. The reason why such a simple latitudinal progression of the time of rainfall maximum is absent in Europe is to be found in the more variable and complicated patterns of westerlies and jets that prevail in that region (Riehl et al., 1954). During periods of high circulation index a zonal pattern of jet stream and surface winds prevails, with the principal cyclonic belt located at about 55°N. But when blocking action occurs the high westerlies and the jet are bifurcated, one branch passing around the high on its northern margins and the other to the south, with the cyclone tracks following a similar pattern. This irregular alternation of zonal and cellular circulation patterns tends to give a wide latitudinal spread to the cyclone tracks so that the resulting rainfall distribution is complex, both areally and temporally. It is likewise responsible for the important fact that the seasonal concentration of precipitation is not marked.

Winter Half-Year Rainier along Coast. —This is not to indicate that a slight seasonal rainfall concentration is lacking in western-central Europe. If the year is divided into a winter half and a summer half, then only the narrow Atlantic borderlands have more rain in the cooler half. This includes all of western Norway, the British Isles except parts of eastern England, and a coastal belt in France and the Low Countries usually less than 160 km wide. Thus, while

Dunkirk shows a modest maximum in the cooler half-year, Paris has slightly more in the warmer half. Much the larger part of western-central Europe has more rain in the warmer half, but in either case the predominance of one half-year over the other is not great.

This phenomenon of a precipitation maximum in the cooler half-year along windward coasts in higher middle latitudes is relatively common. It seems to be characteristic of the open ocean to the west of Europe and the littoral is influenced by the same controls. The *mP* air masses in the cooler months are usually somewhat more unstable as a result of surface heating and humidifying of what was originally colder, drier air. In addition, the air in moving eastward across the Atlantic in winter has been involved in the Iceland cyclonic circulation, with the consequence that it experiences convergence and lifting and has been made more unstable. As this *mP* air moves against the colder continental air in the cool seasons, fronts are developed which are zones of precipitation. This frontal activity makes for stronger cyclogenesis in the cooler months which adds to the precipitation of that season. The effects of coastal friction as a rain-making process in onshore air flow are likewise at a maximum in the cooler months when the sea air has a lower dew point.

By contrast the continent tends to accent warm-season precipitation processes, for surface heating makes for air-mass instability and convectional overturning. Conversely, the colder land surface in winter has the opposite effect of stabilizing maritime air moving inland.

Fall Maximum and Spring Minimum in Maritime Western Europe.—A somewhat more refined four-season analysis of precipitation (Alt, 1932; Beelitz, 1932) indicates that the previous biannual treatment, while useful, is not sufficiently detailed to reveal the meaningful seasonal variations in rainfall in western-central Europe. Thus, throughout much of maritime western Europe fall, more often than not, is the rainiest season and spring the driest. Only some of the extreme westerly projections of Europe, such as western Scotland and Ireland and northwestern Iberia, show a truly winter maximum, while the Biscay borderland is the one important littoral area where summer instead of spring is the driest season (Figs. 14.12, 14.13). It is noteworthy that the belt of fall maximum on the mainland is widest in those parts which face the open Atlantic, such as France south of Normandy. Farther north in the lee of the British Isles the belt of fall maximum narrows markedly and north of the Zuider Zee disappears almost entirely until the open coast of Norway is reached. Even parts of eastern Ireland and Great Britain lose their fall-winter maximum and show a slight preponderance of rain in summer. It is noteworthy also that there is a tendency for the fall maximum to be at-

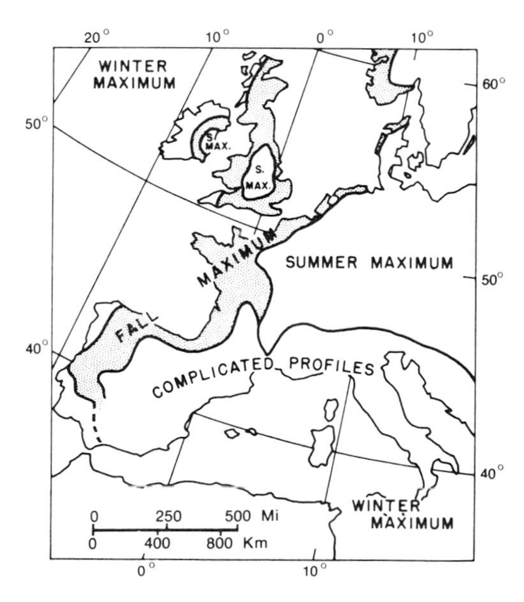

Fig. 14.12. Regions in Europe having a fall maximum of precipitation are shaded. (After Alt, 1932.)

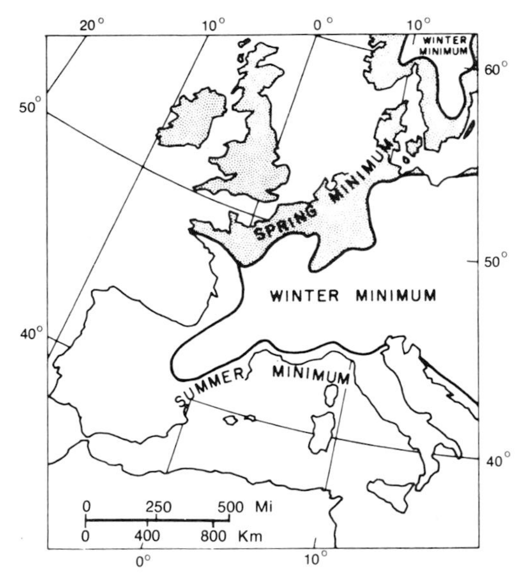

Fig. 14.13. Regions in Europe having a spring minimum of precipitation are shaded. (After Alt, 1932.)

tracted inland by the Baltic Sea, and here too the spring minimum is very conspicuous.

The transition from a maritime winter maximum of precipitation over the eastern Atlantic and the extreme oceanic margins of the continent, to a continental summer maximum a short distance inland, is attained by stages represented by either two or three of the following belts of seasonal concentration: fall-winter, fall, fall-summer, and in that order, from west to east. Most writers on this problem point out that in fall and perhaps early winter the sea surface is warmest relative to the land, so that the maritime air masses have a high moisture content, compared with spring, and are potentially more unstable. In spring when the ocean is coldest relative to the land, the maritime air has a lower moisture content. Flohn (1954) suggests that, since the rainfall accompanying west and southwest weather situations exceeds that of any others, the season of their maximum activity is of greatest importance. As long as the sea is still markedly warmer than the land, in fall and early winter, the west weather is of maximum frequency and intensity. As the temperature difference between land and sea weakens in late winter and spring, west and southwest weather situations decrease in frequency and intensity, and rainfall does as well. It is in these latter seasons that dry continental air breaks through from the east and north more frequently.

Seasonal frequency of blocking action, which is at a maximum in spring and a minimum in fall, likewise affects seasonal rainfall concentration. Since the effect of blocking action is markedly to reduce precipitation over most of Atlantic and interior Europe, it is to be expected that spring, the season of maximum blocking, will tend to be drier than fall unless other controls intervene. It is significant in this respect that the continental axis of high pressure in the vicinity of latitude 50°N is strongest in May and weakest in December.

Flohn (1949) has suggested that the fall maximum in rainfall found in western Europe may, in part and indirectly, be a consequence of hurricane activity which is likewise concentrated in late summer and fall. As these tropical storms move into the middle latitudes, through their advective effects upon cold and warm air masses, they tend to activate the middle-latitude frontal zones and hence increase cyclogenesis in the fall season.

One finds additional explanation of the characteristics of seasonal rainfall in a more detailed analysis of the nature and frequency of the large-scale weather situations and weather singularities of western Europe (Brooks, 1946; Lamb, 1950; 1953). It must never be lost sight of in a search for explanations of the somewhat drier springs and wetter falls, that this seasonal accent is only relative, for spring is by no means a genuinely dry season and fall an excessively wet one, and from year to year they vary greatly in character. C. E. P. Brooks (1946) recognizes 21 singularities or "episodes in seasonal progression of weather," which affect Britain and he finds that these singularities can be classified into four seasonal groups as follows:

1. October to early February. Characterized by stormy periods with minor anticyclonic interludes.
2. February to May. Characterized by cold waves associated with northern anticyclones.
3. The European "summer monsoon," consisting of invasions of Europe by *mP* and *mT* air, alternating with some regularity. This usually begins in June.
4. September and early October. Characterized by spells of warm-anticyclone control.

In the very generalized seasonal grouping of large-scale weather situations as noted above, it is evident that fall and the first part of winter are a period of maximum storminess, with anticyclones less frequent. This is also the period of minimum blocking activity. Spring, on the other hand, is a season when northern anticyclones exert a maximum influence, resulting in numerous spells of cool dry weather. Blocking activity also is at a maximum.

From his study of weather over the British Isles and surrounding regions for the 50-year period 1898–1947, Lamb (1950) classifies that region's weather into seven large-scale weather situations. For each of the seven weather situations he shows its monthly frequency over the half-century when it persisted for long spells of 25 days or more (Fig. 14.14). He likewise shows the distribution of the daily occurrence for each of the seven weather situations over the same period (Fig. 14.15). One must be wary in using these seasonal distributions of large-scale weather situations to explain seasonal rainfall, for the same weather type does not produce similar local weather in all parts of Britain or western Europe. Moreover, a weather situation is not always responsible for similar weather in different seasons. But in spite of these shortcomings and complications certain generalizations appear to be reasonably safe.

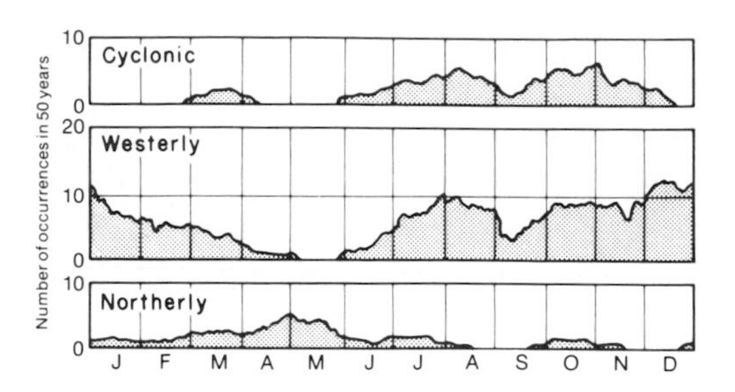

Fig. 14.14. Profiles show the number of occurrences by months, over a period of 50 years, of three important weather situations in the British Isles, persisting for long spells of over 25 days. (After H. H. Lamb, 1950; 1953.)

Using Lamb's (1950) seasonal frequency curves for long spells of weather, it is clear that persistent cyclonic weather, which is one of the better rain-bringers, reaches one of its principal low points in spring, and one of its major high points in fall. This contrast

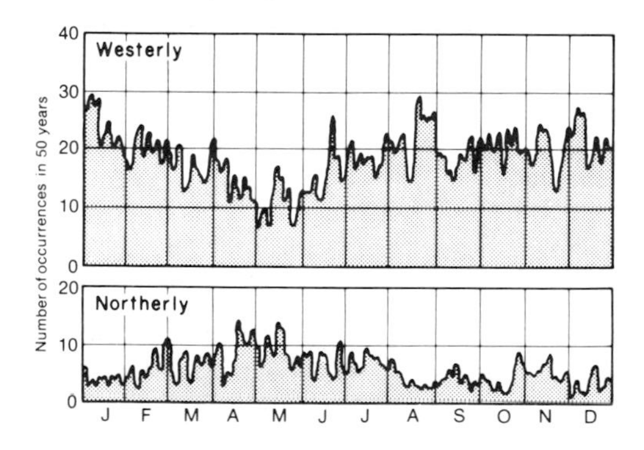

Fig. 14.15. Frequency curves for the daily occurrences, over a period of 50 years, of two common large-scale weather situations (westerly and northerly) in the British Isles. (After H. H. Lamb, 1950; 1953.)

between spring and fall in terms of cyclonic weather is less conspicuous on the frequency curves for daily occurrence of the cyclonic weather situation. Westerly-type weather, which is much the most numerous of the large-scale weather situations, characterized by a succession of depressions and ridges, is likewise a good rain-bringer, and for this situation there is a marked spring minimum on the periodicity curves, both for long spells and for daily occurrence. Fall, on the other hand, has a high frequency of this rainy westerly type of weather. By contrast, the northerly weather situation, which is a dry type, is at a maximum in spring and a minimum in fall. Clearly the wetter falls and drier springs of Atlantic Europe appear to be related to the seasonal frequency of some of the more important weather situations.

In the preceding discussion certain broadscale regional variations in the annual march of rainfall have been noted and explained. It is entirely possible to develop the regionalization of annual rainfall profiles in much greater detail, but since this has already been done by Beelitz (1932), no attempt will be made here to modify or supplement his study in which six principal types, 26 tertiary types, and 80 subtypes are recognized, described, and illustrated with representative profiles.

Weather-Type Climatology of Western-Central Europe

Thus far in the descriptive analysis of the climates of western-central Europe, principal use has been made of seasonal and monthly averages of two climatic elements, precipitation and temperature. Yet if climate is recognized as a summation or composite of large-scale weather situations, and also of briefer weather types, it must be admitted that use of monthly averages of the individual climatic elements is not a completely satisfactory way of analyzing the geography of individual weather situations and weather types, and of the regional patterns of temperature and precipitation that each produces. This is especially true in a region like western-central Europe where extensive and persistent weather situations, together with the briefer weather types which comprise the former, are largely responsible for the climate. These weather situations and weather types have little relationship to the calendar months, so that summaries of individual climatic elements made by such a coarse and arbitrary time subdivision have the effect of effacing the very important nonperiodic weather element.

In spite of the fact that folklore indicates there has long been a lay consciousness of a tendency toward repetition of certain general weather situations at particular times of the year, the weather scientist until recently has been skeptical about such a regular recurrence of weather episodes. The autumn Indian Summer and the January Thaw of the Anglo-Americans, and the June *Schafskälte* (sheep's cold) of the Germans, are two illustrations of such weather episodes. These spells of characteristic weather which have a tendency to recur in most years on or about the same dates are designated as singularities. They are responsible for the major irregularities in the annual solar-controlled march of temperature, rainfall, cloudiness, and other weather elements. A chronological table of the year's weather singularities is spoken of as the weather calendar. The various singularities comprising the calendar result from distinctive types of broadscale pressure systems with their associated winds and air masses, designated by the Germans as *Grosswetterlagen*, a term which will be employed in the subsequent discussion since there is no very satisfactory short translation. Thus a *Grosswetterlage* is a fairly persistent synoptic situation which determines the form and sequence of weather events for a period of several days or even weeks. During such a period the weather is generally consistent, although usually it is not without some change.

It should be stressed that a calendar of *Grosswetterlagen* is not sufficiently reliable to be a useful tool in forecasting. It is only a record of past weather events, although, to be sure, it has some value in judging the likelihood of future occurrences. Not all singularities are equally probable, and even the most reliable ones vary considerably in their times of beginning and ending and consequently in their duration. Still, the calendar of singularities is probably the best tool now available which offers some assistance in the scheduling of future out-of-doors events whose success is dependent upon the weather. The distribution of *Grosswetterlagen* is neither geographically uniform nor temporally regular in occurrence, but neither is it accidental or without plan.

Since it is in western-central Europe that the greatest amount of research and writing has been done on *Grosswetterlagen*, singularities, and weather calendars, it is for this area that a regional treatment of climates based on such tools would seem to be most possible. Yet even for Europe there is still lacking an objective, quantitative, regional climatology based upon weather situations. The difficulties still to be overcome in preparing such a climatology are very great. There is not as yet a widely accepted standard classification of *Grosswetterlagen*, different writers recognizing different types and different numbers of them. Numerous transitional and hybrid types make classification and standardization difficult. What is much needed as one of the essential first steps is a statistical analysis of all of the *Grosswetterlagen* in terms of their effects on temperature and precipitation, and the recording of their distribution on maps. Flohn and Huttary (1950) have provided a sample of what can be done along this line as it applies to the Vb-Lagen.

The broadest and most fundamental subdivision of *Grosswetterlagen* is into cyclonic and anticyclonic types. C. E. P. Brooks (1954) considers that cyclonic weather is the most fundamental type in Britain and western Europe, while, on the other hand, Flohn (1954) asserts that in more continental central Europe the anticyclonic *Grosswetterlagen* are the more regular and significant. He describes anticyclonic weather as being to a large degree self-supporting, independent, or autonomous, for the clear skies, slight air movement, and general subsidence make for weather which is largely regional and diurnal. It is not so much affected by outside sources. By contrast, cyclonic weather, which emphasizes advection and usually has a cloud cover, is greatly affected by conditions outside the area. Instead of being diurnal, it is dominantly nonperiodic in its weather changes.

Since *Grosswetterlagen* usually are of a duration shorter than a calendar month, and since their beginnings and endings have no relationship to the monthly period, it is to be expected that their annual march by months, while useful, provides only a coarse measure of their temporal distribution (see Table 14.2). A finer time screen is required if the facts of distribution are to be most serviceable in the explanation of regional climatic characteristics. From Hess and Brezowsky's (1952) detailed catalog of 18 *Grosswetterlagen*, Flohn (1954) recognizes eight fundamental types which he believes fairly well represent the basic units of central Europe's weather. For each of these eight he shows the monthly and annual frequency, and for the three most important ones the annual march by daily frequencies for the years

Table 14.2. Frequency of *Grosswetterlagen* for Central Europe (in percent) (Hess and Brezowsky, 1952)

	J	F	M	A	M	J	J	A	S	O	N	D	Year
High-pressure situation	19	19	14	12	14	14	16	16	*25*	18	18	20	17
West and southwest situations	35	33	33	29	24	28	38	*45*	30	37	38	41	34
North and northwest situations	16	19	20	24	27	*35*	31	25	21	18	17	15	22
East and northeast situations	12	15	15	17	*21*	14	9	9	12	10	7	9	12
South and southeast situations	10	8	10	8	4	1	0	1	6	10	*12*	10	7
Trough situation and central low	7	7	*8*	*8*	7	4	6	4	6	6	*8*	4	6

1881–1947. These eight, subsequently simplified to six, are briefly characterized as follows:

1. High-pressure situations. These may be warm highs, less frequent cold highs, and also the more numerous moving highs of the kind that exist between cyclones. Such situations are characterized by fair weather with a strong diurnal periodicity. In the restless course of the weather these are the quiet centers. For the year as a whole this type prevails 17 percent of the time, with its greatest frequency (25 percent) in September, but showing seasonal strength throughout fall and winter, and general weakness in spring (April, 12 percent).

2. West and southwest situations. These represent the most important weather types of central Europe, their prevalence being 34 percent for the year as a whole. The mild, overcast, rainy weather associated with a deep penetration of maritime air is quite opposite from that associated with anticyclonic control. Depressions are numerous, and advective rather than convective processes dominate the weather. Warm fronts are more important than cold fronts. There is a maximum occurrence of fronts in late summer, fall, and early winter, and a minimum in spring.

3. North and northwest situations. In these situations the air flow from the north is commonly associated with a blocking high in the eastern Atlantic. They are responsible for some of the coolest weather in spring and early summer, and the arrival of the northerly air in June, often designated as the summer monsoon, is one of the most striking singularities in central Europe. During the warmer seasons this northerly maritime air, due to surface heating plus frictional drag, results in showery rain, with blue sky occurring between the shower episodes. The combined types are at a maximum in June and July, at which time they contribute to the primary rainfall maximum of midsummer.

4. East and northeast situations. At such times central Europe is under the control of continental air masses, which, while they bring the lowest temperatures of winter, may be relatively warm in summer. Since easterly flow represents a reversal of the prevailing westerly circulation, its occurrence is not so frequent, representing only 12 percent of the year's weather. Strong subsidence with stable stratification characterizes the easterly flow, so that spring and late winter when east and northeast *Grosswetterlagen* are most prevalent are the driest part of the year.

5. South and southeast situations. These *Grosswetterlagen*, while not frequent (7 percent of the year), are accompanied by extreme air-mass and weather conditions. They are associated with southerly and southeasterly air flow from the western flanks of a warm high over eastern Europe. Abnormally high temperatures result. The föhn in Germany is of this origin. Uncommon in summer, this severe type is most frequent in the period October to April.

6. Trough situation and central low, Vb-situations. These situations characteristically are associated with severe storms giving widespread bad weather and heavy precipitation leading to flooded streams and deep snows. The high meridional trough results in great southward surges of cold air which reach well into the subtropics and even the tropics. Well-developed fronts are the result. The trough aloft is frequently associated with a deep low (*Centraltief*) at the surface. Such is van Bebber's (1882) Vb-type of disturbance which follows a course from the Adriatic into central Europe. In storms following this track there is an upgliding of warm Mediterranean air which produces abundant rainfall. In central Europe storms of this variety may account for 34 percent of the fall–spring rainfall, 26 percent of that in summer, and 24 percent of that in winter. (Flohn and Huttary, 1950). The region of greatest Vb rainfall lies along the eastern Alps and from thence in a belt that stretches northeastward over Poland and the Danzig area.

Although much is known about the variety of *Grosswetterlagen* that produce the climates of central-western Europe, no one yet has succeeded in expressing quantitatively the regional characteristics of climate in terms of the sequence and variety of weather situations. Nevertheless, what may be considered to be preliminary steps in the direction of a *Grosswetterlage* type of regional climatology are already available (Lauscher, 1954).

CHAPTER 15

Mediterranean Borderlands

GEOGRAPHICAL ELEMENTS.—Although the Mediterranean Basin has given its name to one of the earth's most distinctive climatic types, the region as a whole is not a good model of subtropical summer-dry climate (*Cs*) in its simple or pure form. Only the southern and eastern parts can so qualify. The northern parts of the basin most emphatically are transitional, and in some places even atypical, in *Cs* climatic qualities. It is, rather, the extensiveness of subtropical dry-summer mesothermal climate within the Mediterranean Basin, and not the purity of the type, which has led to the widespread adoption of the regional name for this unique climate. Its extensive development, however, is a uniqueness in itself, for the type is characteristically restricted to the western margin of a continent. The deep inland penetration of this *Cs* type in the region of the Mediterranean Basin is nowhere duplicated in other subtropical latitudes, and its widespread existence there appears to be a consequence of the warm Mediterranean Sea, which in the cooler months acts to make unstable and showery the cool air masses invading from the north, and to provide an environment favorable for cyclone genesis and transmission. Actually the most extensive development of pure *Cs* climate, with a single maximum of rainfall in winter and a minimum in summer, is confined largely to the eastern half of the Mediterranean Basin, 1,500 to 3,000 km from the Atlantic coast of Africa and Iberia.

The latitudinal spread of the Mediterranean Sea is from about 31°–33°N in Tripoli and Egypt, to 45° in the Adriatic and Black seas. On the north, therefore, the sea influence extends well beyond subtropical latitudes, so it is not surprising that most of the European Mediterranean borderlands, lying largely north of 40°N, do not have a typical *Cs* climate. Likewise of great significance climatically is the fact that the Mediterranean Sea is not a compact body of water but instead is divided by large peninsulas and islands into a number of smaller seas and bays. The effect of such fragmentation is to complicate the seasonal pressure and circulation systems and the cyclogenesis patterns, resulting in a complex climatic distribution.

Moreover, the great Mediterranean peninsulas and islands are prevailingly hilly and mountainous, with the grain of the relief forms having various directional trends. Some have extensive uplands like Iberia, Anatolia, and the Atlas of northern Africa, while highland-enclosed basins are conspicuous in Italy and the Balkans. Highlands like the Pyrenees, Alps, Dinaric Alps, and Caucasus define the northern rim of the Mediterranean Basin, and in some instances form significant climatic divides. In winter they act to obstruct in some longitudes a free flow of northerly air into the Basin, but elsewhere there are important gaps in the discontinuous highland rim which serve as important funnels for cold-air invasions.

The Mediterranean Sea is an extensive and deep reservoir of water in which temperature changes little with depth. It has a pronounced negative precipitation-evaporation balance, which is compensated for chiefly by an inflow of Atlantic surface waters through the shallow Straits of Gibraltar. From October to March the temperature of the surface water is higher than that of the air, and this differential amounts to 3° to 4°C in midwinter (Biel, 1944). The Black Sea is much colder. By June temperatures of surface water and coastal air are about equal, but for the next three months the sea is cooler.

PRESSURE AND WINDS.—During the cooler months the average sea-level pressure chart shows a trough over the Mediterranean oriented in a northwest by southeast direction and separating the Eurasian thermal high to the east and north from the dynamic Azores high to west and south (Baum and Smith, 1952). While partly a consequence of the relative warmth of the sea surface, the winter trough in some measure is a statistical feature, since the Mediterranean is an important route of atmospheric depressions in the cooler months. In this respect it is noteworthy that the principal centers of cyclogenesis are likewise centers of lowest average pressure. Local pressure centers are a conspicuous feature of the cooler months, with regional cells of high pressure characterizing the cold peninsulas, and local lows the intervening warmer seas (Biel, 1944).

Winter air flow over the general Mediterranean area is strongly convergent as would be expected from the pressure distribution. A map of streamlines shows conspicuous convergences around the West Mediterranean low, the Cyprus low, and the Black Sea low. It is weaker over the Adriatic. In conformity with the general Mediterranean convergence, northern coasts experience much northerly air flow. The local anticyclonic and cyclonic circulations characteristic of the peninsulas and the intervening seas result in a prevalence of warm, moist, southerly winds along west- and south-facing coasts of peninsulas, while colder and drier northerly winds characterize the eastern and northern sides, with resulting contrasts in temperature and rainfall (Conrad, 1943).

In summer the Eurasian thermal high has disappeared. Concurrently the Azores high has been displaced northward and eastward and is intensified so that most of the Mediterranean is dominated by it. From the five-day normal sea-level pressure charts, the episodes in the change from winter trough to summer high can be observed. By mid-April a ridge appears over Mediterranean Africa. The trough disappears during the last 10 days in May and the ridge shifts northward. It is not until after June 25, however, that the Azores high expands eastward to dominate the whole Mediterranean area (Bryson and Lahey, 1958). This fact has important climatic consequences, for June has many of the weather characteristics of a spring month, and the genuine summer-drought conditions are postponed until nearly July. A weak summer low is positioned over northern Africa and a deeper one is centered over northwestern India-Pakistan and western Asia. As a consequence the pressure gradient over the Basin declines from west to east and from north to south. Surface pressure over the sea is slightly higher than over the heated peninsulas of Iberia and Anatolia where local low-pressure centers prevail. Weak local highs also exist at times over the Black, Aegean, Adriatic, and Tyrrhenian seas. General subsidence with divergent air flow characterizes much of the Mediterranean area in the warmer months, so that the air is stable and the weather element weak (Conrad, 1943).

Without doubt the Mediterranean Basin is the earth's best-known area of local winds, many of which have acquired special names (Biel, 1944). This fame is partly a consequence of the fact that local winds of contrasting character are unusually well developed in the region. In part it may also be due to the publicity given to the topic by the highly literate peoples who have occupied the region for so many centuries. The prevalence of local winds stems from the fact that the Mediterranean Basin, an area of rugged terrain, is for much of the year characterized by air-stream convergence and even more important, by local convergences in the form of moving cyclonic storms. Such convergence results in the advection of air from contrasting source regions lying to the north and south, the tropical Sahara for hot, dry air, and the lands to the north for cold air, especially in winter. Even though the local names for special winds are very numerous, they actually may be divided into a few classes, such as sirocco, bora, etesian, and föhn.

Large-Scale Weather Situations and Atmospheric Disturbances

Omitting the hills and mountains where orographic effects are prominent, rainfall over the Mediterranean lands is almost exclusively associated with extensive atmospheric disturbances of the vortex and wave variety. The Basin is one of the earth's important concentrations of cyclonic activity. This activity varies both temporally and regionally, but middle and late summer is much the most inactive season. This is not unexpected since the Azores high dominates the Mediterranean area during the warm months. It is not unusual that the Mediterranean area should be a prime region of cyclonic activity, since it represents for more than half the year a zone of convergence for air streams derived from unlike source regions to the north and south. Not infrequently, also, the southern branch of the European jet stream overlies Mediterranean latitudes, acting to intensify disturbances and to steer them eastward along the main axis of the sea.

High-Index Situations.—It has long been known that cyclonic activity in the Mediterranean area is intimately connected with weather conditions prevailing over the eastern Atlantic and in Europe north of the Alps. During periods of high-index circulation and west steering, occurring mainly in fall and winter, strong westerly flow occurs both at the ground and aloft. At such times the main west-wind belt and its jet stream are located north of the Mediterranean, and over the latter area the west winds are relatively weak. Since the air trajectory is W-E and paralleling the long axis of the sea, there is less likelihood of invasions of genuinely cold air from the north, and disturbances are few. On such occasions the Mediterranean is not necessarily without the passage of cold fronts, but these are only brief interludes in a prevailingly durable high-index, fair-weather condition (Staff members, Chicago, 1944).

Low-Index Situations.—Most of the disturbed weather, with associated cloud and rain, occurs, however, during periods of low circulation index, of which there are two main types in the Mediterranean area. These are (1) prevailingly northwest flow pattern with trough or ridge steering and (2) southwest steering and flow pattern. It is during those periods dominated by the first of these two large-scale weather situations that the most widespread and prolonged cyclonic activity occurs.

Northwest Flow Pattern.—Low circulation index with northwest flow and trough or ridge steering develops when the Azores high expands in the direction of the British Isles, with the Iceland low displaced east of its mean position or split into two centers so that one upper trough lies over the Baltic or North Sea and the other near the southern tip of Greenland, with the subtropical high projecting northward between them. The isobars aloft have a general northwest-southeast direction, and outbreaks of polar air reach the Mediterranean following the course of the isobars on the rear of an eastward-moving upper trough (Fig. 15.1). In a study of upper-air conditions associated with cyclogenesis over the sea just to the west of Italy, Gleeson (1954) found that at the 500 mb level there is a trough located immediately west of the cyclogenetic area with a northerly flow of air invading the Mediterranean region (Gleeson, 1954). Hermann Flohn (1948) similarly has noted a strong correlation between cyclogenesis in the Mediterranean and a high-level trough over Italy and to the north and west which leads to an invasion of northerly air. Such periods of low circulation index and trough or ridge steering, with a blocking high situated over the easternmost Atlantic, likewise appear to be the occasions for a bifurcated jet stream, with one branch located over subtropical latitudes in the eastern Atlantic and North Africa. This in turn favors cyclogenesis and the eastward routing of disturbances in these subtropical latitudes.

Most of the cool air reaching the Mediterranean from the northwest must enter across the Iberian Plateau, or more easily through the corridor in southern France bounded by the Pyrenees and the southwestern Alps (Fig. 15.2). In the latter region the two mountain areas form a funnel open to the northwest which channels the northerly air through a constriction some 400 kilometers in width into the Gulf of Lions. The Gulf of Genoa and Po Valley to the east are sheltered from these invasions by the Alps. Thus, when strong northwest winds are moving southward west of the southern Alps, while relatively calm conditions exist over the sea just to the east of this current, the situation is favorable for the development of vortex disturbances over the Gulf of Genoa and adjacent Italy (Staff members, Chicago, 1944). The flow of northerly air through the constriction north of the Gulf of Lions is not steady and continuous, but instead is in the

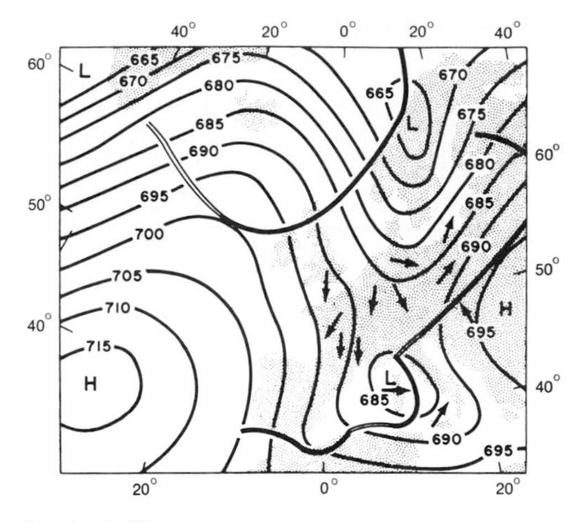

Fig. 15.1. Three-kilometer pressure chart showing surface fronts, under conditions of a northwest flow pattern. Northerly air is entering the Mediterranean Basin by way of southern France and a disturbance is developing over the adjacent sea. (After map in University of Chicago, Department of Meteorology, *Misc. Repts.*, No. 14.)

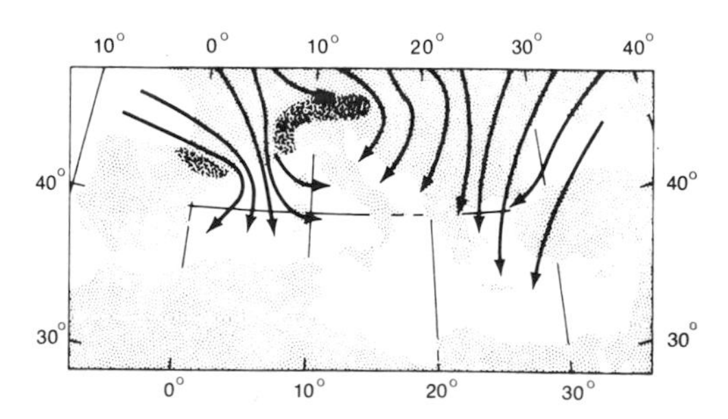

Fig. 15.2. Common routes by which cold northerly air enters the Mediterranean Basin, thereby inducing cyclogenesis.

form of surges which form secondary fronts to the rear of the main cold front. Weak troughs are associated with these secondary outbreaks of cold air, but marked air-mass changes are lacking.

A second major area where cold northerly air invades the Mediterranean is by way of the topographic depression between the Balkan Highlands and the Caucasus, and even across the Anatolian upland and the Balkans, this cold air being derived mainly from Russian and Balkan sources. Italy and the eastern Mediterranean around Cyprus thus become regions of cyclogenesis associated with this more easterly invasion route of cold air.

Flohn (1948) points out that conditions favoring cyclonic weather in the central Mediterranean are associated with two high-level trough situations, one over eastern France and north Italy, and the other over the Balkans. The former leads to cold invasions of unstable maritime arctic air from the northwest, and the latter to inroads of cold continental Russian air. North Italy and the central Mediterranean feel both types of air. The combined frequency of the Italian and Balkan troughs shows a main maximum in March-May, a sharp minimum in July-August, and an abrupt rise again in September-October, with little variation throughout the winter. If the frequencies of the Italian trough and the Balkan trough are drawn separately, the former shows a strong maximum in May-June and a slight secondary maximum in October, with a main minimum in winter and a slight one in August (Fig. 15.3). The Balkan-trough weather situation is simpler in annual distribution, with a two-pronged maximum in winter (March and December) and a broad minimum in summer.

Flohn (1948) finds a close correlation between rainfall frequency in the Italian-Albanian area and the rate of recurrence of cyclonic weather situations associated with

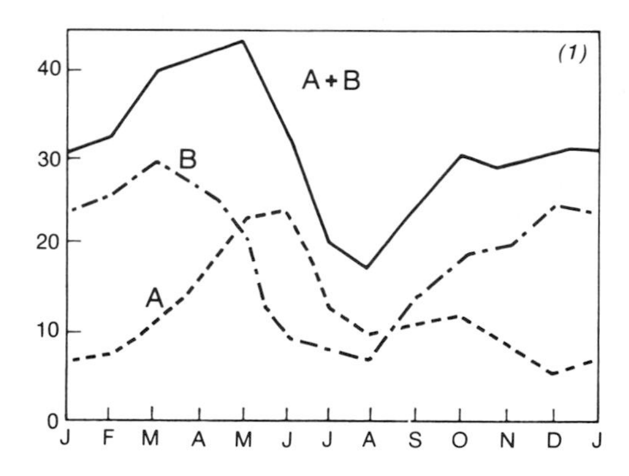

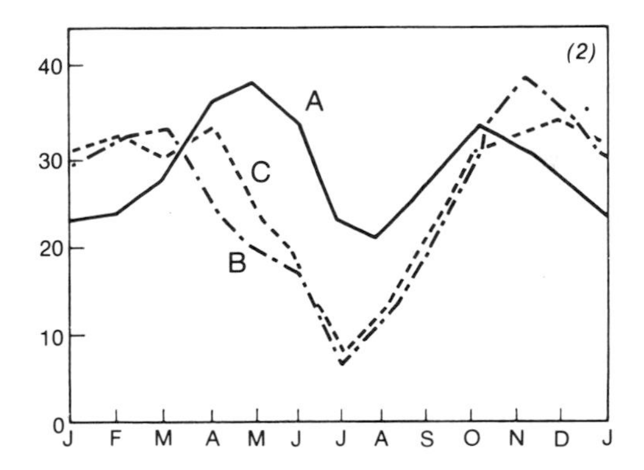

Fig. 15.3. 1. Frequency (percent of all days, period 1881–1943) of cyclonic *Grosswetterlagen*: (A) trough over Italy; (B) trough over the Balkans. 2. Frequency (in percent, 1881–1943) of precipitation: (A) for northern Italy; (B) for southern Italy; (C) for Albania. (After Flohn, 1948.)

Italian and Balkan upper troughs and the invasions of northerly air which accompany them. Thus, northern Italy, which feels both types of cold invasions, shows one maximum rainfall frequency in April-June and another in September-October, paralleling the frequency of the combined Italian-trough and Balkan-trough weather situations. In southern Italy and Albania, on the other hand, where the cold invasions are principally from the Balkans and Russia, and are at a maximum in winter and a minimum in summer, the rainfall frequency shows a similar uncomplicated distribution, with a minimum in summer and a general maximum in winter. It appears, therefore, that the yearly march of rainfall frequency in Italy and the Adriatic is rooted in the yearly march of cyclonic weather situations.

Deep depressions are likewise formed or regenerated in the eastern Mediterranean in the vicinity of Cyprus during the cooler seasons (El-Fandy, 1946). These disturbances in some instances may be the result of regeneration of depressions that have moved eastward from the central Mediterranean region, or farther west. Many of the latter, however, move northeastward toward the Balkans or the Black Sea and hence miss the eastern Mediterranean. Usually the depressions degenerate as they move eastward, for the air from the Sahara in winter is dry and not very warm, so that the only apparent source of energy for disturbance development is fresh cold air with a short sea trajectory. This is supplied in the cooler months by *cP* air from the cold anticyclone over Russia. El-Fandy speaks of the areas surrounding the Black Sea as being the "North Pole" for the eastern Mediterranean. Surges of cold air from this region flowing through the Marmara depression or across Anatolia cause active cyclogenesis in the region around Cyprus, or regenerate weak depressions moving in from the west. On those occasions when a succession of cold northerly surges enters the Basin from the region of the Black Sea, the system of isobars and fronts in the eastern Mediterranean resembles somewhat the pattern of a spider's web. Showers accompany the cold surges, but many times they appear to be instability showers in the warmed and humidified cold air behind the front. These cold fronts extend on into Iraq and Iran, some of them even reaching India. It is these so-called Cyprus disturbances which cause the single cool-season maximum of precipitation in the eastern Mediterranean.

Southwest Steering and Flow Pattern. —The second of the two principal largescale weather situations in which low circulation index prevails is that associated with a southwest flow pattern. This characteristically develops when the Iceland low is displaced southward to a position west or south of the British Isles, from its mean location north or northwest of Britain, and there is a division of the Azores high into two cells, one of them over the Mediterranean (Fig. 15.4). The above-surface trough associated with the displaced Iceland low is then located far to the south, as is likewise the air stream from the southwest along the trough's eastern side. With this situation most of the disturbances crossing the coast of Europe enter from the southwest. Associated with the trough aloft, which is deep and extends to low latitudes, there is almost always cyclogenesis. The polar front lies in the eastern or leading part of the trough and along it wave cyclones move from southwest to northeast as carried by the southwesterly

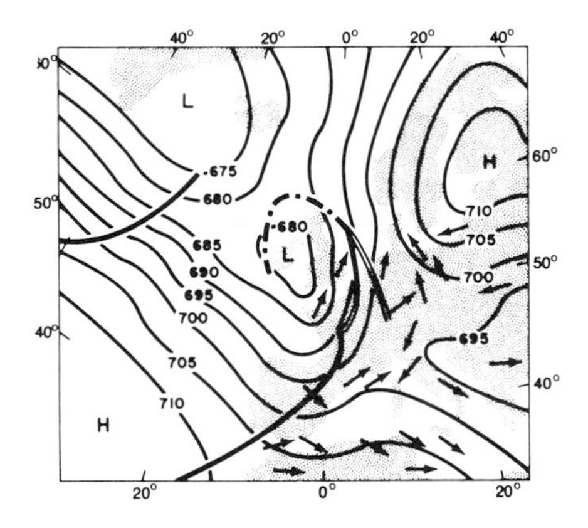

Fig. 15.4. Three-kilometer chart with surface fronts under conditions of a southwest-steering situation in Europe. The trough and the surface front are displaced well to the south and there is a flow of southwesterly air along the east side of the upper trough. (After map in University of Chicago, Department of Meteorology, *Misc. Repts.*, No. 14.)

current. Weather conditions in the Mediterranean Basin will depend upon how far inland the southwesterly flow prevails. If pressure is weak over the Mediterranean so that the southwesterly flow extends well inland, a system of cyclones may advance eastward over the Mediterranean, producing cloud and precipitation as a result of convergence and overrunning in the warm air ahead of the cold front, as well as cumulus activity in the cool air behind the front as it is warmed over the water. But with a southwest flow pattern the unsettled weather is confined mainly to the western Mediterranean and is of brief duration.

As will be seen from the previous discussion a great majority of the disturbances in the Mediterranean develop as secondary depressions when the primary disturbance lies farther north over western or central Europe (Great Britain Air Ministry, 1937). The secondaries develop chiefly during periods of low circulation index. However, not all Mediterranean disturbances originate as secondaries along cold fronts. Some enter the area as primary systems moving south or southeastward across France, or eastward across Iberia, or through the Gibraltar area, while other independent depressions form over the Mediterranean itself.

Distribution of Cyclogenesis in the Mediterranean Basin.—A comprehensive analysis of the temporal and areal features of cyclogenesis in the Mediterranean area is that by Gleeson (1954). Unfortunately, four months, June through September, are omitted, so that analysis of warm-season cyclogenesis is lacking. It should be stressed, also, that this study refers to a region bounded by the 30° and 50°N parallels and by the 20°W and 50°E meridians, so that it covers a somewhat more extensive area than the Mediterranean region proper. For a ten-year period (1929–1939), 442 cyclones were tabulated for the October-to-May period and they were temporally distributed as follows:

O	N	D	J	F	M	A	M
42	57	62	55	55	63	63	47

The decline in numbers in both October and May suggests the existence of a primary summer minimum, but there is evidence likewise of a slight midwinter secondary minimum separating two maxima, one in December and the other in March-April.

The distribution of the point of origin of the 442 cyclones observed by Gleeson (1954) is represented in Figure 15.5, location being made within a grid composed of 3° rectangles. Much the most conspicuous center of cyclogenesis is in the central Mediterranean extending from northern Italy to Tunis and Sicily. Its focus is the Tyrrhenian and Ligurian seas and central and northern Italy. Significantly this major concentration is positioned at the point where surges of cold air are channeled into the Mediterranean from the north, especially through the Rhone and Carcassonne gaps in southern France, but also through the Balkans. This feature has been described in an earlier section. Cyclogenesis in this center is at a maximum in the winter months, but is active in spring and fall as well. The disturbances here generated most frequently move in a southeasterly direction parallel to the Italian peninsula.

A second principal Mediterranean center of October-to-May cyclogenesis is situated in northwest Africa south of the Atlas Mountains, a location which appears much more unusual than the first. Depressions originating here develop during periods of strong northwesterly flow on the rear of an upper trough which extends far southward over North Africa from western Europe (Hare, 1943). At such times there is usually a primary disturbance farther north over the British Isles. The disturbances appear to develop at all times of the year, including sum-

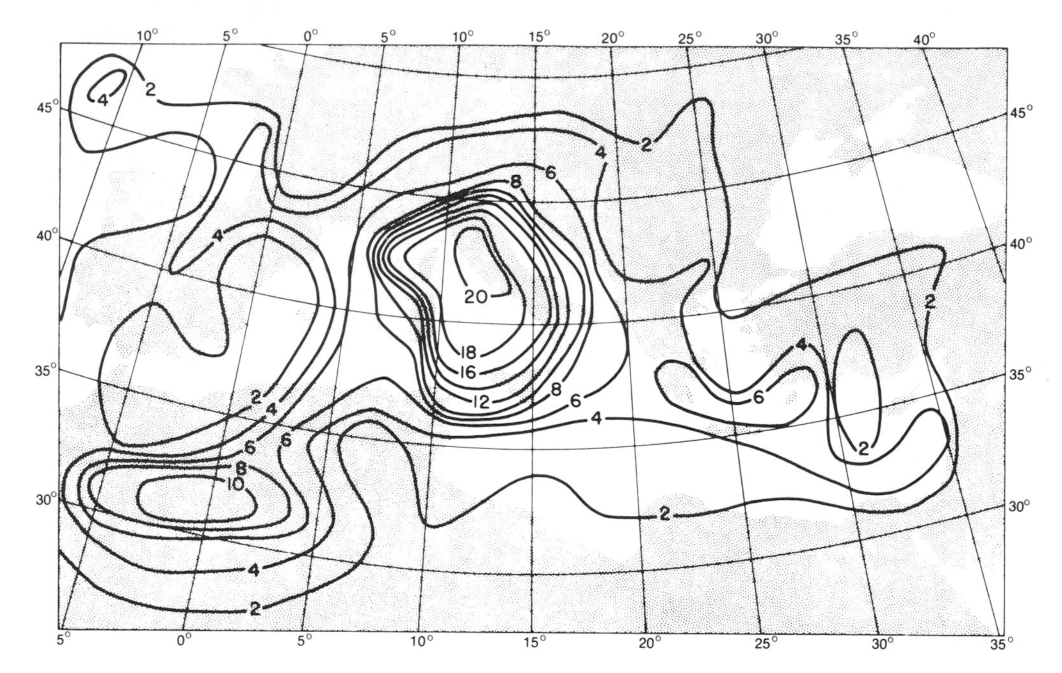

Fig. 15.5. Frequency of cyclogenesis in the Mediterranean Basin. The numerals on the isolines refer to the number of cyclones observed as originating within the area bounded by the 30° and 50°N parallels and the 20°W and 50°E meridians over a 10-year period, 1929–1939, for the months October to May, inclusive. Location was made within a grid composed of 3°-rectangles. Two main centers of cyclogenesis are to be observed. (Data furnished by Gleeson, personal communication.)

mer, but are most numerous in spring which is the season of maximum northwesterly flow. These North African depressions move eastward and northeastward toward the coast of Tunis where they deepen and become significant rain-bringers to the coast of northern Africa and the central Mediterranean.

A third weaker and more diffuse center of cyclogenesis is located in the eastern Mediterranean which feels the effects of cold surges arriving by way of the Marmara gap and Aegean Sea. Not shown are the more numerous disturbances from the west which experience regeneration in this area.

Summer Disturbances.—Summer disturbances in the Mediterranean area are so commonly unmentioned in the climatic literature that they appear to warrant further comment. Scully (1951) notes two principal sources of summer transient low-pressure cells. One of these is the thermal summer low in northern Africa inland from the Atlas Mountains. Occasional small cyclonic circulations become detached from this low and move either eastward or northward. They intensify over the water but remain small (80–160 kilometers or less in diameter) with tight gradients. They do, however, produce strong winds, low cloud, and even showery, thunderstorm rainfall. Oliver and Fowler (unpublished) indicate that depressions may occur in summer and early fall in northwest Africa upon occasions of southwest steering as described earlier.

A second source of small transient lows in summer is the Gulf of Lions, where they develop at the times of northerly flow. These seem less unusual than the disturbances originating in North Africa, since they are simply weak counterparts of those that develop under similar circumstances and in the same area in the more favored cooler seasons. Weak summer lows originating in the Gulf of Lions may recurve toward the Gulf of Genoa and back into northern Italy; others move across Italy into the Adriatic, or southward into the Tyrrhenian Sea, weakening south of Malta, and intensifying again over the Aegean Sea.

SEASONAL WEATHER.—While it is true the Mediterranean area as a whole has more disturbed weather and rain in the winter half-year than in the summer half (less true in the northern parts), and although summer, except in the north, is a relatively dry period, it is scarcely correct to state that disturbances are lacking in the summer season. To be sure the synoptic chart from July through August looks rather indifferent much of the time. Still, it is unusual if some part of the Mediterranean Basin does not show at least a weak depression either within the Basin or on the point of entering, even in midsummer. In the 300 days (approximately) associated with each of the months during the ten years 1926–1935, or about 3,600 days in total, the percentages of depression-free days for the entire Mediterranean were as follows: January, 2; February, 5; March, 2; April, 2; May, 1; June, 6; July, 17; August, 25; September, 14; October, 8; November, 2; December, 2 (Great Britain Air Ministry, 1937).

The depressions which occurred in the Mediterranean during the period 1926–1935 numbered between 800 and 900. The average number per month in the four seasons (excluding those south of the Atlas) was approximately as in Table 15.1. The data, while they indicate that summer has the fewest disturbances, also show that they are not absent during that season. Especially the northern Mediterranean, which lies north of the normal subtropical latitudes, experiences more summer disturbances than other parts, as is indicated by its moderate amount of summer rainfall. But while summer disturbances are few farther south, even there it is not true that they are completely absent. It would seem, then, that the depressions that do exist in summer are relatively weak and of limited extent, so that they are not conspicuous on the synoptic chart, while at the same time they produce only a small amount of cloud, and still less precipitation, because of the general subsident anticyclonic character of the summer air. In addition, the land-sea temperature contrasts in the warmer months are not such as to develop instability in air currents entering the Basin from the north. This effect is largely absent in summer.

TRACKS OF MEDITERRANEAN DISTURBANCES.—Most writers in analyzing the movement of depressions in the Mediterranean area are inclined to follow van Bebber (1882) and Weickmann (1922) in the delineation of the routes followed. Such a system of tracks or routes admittedly has the advantage of being simple, and the routes definite and precise (Fig. 15.6). This definiteness and preciseness of pattern, however, does a measure of injustice to what in reality is a very complex situation. It is an error to conceive of a depression track as a definite thing, for it is not, and in allotting depressions to individual tracks the personal factor plays an important part. Moreover, in their progress eastward, most depressions do not follow a single track but switch back and

Table 15.1. Average number of depressions in the Mediterranean Basin by season (Great Britain Air Ministry, 1937)

	Winter	Spring	Summer	Fall
West Mediterranean	8	9	5	6
East Mediterranean	6	7	3	4

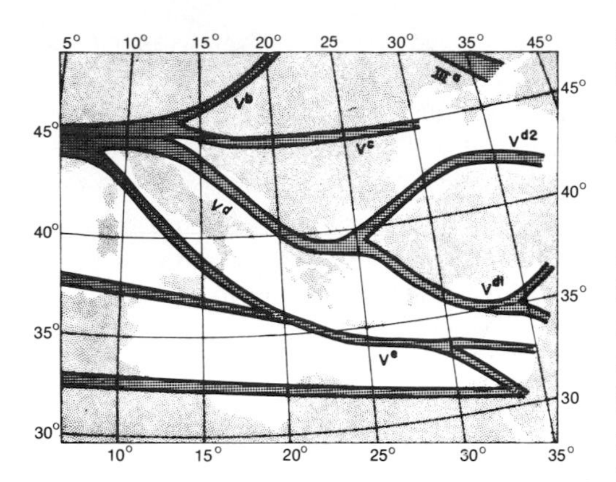

Fig. 15.6. Main routes followed by cyclonic disturbances in the Mediterranean region. (After van Bebber, 1882, and Weickmann, 1922.)

forth from one to another. The number of individual depressions which follow one or the other of the main tracks with even a moderate degree of consistency is only about 50 percent of the total. " . . . the seasonal characteristics of the weather at any place cannot be summarized as a compound of ingredients furnished in certain proportions by depressions following the main tracks. The number of depressions which change from one track to another (it may be several times in the course of their passage through a region), the number of apparently 'retrograde' depressions, and of depressions which form, or fill up, locally, the great variation of intensity of depressions, and the frequency of disturbed weather due to small local systems not discernible as depressions in the ordinary synoptic chart combine to render such a simple synthesis impracticable" (Great Britain Air Ministry, 1937).

There is a tendency for the disturbances to skirt the high mountains and plateaus, and when they do cross them they show signs of deterioration. Conversely, they are attracted by water bodies such as the Tyrrhenian, Adriatic, and Black seas. As expected, the belts of easterly-moving disturbances shift latitudinally with the seasonal course of the sun so that in the summer the chief concentration is in the northern parts.

Temperature

Almost all of the Mediterranean climate of this area, although located in close proximity to extensive seas, is of the warm-summer type. The only exceptions are the moderate elevations, and also those littoral areas in northwest Africa and western Iberia which front upon the cool Atlantic. Even though most of the Mediterranean lowlands front upon salt water, that water is unusually warm even to great depths.

Both summer heat and winter cold are tempered somewhat by the adjacent sea. Omitting highlands, the coldest spots are adjacent to gaps in the northern highlands through which cold northerly air is funneled southward, such as the northern Aegean shores, the northern coast of Anatolia west of Cape Ince, and southern France. Marseilles in southern France, which feels numerous northers, is 2°C (3°–4°F) colder in winter than Genoa farther north but protected by the mountains.

Along the European shores there is a general decrease in winter temperatures from west to east as the Atlantic influence is left farther behind and the continental influence of eastern and central Europe becomes stronger. It is chiefly *mP* air that invades through France, but it is colder *cP* air that enters the eastern Mediterranean. Accordingly, the freezing isotherm in January is 10° farther south in the Balkans than in Italy.

As a general rule, also, the east coasts of the Mediterranean peninsulas are colder in winter, and likewise drier, than the west coasts (Fig. 15.7). The latter are on the windward sides and feel to a greater degree the tempering effects of the water. In addition, each of the colder peninsulas in winter tends to develop an anticyclonic circulation

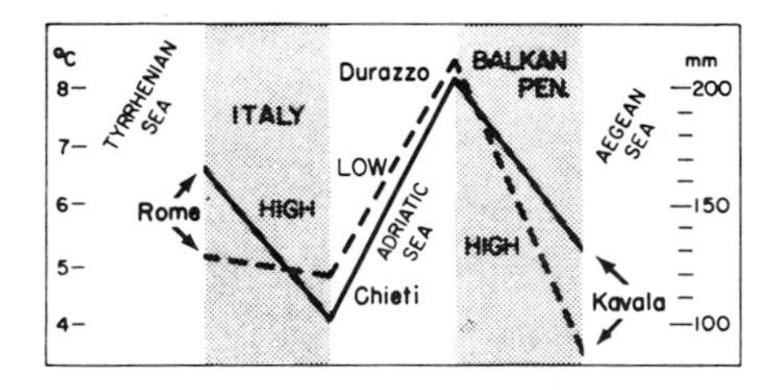

Fig. 15.7. Temperature (solid line) and rainfall (broken line) contrasts between the east and west sides of the Mediterranean peninsula. Note that east side locations are usually cooler and drier.

and each warm intervening sea a cyclonic one. This arrangement operates to bring warm, moist southerly air in over the western coasts, and cooler, drier, northerly land air over the eastern littorals (Conrad, 1943). Southerly ocean currents along west coasts, and northerly along east coasts, further tend to accentuate the temperature contrasts. Of the three great peninsulas, the winter temperature contrasts between east and west coasts are greatest in the case of the Balkan Peninsula, including Greece, its most southerly extension, while winters are also colder on its east side than in comparable locations on the other two (the January mean at Perpignan, France, is 7°C (44°F); at Ancona, Italy, 6°C (42°F); at Varna, Bulgaria, 2°C (35°F). Also Varna has only half the precipitation in January that is recorded by the other two stations. The colder, drier winters in the eastern Balkans reflect the latter's greater exposure to the continental thermal anticyclone and its decreased protection from northerly air by highland barriers. Even the Aegean side of Greece is 3°–4°C (5°–7°F) colder in winter and has only half as much precipitation as does the western side of that country (Philippson, 1947).

PRECIPITATION

ORIGIN AND NATURE OF THE PRECIPITATION.—Omitting the effects of orography, the precipitation of the Mediterranean area is to an overwhelming degree associated with extensive disturbances within the zonal westerlies. But a large proportion of the rain that falls over the sea and along its southerly margins, develops in the cold air lying to the rear of the cyclone's main cold front, this air having been made convectively unstable as a consequence of additions of moisture and heat at its base in its passage over the warm sea surface. In this unstable cold air swelling cumuli and cumulo-nimbus are wisespread, and showery rainfall and squall winds are characteristic. The flow of northerly air through the gaps in the highland rim is not steady, but occurs, rather, in the form of a series of pulsations or secondary outbreaks, which are in the nature of speed convergences and not associated with marked air-mass change. Each new surge of fresh polar air produces rainfall through lifting the more modified polar air which it displaces (Staff members, Chicago, 1944). Clearly, an important part of Mediterranean rainfall owes its origin directly to the effects of the sea surface on polar air masses located on the rear side of depressions. Flohn (1949) states that in the Mediterranean area cold fronts are of first importance in producing rainfall, while warm fronts in general are of less consequence. But this varies with region and season as is shown by Reichel (1949) (see Table 15.2). His data refer to frequency of rainfall expressed as a percent of all weather observations taken by ships operating between Gibraltar and Port Said.

Reichel (1949) tabulates the count of rain frequency according to seasons and regions, and also notes whether the rain occurrence was associated with air flow from a northerly or southerly direction. His data show, for the year as a whole, that the extreme west, in the vicinity of Gibraltar, is the only part of the Mediterranean area where there is a predominance of rain occurrence with southerly air flow, or the front side of a depression. From about the Algeria-Morocco boundary eastward northerly flow was associated with a greater rain frequency, with a maximum of this predominance (N/S = 1.7) in the Straits of Sicily. Farther east the predominance of rain with north winds is less emphatic. It may be said therefore that, except in the extreme western Mediterranean where the northerly air masses have had only a short trajectory over water, rear-side rains predominate over front-side, even in the southern Mediterranean. It is the extreme western Mediterranean that feels the effects of low-index southwest steering, and the disturbances that move from southwest to northeast along the front oriented in the same direction. Here in all seasons southerly flow is associated with greater rainfall frequency.

Further insight into the origin of the rainfall is gained by separating the rainfall-frequency data according to whether the precipitation at the time of observation was of the unstable type (showers and thunderstorms) or of the stable type (gentle and long-continued). By combining showers with thunderstorms to form one group, and widespread rainfall with drizzle to comprise another,

Table 15.2. Rainfall frequency (in percent of all weather observations) over the Mediterranean Sea (Reichel, 1949)

	Sea area																			
	Both sides of Gibraltar		East of Gibraltar		North of west Algeria		North of east Algeria		Straits of Sicily		Southeast of Malta		North of Cyrenaica		Southeast of Crete		North of Egypt		Number of maxima	
	SEPARATED FOR NORTH (N) AND SOUTH (S) WIND DIRECTIONS																			
	N	S	N	S	N	S	N	S	N	S	N	S	N	S	N	S	N	S	N	S
Winter	22	53	21	21	33	19	55	42	39	21	43	31	35	48	25	34	28	22	5.5	3.5
Spring	17	46	13	20	18	19	32	23	29	11	25	17	18	6	13	4	5	6	5	4
Summer	2	3	1	3	5	4	8	3	4	1	2	—	3	1	2	—	—	—	6.5	2.5
Fall	15	36	13	16	31	25	37	22	26	26	14	27	9	20	14	8	6	6	4	5
Year	14	35	12	15	22	17	33	23	25	15	19	19	17	16	14	11	10	8	6.5	2.5
N : S	0.4		0.8		1.3		1.4		1.7		1.0		1.1		1.2		1.2		—	
	SEPARATED FOR UNSTABLE (U) AND STABLE (S) FORMS OF PRECIPITATION																			
	U	S	U	S	U	S	U	S	U	S	U	S	U	S	U	S	U	S	U	S
Winter	25	50	9	32	23	30	61	38	40	22	47	28	55	30	39	19	32	20	6	3
Spring	31	32	8	26	14	24	24	32	16	24	13	19	12	12	10	7	5	7	1.5	7.5
Summer	1	4	1	4	4	5	6	6	3	4	2	1	1	3	2	0	0	0	3	6
Fall	19	31	12	18	25	33	27	32	29	24	19	23	17	19	14	9	9	4	3	6
Year	19	29	8	21	17	23	31	28	23	19	21	18	21	15	17	9	12	8	6	3
U : S	0.6		0.4		0.8		1.1		1.2		1.2		1.4		1.9		1.5		—	

there is obtained an approximate division into rear-side and front-side rainfall, although there will be some general rains in the cool air and a few thunderstorms on the front-side.

From Table 15.2 it is clear that in the western parts, eastward to about mid-Algeria, the stable forms of precipitation (light, general rain) predominate. This is not unexpected in view of the prevalence of southerly air at the time of rainfall and the importance of low-index southwest steering. But from eastern Algeria eastward the showery, unstable rear-side type of rain associated with cool northern air predominates, with a maximum frequency of this precipitation type in the vicinity of Crete. This is the case not only for the year as a whole, but also for the winter season, when the northerly air is coldest and the sea surface relatively warm. In the other seasons the stable type of gentle, general rains associated with the upgliding of warm air is more frequent, especially in spring. The stable rainfall types strongly predominate in the vicinity of Gibraltar, where southerly Atlantic air prevails. Characteristically, the dry southerly air derived from the Sahara provides a relatively poor source for rainfall until it has had a sufficiently long trajectory over the sea surface to have absorbed considerable moisture. Warm fronts, as a consequence, are diffuse and weak in the southerly parts of the Basin.

From what has been said above it does not appear unusual that thunderstorms should be fairly common over the sea during the cooler months in the northern parts of the Basin. Summer thunderstorms, by contrast, are more characteristic of land areas, and while they are infrequent along the southern or African margins, they are sufficiently numerous in the European borderlands to contribute significantly to the warm season's rainfall of those regions.

Spatial Distribution of Rainfall

Topographic irregularities within the Mediterranean area make for great regional and local differences in rainfall. As a rule the western or weather side of each major peninsula shows significant contrasts with the east side, not only in temperature, but likewise in rainfall (Fig. 15.7). Thus, western Iberia has double the amount of annual rainfall of the east coast and the same ratio holds true in Greece. In Italy the coastal contrasts are somewhat less marked. The explanation for rainfall contrasts between east and west coasts is much the same as that for temperature contrasts described in an earlier section.

The seasonal variation in rainy days is especially marked. On the average the Mediterranean area has 90–100 rainy days a year, approximately 10 in summer, 20 in fall, 40 in winter, and 25 in spring (Reichel, 1949). This distribution is understandable in terms of the annual march of days with rain-bringing weather situations described previously. For the western Mediterranean Reichel selects five weather situations (NH_z, N, NF_z, HO, and S) which are responsible for most of the rainfall in that region, and shows the annual march of these situations in combination (see Table 15.3). The same is done for the central Mediterranean, employing eight of Hess and Brezowsky's (1952) weather situations (HN_z, NH_a, N, HO, TJ, TB, TK, and W_s), which are the principal rainfall generators in that region. For the eastern Mediterranean a monthly count is made of the number of depressions traveling on tracks V^c, V^{d1}, and V^{d2} as described by Weickmann (1922) (Fig. 15.6). For comparison, there are also provided data on the average number of days per month on which outbreaks of cold northerly air move into the Mediterranean through the gap between the Pyrenees and Alps, for this situation is fa-

Table 15.3. Statistical analysis of Mediterranean weather situations (Reichel, 1949)

Large-scale weather situations that are good rain-bringers (number of days)	J	F	M	A	M	J	J	A	S	O	N	D	Year
Western Mediterranean	4	3	6	5	5	4	2	2	3	5	5	3	47
Central Mediterranean	5	5	8	7	6	5	4	3	3	5	6	5	62
Eastern Mediterranean	5	7	8	6	4	3	1	1	2	4	6	7	54
Days with passage of cold air through gap between Alps and Pyrenees in southern France	16	13	9	9	13	10	6	6	7	13	12	11	125

vorable for cyclogenesis in the central Mediterranean. All four of these weather conditions favorable for rain show a minimum in summer, thereby coinciding with the season of least rainfall. In both the western and central Mediterranean it is noteworthy that summer, although it is the season with a minimum of large-scale weather situations favoring rainfall, is by no means lacking in such situations, so that summers are far from being dry as they are in the eastern (and southern) Mediterranean where the count of summer disturbances is only one a month for July and August. Cold outbreaks through southern France are at a maximum in January–February but are also frequent in May and October–November. Spring and fall commonly have somewhat more large-scale weather situations favoring rainfall than does mid-winter, this fact being reflected also in the bimodal annual march of rainfall amounts so characteristic of the northern Mediterranean margins (Flohn, 1948). Not uncommonly spring shows a greater frequency of weather situations favoring rainfall than does fall, although this situation is reversed as it applies to amounts of rainfall. This reflects the fact that the sea surface is warmer in fall than in spring, so that it is in fall that the sea creates the greater instability in invading cold air masses. Rainfall intensity is therefore greater in fall even though rainy days are more numerous in spring.

Seasonal Distribution of Rainfall Frequency.—In summer, except where terrain makes for irregularities and steep gradients, the isolines of rainfall frequency show a strong zonal arrangement, their general trend being east-west and with the values declining from north to south (Fig. 15.8). This arrangement points to the fact that in the warmer months the great planetary controls largely dominate the weather, and the local effects of land and sea temperature contrasts are at a minimum.

The highland rim on the north provides a climatic divide which sharply separates the Atlantic area with its 40–45 rainy days in summer, from the Mediterranean Basin where, even in its rainier north, fewer than 22 days with rainfall are encountered. In reality the summer months June, July, and August do not comprise a good seasonal unit, for June is more like a spring month with a fairly large number of rainy days. It is particularly in mid- and late summer and early fall that the Azores high and Indian low give rise to an air flow that is genuinely conducive to drought. August, therefore, has many fewer rainy days than June.

In winter, by contrast with summer, the zonal or east-west arrangement of rainy-day isolines is weak, and there is substituted a more confused pattern characterized by isolated regions of maxima and minima (Fig. 15.9). At this season the planetary controls

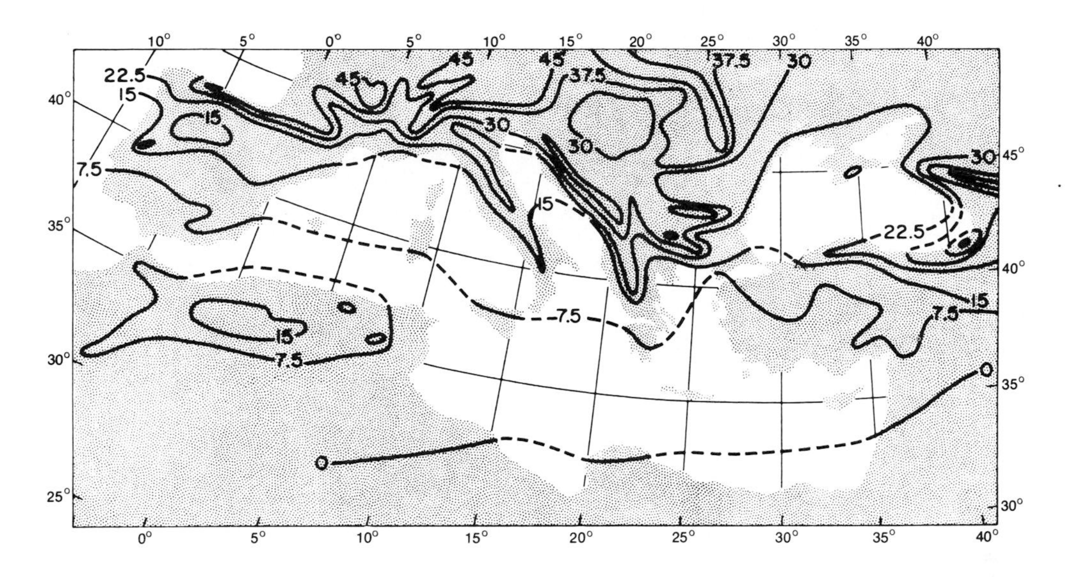

Fig. 15.8. Average number of days with rainfall in summer in the Mediterranean Basin. (After Reichel, 1949.)

are relatively weaker, while the Mediterranean Sea, through its modifying effects on cold northerly air masses, develops a dynamic system of its own which favors rainfall. As a general rule, the higher values of winter-rainfall frequency are positioned over the sea and along the western sides of the large peninsulas. Two conspicuous regions of concentration are (1) the central Mediterranean from the Gulf of Genoa to the coasts of Algeria and Tunisia, and (2) the extreme eastern end of the sea in the vicinity of Cyprus, both areas being centers of cyclogenesis associated with routes of cold-air invasions. Over the Black Sea the eastward-traveling depressions give numerous rainy days (35–45) in winter to the eastern parts, but many fewer (20–25) to the northwestern shores. The relatively few rainy days in winter over eastern Iberia, the Po Plain, the southeastern Balkans, and eastern Anatolia are associated with locally developed cold anticyclones.

For the sake of emphasis, it bears repeating that within this region of *Cs* climate the cool-season maximum of precipitation is not exclusively the result of planetary controls which bring the storm belt of the westerlies over these subtropical latitudes at the time of low sun. This, the earth's most extensive area of Mediterranean climate, owes its existence in considerable part to the effects of a great expanse of water whose surface during the cooler months is warmer than the air entering the Basin from the north, a condition which not only favors perturbation generation and propagation, but likewise acts to make unstable the invading cold air masses. Obviously this influence of the sea surface is absent or greatly weakened in summer.

On the map of rainfall frequency for the year as a whole certain features are noteworthy (Fig. 15.10). (1) Isolines of rainfall frequency are relatively zonal in the south, but much more complicated and with a stronger meridional alignment in the northern parts. (2) Mediterranean lowlands have distinctly fewer rainy days than do those of

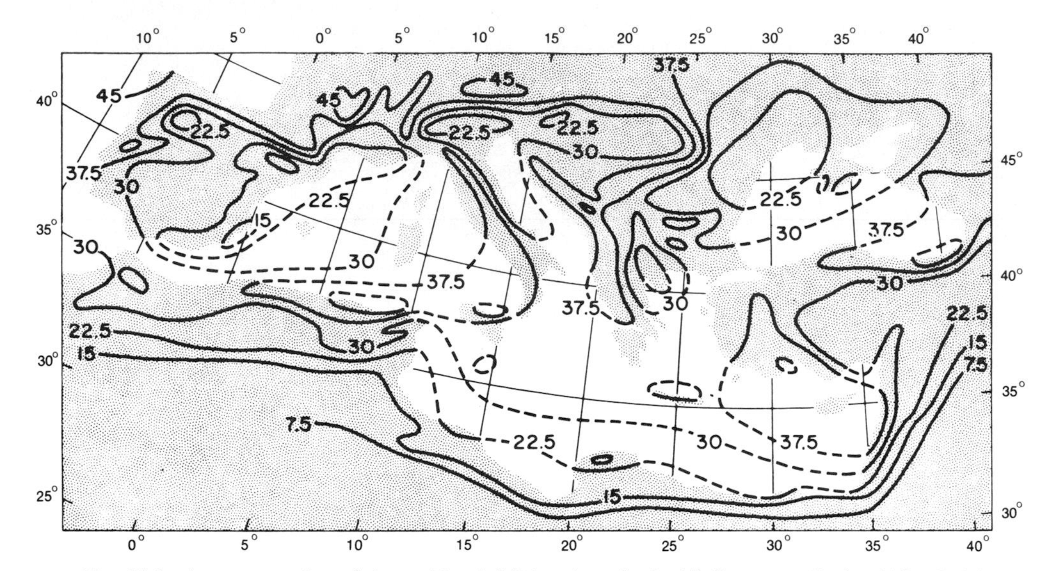

Fig. 15.9. Average number of days with rainfall in winter in the Mediterranean Basin. (After Reichel, 1949.)

Europe north of the Alps and Pyrenees. But although the former have only 90 to 100 days with rain per year, they nevertheless have a total annual amount of rainfall which is greater than that of central Europe where rainy days are more numerous. This reflects the greater intensity of rainfall in the Mediterranean area, which in turn is associated with the greater amount of instability shower rain falling from cool northerly air modified over the warm sea surface. (3) The eastern sides of the two broad peninsulas, Iberia and

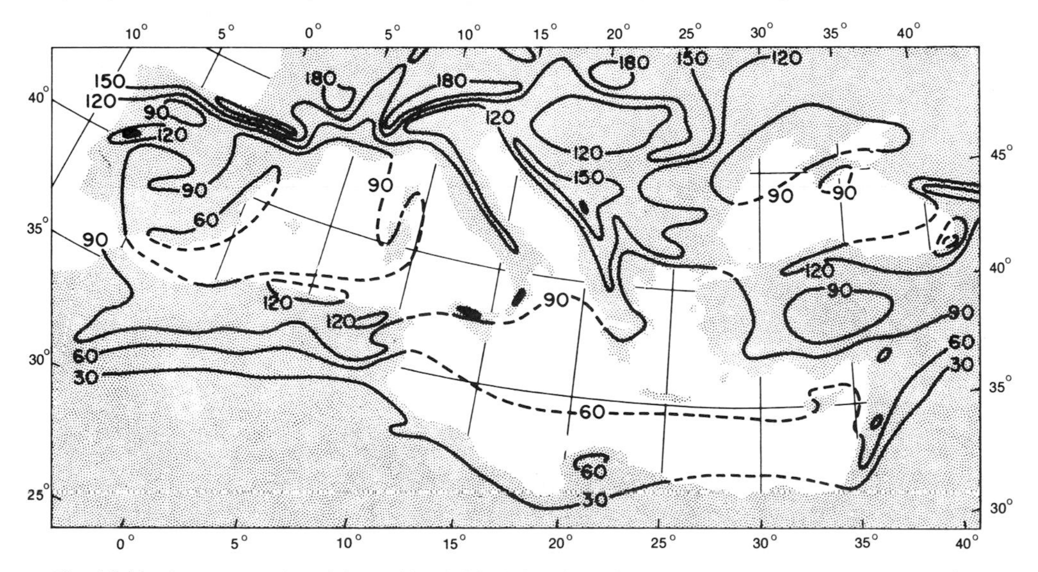

Fig. 15.10. Average number of days with rainfall during the entire year in the Mediterranean Basin. (After Reichel, 1949.)

the Balkans, have markedly fewer rainy days than the western sides.

If one classes as genuinely *Cs*, or Mediterranean, climate only those parts of the Mediterranean Basin which have a single minimum of rainfall frequency in one of the summer months and a single maximum in a winter month, it is discovered that only the southern and eastern parts of the Basin qualify for inclusion. Moreover, the African borderlands east of Tunisia, while they qualify in terms of seasonal concentration, would be excluded because of a total annual rainfall deficiency. Genuine *Cs* is limited to the African lands eastward to Tunisia, Sardinia, Sicily, Italy south of Rome, Greece, most of the Anatolian margins, and Syria-Lebanon-Israel. Iberia, southern France, most of Italy, and the Balkans are excluded. It is noteworthy that much of southern Europe which is ordinarily considered to have a Mediterranean climate, is characterized by complicated annual rainfall profiles, many of which show a double maximum and minimum of rainy days.

SEASONAL DISTRIBUTION OF RAINFALL AMOUNTS. — Complementing Reichel's (1949) study of seasonal rainfall frequency in the Mediterranean area is that by Huttary (1950) which analyzes the seasonal distribution of rainfall amounts. By emphasizing seasonal percentage distribution rather than absolute values, local orographic effects are minimized, while more extensive controls are stressed.

Winter has more than its proportionate share (25 percent) of the annual rainfall over most of the Meditteranean Basin and the proportion increases from northwest to southeast (Fig. 15.11). In east-central Iberia, southern France, northern Italy, and the interior Balkans and Anatolia, regions of moderately prevalent anticyclonic control, the winter proportion falls below 25 percent. It rises to more than 50 percent in northern Africa east of Tunis, and in Israel, Lebanon, and Syria. According to Huttary (1950) the decisive control producing this strong winter concentration on the southeastern littorals is the winter frequency of depressions following Weickmann's (1922) V^d and V^e tracks. A corollary cause is the expulsions of cold air from the Balkans and Russia into the eastern Mediterranean which reach their maximum frequency during the winter months.

In summer little of the Mediterranean Basin shows a seasonal concentration of rainfall amounting to over 25 percent (Fig. 15.12). A small area in southern France and northernmost Italy so qualifies, as does a more extensive area in the southernmost Balkans. Nevertheless, a moderate amount of summer rainfall does occur over the northern parts of the Mediterranean Basin, this being associated with depressions following the V^c route across northern Italy and the Balkans. South of about parallel 40°N the stable air of the Azores high and its etesian winds tend to stifle the rain-making processes.

Although spring and fall resemble each other in cyclonic activity, the rainfall patterns of the two seasons show some important contrasts. These stem largely from the reversal in temperature contrasts between land and sea in the two seasons. Thus, in fall, as the lands assume a negative temperature anomaly with respect to the sea, the latter acts to make unstable the invading cold air masses from the lands to the north. As a consequence, the water surface increasingly affects precipitation, and the coasts become wetter than the interior. Accordingly, in fall the coasts of eastern Spain, southern France, western Italy, and the eastern and southern borderlands of the Black Sea show relatively strong rainfall concentrations. In spring, by contrast, as the lands heat rapidly and be-

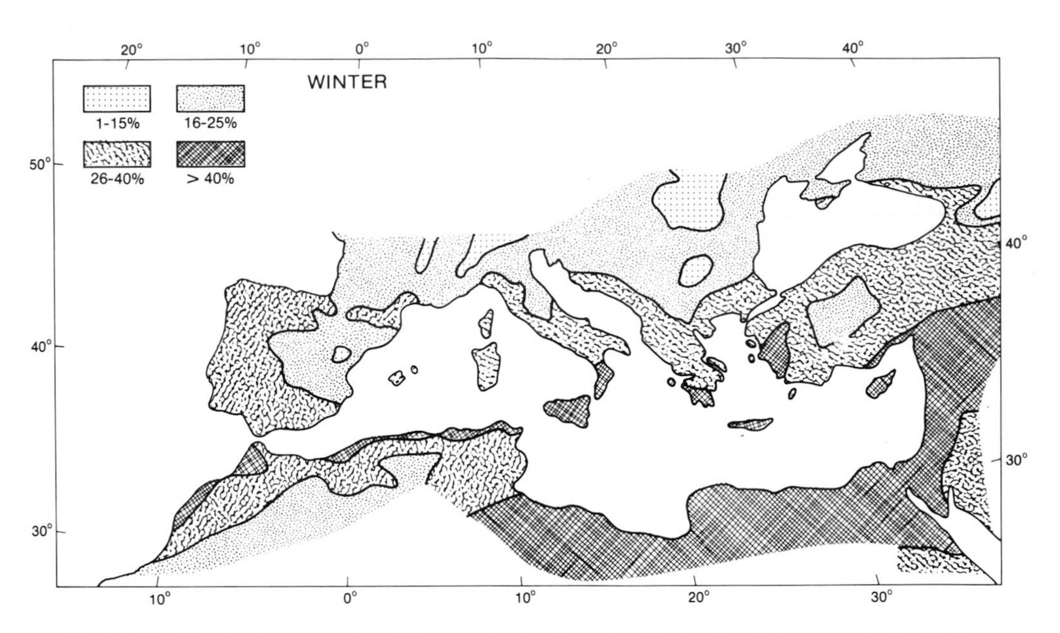

Fig. 15.11. Showing the proportion of the annual rainfall received during the winter season in the Mediterranean Basin. Note that parts of the northern Mediterranean Basin receive even less than winter's normal share, or 25 percent. (After Huttary, 1950.)

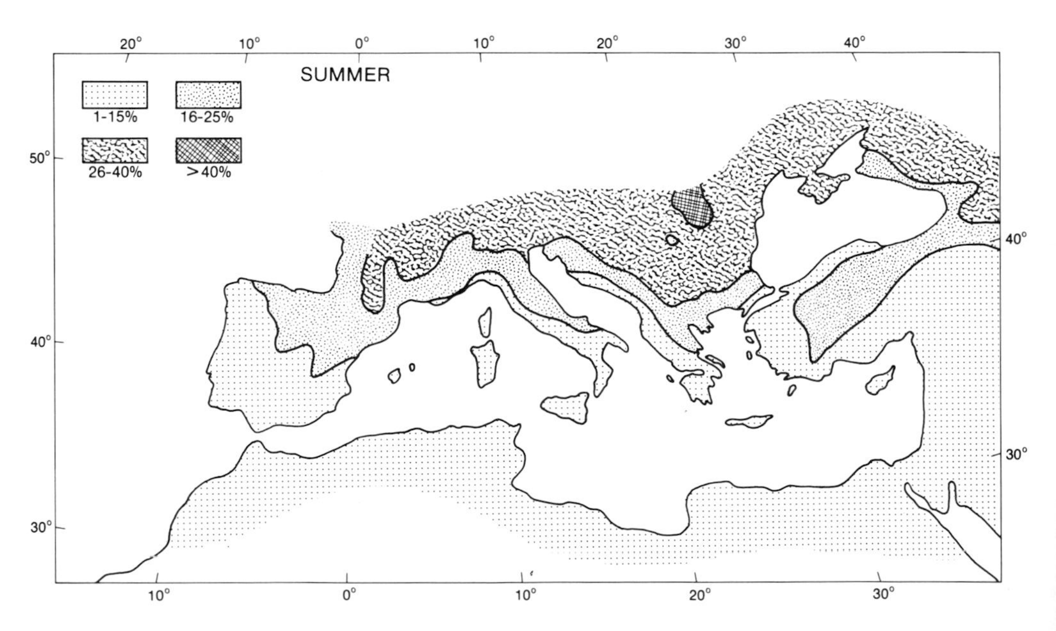

Fig. 15.12. Showing the proportion of the annual rainfall received during the summer season in the Mediterranean Basin. (After Huttary, 1950.)

come warmer than the sea, the latter's effects upon rainfall decrease. It is at this season that interior Anatolia, and to a less degree Iberia and the Atlas, show rainfall concentrations (Fig. 15.13).

If the principal types of seasonal rainfall concentration are combined, and represented on one map (Fig. 15.14), certain generalizations are valid (Huttary, 1950). The simple winter rainfall type is found along the Atlantic margins of Iberia and Morocco, while within the Mediterranean Basin it is confined principally to the sea margins of northern Africa and western Asia. Europe has only a restricted development of the simple winter-maximum regime, and where it exists it is confined to the southern extremities of the three great northern peninsulas, together with Sicily and Crete. Obviously most of Mediterranean Europe has a much more complicated seasonal rainfall concentration than the standard *Cs* type suggests. A single spring concentration is confined largely to the central Anatolian upland and to the inner Atlas, while bands of winter-spring concentration form transition belts around these areas.

THE AREA WITH A TWINNED RAINFALL MAXIMUM.—The bimodal annual rainfall profile, in which the peaks commonly coincide with fall and spring (Beelitz's [1932] "equinoctial rains") is so widespread that it appears to warrant special comment (Fig. 15.15). Such a profile is a feature not only of extensive areas in the northern parts of the Mediterranean Basin, but of the upper Danube plains of central Europe as well. In the west it is characteristic of large parts of interior and eastern Spain and southeasternmost France, most of the northern half of Italy, much of the Balkans north of Greece, parts of the Hungarian Plain, and interior Anatolia. Within this extensive area there are considerable variations in the annual rainfall profile, to be sure. In the southern or more Mediterranean parts, as in Spain and

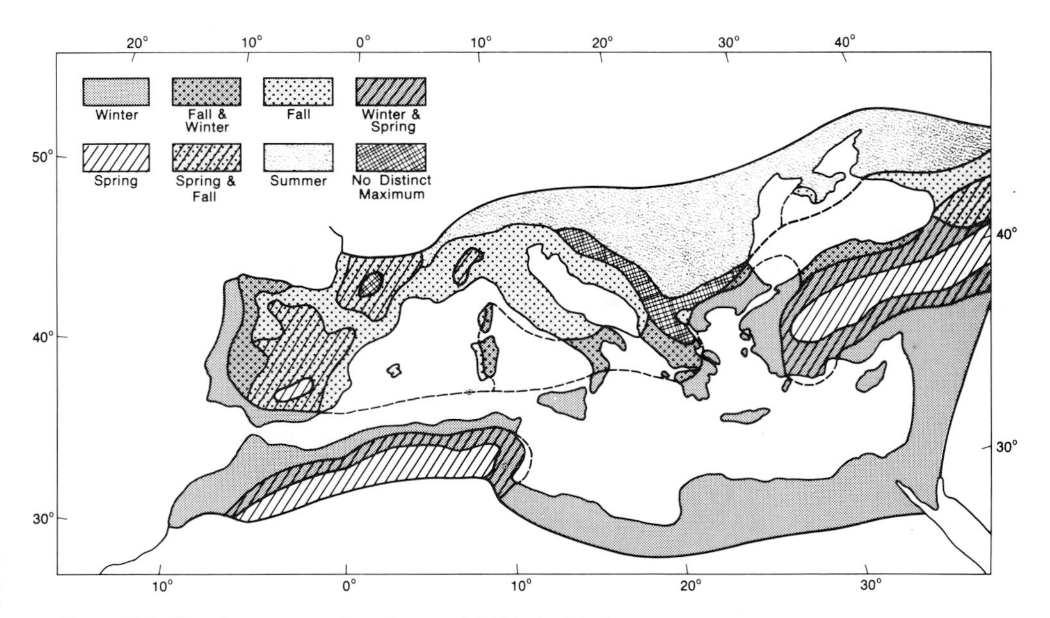

Fig. 15.13. Showing seasons of maximum rainfall in the Mediterranean Basin. Much of the northern basin does not have a simple winter maximum so typical of the Mediterranean type of climate. (After Huttary, 1950.)

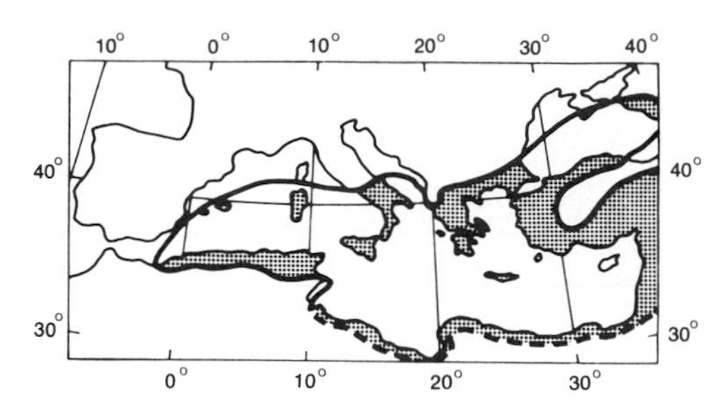

Fig. 15.14. Showing those parts of the Mediterranean Basin having an annual march of rainfall typical of Mediterranean climates, with the maximum falling in a winter month and the minimum in a summer month. (After Huttary, 1950.)

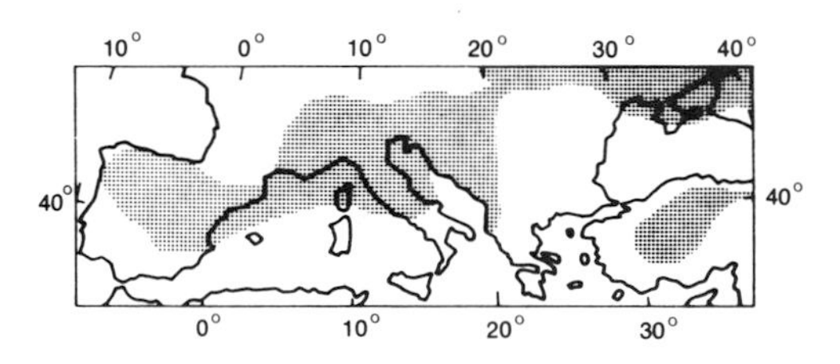

Fig. 15.15. Showing those parts of the Mediterranean Basin having a complicated annual march of rainfall, usually with two maxima and two minima. This is sometimes designated as the region of equinoctial rains. (After Huttary, 1950.)

peninsular Italy, the primary minimum is usually in summer while the winter dry period is a secondary. The reverse of this is the case in most of the Po Valley and the northern Balkans, for there winter is the season of primary minimum, with the secondary in summer. In some parts, also, while the rainfall profile exhibits a double maximum, the peaks more closely coincide with the extreme seasons than with spring and fall.

The biannual type of rainfall variation which characterizes this extensive region does not have a single and simple cause. To be sure, a large part of the area does lie between about parallels 40° and 45° N, that is, along the northern margins of the Mediterranean Basin and consequently in an intermediate position between Mediterranean and continental climates. It therefore experiences the drying effects of the continental high in winter and of the subtropical high in summer, resulting in the two drier seasons. In interior eastern Iberia and interior Anatolia, located farther south, the summer drought has a similar origin, but the winter depression in the rainfall profile is less the result of the general continental high than it is of local thermal anticyclones which develop over these extensive uplands (Zimmerschied, 1947–48; Lautensach, 1951).

Just as there exists nowhere else such an extensive area of dry-summer subtropical or Mediterranean climate, or such a deep penetration of this climate into the interior, so is there lacking elsewhere a similar extensive area characterized by equinoctial rains on the poleward margins of Mediterranean climates, for this is the only part of the earth where Mediterranean climate makes extensive contact with continental climates. Here then is a unique situation where, over extensive longitudes, summer-dry climate passes over into summer-maximum climate, a transition region of equinoctial rains being the result. Significantly, in the Balkans, even where the double maximum is conspicuous, the early summer or June maximum, so characteristic of interior continental climates, is usually much stronger than that of fall. Farther south, however, as for example in northern peninsular Italy and eastern Spain, the situation is reversed and the fall maximum is the stronger.

The Black Sea Borderlands.—In this general transition area between Mediterranean and continental climates is the Black Sea and its borderlands, a region of great climatic variety and complexity. As is the case with other local seas within the Mediterranean Basin, here likewise there are exhibited climatic contrasts between the drier and colder west and northwest leeward shores, and the milder and much wetter southern and eastern windward shores, but here the contrasts are greatly magnified. Even parts of the north shore, where the hills and mountains offer protection from invasions of northerly air, as for example in southernmost Crimea, bear some of the hallmarks of subtropical climate. At the extreme eastern end of the sea, in the vicinity of Poti and Batum, the year-round precipitation not only is unusually heavy (about 2,000 mm), with no dry season, but there is also a strong late summer-fall maximum, and the winters are genuinely mild for the latitude. Emphatically this is a subtropical humid climate (*Cf*). Somewhat similar conditions are to be found along wetter parts of the Anatolian Black Sea coast. Those sections of this littoral where the coastline is aligned in a WSW-ENE direction, or even E-W, so that the WNW winds, including those of summer, have an onshore component, experience rainfalls of (1,000–1,500 mm), relatively well distributed throughout the year and with maxima in both winter and summer. By contrast, those other stretches of coast having a WNW-ESE direction, so that the winds parallel the coast or are obliquely offshore, experience only (650–750 mm) of rainfall and the warmer months are dry, so that the climate is distinctly Mediterranean in character (Lembke 1940). This latter condition is particularly conspicuous just eastward of Cape Ince in the vicinity of Sinope, Bafra, and Samsun where the dry season may extend over a period of three to five months (Beelitz, 1932).

Subtropical Southwestern Asia (Asia Minor, Near and Middle East)

The area here included is that part of southwestern Asia reaching from the Mediterranean and Black seas on the west to India-Pakistan on the east. It is a transitional area, chiefly of dry climates, lying between the summer-dry climates of the Mediterranean and the summer-wet tropical and subtropical climates of India. Actually it is much more of an extension of the former than of the latter, and bears far more of the hallmarks of summer-dry than of summer-wet climates.

Extending as it does through about 30° of latitude, from about 13°N in southern Arabia to 43°N in the Caucasus region, and characterized by complex terrain and an interdigitation of land and water surfaces, homogeneity of climate is not to be expected. Its southern parts, chiefly Arabia, lie constantly under the influence of those aridifying controls that produce the Saharan climate. The same is true of southern Iran and southern West Pakistan. It is principally north of 30°N that the Mediterranean, Black, and Caspian seas become important climatic controls as they act to humidify the westerly and northerly air streams that enter southwestern Asia and likewise serve as the breeding grounds and propagators of those disturbances of the cooler months that are responsible for most of the region's precipitation. This fact that on its western and northwestern frontiers lies one of the earth's great regions of seasonal cyclogenesis and cyclone transmission is of primary importance in determining the weather and climate of southwestern Asia.

Seasonal Circulations and Their Associated Weather

Winter.—During the cooler months the entire region lies within the very broad stream of anticyclonic continental air from interior Eurasia. This northerly flow is interrupted during the cooler months by depressions invading the area from the west via the Mediterranean and Black seas. From October to April around 30 such disturbances reach Mesopotamia or the Persian Gulf and it is estimated that 20 to 25 traverse Iran toward India. During these cyclonic episodes southerly air, part of it Mediterranean in origin, is found on the southern sides of these convergent centers (Bauer, 1935).

Obviously little rainfall is likely to occur during the period of normal undisturbed northerly flow, except in a few special locations. The most conspicuous of these is the relatively abundant cool-season rainfall on the slopes and plain bordering the southern margins of the Caspian Sea. Here, although depressions are few, a major flow of cold northeasterly air meets a westerly flow over the Caspian along a convergence extending north-south over that water body. As a result, there is a strengthened northerly flow over the Caspian which is obstructed in its continued southward movement by the highlands at the south end of that sea. In its long trajectory over the Caspian the cold air is greatly humidified and warmed, so that when it is forced to ascend the Elburz highlands heavy rainfall results. Thus, Ensli on the south coast has over 1,000 mm of annual rainfall and Lenkoran on the southwest coast about the same amount, while much heavier rainfalls characterize the slopes to the rear. A fall-winter maximum is conspicuous at both stations, with summer the driest season, the period of heaviest rainfall in the autumn months coinciding with the season when the relatively greater warmth of the sea surface makes for a maximum modification of the *cP* air masses. In its late fall and winter trajectory across the Caspian Sea from north to south, the temperature of the surface air may be increased by as much as 14°C (25°F) and its vapor pressure multiplied three times (Bauer, 1935). No doubt some of the winter rainfall along the north coast of Anatolia facing the Black Sea is likewise due to the lifting of undisturbed northerly winds which have been modified in a similar fashion in their trajectory over the Black Sea.

However, by far the largest part of the cool-season rainfall over the general region of southwestern Asia originates in cyclonic storms which cross the area from west to east. Most of Arabia is too far south to be affected by a majority of these depressions, and as a result rainfall amounts, annual as well as cool-season, decrease from north to south (Fig. 15.16). There is a similar decrease from west to east except where highlands interrupt. This is to be expected since the major cyclonic activity is concentrated to the west in the vicinity of the Mediterranean and Black seas, while the depressions tend to fill and weaken as they progress inland.

The rain-bringing depressions which affect southwestern Asia in the cooler seasons enter the region by both Mediterranean and Black Sea routes (Weickmann, 1922). The most northerly route is that designated as track III[a] (Fig. 15.6). These disturbances, coming from the Baltic region and moving toward the southeast, chiefly influence the Black Sea area, although in winter their rainfall effects may be felt over all of Asia Minor. On the average this track is followed by about 14 depressions a year. Its path is marked by a col through the continental axis of high pressure. Track III[a] storms are asso-

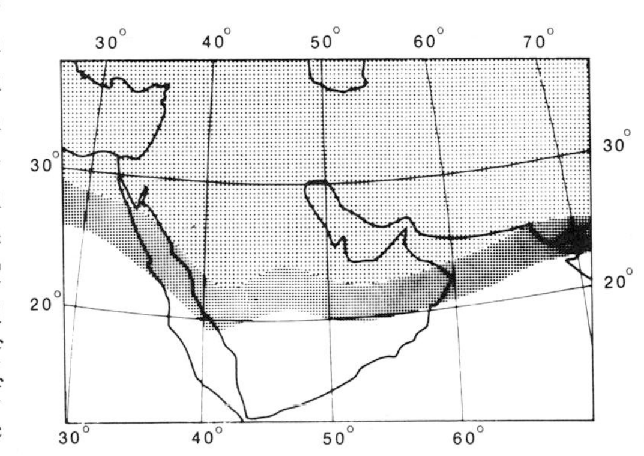

Fig. 15.16. Showing the approximate southern limits (dark shading) of, and the area covered by (light shading) cyclonic winter rains in southwestern Asia. (After Bauer, 1935.)

ciated with outbreaks of cold air from the north, which surge southward on the rear of the depressions. Such lows have a tendency to become stationary or move slowly in the eastern Black Sea area, giving rise there to heavy rains lasting for several days. Storms also move into the Black Sea basin over routes V^{c} and V^{d2}, each of these routes carrying a similar number of depressions as III^{a}. On the whole, disturbances following these more southerly tracks have greater effects on Asia Minor's weather than do those on track III^{a}. Unfortunately weather stations are too few in southwestern Asia to be able to plot the tracks of storms over that extensive region. During the winter months alone, about 20 cyclones follow the V^{d1} and V^{e} routes into the eastern Mediterranean and enter Asia Minor. Thirty to 35 depressions entering the area from the west by any and all routes reach the Persian Gulf, 20 of them in the winter months. Somewhat fewer travel across Iran (Bauer, 1935). The fact that southwestern Asia has funneled into it numerous storms traveling eastward along the various tracks concentrated in the eastern Mediterranean and the Black Sea, causes this region to be the most extensive of the earth's dry areas with a winter maximum of rainfall.

SUMMER.—In the warmer months there are two primary circulations affecting southwestern Asia. Much the more important is the northwesterly flow which has its sources in the Azores high and is directed toward the centers of low pressure over southeastern Arabia and India-Pakistan. These winds are the hot, dry, and stable etesians (Schneider-Carius, 1947–48). The onset of the dry northerly winds of summer in Iraq is almost as abrupt as is the onset of the summer monsoon in India, and significantly the two phenomena begin almost simultaneously (Frost, 1953). The etesians, however, are quite unlike the southern or maritime branch of the monsoon that reaches India and southeastern Asia. Rather they are trade-like, and just as the drought in the eastern oceanic trades is fundamentally due to a low temperature inversion layer resulting from subsidence, through which it is difficult for convection to penetrate, such is also the case with the etesians. In southwestern Asia, therefore, the etesians represent a far eastward extension of the eastern oceanic trades of the Atlantic. In spite of high surface humidity along coastal areas, and the intense surface heating over the land, convection currents cannot reach sufficient altitude to develop shower activity. Over the sea this convectional exchange layer is only about 900 meters thick and over the lands not much more than double this figure. Thus, over 90 percent of the area, where the etesians are dominant, drought grips the land in summer.

The second principal air stream which in summer affects southwestern Asia, but only its southernmost parts, comprising about one-tenth the total area, is the so-called Indian monsoon air. Although this southwesterly maritime current is moist, it is relatively impotent as a rain-producer because of its shallowness, in these parts rarely reaching more than 900 meters in depth. Aloft, the dry etesian air acts as a lid over the humid southwesterly current, stifling the formation of clouds and precipitation.

There are only three areas with noteworthy precipitation in summer (Fig. 15.17). One of these is the Yemen Highlands in the extreme southern portion of Arabia which lies within the realm of the southwest monsoon. Although this humid surface air may penetrate the continent for a few hundred kilometers, only in elevated Yemen is it regularly thick enough to produce a moderate amount (250–500 mm) of summer rainfall, most of it in the form of heavy showers. On occasions when the southwesterly flow is unusually strong, and as a consequence the etesians are forced to retreat farther north, significant summer rains may fall farther inland. Bauer (1935) estimates that the windward slopes of the Oman Highlands in Arabia may receive 50 to 150 mm of summer rain from the southwesterly circulation.

A second area in southwestern Asia with appreciable summer rainfall is eastern Afghanistan and adjacent parts of Pakistan, which are also within the domain of the Indian monsoon. Here southeasterly air currents from up the Ganges Valley, as well as the southwestern branch of the monsoon from the Arabian Sea penetrate sufficiently deeply to produce a modest rainfall on the southern and eastern slopes of the highlands west of the Indus Valley. But the shallowness of the humid southerly currents precludes much rainfall, and amounts decline quickly to the west as the etesians increasingly dominate.

The third and largest of the areas within southwestern Asia which receive a significant amount of summer rainfall, lies in the northwest and includes Anatolia, Armenia, northern Iran, the Caucasus, the south and east coasts of the Black Sea, and the south coast of the Caspian. The summer's rainiest area in all of southwestern Asia is along the east coast of the Black Sea where at Batum some 650 mm of rain fall in the three summer months. Elsewhere the warm-season rains are more modest.

In part, this summer rainfall in the northwestern sections is a consequence of the fact that here the etesians are more humid than

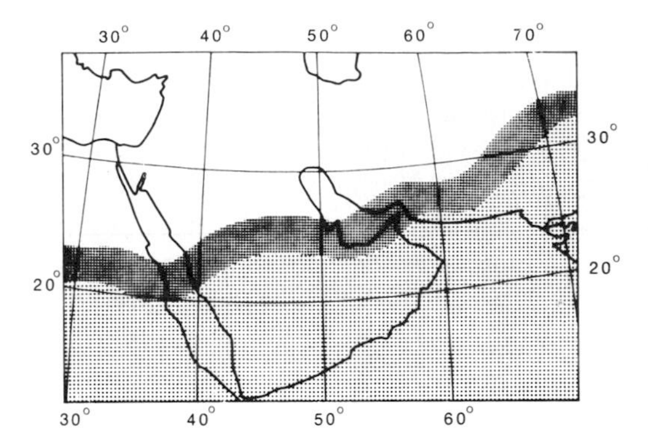

Fig. 15.17. Showing the northern limits (dark shading) of, and the area covered by (light shading), summer rains associated with the southwest monsoon in southwestern Asia. (After Bauer, 1935.)

farther south and have not as yet attained their genuinely stable characteristics. Moreover, in these latitudes which lie north of parallel 35°N, and parts north of 40°, cyclonic storms are still operative even in summer, especially along the tracks entering the Black Sea. There is no doubt, however, that the Black and Caspian seas themselves play an important part in making summer rainfall possible along their southern margins. Air flow is prevailingly from the northwest across the Black Sea and is NNW over the Caspian, and these trajectories result in a sufficient humidification of the northerly currents so that when they are forced upward by highlands, or in vortex or convectional systems, they yield moderate precipitation. South of the north-facing highlands precipitation drops off very rapidly, however.

PART VI
Anglo-America

CHAPTER 16

Temperature and Precipitation

Since the general controls and major lineaments of the climates of Anglo-America are dealt with in a number of sources, both general and specific, these more familiar materials need not be repeated here. Instead an attempt will be made to supplement these sources and likewise to emphasize the unusual in climate.

Other than Eurasia, North America is the only broad continental area in the middle latitudes. It is to be expected, therefore, that important climatic resemblances will exist between these two great continental areas, growing out of similarities in latitude, and in large longitudinal spread as well. But while many features of climatic arrangement are similar, there are, as well, important climatic contrasts between Anglo-America and Eurasia, some of which have been set forth in the earlier sections on Europe and Asia. The high cordillera which closely parallel the west coast of North America greatly restrict the logitudinal extensiveness of marine influence there, in contrast to the situation in western Eurasia, so that the transition from humid maritime climates to dry climates is very abrupt and without intervening humid continental climates, except at high altitudes.

In addition, while North America has a warm-water drift paralleling its west coast poleward of about 50°N, this current is not comparable in extensiveness and warmth to the North Atlantic Drift off the European coast. Thus, the air-temperature anomaly in January over the Gulf of Alaska is about 11°C (20°F) while a maximum amonaly of 28°C (50°F) is characteristic of the northeastern Atlantic off Norway. Decreased winter mildness is, therefore, characteristic of the North American Pacific littoral. Moreover, the warm current in the northeastern Pacific is unable to extend its influence as far poleward as does its counterpart in the eastern North Atlantic. The former is really a subordinate eddy in the Gulf of Alaska and is blocked by the westward bending of the continental coastline and the Aleutian barrier from flowing northward into the Arctic Ocean. Consequently, while Europe feels the warming effects of the North Atlantic Drift as far north as North Cape (71°–72°N), and eastward along the Lapland shores of the Arctic Ocean, the warm current in the North Pacific gets no farther north than about 60°. Drift ice extends southward in the Bering Sea all along the Alaska coast to the Aleutian Islands, so that western Alaska, unlike Norway in similar latitudes, is scarcely marine in winter.

Moreover, North America's windward littoral is to a much less degree insular and peninsular, and has fewer large-scale indentations in the form of bays and seas which further tend to reduce the marine qualities of climate. Especially does it lack anything comparable to the Mediterranean Sea, which in the lower middle latitudes carries marine influences more than 3,000 km inland from the west coast. No such region of cyclogenesis and cyclonic concentration exists in comparable latitudes and longitudes in North America, so that there is lacking a similar extensive development of summer-dry subtropical climates (*Cs*).

On the other hand, lacking east-west terrain barriers such as exist in Eurasia, eastern and central North America is freely open to invasions of tropical warmth and humidity from the Gulf of Mexico to the south, and of arctic cold from the north. Here climatic divides are lacking and climatic change in a north-south direction is gradual. In no other continental area of the earth do strongly contrasting air masses meet so freely as they do in central-eastern Anglo-America. Corollary to this is the fact that no other continental area can match Anglo-America east of the Rockies as a region of cyclogenesis and vigorous cyclonic activity. Thus, the Gulf source of heat and moisture and the unusual cyclonic activity result in humid climates extending inland for nearly 2,500 km from the lee coast, a much greater depth than is the case in eastern Asia. Accordingly, winter-dry climates which are so prevalent in eastern Asia are completely absent in eastern Anglo-America.

Temperature

Seasonal Temperatures.—Most of the larger lineaments of temperature are relatively standard in character. It is noteworthy, however, that the cold pole of the continent in winter is symmetrically located about midway between the Atlantic and Pacific oceans along meridians 90°–105°W. This is in contrast to the situation in Eurasia where the cold core of the continent is asymmetrically positioned in the northeastern part of the land mass. The fact that the center of winter cold in Anglo-America is about equidistant from both oceans suggests that the marine influence from the west, or windward side, is blocked by mountains.

From the January cold centrum over the arctic pack ice, two wedges of cold air extend southward, one over the continent proper and the other over Greenland. The latter is the colder by 11°C (20°F), but the average January temperature of −40°C (−40°F) is for an ice plateau which in places extends to over 3,000 m in altitude. Greenland is not the primary source of the cold surges that generate North American and North Atlantic cyclones in winter (Hare, 1956), for the air that descends from the Greenland Ice Cap is so warmed by compression that when it enters the Atlantic circulation it is much less cold than *cP* air from Canada.

Abnormal temperatures in fall and early winter are to be observed in certain subarctic locations which are favorably situated with respect to open water (Hare and Montgomery, 1949). Thus, western Greenland is relatively less cold by reason of open water in Baffin Bay and Davis Strait to the west. Accordingly, in December temperatures are 14°C (25°F) higher at windward Disko Island off Greenland than at leeward Clyde River on the Baffin Coast only 600 km distant. This differential diminishes after December as freezing greatly restricts the area of open water. It is possible that subsidence from off the Greenland Ice Cap likewise decreases winter severity along that island's west coast.

Hudson Strait separating Baffin Island from Labrador Peninsula is another area of open water which is responsible for areas of early-winter mildness along its lee margins. Hudson Bay remains unfrozen until about January, so that its eastern land margins are distinctly less frigid in fall and early winter than are the opposite western margins. This effect reaches a maximum in November, after which time freezing reduces the area of open water. Thus Port Harrison on the eastern side of Hudson Bay is 6°C (11°F) warmer in November and 5°C (9°F) in December, but 2°C (3°F) colder in January, than Churchill on the western side of the Bay.

By July the cool pole of the continent has shifted to the northeast with Baffin Island at its core. Isotherms assume a diagonal northwest-southeast course, so that arctic Canada divides into a warmer west and a cooler east, the latter being influenced by the abundance of water surface, including ice-melt water and icebergs.

Greenland in summer is a genuine cold source, for here the air originates over a permanent ice cap. Feeding into the atmospheric circulation over the Atlantic, this cold air is effective in generating disturbances which cause the North Atlantic to be much stormier in summer than the North Pacific which has no comparable cold source.

The strongest July plus anomaly of temperature in North America is in the Great Basin area where latitude, basin configuration, and anticyclonic circulation aloft, providing clear skies and intense insolation, permit of strong surface heating. Between the cool California coast with its cool current and fog, and the intense summer heat just inland from the mountains, there exists the steepest summer sea-level temperature gradient anywhere in Anglo-America.

TEMPERATURE CONTINENTALITY.—Continentality as a function of temperature may be expressed in a variety of ways. V. Conrad's (1946) formula, $k = [1.7A/\sin(\phi + 10°)] - 14$, in which k is the coefficient of continentality stated in percent, A is the average annual temperature range in C°, and ϕ is the geographic latitude, yields a coefficient of continentality of about 100 percent for Verkhoyansk in northeastern Siberia and zero for Thorshavn in the Faeroe Islands, representing the two extremes. There are two published maps showing distribution of continentality according to Conrad's formula; for western United States (D'Ooge, 1955) and for New England (Fobes, 1954). Subsequently a map has been prepared by the author covering all of the United States and much of Canada (Fig. 16.1). Among the noteworthy features revealed by the last map are the following. Continentality is much weaker on the west coast than on the east, and while the Gulf of Mexico coast is somewhat less continental than the Atlantic margins, it is markedly more so than the Pacific littoral. Five to 10 percent are typical of the Pacific margins, 25 to 30 percent along the Gulf, and 30 to 40 percent along the Atlantic coast. A very steep east-west gradient in the continentality index characterizes the western margins of Anglo-America, much steeper than along the Gulf or Atlantic littorals. This is attributable not only to the stronger and more continuous marine influence on this windward side, but likewise to the fact that there the paralleling mountain ranges rapidly weaken the oceanic influence inland from the coast. A core area of maximum continentality as delimited by the isoline of 60 percent is situated in central Canada west of Hudson Bay. From this core area the roughly concentric isolines loop southward over central United States as far as Texas. But for the modifying effects of the Great Lakes, in whose vicinity the isolines bend sharply northward and westward, thereby producing a major dent in the eastern flanks of the wedge of strong continentality, the breadth of the latter probably would be much greater in northern United States. It would appear as

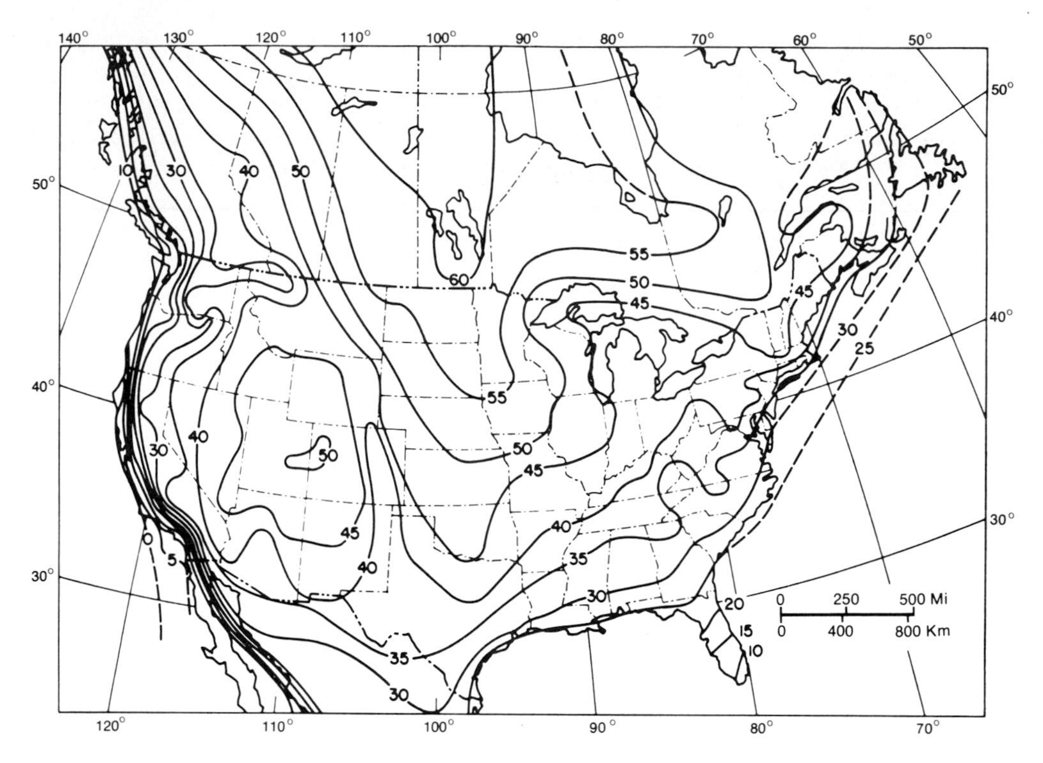

Fig. 16.1. Isolines of equal continentality for Anglo-America, using V. Conrad's formula.

though the Great Lakes project an outlier of relatively low continentality into a region that in their absence probably represents the center of maximum continentality in North America. The *local* effects of the Great Lakes on temperature continentality are made remarkably clear on a map showing isopleths of continentality coefficients at 1 percent intervals (Kopec, 1965). Here the striking feature is the bunching of isolines in a pattern outlining the individual lakes, with the isolines paralleling the shorelines. Especially steep gradients are to be observed around Lake Superior, at the southern end of Lake Michigan, along the eastern margins of Lake Huron-Georgian Bay and northeast of Lake Ontario. In the western intermontane region continentality is only moderate over the narrower northern parts, actually no higher than it is along the Atlantic seaboard. In the broader sections south of about the 45° parallel, however, continentality increases, reaching a high of 50 percent in western Colorado. A major trough of reduced continentality is positioned just to the east of the Continental Divide.

TEMPERATURE VARIABILITY.—Interdiurnal variability of average daily temperatures, or the difference in the average temperature of succeeding days, reflects not only continentality, but in addition the nonperiodic effects of moving highs and lows. Clearly this term will be weak in regions of great temperature uniformity such as marine climates, but is likely to be strong in regions of steep sea-level temperature gradients, such as interior North America and Asia in winter. But equally important is the irregular cyclonic-anticyclonic control which is instrumental, through advection and cloud-cover variability, in causing large and rapid temperature changes of a nonperiodic nature. Supan (1927) has developed a table which shows that the Northern Hemisphere in January has two far-separated areas of maximum interdiurnal temperature variability, one in interior North America, including north-central United States and south-central Canada, and another in western Siberia. The North American center is somewhat the stronger of the two, in spite of the fact that it has less severe winter cold. On the other hand, the perturbation control is better developed here, and this in conjunction with the warm Gulf source to the south and the cold Canadian source to the north, with no terrain barriers between, makes for extraordinarily large aperiodic temperature changes associated with advection.

For the most part studies in interdiurnal changes have been applied to the average daily temperature. This term has been mapped by Greely for the United States in January, and subsequently described by Ward (1925). More recently Calef (1950) has prepared maps showing the interdiurnal variability of the daily minima in January and the daily maxima in July (Figs. 16.2, 16.3). In major lineaments the January maps of interdiurnal variability of average daily temperatures and of the daily minima show strong resemblance. However, the variability of the minima over eastern United States is considerably weaker than that of the mean, which indicates that the greater temperature variability of that region in winter is more a consequence of variability of the daily maxima than of the daily minima. The greatest contrast between the two maps is in New England and northern New York State where the January daily temperature minima vary much more than do the daily averages. In fact, the highest variability of daily minima anywhere in the United States is in northern New York and New England. This situation reflects the region's intermediate position

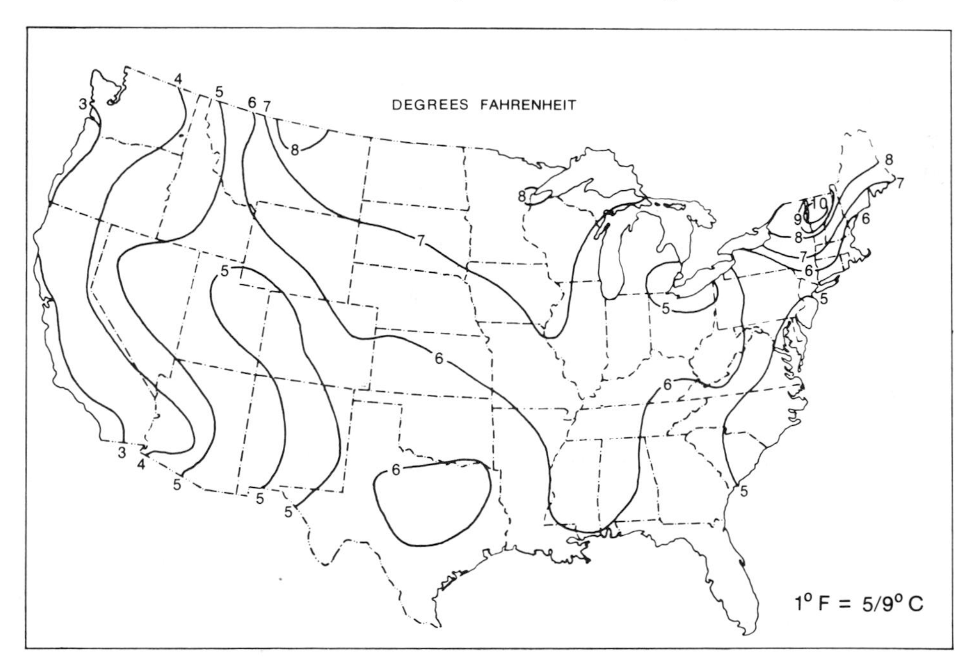

Fig. 16.2. Interdiurnal variability of daily minima in January for Anglo-America. (From Calef, 1950.)

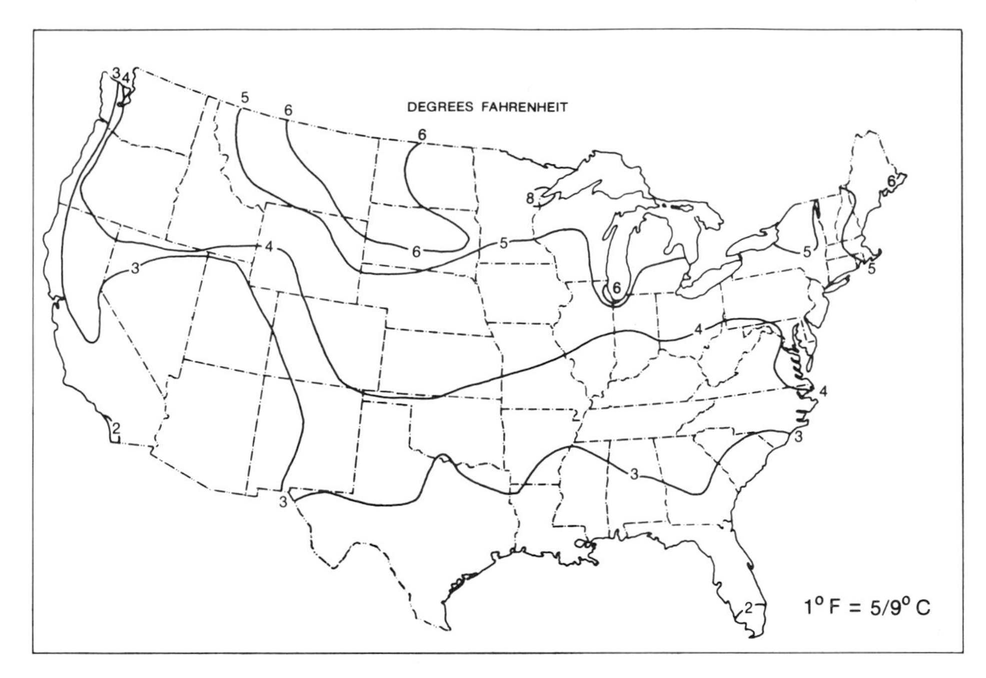

Fig. 16.3. Interdiurnal variability of daily maxima in July for Anglo-America. (From Calef, 1950.)

between the mild Atlantic source to the south and east and the cold continental source to the northwest, as well as the high concentration of cyclone tracks.

Variability of Monthly Mean Temperatures has been mapped for Anglo-America and likewise for the Northern Hemisphere in the form of the standard deviation from the long-term mean (Sumner, 1953). Quite clearly, the reasons why successive Januarys vary so greatly in temperature over parts of North Amierica are quite different from the causes for variations in daily means or daily extremes. Monthly variability must stem from a change in general circulation patterns from one year to the next, as reflected in the strength and prevalence of certain *Grosswetterlagen*, or in air masses. Sumner's maps reveal a much larger standard deviation of mean monthly temperatures for winter than for summer months (Figs. 16.4, 16.5). This in turn reflects the higher pitch at which the general circulation operates in winter, so that modest variations in airflow patterns are able to produce relatively large temperature modifications. The weaker latitudinal temperature gradients and atmospheric circulation in summer have the opposite effect.

In the winter months the areas of largest deviations from the normal average temperature are fairly coincident with areas of lowest mean temperature. Mild marine areas have small deviations, while strong continentality appears to coincide with large variations. Even the eastern Great Lakes become centers of somewhat lower-than-average deviation in winter. The geographical area with the strongest deviations in the winter months extends from the Yukon Valley in central Alaska southward and eastward following the band of highlands to western Montana, with the highest values at or near the eastern base of the Rockies. Here föhn effects are strong, and the variability in chi-

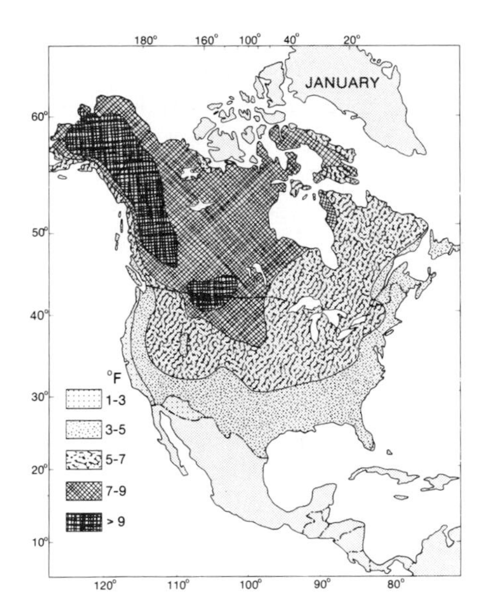

Fig. 16.4. Standard deviation of mean monthly temperatures (1°F = 5/9°C) for January for Anglo-America. (After Sumner, 1953.)

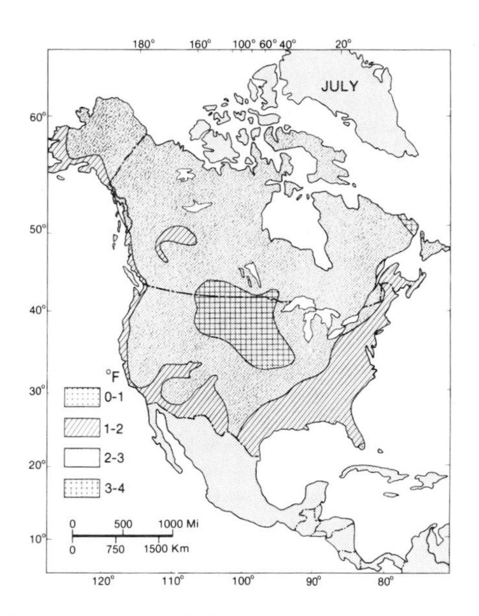

Fig. 16.5. Standard deviation of mean monthly temperatures (1°F = 5/9°C) for July for Anglo-America. (After Sumner, 1953.)

nooks from year to year, which in turn is related to changing circulation patterns and the associated routes of cyclonic storms, largely determines the intensity of winter cold.

Southern Alberta east of the mountains is particularly unusual in its winter temperatures. Less severe average winter cold characterizes the western part of the Canadian plains than regions farther east. Thus Calgary and Banff have average winter temperatures which are about 6°C (13°F) warmer than Winnipeg farther east, and 8°C (11°F) warmer than Qu'Appelle. Chinook prevalence along the eastern base of the Rockies chiefly explains this contrast. But variability of winter temperatures, both between different years and within a single winter, are still more characteristic of this region. Probably no other region on the earth equals the plains of southwestern Alberta in their winter temperature variability (Stupart, 1905).

Within any single winter month variability is closely related to the eastward progression of cyclonic storms and the chinooks which they generate. Thus, a cyclone crossing northern or central Alberta will have the effect of shifting the wind from north or northeast to a chinook southwesterly flow of subsident character from over the mountains, with a resulting rise in temperature of 11° to 22°C (20° to 40°F) within a few hours.

The temperature variability between different winters is equally phenomenal. Stupart cites the case of Edmonton where over a period of 3 years the January average was as high as −6°C (22°F) and as low as −25°C (−13°F), a difference of 19°C (35°F). November in one year had an average temperature of −18°C (0°F), while four years later it was 3°C (38°F). Havre, Montana, has recorded a January mean as low as −25°C (−13°F) and as high as 1°C (34°F).

The great inter-January variability is to a large degree a consequence of the varying positions of the mean tracks of storm centers in different years. In those winters when the disturbances follow southerly tracks predominantly, chinooks are largely absent in Alberta, and northerly winds prevail with accompanying low temperatures. But when, in other winters, the disturbances move on northerly tracks across northern British Columbia, chinooks are frequent in Alberta, southerly flow is prevalent, and higher average temperatures are the rule. Even apart from chinooks themselves, the advection associated with depressions following routes to the north or to the south of Alberta would operate to cause a high degree of temperature variability.

TEMPERATURE SINGULARITIES.—The reality of climatic singularities is still a controversial topic, statisticians being inclined to doubt their statistical significance, while many meteorologists, notably those in Europe, are proponents of their validity. Even the most pronounced singularities, however, occur only in 50–60 percent of all the years on record, and in any particular year a singularity may be absent, or out of place in terms of time. Consequently it is not a very useful tool in long-range forecasting. In Anglo-America singularities have been given much less attention than in Europe. In part this reflects the fact that instrumental weather records are too brief in much of the New World to permit a long-time analysis of daily values of temperature and precipitation.

The concept of temperature singularities in this country is not new (Marvin, 1919). January Thaw and Indian Summer are the two which are most universally recognized by laymen, so that they have become essential elements of American folklore. For the Atlantic seaboard the January Thaw was tentatively verified through statistical analysis 50 years ago by Nunn (1927) and recently has been more rigorously tested by Wahl (1952). Nunn found that the records of daily mean temperatures disclosed a marked crest of temperature for the three-day period January 21 to 23 at Boston, Philadelphia, New York, Baltimore, Pittsburgh, Raleigh, and Atlanta. There was an abrupt rise of 2°–3°C (3°–4°F) starting about January 19 or 20 and culminating on 21–22, with a sharp drop on January 24 and the days immediately following. More recently Wahl has confirmed this same feature for Washington, New York, Boston, and Blue Hill. Wahl's analysis proceeds to show the synoptic causes for the January 21–23 warmth. During the warm spell the Atlantic subtropical high and its southwesterly circulation characteristically dominate the Northeast. By contrast, the average normal sea-level pressure map for January 27 shows a cold high over the Middle West and the Atlantic seaboard, with accompanying northwest winds. Thus, a large-scale change in the circulation pattern seems to account for the January Thaw temperature singularity in the northeastern states.

Wahl (1954) has likewise studied a temperature singularity common to central North America occurring on or about October 15. As of about that date the frequency of cold days and snow probability suddenly increases in the midwest (Denver, North Platte, St. Paul). This increase is associated with a rapid change in the large-scale circulation patterns. The composite five-day sea-level pressure pattern for October 11–15, just prior to the cold period, shows a large subtropical anticyclonic cell over the eastern half of the continent which produces an Indian Summer type of weather with southerly air currents, and one in which northerly invasions are blocked. Within the next five days the high-pressure cell rapidly declines in size and strength, so that cold northerly air is able to invade the central part of the country, accompanied by lower temperatures and a greater probability of snow.

Other climatic consequences of singularities in the annual variation of synoptic patterns and indices have been noted by Bryson and Lahey (1958).

Precipitation

Precipitation Amounts

Within Anglo-America precipitation is concentrated in two extensive and far separated regions, the Pacific margins of the continent, and most of eastern-central North America, together with smaller scattered areas coincident with higher elevations in the drier West. The heavy precipitation along the continental western margins seems normal for the latitude and the windward location. While it is probably true that dynamic processes usually are more important than advective considerations in explaining the distribution of rainfall amounts, nevertheless a substantial amount of water vapor moving into an area is a prerequisite for, but not in itself a sufficient condition to explain, abundant precipitation. For the single year 1949 the net inflow of water vapor across the Pacific coast of Anglo-America represented the maximum across any boundary of the continent, with the Gulf of Mexico boundary next in importance (Benton and Estoque, 1954). Along the west coast rainfall definitely decreases in amount south of about latitude 38° or 39° as the effects of the stable air from the eastern limb of the subtropical high are increasingly felt.

Less than 500 mm of rainfall characterize most of the great intermontane area which lies in a rainshadow with respect to humid air streams from western, and also from southern and eastern, sources. But this deficiency of rainfall likewise extends east of the Rocky Mountains for several hundred kilometers over the Great Plains area of the United States and Canada, the dry belt broadening toward the north where it merges with the low rainfalls characteristic of most of the subarctic and arctic lands. The location of a meridional dry belt to the east of the Rockies has been attributed to the dynamic deformation of the westerly flow by the highlands resulting in a ridge of high pressure over the mountains with anticyclonic flow curvature and northwesterly flow persisting for some distance east of the barrier (Boffi, 1949; Riehl et al., 1954). Under this high-level anticyclonic flow, depressions are weak and the rain-making processes damped. Thus a strong westerly flow across the Rockies is conducive to drought, so that on occasions when this anticyclonically curved air stream extends eastward of its normal position, abnormally dry summers in the prairie states result (Borchert, 1950). Supplementing the above explanation is the fact that surface air from the Gulf of Mexico has had a long trajectory over land before reaching most of the Great Plains, so that it is not a good source for abundant precipitation. Moreover, the streamlines of Gulf air tend to bend eastward after leaving the Gulf, so that under normal conditions they carry only a modest amount of moisture to the Plains region. It is difficult to say whether a station such as Denver near the base of the Rockies receives most of its annual precipitation of about 360 mm from Pacific or Gulf moisture sources, for while most summer thunderstorms there appear to occur in Pacific air, the fewer storms that do originate in Gulf air result in more rainfall per storm.

A noteworthy feature of rainfall in Anglo-America is the depth to which moderate rainfalls penetrate inland from the Gulf and Atlantic. Southward from about 50°N the 500-mm isohyet has an approximate north-south alignment along the 100° meridian, roughly 160 km east of it in northern North

Dakota and 200 km west of it in Texas. This means that in North Dakota the 500 mm isohyet is at least 2,200 km from the Gulf or Atlantic moisture sources. There seems little doubt that this unusual extensiveness of humid climates inland from the leeward side of a great continent is largely attributable to the moisture source provided by the warm Gulf of Mexico which forms the southern boundary of this continental area westward to about the 97°W meridian. In all seasons of the year the Gulf provides the prime moisture source for most of the extensive land area east of the Rockies, and a modest one as well for the intermontane region to the west of those same highlands. By contrast, the Atlantic source is of less consequence, for in each of the seasons there is a net loss of moisture from the continent across the Atlantic coast (Benton and Estoque, 1954).

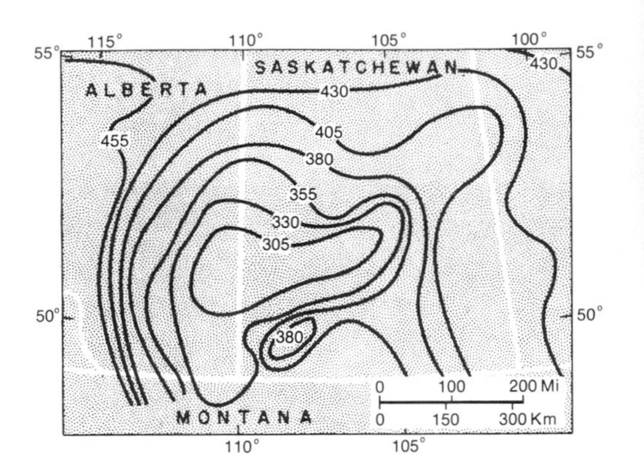

Fig. 16.6. Isolines of annual rainfall which show the general location of the Canadian Dry Belt in Saskatchewan and Alberta. (After Villmow, 1956.)

THE CANADIAN "DRY BELT".—Within the semiarid Great Plains region there are a number of areas having local rainfall deficiencies in excess of the general regional normal, a principal one of which is located in southern Alberta and Saskatchewan. There, within an area of about 130,000 sq km the smoothed isohyets take the form of distorted concentric ellipses around a dry area whose center is approximately the Saskatchewan River basin, where the annual rainfall drops as low as 300 mm (Fig. 16.6). The distinctive climatic character of this local area was first recognized by Captain John Palliser during his explorations in southern interior Canada during the mid-nineteenth century, and it was described in his reports of 1859 and 1860. Subsequently the region went by the name of Palliser's Triangle. A. J. Connor (1938) recognizes the climatic uniqueness of this local area and refers to it as the "Dry Belt," noting that it defies explanation. Later on it has been the object of a more intensive study by Villmow (1955), who finds that the "Dry Belt" is not only distinguished from surrounding areas by a lower annual rainfall, but likewise by (1) less rain in the wettest months, (2) a higher degree of rainfall variability and a longer duration of droughts, and (3) a higher ratio of evapotranspiration to precipitation.

The origin of the Canadian "Dry Belt" is not entirely clear. Villmow (1955) has found that the frequency of cyclones is least, and of anticyclones greatest, in this local area, which is at least suggestive. In addition the "Dry Belt" experiences a stronger mean westerly flow than other parts of the northern Great Plains in summer, a condition which is conducive to drought. And while it is not an ultimate explanation, computation of divergence employing the Bellamy method indicates that during the period April to September inclusive, a condition of positive divergence prevails over the "Dry Belt" as compared with nearly neutral conditions both to the north and to the south. It is possible, also, that the somewhat lower elevation of the Saskatchewan River basin may have some effect on reducing the rainfall.

More recently Dey and Chakravarti (1976) have emphasized summer droughts as noteworthy features of the Canadian prairies. The *average* annual precipitation over much of this region is so scanty that even small negative departures from the normal result in crop injury or failure. Records indicate that this region has a highly variable precipitation both in space and in time, causing widespread damage to crops. Moreover, it is during the warm summer months when precipitation is at a maximum, and the agricultural need for water is also, that variability is greatest. In southeastern Alberta and southwestern Saskatchewan summer (May–August) rainfall variability ranges from 35 to over 40 percent. From this core area of greatest frequency of dry spells, which is the heart of the Palliser Triangle, drought frequency decreases to the north, east, and west.

The synoptic conditions favoring summer drought belong to both lower and upper atmosphere levels. During 33 dry spells that occured between May through August, 1941–1970, it was found that the North Pacific High was thrust farther eastward than normal, resulting in low-level subsidence and drought. Such a northeastward displacement of the North Pacific High causes a northwesterly flow of air along the coastline of British Columbia leading to an upwelling of cold water and further stabilizing the air entering the continent from the Pacific. During the dry spells, with anticyclonic dominance, frontal systems were poorly developed over the prairies, or were displaced farther north than normal.

Upper level synoptic patterns also play a part in producing the dry spells. Widespread dry conditions and dry spells are related closely to the presence of a quasi-stationary upper-level ridge (500 mb level) with its north-south axis situated over western Canada. This ridge operates as a block, shifting the jet stream, cyclone tracks, and associated boreal air masses northward and, thus, beyond the prairies. Also, the anticyclonic circulation under the ridge induces subsidence and divergence. Therefore, when the meridional flow pattern associated with a midtroposphere ridge is present, dry spells are very likely (Dey and Chakravarti, 1976).

SOUTHERNMOST TEXAS.—In a previous section on the Caribbean lands, including Mexico, comment was made concerning the somewhat unusual area of rainfall deficiency in the west-Gulf borderlands, both in southernmost Texas and adjacent northern Mexico. Especially unusual is the fact that the deficiency, compared with the littorals to the south and also to the north and east, is most marked in the warm season when the tropical easterly circulation reaches its maximum development in the Gulf of Mexico. In summer the isohyets are aligned in a NW-SE direction across Texas, so that the subtropical windward littoral just north of the Rio Grande is drier than the Great Plains in the western Dakotas some 2,000 km inland. For a genetic treatment of this problem area, reference should be made to the earlier discussion on Mexico in Chapter 5.

As indicated previously, the N–S belt with under 500 mm of rainfall east of the Rocky Mountains broadens northward to include most of Canada north of the 50° parallel. It is only in the eastern subarctic that rainfall exceeds 500 mm, even reaching 750 mm in places. In part this results from a converging of storm tracks in the east, a closer proximity to local sources of moisture (Hudson Bay and the Atlantic Ocean), and the reactivation of perturbations as they come within the destabilizing effects of the cyclonic circulation around the western margins of the Iceland low. Precipitation actually drops below 250 mm over most of the Arctic archipelago where low temperatures preclude much atmospheric moisture.

THE PRAIRIE WEDGE.—Within humid-subhumid interior United States, where the normal annual rainfall for the most part amounts to 750 to 1,000 mm there existed that easternmost projection of the North American grasslands, sometimes designated as the Prairie Wedge. In its broader western parts where it made junction with the short-grass steppe of the semiarid Great Plains, the original prairie extended from southern Canada through the eastern Dakotas and southward to Oklahoma. Narrowing rapidly eastward where it was constricted on the north and south by forest, prairie vegetation covered most of Iowa, central and northern Missouri, southernmost Wisconsin, and most of central and northern Illinois (Fig. 16.7). This conspicuous eastward thrust of a grassland vegetation into the humid forested area, even beyond the eastern margins of Illinois, presents a problem which has long attracted the attention of workers in the natural, and especially the biological, sciences.

Borchert (1950) has analyzed the climatic peculiarities of the prairie region which may have been responsible for establishing the grasslands so far eastward, and suggested explanations for them. In summary, the climatic distinctiveness of the Prairie Wedge of central North America consists of the following elements. (1) Much less total winter precipitation falls over the grassland area than over that part of the United States farther to the southeast, so that the precipitation gradient is steep along the transitional boundary between grass and forest to the southeast of the prairie. (2) Low snowfall and meager snow cover in winter also characterize the prairie region, so that a relatively steep mean winter-snowfall gradient coincides with the boundary between forest and grassland along the northeastern boundary of the Prairie Wedge. Thus, this grassland occupies a unique position in the winter precipitation pattern of central North America. In its northern parts where winter cold is severe, the grassland lacks the protective effects of a deep and persistent snow blanket such as characterizes the forest area to the north. In its southern parts it lacks the abundant winter rainfall which falls in the southeastern forest area.

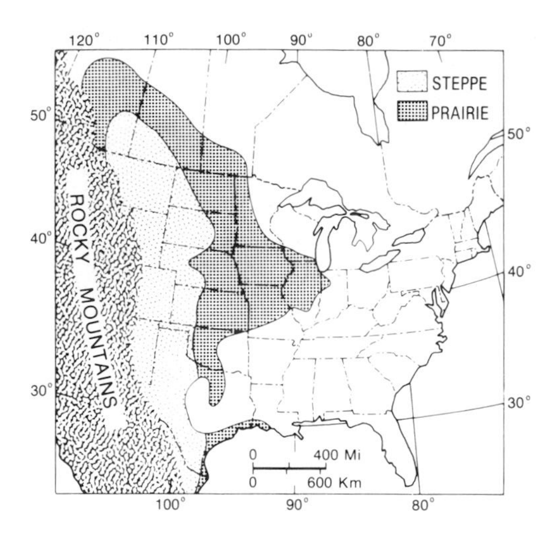

Fig. 16.7. The grasslands of Anglo-America east of the Rocky Mountains, showing the location of the Prairie Wedge. (After Borchert, 1950.)

(3) While the short-grass steppe of the Great Plains receives markedly less summer precipitation than the grass-forest region eastward from the 100°W meridian, the May-to-August summer rainfall of the Prairie Wedge is equal to that of the forested areas bordering it to the north and south. However, the increase in summer rainfall eastward from the short-grass steppe to the prairie is not the result of an increase in rainy days, but rather of an increase in the intensity of the rainfall, so that it is relatively less effective for plant growth.

(4) The prairie region is climatically distinguished from the forested areas in summer by (a) a small but abrupt increase in the number of rainy days along the northern margin of the Prairie Wedge, and (b) a higher concentration of rainfall variability and of regional drought in the prairie than in the forested regions to the north and south. Thus while the prairie region has an amount of summer rain equal to that of the forest regions to the north and south, it is subjected to more severe and frequent summer drought.

(5) During the occasions of summer

drought the prairie region is characterized by large positive anomalies of temperature and of hot winds to a much greater extent than is the forested area. Thus, the injurious effects of drought in the prairie are intensified by searing temperatures under mostly cloudless skies.

Droughts in Anglo-America east of the Rockies have been shown to occur when there is a persistent and strong flow of dry continental air eastward from the base of the Rockies. When this westerly flow that produces the usual drought condition of the Great Plains extends farther eastward than normal, drought conditions likewise are displaced farther east. An abnormally strong westerly flow, in turn, is associated with a Northern-Hemisphere mean circulation pattern which is characterized by (a) a rapid pressure decrease from north to south across the middle latitudes, with a consequent strong mean westerly component of the gradient wind, and (b) also a relatively high pressure over the Great Basin and southern plateaus. Under such conditions the humid, northward-streaming Gulf air is caused to turn abruptly eastward, thereby missing the grassland. Strong westerly flow from the eastern base of the Rocky Mountains across central United States is likewise abetted by a high frequency of summer cyclones traveling on the northern or Alberta track, and of the west winds developed in the southern quadrants of these disturbances. In addition, the Alberta lows characteristically deliver most of their precipitation to the northern forests rather than to the grassland farther south.

Under normal conditions there is throughout the entire year a prevailing westerly, and therefore dry continental, air stream over the western part of the Great Plains from northern New Mexico to southern Alberta and Saskatchewan. But the grassland farther east has a continental air flow for only a part of the year, essentially the cooler months. The pattern of the isolines of monthly prevalence of air-mass continentality projects eastward in the form of a wedge and reflects chiefly the period of strong westerly flow in the cooler months (Fig. 16.8). In drought years, however, the continental air flow from the west may persist across the whole grassland for most or all of the year. Thus, according to Borchert (1950) the location and shape of the prairie appear to reflect the greater prevalence of summer continental westerly flow in this region than in surrounding areas to the northeast, east, and southeast.

Diurnal Concentration of Precipitation

In sufficiently moist air, unstable enough to permit convection, the maximum shower activity over land normally is expected to coincide with the hours of greatest surface heating. The fact that over continental Anglo-America there are extensive areas where, during the warmer half of the year, rainfall has a nocturnal maximum is something of an anomaly. It seems the more unusual because the two most extensive areas of nocturnal summer rains and thunderstorms are strongly continental, being well insulated from oceanic influence by distance or by terrain barriers. The more extensive of the two areas where over 60 percent of the precipitation in the period April to September inclusive occurs during the 12 hours ending at 0800 is centered over Kansas, Nebraska, and Iowa and parts of adjacent states (Fig. 16.9). The second and smaller area is located in southern Arizona (Kincer, 1922). Summer thunderstorms also show a disproportionate nighttime concentration over these areas (U. S. Weather Bureau, 1947). By contrast, warm-season thunderstorms and precipitation in subtropical southeastern United States are strongly concentrated in the afternoon and evening.

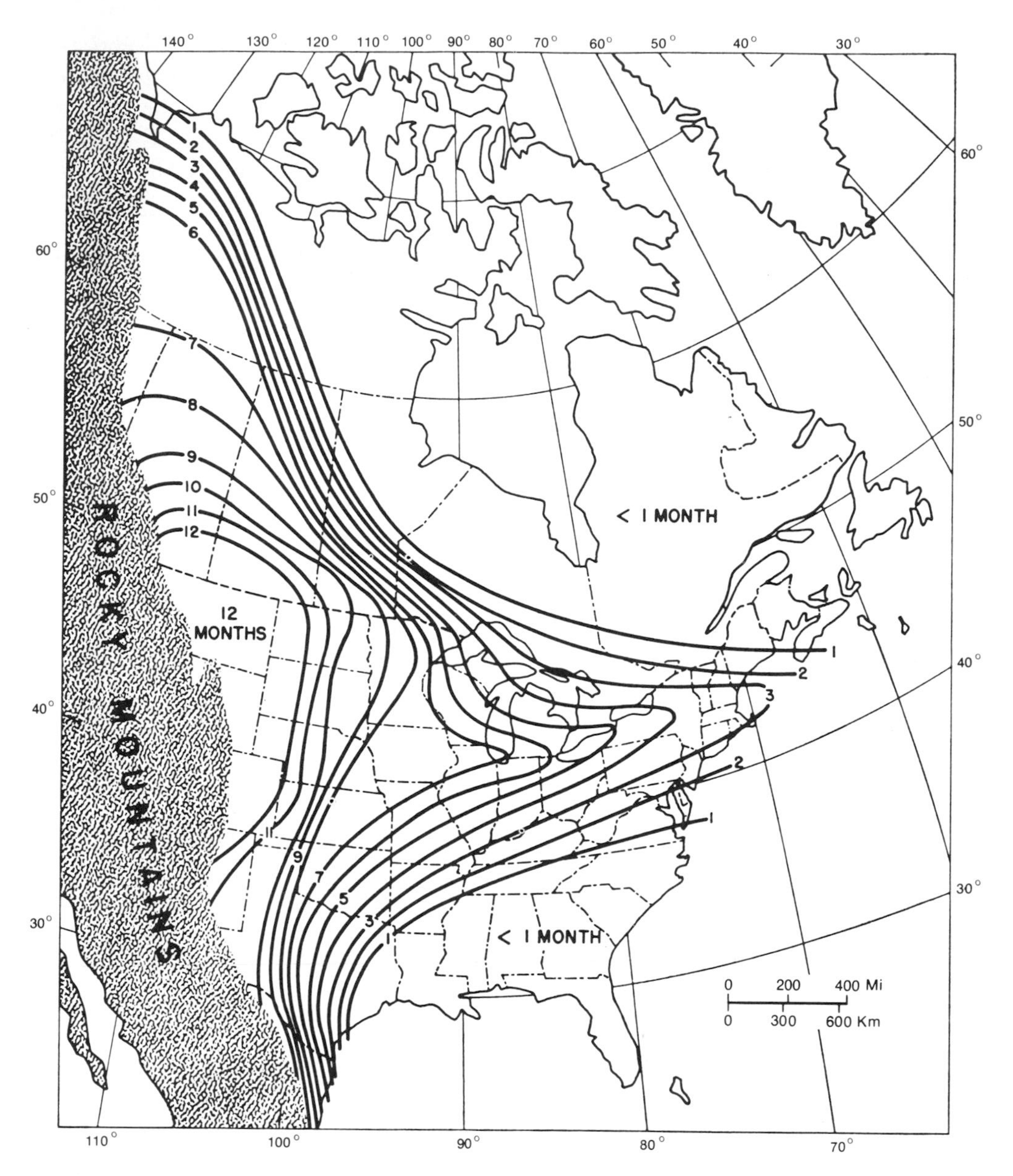

Fig. 16.8. Average number of months per year with a mean transport of air from the eastern base of the Rocky Mountains. (After Borchert, 1950.)

A number of explanations have been proposed for the nocturnal rainfall and thunderstorms over interior United States. Means (1946) is of the opinion that the advection of warm air in the lower layers is an important factor. Bleeker and Andre (1951) are critical of this interpretation and contend that other causes must be sought. These are required to explain not only the concentration of thunderstorms and rainfall at night over central

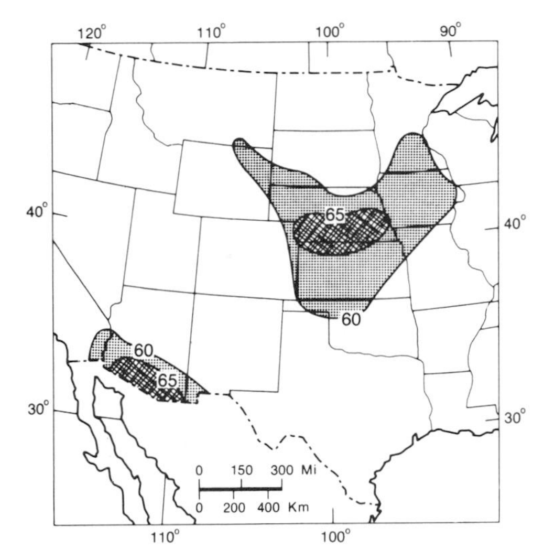

Fig. 16.9. Showing the location of areas in the United States with a preponderance of night (12 hours ending 0800, 75 meridian time) rainfall during April–September, inclusive, in percent. (After Kincer, 1922.)

United States, but also the relatively smaller proportion that occurs during the day. There must be a stabilizing influence during the day to counteract the effects of surface heating, as well as a destabilizing influence at night. These authors contend that an extensive lowland between highlands, or even a lowland bordered on one side by highlands, presents an orographic condition which favors convergence and upward movement of air by night and divergence and subsidence by day. Thus, there is set up a large-scale diurnal oscillation of the atmosphere between the Rocky Mountains and the lowlands to the east.

At night as nocturnal cooling causes the isobaric surfaces to sink they will be depressed by the same amount at each level except toward the mountain slopes where the cooled layer under the isobaric surface is thinner. By reason of the above arrangement the isobaric surfaces will slope downward from the mountain toward the plain and there should be a gradual transport of air toward the adjacent plain. This will produce a low-level convergence over the plain resulting in upward movement.

During the day the opposite condition should result, leading to divergence over the plain and accordingly subsidence and stability. This circulation system is one of far greater magnitude than those small-scale systems known as diurnal mountain and valley winds, or upslope and downslope winds.

To check the conclusions reached in the hypothesis Bleeker and Andre (1951) computed the changes in divergence for 6-hour periods in August 1947 and 1948 for central United States east of the Rockies. Their maps show a definite daytime increase in divergence over central United States and a decrease at night when positive convergence prevails. They conclude that such a diurnal reversal as is suggested in their hypothesis does actually take place.

Sangster (1958) more recently also has computed the vertical atmospheric motion for the area of high nighttime thunderstorm frequency in central United States, and has discovered that there is a significant diurnal variation of vertical motion at the 700 mb level, with the maximum upward motion occurring at about 0000 hours and the maximum downward motion at 0900 CST. He attributes this diurnal variation in the vertical motion of the atmosphere over the region in question to a diurnal pulsation of the mass transport across the southern and northern boarders of the region, the former being the larger. Diurnal variations of the convergence of east-west velocity components, by contrast, do not contribute significantly to the diurnal variation of vertical motion, which concept is in opposition to the mechanism proposed by Bleeker and Andre (1951).

The most common summer synoptic situation in the area under consideration is a quasi-stationary or slowly moving front oriented more or less east-west. South of the front is warm air with small daytime cloudiness and high maximum surface temperatures. To the north, cloudiness is greater and maximum temperatures are relatively low. Thunderstorms occur along and to the north of the front. During the daytime, surface heating and increased eddy viscosity slow down the mass transport toward the convergence, so that upward motion along the front is reduced. By contrast, when at night the eddy motion is weak, there is an increased transport of air toward the front, chiefly from the south, resulting in greater upward motion along the convergence.

Seasonal Characteristics of Precipitation

Areal variations in the way the average precipitation is distributed over the 12 months provide some of the most numerous and most difficult problems in the climatology of Anglo-America. Regional variations in the annual march of precipitation can be depicted graphically either through the use of station profiles constructed with time as the abscissa and inches or millimeters of rainfall as the ordinate, or by means of maps. Representative of such maps are those showing (1) the percent of the annual rainfall occurring in each of the four seasons (Kincer, 1922), (2) percent of change in rainfall amounts from one month to the next (Horn, Bryson, and Lowry, 1957) and (3) phase-angle and amplitude maps as derived by the method of harmonic analysis (Horn and Bryson, 1960).

Maps have an advantage over individual station profiles in that they more clearly and precisely portray the areal arrangement of the features of seasonal rainfall distribution. This is especially true of maps which are a product of the harmonic analysis of precipitation types, where core areas possessing a certain profile form stand clearly revealed as do the rates of change from one core area to another. On the other hand, an objective analysis of seasonal rainfall by means of harmonics always requires two basic maps, phase-angle and amplitude, for each harmonic. The phase-angle of a given harmonic reveals the approximate date or dates on which a single (or multiple) maximum or minimum in rainfall occurs, while the amplitude map exhibits the relative importance or magnitude of a particular harmonic. Since as many as six harmonics may be required to accurately portray the details of the rainfall profiles of an extensive region, it becomes obvious that up to a dozen different maps may have to be utilized, the numerous distributional features of which are difficult to integrate regionally. It is this necessary multiplication of maps in the method of harmonic analysis that argues in favor of the station-profile method for representing the areal qualities of the annual march of precipitation, even though the latter method is open to the objection that it presents less effectively in an objective and quantitative form the areal distribution of the annual march of precipitation.

In the analysis of the regional characteristics of the annual march of rainfall use has been made of data for some 400 stations in the United States covering the 30-year period 1921–1950. These have been supplemented with data from about 200 additional stations covering the period 1931–1952 inclusive. For Canada some 350 stations with rainfall records of varying lengths have been made use of.

CHAPTER 17

Regional Rainfall Types: The West

Based upon similarities in station profiles of annual rainfall, Anglo-America may be subdivided into a number of regional rainfall types, most of which are composed of a number of subtypes. Since it is normal for rainfall types to be separated from each other by transition zones of variable width, any boundaries drawn to separate the individual types and subtypes must be located arbitrarily. Moreover, it is frequently difficult to decide whether a transition area should be divided by a boundary separating two relatively pure types, or whether the transition itself should be recognized and set apart as a distinct and separate regional entity. With this in mind, it bears emphasizing that the scheme of regional rainfall types here presented (Fig. 17.1) is only one of several which might have equal validity. In parts of Anglo-America also, especially in the higher latitudes, weather stations are so few, and for some of them the records are so brief that any attempt to delineate rainfall types must rest upon meager and unreliable evidence.

The North American West (Fig. 17.1, regions 1 and 2), situated between the Rocky Mountains and the Pacific Ocean, is indeed a region of precipitation complexity, both as regards mean annual amounts and seasonal distributions. More than in other parts of the country, the West's terrain features, expressed both in altitude and in directional alignment, play an important role in influencing both spatial and temporal distribution of rainfall. Pyke (1972) recognizes four primary precipitation regimes within the West based on time of maximum. (1) the winter maximum of the Pacific coast region mainly in California and Baja California (Mexico); (2) the late-fall maximum of the Pacific coast region of Oregon, Washington, British Columbia (Canada), and the Alaska Panhandle; (3) the late-spring maximum of the northeastern parts, mainly in the Rocky Mountain areas of Montana, Idaho, Utah, and Colorado; and (4) the summer maximum of the interior southwest, mainly in Arizona, western New Mexico and adjacent southern parts of Colorado, Utah, and California (Fig. 17.2). In modified form the winter and late-fall maxima spread eastward beyond Sierra Nevada and the Cascade Range, while the late-spring and the summer maxima invade westward from their primary Rocky Mountain centers. As a consequence the intermediate intermontane region is one of relatively complex seasonal regimes of rainfall. (It is believed there is some merit in combining rainfall regimes 1 and 2. See Figure 17.1.)

The one main seasonal *minimum*, which prevails over all those sections of the West,

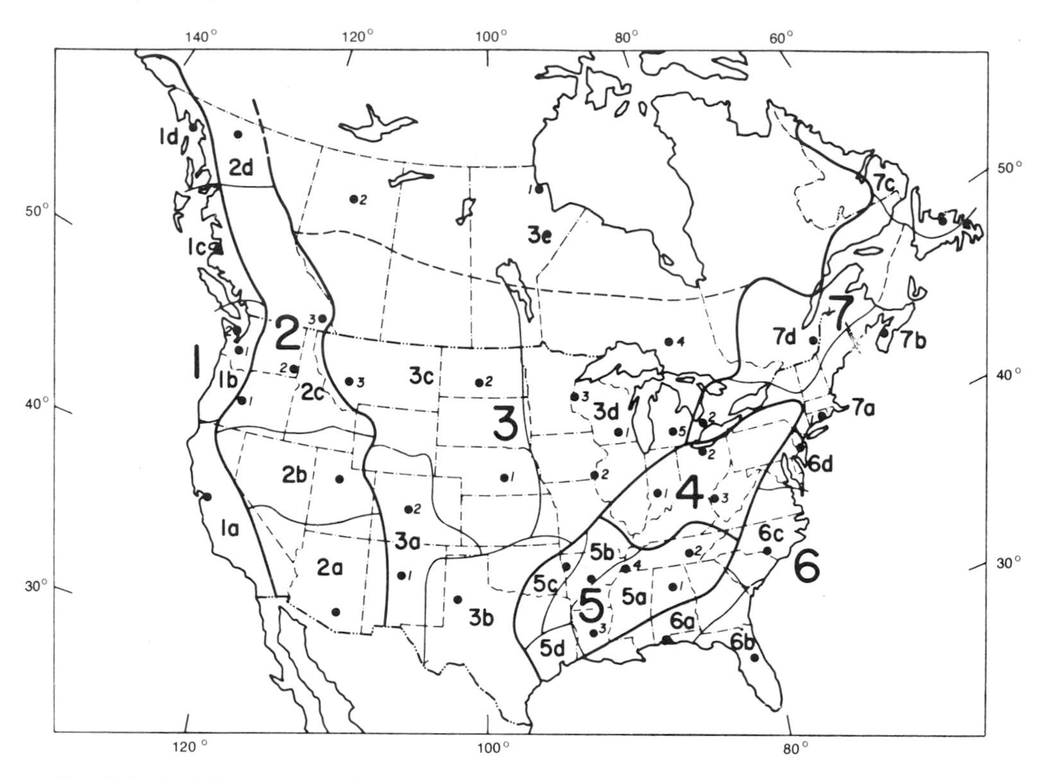

Fig. 17.1. Precipitation regions of Anglo-America based on characteristics of the annual march of precipitation. Where more than one station profile is used to illustrate rainfall characteristics within a subregion, a numeral is added for identification purposes, both on Figure 17.1 and on the individual station profiles.

situated to the west of about 118°W, or south of about 37°N (or both), occurs during the summer in Washington, Oregon, most of California, and northwest Baja California. The minimum occurs in late spring from southeast California eastward to New Mexico and southward over eastern and southern Baja California (Pyke, 1972). This warm-season minimum is strongest in the parts of the Pacific states west of Sierra Nevada and the Cascade Ranges and in Baja California.

TYPE 1. THE PACIFIC COAST REGION

At least three features of precipitation are common to the whole west-facing maritime region of Anglo-America extending from almost 60°N to southern California: (1) there is a strong annual type of variation in the precipitation profile; (2) the conspicuous single maximum occurs in the cooler part of the year while the marked single minimum is in summer; and (3) the month of maximum precipitation is a function of latitude. All three of these features are closely interrelated and all are associated in a cause and effect relationship with the seasonal migration of the North Pacific subtropical high, and of the jet stream and associated cyclonic storms which flank the anticyclone on its poleward side. The single minimum coincides with that period when the subtropical anticyclone extends its influence farthest poleward, displacing the jet and the major tracks of cyclones far to the north. The single maximum coincides with the retreat

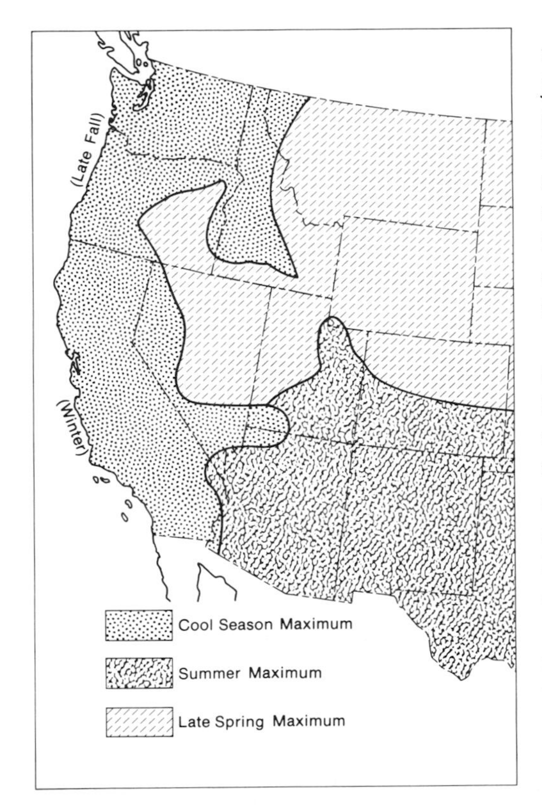

Fig. 17.2. Season of primary-maximum rainfall in the western United States. (From Pyke, 1972.)

southward of the high and the advance toward lower latitudes of the jet stream and storm belts.

Except in Alaska west and north of the Panhandle, the Pacific coast has more precipitation in the winter half year than in the summer half. At Anchorage and Matanuska the reverse is true. But in the Alaska Panhandle only about 60 percent of the annual total is concentrated in the winter six months, so that in these higher latitudes summer is relatively wet. Southward the degree of cool-season concentration becomes progressively greater—two-thirds in northern coastal British Columbia, three-quarters in the southwestern part of the same province, about 80 percent in the Puget Sound-Willamette Valley, and 85 percent in California.

Fall and winter are the seasons of maximum rain in the Alaska Panhandle and adjacent British Columbia, while winter is emphatically the wettest season in southern British Columbia and southward, although autumn is almost everywhere wetter than spring. To an unusual degree the annual profile or curve of rainy days (.254 mm and over) bears close resemblance to that of rainfall amounts.

The cool-season maximum precipitation regime of maritime Pacific North America (Fig. 17.1, Type 1) is the most important and extensive producer of precipitation within the West. In most areas the season represents a broad maximum in time, which begins in early fall and ends in late spring. Along the Pacific coast this latitudinally-extensive cool-season regime never fades out except in maritime mainland Alaska. The cool-season rains originate almost exclusively in the extratropical disturbances of the northern Pacific ocean, which in turn are closely associated with a widely fluctuating jet stream and with large synoptic trough patterns of the upper troposphere.

Strong Summer Minimum Shifted Far Poleward.—One of the most unusual features of the annual rainfall profile is that the relatively dry summer extends so far poleward in Pacific Anglo-America, much farther than in Atlantic Europe and in Chile. Even along the northern British Columbia littoral the summer rainfall is only 40 percent that of winter and in the vicinity of latitude 50° this is reduced to 20–25 percent. The whole Puget Sound-Willamette Valley region in the United States, and the Georgia Basin in British Columbia, Canada, qualify as having the dry summer typical of a Mediterranean climate (*Cs*), but winter temperatures there are too low and annual rainfall totals generally too large to qualify as typically dry-summer subtropical. Still, over most of this region the moderate annual precipitation is sufficient to develop a relatively luxuriant forest cover. But in a more restricted part of the larger region, mainly the borderlands of the Strait of Juan de Fuca and southeastern Vancouver Island, annual rainfall generally is below 800 mm and a few localities drop below 500 mm. Here the July–August rainfall declines to less than 7 percent of annual total, and the original vegetation cover is an oak-grassland transition type, while its mature soils are part of the brown-podzolic group (Kerr, 1951). In this restricted subhumid region a dry-summer subtropical climate is to be found 10°–15° poleward from subtropical latitudes, and the serious rainfall deficiency in summer has stimulated the development of irrigation. Even dangerously dry conditions may occur on occasions when the thermal trough type of pressure distribution results in the advection of hot, dry air from the southeast into the lowlands of the Pacific Northwest (Chapman, 1952).

The abnormally low summer precipitation which is characteristic of the British Columbia-Washington-Oregon lowlands west of the Cascade Range, has its origin in the unusual poleward displacement along the coast of an arm of the North Pacific subtropical cell of high pressure, and of the cool coastal waters associated with the circulation pattern which it generates (Kerr, 1951). While the center of the anticyclone over the ocean is positioned at about 38°N, the isobars on its eastern side bulge in a northeasterly direction, so that along the coast the highest pressure is somewhat poleward of 50°N. The northerly winds with anticyclonic curvature blow steadily along this coast in summer, developing a cool ocean current whose temperatures are lowest near shore where active upwelling occurs. Since the northerly air flow circulating around the northeastern side of the Pacific high is warmer than the underlying water until it reaches latitude 38° or 40°N, stabilization of the lower air results, while stability aloft is maintained by anticyclonic subsidence (Patton, 1956). A temperature inversion exists about 20 percent of the time in summer as far north as Tatoosh Island, 48°N, and somewhat less frequently at higher latitudes.

The northward-displaced anticyclone is not only effective in stabilizing the air, but it also acts so as to block depressions from penetrating to the coast except infrequently. On occasions when the northern spur of the anti-cyclone weakens and is displaced southward in summer, depressions do reach the coast and bring some rainfall to British Columbia, Washington, and Oregon, and at rare intervals even as far south as central California. Variations in the strength and position of the tongue of high pressure create the synoptic conditions so typical of the coastal region north of 45° or 50°N in the warmer months. Progressively farther north as the anti-cyclone becomes weaker and more variable in position, disturbances are able to reach the coast more frequently, resulting in increased warm-season rainfall (Patton, 1956).

The lowest water temperatures near the coast appear to be at about 40°N and the maximum temperature contrast between air and water is in approximately the same latitude. Maximum anticyclonic subsidence on the southeastern rim of the anticyclone is at about 35°N. The combined effects of subsidence and the cool ocean current are able to hold summer precipitation to near zero from 30° to about 38°N. But northward from 38° or 40°N, as surface temperature contrasts and anticyclonic subsidence wane, some summer precipitation occurs, but it continues small considering the latitude.

The Seasonal Southward Progression of the Cold-Season Precipitation Maximum.— A distinctive feature of the time of the cold-season rainfall maximum along the North American west coast is that it is a function of latitude; it occurs increasingly later in the year with decreasing latitude. Thus at Anchorage and Valdez (61°–62°N) in Alaska, the maximum rainfall is typically in September. At Juneau (58°N) and at Ketchikan (55°N) it occurs in October; at Vancouver (49°N), November; at Aberdeen (47°N), December; at Eureka (41°N), January; and at Los Angeles (34°N) and northernmost Baja California (Mexico) (32°N), February. But farther south in Baja there is a reversal of this southward progression of the cold-season precipitation maximum, and January or December become the wettest months. This southward progression of the time of the cold-season maximum has both direct and indirect causes. Clearly, it is directly related to the southward shift of the mean jet stream and the associated similar shift of the Pacific cyclone track and its deepening disturbances, following the retreat of the sun. Thus, in late summer and early fall the jet and its cyclonic storms are concentrated at about latitudes 60°–65°N, but in February they are south over California. These occurrences in turn are linked with the seasonal change of the sea-surface temperatures between fall and early spring. An additional factor is the blocking highs over north-central Canada which exert their strongest effects in late winter and early spring in permitting the maximum penetration of the cyclone belt into southwestern United States. Pyke (1972) has proposed also that the southward progression of the North Pacific cyclone track along the west coast of North America may be linked with seasonal fluctuations of ocean temperatures in the equatorial Pacific. These make themselves felt through changes in the subtropical jet and variations in the Hadley circulation. Although the above relationships are complicated, Pyke has hypothesized as follows: "Heavier precipitation in southern California and adjacent areas should tend to be positively correlated with both the intensity of the subtropical jet stream across Baja California and the strength of the eastern north Pacific Hadley Circulation, positively correlated with the intensity of the Hadley Anticyclone over the central north Pacific Ocean, negatively correlated with the strength of the eastern Pacific Walker Circulation, positively correlated with the sea surface temperatures of the eastern equatorial Pacific, and negatively correlated (perhaps with some time lag) with the intensity of the eastern Southern Hemisphere Pacific trade wind circulation."

The seasonal shift of the jet poleward in spring and summer does not seem to occur in a steady manner as in the southward movement in fall and winter, but instead one jet appears to wane in the south in late winter, while another one develops near the Arctic Circle. As a consequence there is no secondary spring maximum of rainfall in Washington and Oregon as might be expected if the jet and its associated storms moved steadily northward between late winter and late summer (Riehl et al., 1954). As noted in an earlier section, no such simple latitudinal progression of the month of maximum rainfall occurs in western Europe where the pattern of jet stream and its associated cyclonic disturbances is a very complicated one, in part a consequence of the frequent blocking action by anticyclones.

It is worthy of mention that the southward progression of the late fall-winter precipitation maximum is scarcely smooth; rather it is characterized by weak double peaks, singularities, and other irregularities. In addition there are irregularities in an east-west direction, most of the latter probably being a consequence of terrain. Thus the eastern slopes of major mountain ranges tend to experience somewhat earlier precipitation maxima than do coastal stations in the same latitude, while western slopes are likely to have later maxima. East of the Sierra Nevada and the Cascade Ranges in the western intermontane parts, the prime winter maximum still prevails, but usually there is a secondary peak as well so that the annual rainfall regime is more complex. This is in the nature of a transition region.

BAJA CALIFORNIA.—An extension of the southern California and Arizona precipitation regimes southward into Baja California is not unusual. Throughout nearly the entire length of the peninsula the Pacific side shows a strong primary winter precipitation maximum, like California, while the Gulf side resembles Arizona in having a dual maximum, a sharp primary one in August-September, and a secondary in December. The moisture source for the summer maximum is probably the Pacific Ocean and the Gulf of California and less likely the Gulf of Mexico. A portion of the summer rainfall here, especially in the south, is derived from tropical disturbances, including occasional hurricanes, which originate off the west coast of Mexico and move northward to the southern part of Baja California. The late summer regime consists almost entirely of convective air-mass precipitation mostly associated with thunderstorms.

Noteworthy is the reversal in Baja California of the southward progression of the season of the coastal winter precipitation maximum. South of the Mexican border the cool-season maximum regresses from about mid-February back to around January 1. In addition, there is an increasingly earlier cutoff of the late winter and early spring precipitation toward the south. These phenomena appear to be associated with the seasonal changes in the position and intensity of the eastern subtropical Pacific anticyclone, with variations in the sea-surface temperature of the ocean west of Baja California, and with the increase in intensity in late winter and early spring of the subtropical jet stream over northern Baja California (Pyke, 1972).

Amounts of Precipitation.—In a region of such diverse terrain, great areal variations in the amounts of annual precipitation are to be expected. Locally the amount is controlled by exposure to the open ocean and the abruptness of the mountain slopes to the rear. South of about 40°N there is a rapid decrease in the annual precipitation as the stabilizing effects of the eastern limb of the subtropical high become increasingly dominant, and coastal deserts prevail south of latitude 32°.

Along the Canadian littoral precipitation varies from a high of about 5,000 mm at Swanson Bay to 600 mm at Victoria in the lee of Vancouver Island. The relatively modest rainfalls that characterize parts of southeastern Vancouver Island are duplicated in the state of Washington along the Strait of Juan de Fuca and on the shores of Puget Sound behind the Olympic Mountains (Fig. 17.3). Thus, Coupeville records only 430 mm, Sequim 400 mm, Port Townsend 460 mm, and Port Angeles on the Strait of Juan de Fuca 600 mm. As noted previously, the relatively small annual amounts, together with the strong concentration in the cool season, make for a genuninely summer-dry climate at latitudes of 45°–50°N.

This relatively modest annual precipitation in the vicinity of northern Puget Sound and the straits of Georgia and Juan de Fuca is usually attributed to the rainshadow effects of the Olympic Mountains and Vancouver Island. It is possible, however, that local air streams moving through these restricted channels, associated with coastline alignment, may be an auxiliary factor in producing the reduced rainfalls. South of Tacoma to about the Umpqua River in Oregon, the interior valley has increased annual rainfall amounts of about 1,000 mm. Here the lower and less continuous nature of the coastal mountains permits the freer entrance of Pacific air. In the lee of the higher Klamath Mountains of southern Oregon and northern California rainfall amounts again decline, so that Medford, Oregon, records only 400 mm. But in the latitude of southern Oregon the effects of the Pacific high are becoming

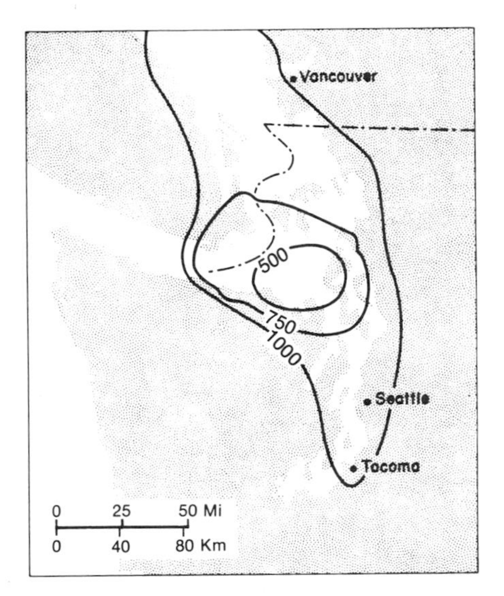

Fig. 17.3. Generalized isohyets showing the location of areas of atypical modest annual precipitation (in mm) in the northern Puget Sound region.

increasingly strong so that low rainfalls are not unusual.

Origin of the West-Coast Rainfall.—It cannot be doubted that much of the precipitation is associated with eastward-moving occluded depressions and the lifting effects of highlands. The highlands likewise have the effect of delaying the eastward progress of the depressions and of increasing the duration of the cyclonic rainfall. In addition to actual frontal precipitation, instability showers in the cool maritime air following the cyclones make an important contribution to the rainfall totals (Pincock, 1951). This effect is especially strong in the cooler months when the water is so much warmer than the air.

Connor (1938) points out that there is a temperature connection with British Columbia west-coast rainfall as well as an orographic one. He notes that the sharp seasonal increase in precipitation occurs simultaneously with an abrupt drop in seasonal temperatures in the interior valleys in September. Thus, easterly winds tend to predominate along the west coast in winter as the colder air drains from the interior, so that there appears to be a fairly persistent discontinuity in winter between the cool easterlies of the mainland and along the coast, and the mild *mP* air from the west. This, Connor (1938) designates as an orographic front over which the *mP* air is forced to ascend.

Subtype 1a.—Here in the southernmost, or California, part of Type 1, the distinctive feature of the annual rainfall profile is the fact that one, or more, of the summer months is usually rainless, while the single maximum is in mid-winter (Fig. 17.4). It is the dry-summer type, either Mediterranean (*Cs*) or steppe (*BSs*) in its most standard form. Rainfall declines southward as the dominance of the subtropical high increases, so that in those parts of California west of the Sierra Nevada, the San Gabriel, and the San Bernardino mountains, steppe climate is found both in the southwestern part of the San Joaquin valley and also along the coast south of Los Angeles. Marine *Cs* and *BS* climates with a large amount of summer fog characterize coastal California bordered by the cool current.

One of the most unusual local climates of the California region is that of the San Francisco Bay area (Patton, 1956). In summer this area is one of the coolest within the continental United States, in spite of its subtropical location, July at San Francisco having a mean temperature of only 15°C (59°F) while the average daily maximum of the three summer months is 18°C (65°F). Two mech-

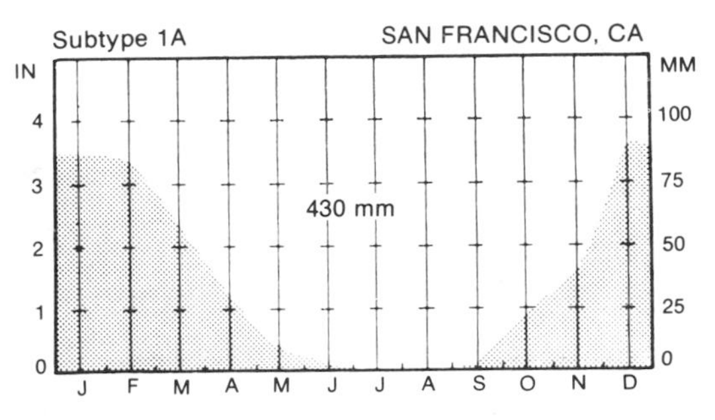

Fig. 17.4. Annual march of precipitation, San Francisco, California.

anisms account for the low summer temperatures of the Bay area, the prevalence of low stratus which reduces insolation, and the effects of strong advection of very cool air from the ocean which is funneled through the break in the coastal hills at San Francisco as it moves toward the strong thermal low over the superheated Great Valley. A delayed temperature maximum is likewise characteristic of many stations.

Subtype 1b differs from 1a in that there is a marked increase in total rainfall and in the number of rainy days (Figs. 17.5, 17.6). Summer is still relatively dry. However, the total rainfall is too great and the winter temperature too low for a normal *Cs* climate and the resulting vegetation cover of large trees is likewise atypical.

Subtype 1c.—Rainfall amounts are variable as determined by terrain and exposure, but on the whole they are larger than in 1b and summers are somewhat wetter as the effects of the subtropical anticyclone weaken in the higher latitudes (Fig. 17.7). The precipitation profile shows a rapid increase during the fall months, with a maximum reached in November-December rather than in midwinter as is the case farther south.

Subtype 1d shows a still earlier rainfall maximum with a sharp primary peak in October and a minimum in June (Fig. 17.8). Spring, the season of least rainfall, is far from being dry.

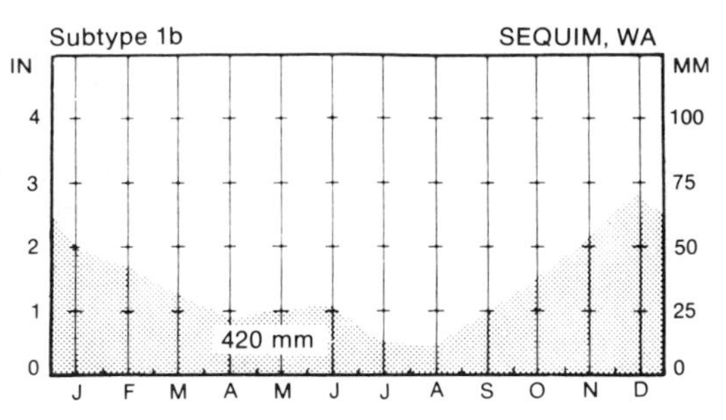

Fig. 17.6. Annual march of precipitation at a station located on the southern side of the Strait of Juan de Fuca where total annual rainfall is strikingly modest and a dry-summer climate prevails.

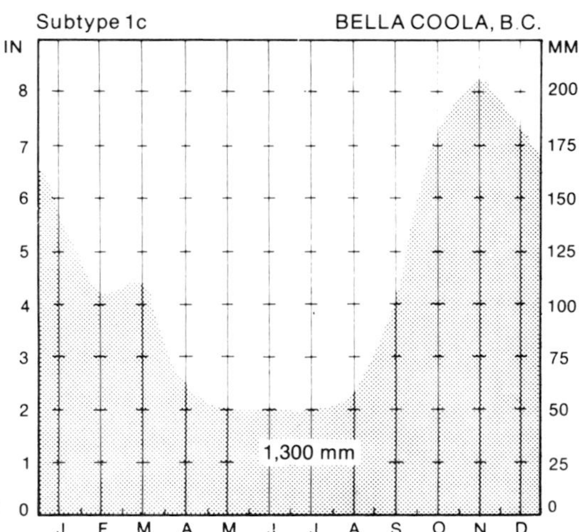

Fig. 17.7. Annual march of precipitation at a station located at about 50°N on the coast of British Columbia, Canada. Here the primary maximum is in the late fall and early winter.

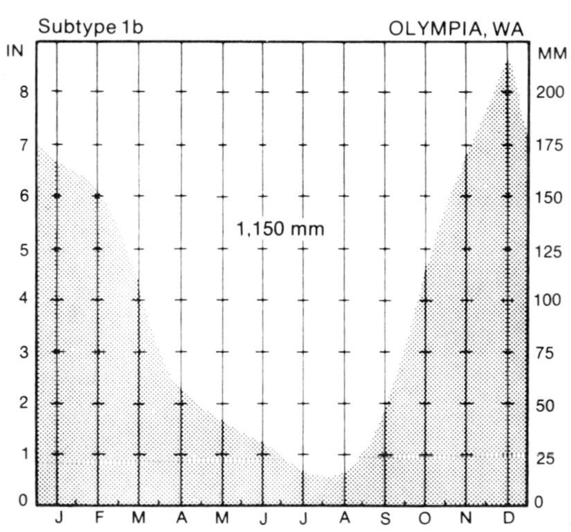

Fig. 17.5. Annual march of precipitation at a station located at the southern end of Puget Sound. There rainfall is fairly abundant, and a strong cool season maximum prevails. Summer is relatively dry.

TYPE 2. THE INTERMONTANE REGION

Situated as it is between the Rocky Mountains to the east and the Pacific Coast moun-

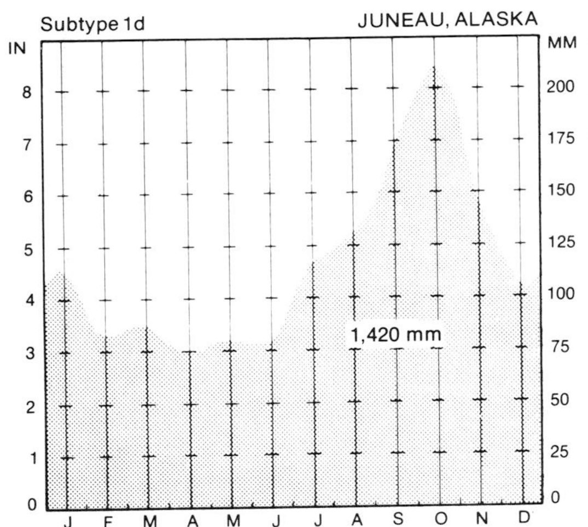

Fig. 17.8. Annual march of precipitation at a coastal station in Alaska at about 58°N. Here there is a strong fall maximum, with October the wettest month.

tains to the west, this extensive intermontane region lies in the rainshadow of highlands. Except at higher elevations it is, therefore, a region of modest precipitation, much of it being classified as steppe or semiarid, although some parts are genuinely arid and others, chiefly in the north or at higher elevations, subhumid. In the southwest there are stations with as little as 130 mm of annual rainfall, although for the entire region 250 to 500 mm is a common range.

In spite of numerous localisms imposed by relief, a dominant climatic feature is the biannual rainfall variation. One maximum very consistently occurs during the winter season. In some parts this is the primary maximum and in others the secondary. The second or warm-season maximum is more variable in its time of occurrence, for in some parts it comes in early summer or even spring, while in other sections it is a mid- or late-summer phenomenon. It appears that Type 2 is in the nature of a transition located between the area of strong winter maximum characteristic of the Pacific Coast, and the marked summer maximum typical of the great interior region to the east of the Rockies. It has the qualities of a hybrid, therefore, which bears the earmarks of both parents. At the time of the cool-season maximum the precipitation is derived from the numerous depressions moving eastward from the Pacific coast which result in general widespread rains. The warm-season rains usually are more local and showery.

Subtype 2a, situated in the southernmost part of Type 2, which approximately coincides with the intermontane region, includes Arizona and adjacent parts of the bordering states of New Mexico, Colorado, Utah, Nevada, and California. It occupies an intermediate position between the area of strong annual rainfall variation and winter maximum in California (1a) to the west and the New Mexico area of weaker annual variation and summer maximum (3a) to the east. Subtype 2a has the distinction of including the driest parts of Anglo-America, although at higher elevations rainfall may exceed 500 mm.

The biannual variation in rainfall is most strongly developed over central Arizona, and there is a decrease in amplitude in all directions from this center. The strong primary maximum usually is reached in July-August and the secondary maximum in February (Fig. 17.9). Characteristically the annual profile of rainy days resembles that of rainfall amounts. As a rule the winter maximum becomes stronger westward while the late-summer peak strengthens toward the east. In the southern and eastern part of Arizona over 30 percent of the annual rainfall is concentrated in the two months, July-August, while at stations in southeastern California winter precipitation may actually exceed that of summer. The long dry season in April-May-June is followed by a sharp increase from June to July, with a second and shorter dry period occurring in October.

The winter rainy season in Arizona reflects the same controls that provide the

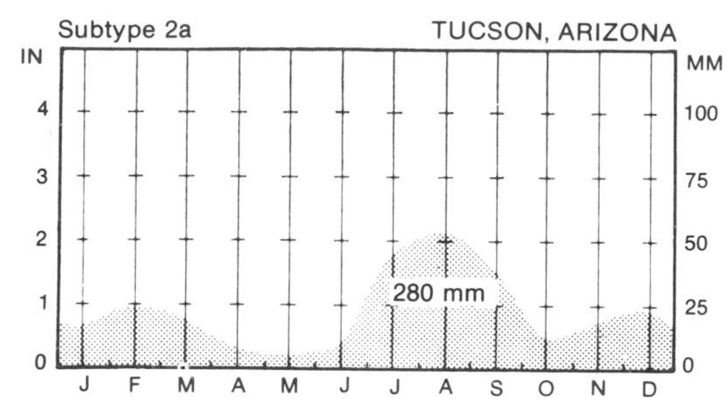

Fig. 17.9. Annual march of precipitation at a desert station in southern Arizona. Here the primary maximum is in summer and a secondary maximum is in winter.

strong winter maximum in California farther west. It occurs during the period when the Pacific subtropical high has receded farthest south and the depressions that come in from the Pacific bring with them moist maritime air from that source (Jurwitz, 1953). More especially it is those depressions which enter the continent south of San Francisco, the southwest type of storm, which give the Arizona region its heaviest winter precipitation. These are widespread, soaking rains and they provide the winter snows in the highlands, fill the irrigation reservoirs, and start the greening of the range in February and March. As the cyclonic belt creeps northward with the sun and the subtropical high exerts increasing control, rainfall declines and the region experiences its spring and early-summer drought.

In the April–June dry period subtype 2a experiences a strong westerly circulation derived from the eastern flanks of a northward-shifting Pacific anticyclone located west of California. Subsidence is strong. This westerly air is further stabilized by a trajectory over cold, upwelled coastal water. Such a circulation is not conducive to precipitation.

The rains which produce the July–August maximum of subtype 2a are of a different origin and type than that of winter. In the hot season the precipitation is characteristically of the local, convective, shower type, commonly associated with thunderstorms. But while these summer thunderstorms may account for 30 to 50 percent of the annual rainfall, they contribute only about 15 percent of the runoff into the reservoirs (Jurwitz, 1953). This is because they are local in character and cover only a small part of a drainage basin.

Whether the Pacific Ocean or the Gulf of Mexico is the prime moisture source for the increased summer rainfall in the Arizona region remains highly controversial, but the weight of recent opinion seems to favor the Pacific. This feature may represent the northern margins of the extensive main summer rainfall maximum characteristic of northern Mexico and the general Caribbean-Gulf of Mexico region. Several decades back Wexler (1943) observed that the North Atlantic subtropical anticyclone begins to strengthen in June and push westward from the Gulf of Mexico over the southwestern states. This anticyclonic circulation, not obvious on the surface charts, but conspicuous at the 700 mb level, moves progressively westward from the Gulf, reaching Arizona in late June and western Texas and New Mexico somewhat earlier. A deep tropical current of air on the southern side of the high accompanies the westward shift of the Atlantic anticyclone, providing conditions that are conducive to shower activity as upward movement is triggered off by the highlands and by the heat of the lowlands (Fig. 18.3). The beginning of the thunderstorm activity in late June and early July, Wexler noted, coincides with the arrival of Gulf air when the Pacific anticyclonic flow is displaced to the north and a humid southeasterly current from the Gulf of Mexico takes its place. As the North Atlantic anticyclone and its southeasterly current weaken in September, the rainfall drops off, but not so rapidly as it rose in early July. A number of other climatologists, following Wexler, were led to conclude that the Gulf of Mexico was a major summer source region of moisture for the interior southwest section of the United States.

But more recently several writers have expressed their doubts relative to whether the Gulf of Mexico is the *main* moisture source for the summer rains of the Arizona region (Pyke, 1972; Hales, 1974). Pyke admits that it may serve as the moisture source region of many weak upper-air disturbances which could trigger isolated, or even organized areas of thunderstorms over the general Arizona region in summer. However, he feels that most of the moisture itself, especially that which reaches the central and eastern parts of this southwestern interior region, must come from the Gulf of California, or likely even the Pacific Ocean to the west of Baja California, where at this time the water temperature is warmer than in spring, and subsidence is weaker. The rapid increase in ocean temperature in these parts is due to the northwestward spreading of very warm water from the North Equatorial Current off the coast of central Mexico. Moreover, Gulf of Mexico air masses would have to cross a large extent of dry country, including moderately high mountains, before reaching Arizona, and en route would lose much of their moisture through condensation-precipitation, and also experience mixing, subsidence, and diffusion. Hales (1974) expresses the opinion that the importance of the Gulf of Mexico as a source of precipitable water for regions west of the Continental Divide is minimal; the highland barriers intervening in a movement of moisture westward from the Gulf of Mexico to Arizona are formidable.

Bryson and Lowry (1955*a*,*b*) have shown that the beginning of the summer rainfall of the Arizona area is in the nature of a singularity. The July rains of Arizona being of the air-mass type, the very abrupt rise in rainfall between June and July suggests a sharp change in air-mass character, the change occurring about July 1 (Fig. 17.15). During June, at 5,000 meters, there is an anticyclonic system over the Mexican highlands with a strong west-southwesterly flow over the Arizona region in which the divergence is nearly zero (Fig. 17.10). At 2,000 meters there is a similar anticyclone flow over Arizona from over the coast of Southern California derived from the eastern end of the North Pacific high. In this westerly flow divergence is evident. Such a synoptic condition is opposed to precipitation. Sometime about July 1 a rapid change in the circulation pattern occurs, when the Pacific anticyclone and the jet are suddenly displaced about 7° northward.

It is a debatable question how far one is justified in further subdividing the large intermontane region based upon variations in annual rainfall profile. Terrain irregularities tend to produce so many mutations in the annual profiles that areal consistency in the annual march of rainfall is not always conspicuous. With this in mind, the boundaries of Subtype 2c have been drawn to include a very large part of the intermontane area in Washington, Oregon, Idaho, and British Columbia, even though there is considerable variation in the rainfall profiles of individual stations (Fig. 17.1).

Subtype 2b.—While in the southernmost subdivision, 2a, the late-summer rains predominate, in Subtype 2c, well to the north, winter commonly is the rainiest season and

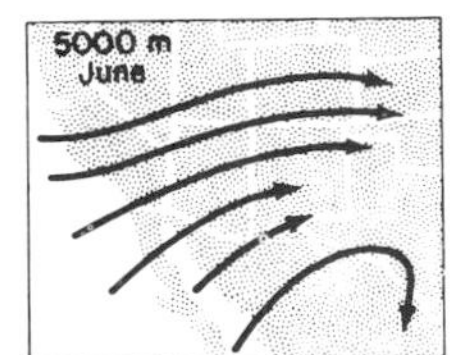

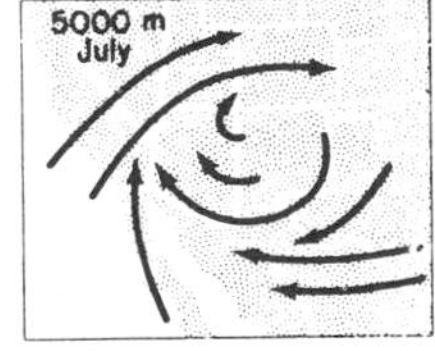

Fig. 17.10. Contrasting mean air-flow patterns at 5,000 m for June and July over the southwestern United States. In June the westerly flow is from the stable eastern margins of the North Pacific anticyclone, while later in the summer, at relatively high levels, there is more of an easterly circulation, some of it probably originating as far east as the Gulf of Mexico region. (After Bryson and Lowry, 1950*a*.)

the secondary summer maximum comes earlier, characteristically in June. Subtype 2b is transitional in character between 2a and 2c. A bimodal profile is conspicuous, there being one maximum in fall-winter and the second in spring with summer the driest season (Fig. 17.11). The origin of the spring maximum can be dealt with most efficiently following the discussion of the more northerly subtype, 2c.

Subtype 2c is not without local differences in annual rainfall variation of some importance, but these differences are not sufficiently regionalized to warrant a further subdivision into genuine subtypes (Figs. 17.12, 17.13, 17.14). As a general rule the winter maximum is the primary one, but more consistently still, the annual variation in number of days with precipitation shows a maximum in winter, even where amounts are equal or higher in a summer month, usually June. This reflects the more general and widespread nature of the winter rains. On the other hand, the rains of summer are more local but heavier. Chapman (1952) has suggested that in southern intermontane British Columbia the rainfall profile is chiefly a matter of elevation and exposure, with higher and more exposed locations having a cool-season primary maximum, while the drier valleys more frequently exhibit a chief maximum in summer. This generalization has numerous exceptions, however. On Chapman's map several river valleys south of 51°N are shown as having summer-dry climates, although this classification is arrived at by including dry April and May as summer months.

The fact that in the biannual rainfall variation characteristic of Subtype 2c, winter rainfall is usually more abundant than in Subtype 2a, and fairly consistently the pri-

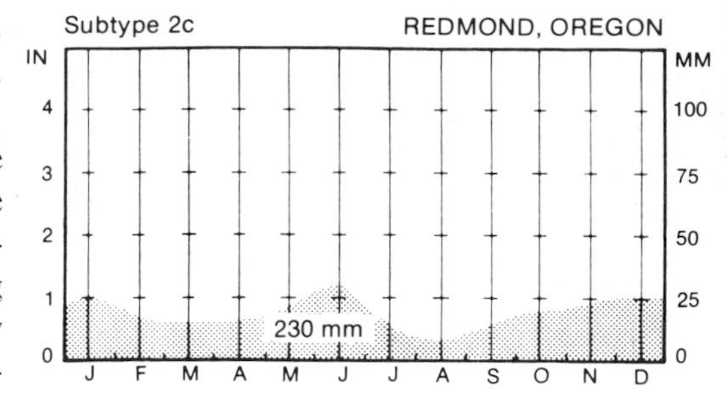

Fig. 17.12. Annual march of precipitation in Redmond, Oregon.

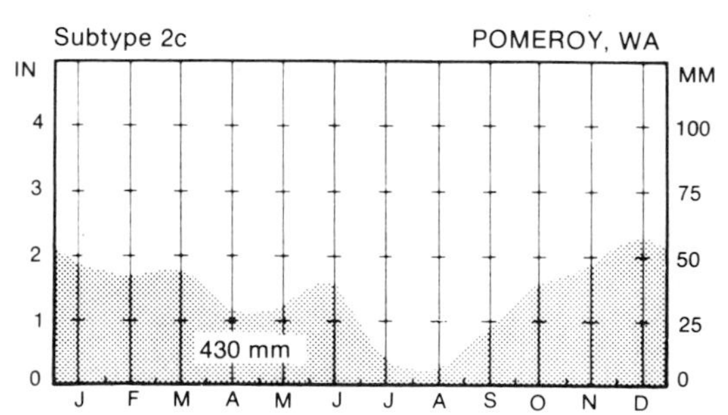

Fig. 17.13. Annual march of precipitation in Pomeroy, Washington.

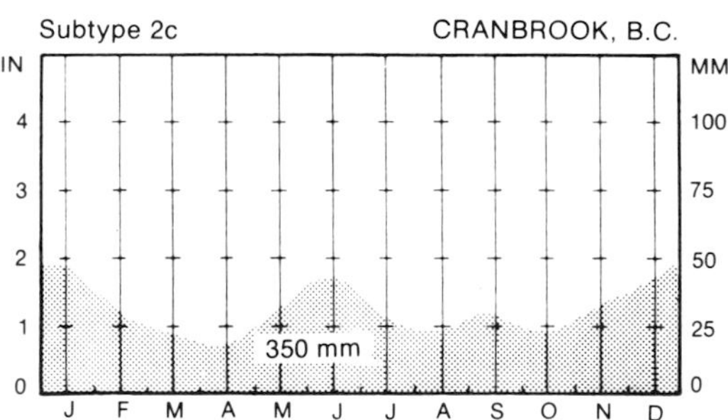

Fig. 17.14. Annual march of precipitation in Cranbrook, British Columbia.

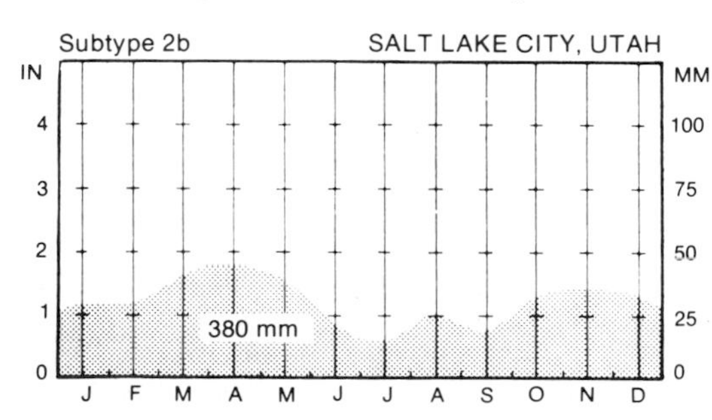

Fig. 17.11. Annual march of precipitation in Salt Lake City, Utah.

mary maximum coincides with the cool season, which is not the situation in the south, does not seem so difficult to understand. Cyclonic control is stronger and winter rains are more abundant with increasing latitude along the Pacific coast of the United States, and it is not unusual that the same is true east of the Sierra Nevada and the Cascade Range as well.

More interesting is the fact that, while in the south (Subtype 2a) the primary warm-season maximum is in July–August, in the north (Subtype 2c) the secondary warm-season maximum arrives in June, while late summer is the season of the primary minimum. Analysis of the mean latitude of the center of the eastern Pacific anticyclone over a period of 20 years shows that it maintains a position at about 34°N during May and June and then early in July shifts abruptly to about 40°N (Bryson and Lowry, 1955*b*). While it is in its May–June southerly location the northern plateau region (2c) escapes the subsident effects which are so obvious in California and Arizona. By contrast, in May and June interior Washington–Oregon is characterized by low pressure at the surface and an upper trough at the 500 mb level, the effects of which are to induce a flow of maritime southwesterly air from the Pacific Ocean. The short wet season of early summer is the result. But when the eastern Pacific high suddenly shifts northward and eastward in early July, the circulation pattern quickly changes and Subtype 2c comes under increasing anticyclonic control with associated subsidence. This initiates the July–August dry season in the north, which, significantly, coincides with the rapid increase in rainfall farther to the south in Arizona, which region has now escaped from the subsident westerly flow originating in the eastern quadrant of the Pacific high and instead begins to feel the effects of invading moist air from the subtropical eastern Pacific and the Gulf of California (Fig. 17.15). The anticyclonic condition is maintained in the north until the southward movement of the jet stream and its cyclones in fall once more increases the number of rainy days and the total amount of precipitation. Substantiating the above analysis is the observed fact that wet Julys in Subtype 2c are characterized by a weaker high and continued lower surface pressure, while dry Julys show a pushing inland of the Pacific anticyclone. The same generalization applies to June. At upper levels unusually wet Junes and Julys are characterized by a well-developed trough oriented in a north-south direction over the coast states with a resulting southwesterly flow of air from the Pacific. By contrast, in dry Junes and Julys the upper trough is absent and straight isobars and a westerly flow prevail.

In the transitional Subtype 2b, the non-winter maximum occurs a month or two earlier than in Subtype 2c, in April or May instead of June. Conceivably this may reflect the earlier arrival in this more southerly location of the same controls which produce

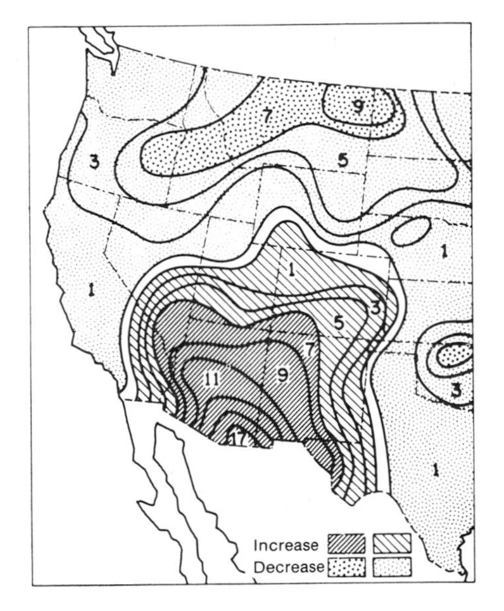

Fig. 17.15. Change in precipitation amounts from June to July, expressed as a percent of the mean annual precipitation. (After Horn, Bryson, and Lowry, 1957.)

the June maximum farther north in Subtype 2c, and simultaneously an earlier weakening of those controls which cause the spring drought in Arizona (Subtype 2a) just to the South.

Subtype 2d is differentiated from 2c because of its strong primary maximum in summer, while the winter secondary maximum is distinctly weaker (Fig. 17.16). At the same time the summer maximum reaches a peak in July rather than in June. Except for the fact that there is still evidence of a biannual rainfall variation, the profile of 2d strongly resembles that of the great subarctic region of interior Canada to the east of the Rocky Mountains where continental controls are responsible for the peak in the modest rainfall coming in midsummer.

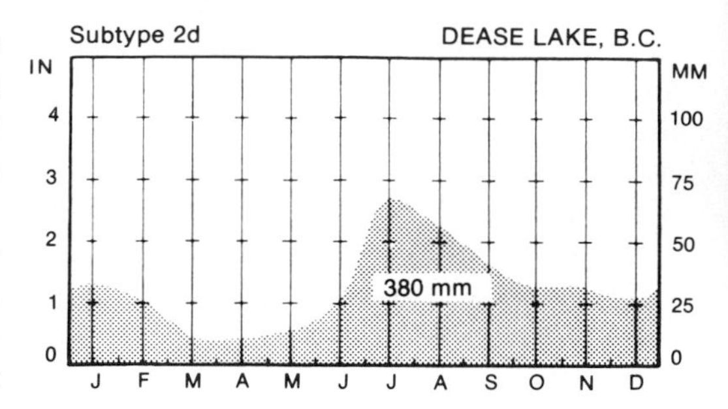

Fig. 17.16. Annual march of precipitation in Dease Lake, British Columbia.

CHAPTER 18

Regional Rainfall Types: The Interior

Type 3. The Interior Region

The immense area included within this type embraces much of the central and northern interior of Anglo-America where continental controls operate consistently and at great strength. Although within Type 3 annual rainfalls vary considerably, modest-to-moderate amounts are characteristic. As a rule precipitation decreases toward the west, with the consequence that large sections of the western and west-central parts of the region included within this type are semiarid and subhumid.

Over this region a marked annual variation in rainfall is characteristic, with a single primary minimum in winter and a maximum coinciding with the warm season. Such a rainfall profile is in accord with the strong continentality. As might be anticipated, there are sufficiently strong regional variations in rainfall profiles within this extensively developed general type to warrant the recognition of several subtypes, and in addition a number of local modifications of some importance.

Subtype 3a.—Located largely in New Mexico and Colorado, this subtype exhibits an annual rainfall variation with a prime maximum in late summer (July–August) and a single minimum in mid-winter (Figs. 18.1, 18.2). In its time of maximum it resembles Subtype 2a in Arizona, but unlike the latter it shows little or no trace of the secondary winter maximum which gives the Arizona region its biannual type of profile. Here the cold season is very dry, for the Pacific winter depressions which are responsible for the important widespread rains in the Arizona region are greatly weakened by the time they arrive farther east of the mountains in New Mexico and Colorado.

It is worthy of mention, however, that two of the three major cyclogenetic regions of Anglo-America are located in the western Great Plains to the lee of the Rockies—one in eastern Colorado and the other in southern Alberta. Genesis is year-round in Alberta; it is largely absent in summer in Colorado. Significantly, in both regions the eastward down-slope from the mountains is steeper than elsewhere along the Rocky Mountain front.

The July–August precipitation maximum in Subtype 3a (Figs. 18.1, 18.2) probably has its origin in the large-scale addition of atmospheric moisture that occurs in mid-summer. As the Bermuda high pushes westward in July it carries with it on its southern flanks a stream of warm humid air from the

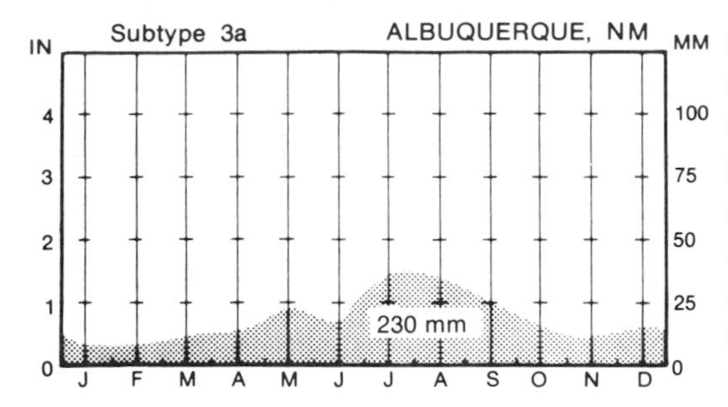

Fig. 18.1. Annual march of precipitation in Albuquerque, New Mexico.

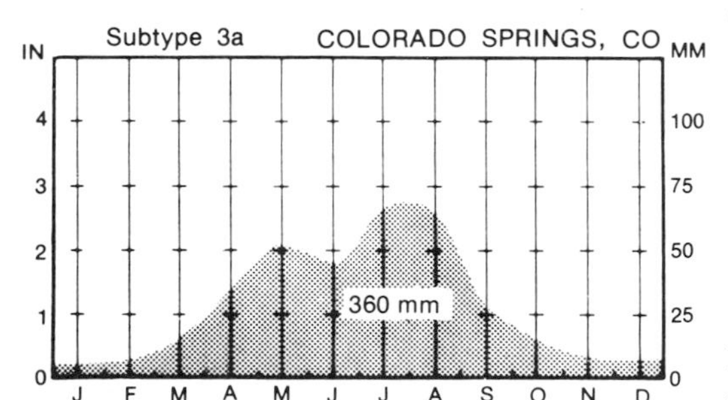

Fig. 18.2. Annual march of precipitation in Colorado Springs, Colorado.

Gulf of Mexico which is drawn in over the New Mexico–Colorado region and furnishes the moisture for a greatly strengthened shower activity (Fig. 18.2). Thus, New Mexico becomes a major center of summer thunderstorms, second only to that in Florida and the Gulf states (Fig. 18.6). In the rainfall profiles of many stations there is also a hint of a secondary May maximum, a feature which becomes very conspicuous in the Texas area (Subtype 3b) just to the east.

Subtype 3b, chiefly characteristic of Texas, but a feature also of easternmost New Mexico and parts of Oklahoma, has an unusual profile characterized by a striking double maximum in the warm season (Fig. 18.3). Clearly the continental type of annual rainfall variation still dominates, for much more rain falls in the warmer months than in winter, and the cold-season primary minimum is particularly well developed. But instead of a single warm-season maximum such as prevails over the Great Plains farther north, the Texas subtype 3b has one marked maximum in May and another in September and there is a conspicuous secondary minimum in July–August. This double maximum about four months apart stands clearly revealed in the phase-angle and amplitude maps of the third harmonic (Horn, Bryson, and Lowry, 1957). In terms of its midsummer secondary minimum this subtype resembles another and more extensively developed one in the upper Mississippi Valley-Great Lakes region, although in the latter the secondary minimum is not so striking or of such long duration. It is entirely possible that the origin of the secondary summer minimum in the two regions has some causal features in common.

As spring advances and rapid surface heating in these subtropical latitudes takes place, the annual rainfall curve trends sharply upward, especially during March and April. Surface air flow is prevailingly from the Gulf of Mexico to the heated land, so that the humid maritime air experiences surface warming and increased instability. At the same time frontal disturbances remain numerous. But for some reason the steep ascent of the rainfall profile is abruptly checked after May and a decline sets in at the very time when surface heating is at a maximum and the inflow of surface maritime air strong. It would appear, therefore, that the

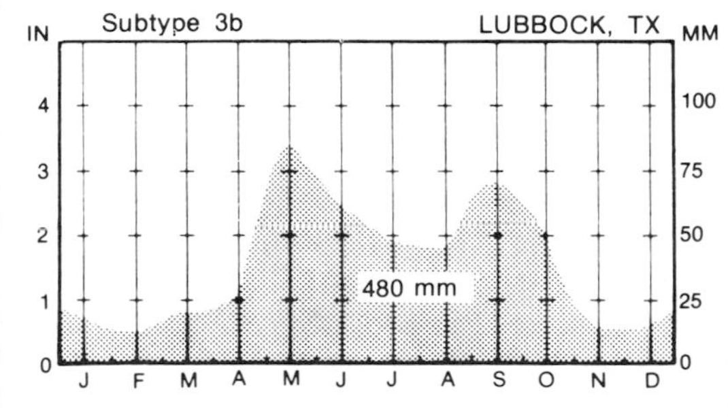

Fig. 18.3. Annual march of precipitation in Lubbock, Texas.

secondary minimum of summer must have its origin in dynamic processes aloft.

At the 750–500 mb level during the summer months anticyclonic flow prevails over the Texas region. This situation is substantiated by Wexler's (1943) mean isentropic chart for August, 1935, and by the average summer isentropic chart based on the mean of all monthly mean isentropic charts from July, 1934, to August, 1939, prepared by Wexler and Namias (Fig. 18.4). As a result of the torque exerted by the steep summer temperature gradients between warm land and cool sea along the west coast, and also of the blocking effects of terrain barriers, the westerly flow over the United States is distorted, giving rise to a series of standing waves downstream, or eastward, from the Pacific coast and the mountains. In the midtroposphere in summer over the United States the circulation pattern shows a large anticyclonic ridge or cell positioned over the Mississippi Valley and southern Great Plains, and two pressure troughs, one in the southwest and the other along the south Atlantic coast. The pressure cells are associated with wet and dry tongues of air as represented in Figure 18.4. Because summer rainfall depends on the presence of a deep, moist current the mean positioning of this pattern of ridges and troughs with their associated dry and wet currents exerts a controlling effect on mean summer rainfall distribution. Since the circulation pattern varies considerably in different summers, the warm-season rainfall distribution also varies.

On the Wexler and Namias (1938) charts a large anticyclone is positioned over the southern Great Plains in summer with a tongue of dry northerly air associated with this cell concentrated over the Texas area and the lower Mississippi Valley. This dry subsident air stream has the effect of damping any convection beginning in the unstable surface air from the Gulf. Evidence of this effect is to be seen in the rapidly decreasing number of thunderstorms from east to west along the Gulf coast in August as the anti-

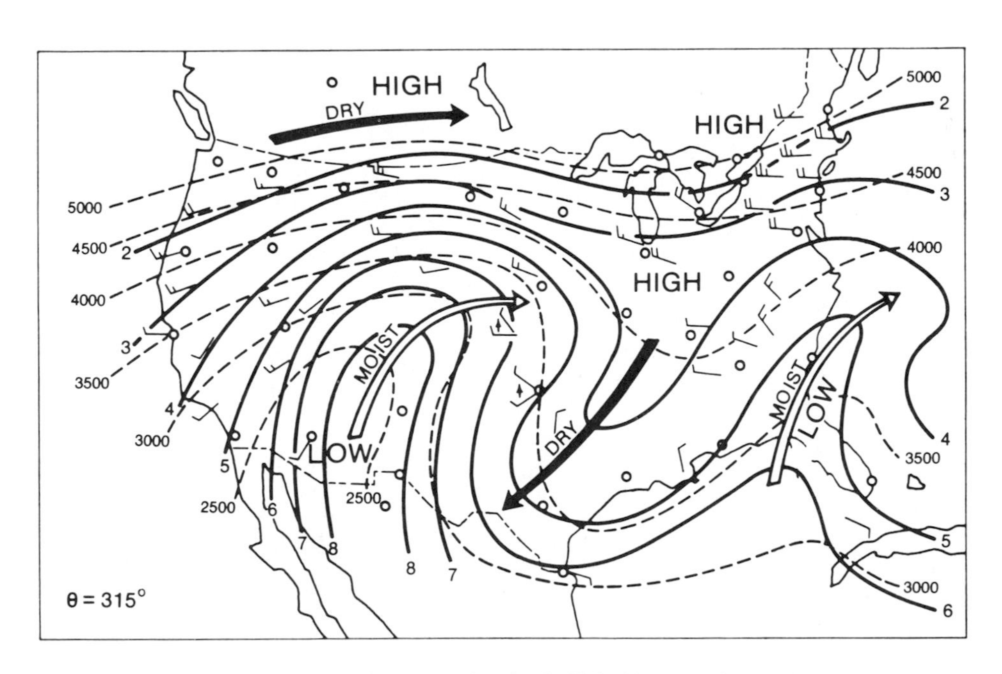

Fig. 18.4. Average summer isentropic chart for the United States. (After Namias, 1972.)

cyclonic dry tongue is approached, and by a decreased percentage of the annual rainfall occurring in August and in the summer months in general (Wexler, 1943) (Figs. 18.4, 18.5, 18.6).

Subtype 3c.—Throughout this very extensive area embracing most of the central and northern Great Plains, and eastward to Minnesota and Iowa, the annual rainfall profile is a fairly simple one. The meager or modest total rainfall is strongly concentrated in early summer, June usually being the wettest month, although a few stations may show a May maximum (Fig. 18.7). The crest of the profile is characteristically sharp. It is fairly common for the profile to show a halt in its rapid decline in August–September, so that either a bench or even a hint of a slight secondary September maximum may be evident. This appears to be a rudiment of what farther east in Iowa, Minnesota, and Wisconsin, and also over the southern Great Plains in Texas, is a modest second summer maximum in August–September.

The fact that precipitation reaches a maximum in early summer, or even late spring, rather than later in the summer when temperatures are higher and the advection of Gulf moisture into the deep interior of the continent presumably is at a maximum, requires comment. Early summer is also the time of an important secondary, or sometimes even primary, rainfall maximum in the northern intermontane region to the west of the Rockies, and to a degree the June maxima to the east and west of the Rockies are interrelated. The primary summer moisture sources of the two regions are not identical, however, for while the Pacific Ocean probably represents the more important moisture source for the intermontane region's summer rains, the Gulf of Mexico very likely sup-

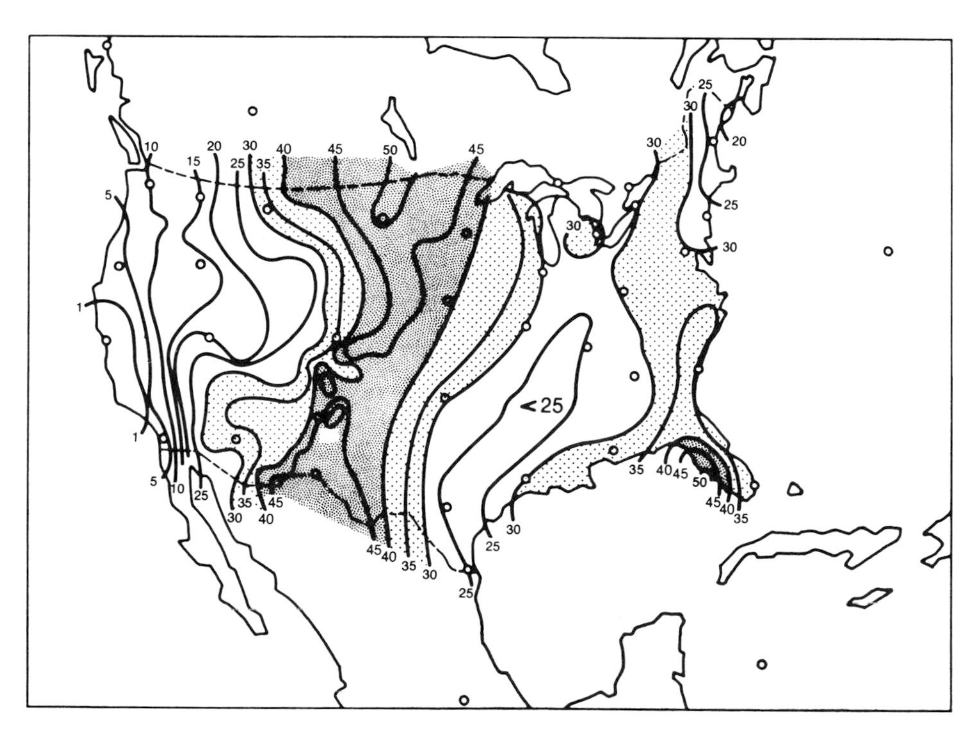

Fig. 18.5. Percentage of total annual precipitation occurring during summer in the United States. (After Kincer, 1922.)

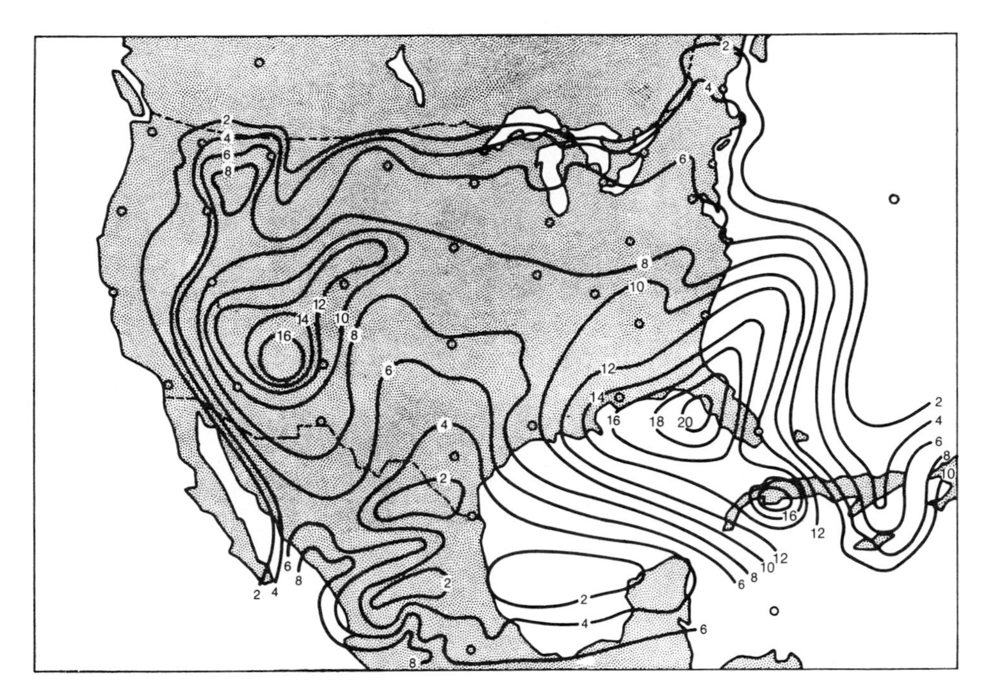

Fig. 18.6. Average number of thunderstorms within the United States in August. (After Wexler, 1943.)

plies a larger proportion of the warm-season moisture for the northern Great Plains.

While over most of the Great Plains June is wetter than July, the core area where June rainfall exceeds that of July by 25 to 50 mm, and in restricted areas by more than 50 mm, is in eastern and central Montana and adjacent parts of the Dakotas and Wyoming (Fig. 18.8). It is especially to this core area that the following comments apply. The fact that rainfall drops off in July, even though air temperature and moisture advection are then at a maximum, appears to be related to the fewer occasions of disturbed weather in mid- and late summer as compared with June. Thus, while June is usually considered to be the first month of summer, it exhibits in its frequency of disturbances many of the characteristics of spring and even of winter. Between June and July, however, there occurs a marked northward shift of the circulation features, including the principal storm tracks. Within a rectangle with dimensions 10°E-W by 5°N-S, centered on the area in question, for the 40-year period 1899 to 1938, it was found that the number of days with low centers was 35 percent greater, and with high centers 32 percent less, in June than in July (U.S. Weather Bureau, 1957). Further evidence of the greater prevalence of cyclonically disturbed weather in June as compared with July is to be found in the larger deviation of the value of the west-east and north-south components of the geostrophic winds at the 500 mb level in June (Lahey et al., 1958). It may likewise be significant that in about mid-June there occurs over the northern Great Plains the greatest fall in sea-level pressure recorded for any 5-day period throughout the entire year (Lahey, Bryson, and Wahl, 1957). Summarizing, it appears that the early summer rainfall maximum in the northern Great Plains is a consequence of the normal continental influence

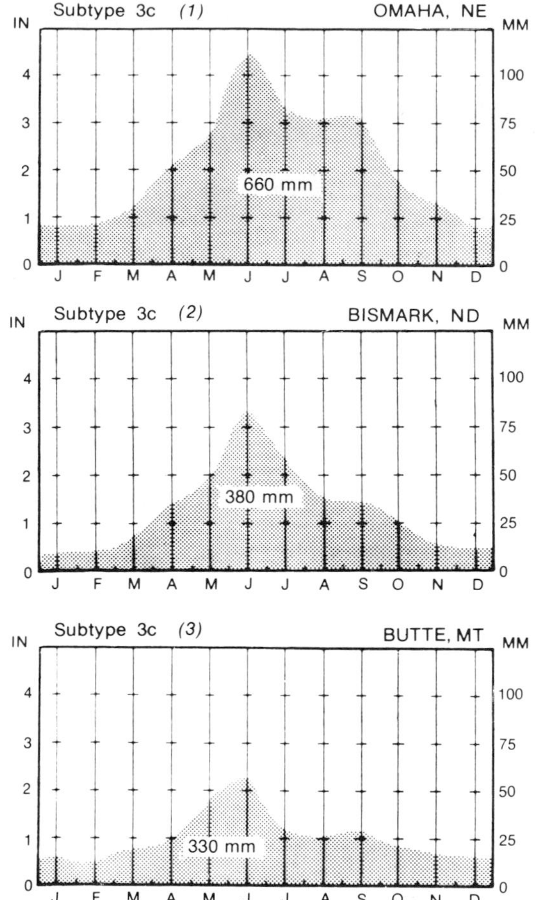

Fig. 18.7. Annual march of precipitation in Omaha, Nebraska; Bismarck, North Dakota; and Butte, Montana.

resulting in increasing temperatures and increasing advection of Gulf moisture, in conjunction with a persistence of spring storm activity into early summer. In this respect, it is significant that north of about 55°N the rainfall maximum occurs in midsummer rather than in June.

Subtype 3d.—In an extensive area in the upper Mississippi Valley–upper Great Lakes region the typical continental type of annual rainfall variation is significantly modified by reason of the profile having a mid- or late-summer secondary minimum (Fig. 18.9). In addition the total annual rainfall is greater than in the subtype to the west, 650 to 1,000 mm being characteristic, the additional 250 to 400 mm being largely attributable to added cyclonic rainfall in the cooler months. Thus, the annual rainfall profiles in Subtype 3d are less sharp than are those for Subtype 3c and this feature increases in prominence eastward.

More striking, however, is the double maximum in summer, with one peak usually in June (occasionally May or July) and the second most commonly in September (occasionally August). If throughout the extensive area of a twinned summer maximum the two peaks were consistently three months apart, in June and September, the boundaries of the area would be revealed on the amplitude and phase-angle maps of the fourth harmonic (Horn, Bryson, and Lowry, 1957). As it is, the fourth harmonic amplitude map only incompletely defines the area in question. It bears re-emphasizing that this twinned peak of rainfall in the warmer part of the year is a feature typical not only of the upper Mississippi Valley. In modified forms it is characteristic as well of the area lying to the

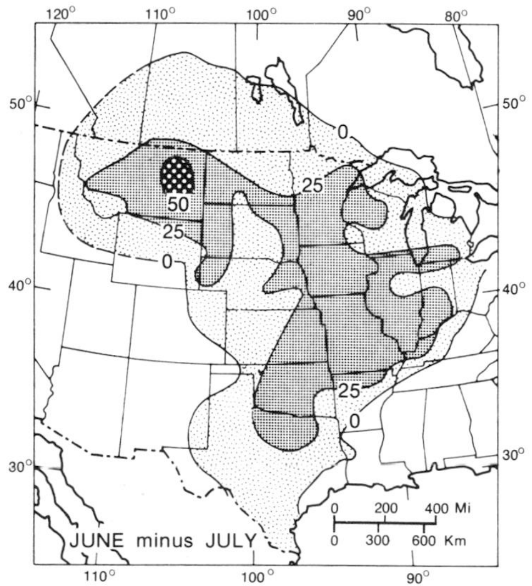

Fig. 18.8. That part of the mid-North American region where June's precipitation exceeds July's.

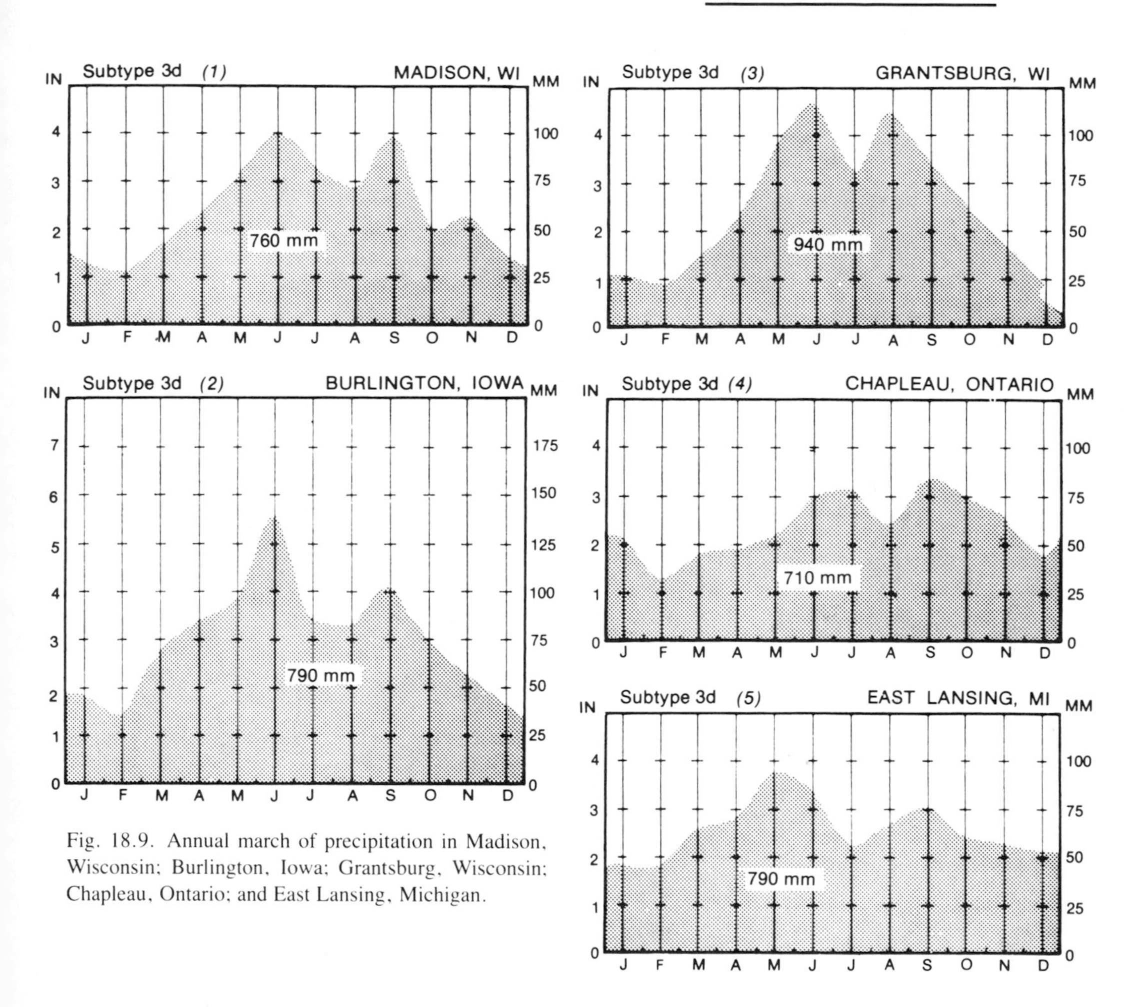

Fig. 18.9. Annual march of precipitation in Madison, Wisconsin; Burlington, Iowa; Grantsburg, Wisconsin; Chapleau, Ontario; and East Lansing, Michigan.

south and southwest of Subtype 3b, especially in the western parts of Type 5, and as described earlier, is present as a dominant feature in Texas (Subtype 3b), although in the latter area the peaks are 4 months apart. In eastern Kansas and Nebraska it is present only in the rudimentary form of either a bench or a slight node in August or September. It is highly noteworthy that the positioning of the mid-continent summer ridge of pressure aloft and its associated dry current of northerly air as described by Wexler (1943), has its analogue in the negative anomaly of August rainfall (Wexler, 1943, fig. 6), and in the extensive region having a twinned peak of warm-season rainfall which extends in a NE to SW direction from south-central Canada in the vicinity of the Great Lakes to southern Texas (Figs. 18.4, 18.5, 18.6). The mid-continent area experiencing an increase in rainfall from August to September is similarly located (Horn, Bryson, and Lowry, 1957).

The core area of Subtype 3d can be thought of as located in southern Wisconsin, northern Illinois, and eastern Iowa where the two rainfall peaks are symmetrical and characteristically fall in June and September, and where the July–August secondary minimum is rather striking. In all directions from this core area there are significant modifications. Toward the southwest the profile is less symmetrical as the September peak wanes (Fig. 18.9). Northward in Minnesota and northern Wisconsin the two peaks more often coincide with June and August, and still farther north in Canada with July and September. Eastward in Michigan the annual profile is a flatter one, with more winter precipitation, and the broad early-summer peak is separated from the September maximum by only a shallow dip in July and August.

It should be emphasized that even within the core region, the majority, but by no means all, of the weather stations have a twinned summer maximum. The primary early summer (June) maximum is rather consistently present; the late summer (September or August) one is less so. It may be noted also that over a span of one-half to three-quarters of a century the number of stations with a dual warm-season maximum has declined from 91 to 82 percent in Wisconsin, 93 to 82 percent in Illinois, and 87 to 67 percent in Iowa. Cause for this change is problematic. One suggestion is that the recent cooler summers over interior Anglo-America may have caused the jet stream and associated cyclonic disturbances to follow somewhat more southerly routes, resulting in more midsummer rainfall than usual over the region in question. Madison is one of the stations that exhibited a well-defined midsummer decline in rainfall in the twenties and thirties, but this feature is not apparent in the rainfall data for more recent decades. It should be emphasized that even within the core region, the majority, but by no means all, of the weather stations have a twinned summer maximum.

In order to obtain a more detailed representation of the warm-season variation of precipitation, 7-day means were substituted for monthly means in constructing a composite mean rainfall profile of five stations in southern Wisconsin for the 30-year period 1906–1935 (Fig. 18.10). The result is startling, for it reveals the fact that the relatively smooth profile constructed from monthly means is really composed of a series of short wet and dry periods which taken together form a very complex profile. The principal peaks and hollows in this profile are in the nature of singularities. Four relatively wet periods can be noted, the 21st week in late May, the 24th and 25th weeks in June, the 31st week in early August, and the 37th week in mid-September. Notably dry weeks are the 19th in the second quarter of May, the 22nd in late May and early June, the 29th and 30th in the second half of July, and the 34th and 35th in late August.

It is clear from the preceding discussion that any explanation of the double warm-season maximum of precipitation is obliged to take into consideration not only the simple rainfall profile based upon monthly means, but also the more complicated pattern of shorter wet and dry periods as revealed by the composite profile derived from employing 7-day rainfall means. The latter in turn necessitates a consideration of the synoptic patterns and associated frontal zones. Random thermal convection is not sufficient to explain the profile derived from 7-day rainfall means for although 90 percent of the July rainfall at Madison, Wisconsin, is associated with thunderstorm activity (80 percent in June, 79 percent in August, 68 percent in September), these thunderstorms commonly are nocturnal when thermal convection is at a minimum. Nearly all thunderstorms are found to be associated with large-scale synoptic patterns, 53 percent with low-pressure troughs at 500 mb and about 70 percent with clearcut warm tongues (Bryson and Lahey, 1955). It must be concluded, therefore, that summer rainfall in the Wisconsin area, although predominantly of the showery thunderstorm type, is not prevailingly a consequence of random thermal convection, but instead is organized in character and associated with large-scale synoptic situations.

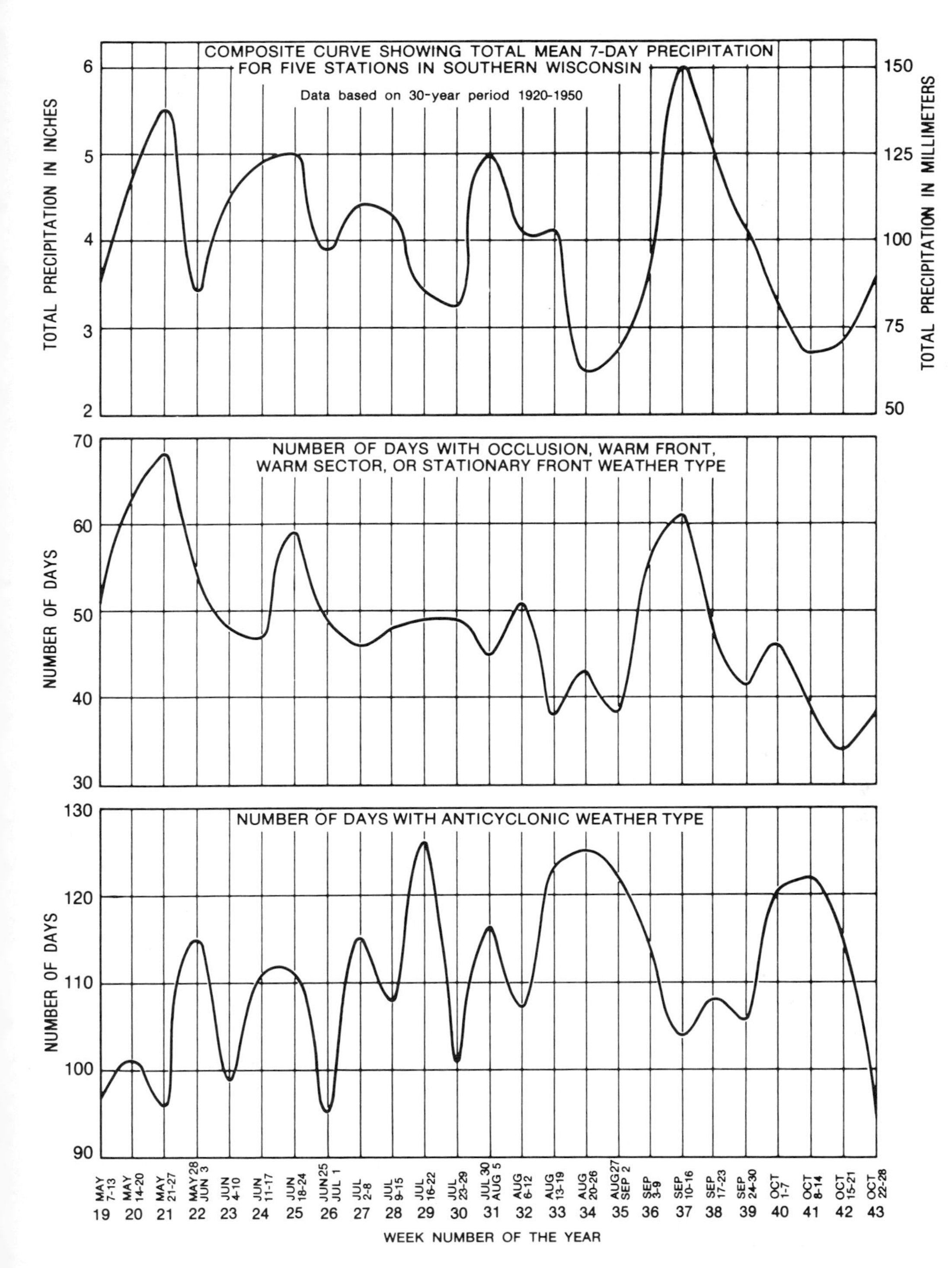

Fig. 18.10. Mean 7-day composite precipitation profile for 5 stations in southern Wisconsin, and similarly constructed profiles showing the frequency of (a) frontal weather situations favoring precipitation and (b) anticyclonic weather favoring drought.

It has been suggested that the double summer maximum of precipitation in the upper Mississippi Valley as revealed by the profile of monthly means may have its origin in the seasonal latitudinal migration of the belt of atmospheric disturbances and fronts following the sun. Such a latitudinal shifting of the zone of convergence between tropical and polar air may be observed on the mean monthly wind-flow charts. But while this simplified explanation no doubt has some validity, it is of little value in understanding the detailed rainfall profile constructed from 7-day means.

There are two ways of viewing this feature of the dual warm-season maxima in the annual precipitation profile constructed from monthly means. One is to focus on the two peaks (June and September) as the positive feature. The other is to emphasize the modest sag in the annual rainfall profile that occurs in midsummer (July–August), which thereby creates the two peaks.

At least a partial explanation for the decreased rainfall in July–August is to be found in the above-surface circulation pattern of summer as described by Wexler and Namias (1938) and referred to in an earlier part of this chapter (Fig. 18.4). The dry northerly tongue of air associated with the anticyclonic cell over the interior of the country reaches its maximum development in July–August, at which time it acts to depress the rain-making processes in its vicinity. It is this dry current aloft which appears to cause a dropping off of rainfall in mid- and late summer throughout an extensive area extending from the Great Lakes southward through the Mississippi Valley to Texas, and is responsible for the twinned peak in summer rainfall in Subtypes 3b, 3d (Figs. 18.3, 18.9).

It is necessary now to turn to the more complicated rainfall profile provided by the 7-day means. In order to obtain some measure of the controls responsible for the shorter wet and dry periods here revealed, which when combined into monthly means of rainfall disclose the twinned summer maximum and the mid-summer secondary minimum, the daily weather over southern Wisconsin was classified into six synoptic types, and tabulations were made of the frequency of occurrence of each type for 7-day periods over the 30 years, 1906–1935. These tabulations were subsequently compared with the composite 7-day rainfall profile for five stations in southern Wisconsin covering the same period of time. The warm front–stationary front is recognized as the type of synoptic condition most conducive to abundant summer rainfall in the upper Mississippi Valley area. Paul Waite (1958) has analyzed the causes for the strong September rainfall maximum in the Madison, Wisconsin, area and concludes that it is generally associated with a warm or semi-stationary front extending east-west or northeast-southwest, with moist Gulf air along the front about 160 kilometers south of the heavier rainfall and with the moist southerly air overrunning a wedge of polar air.

There is a general inverse or negative correlation between the frequency of highs and the 7-day rainfall means, and the correlation is especially strong from mid-August to mid-October (Fig. 18.10). Anticyclonic frequency shows a tendency to increase from May to mid-August, culminating in a broad peak during the weeks 33 to 35 inclusive, from mid-August to early September. This anti-cyclonic peak coincides with a major dry period. Similarly, a striking decrease in the frequency of highs in mid-September is matched by the summer's most conspicuous peak in rainfall. As anticyclonic frequency strongly increases again in late September and early October, the Indian Summer period, rainfall slumps to a sharp minimum. A precise correlation of highs and rainfall is less conspicuous in spring and early summer, although there is a general matching of the rise in frequency of highs from June to August with a general decline in rainfall during the same period.

A reasonably good positive correlation appears to exist between summer wet periods and the frequency of combined warm and stationary fronts, and an even better one between the combined frequency of warm, stationary, and occluded fronts, and warm sectors. Thus, the strong rainfall peaks during the 21st (May 21–27), 25th (June 18–24), and 37th (September 10–16) weeks are matched by peaks in the combined frequency of warm-stationary fronts, occlusions, and warm sectors. The rainfall peak during the weeks 31 to 33 is also fairly coincident with a modest peak of the same synoptic conditions in the 32nd week.

It is worthy of emphasis that in the upper Midwest of Anglo-America throughout the year, circulation patterns, air masses, frontal systems, storm routes, and weather in general are in a state of constant flux (Bryson, 1966; Bryson and Lahey, 1958). The weather element is variably present, to be sure, but still there is an observable seasonal rhythm to the weather episodes. Here the prime interest is in the precipitation element, and almost exclusively that of the warmer part of the year, or from May–June through September.

Sometime in early April, or even March, there occur in the Northern Hemisphere certain relatively abrupt changes in the broadscale circulation patterns which mark the advent of spring. Major synoptic alterations associated with these changes are as follows: (1) a weakening of the cold anticyclone over Anglo-America, a feature that was very prevalent during winter; (2) the more frequent episodes of cyclonic weather; (3) a strengthening of the subtropical anticyclones over adjacent seas, and (4) a reduction in the wavelengths in the upper-air polar vortex circulation coincident with the splitting of the Icelandic Low (Bryson and Lahey, 1958). Following in the train of these noteworthy atmospheric events there occurs a stronger meridional transfer of humid tropical air from the south, induced by a strengthening of the subtropical anticyclone over the North Atlantic Ocean. Cold arctic air continues to invade from higher latitudes with the result that a zone of mean confluence separating unlike *mT* and *cP* air masses is positioned across the Midwest, which acts to stimulate more frontal cyclonic weather and thunderstorms. So spring witnesses an increase in precipitation and the annual rainfall profile shows an upward trend, which usually culminates in May–June, the time of the first peak of rainfall.

June normally witnesses another period of relatively abrupt pressure and circulation changes throughout the Northern Hemisphere, coinciding with the end of spring and advent of summer (Bryson and Lahey, 1958). This first month of summer sees a rapid northward shift of the North Atlantic subtropical anticyclone and a corresponding weakening of the Icelandic and Aleutian low-pressure centers. Rapid disappearance of the arctic-subarctic snow cover in May–June, with a similar change from high to low albedo in high latitudes, no doubt is associated with the northward shift of the subtropical anticyclones. In Anglo-America this results in the anticyclonic circulation around the western end of the North Atlantic high to route southerly *mT* air from off the Gulf of Mexico and subtropical Atlantic more deeply and strongly into eastern Anglo-America, providing a favorable environment for precipitation well into southern Canada. This advected air, although anticyclonic in origin, is derived from the western quadrants of the high where the moisture content is great and inversions are absent or present only at considerable altitudes. In this southerly maritime circulation there is a corresponding shift in average wind speed, from 1–2 meters per second characteristic of the first half of the year, to an average of 3.5–4 meters per second, a feature which is rapidly established by late June. Subsequently it declines slowly for the remainder of the year.

The second of the warm-season precipitation maxima, extending from late August to late September, coincides with a quickening of cyclonic activity and an increased pressure variability in the latitude of the subpolar low (Lahey, Bryson, and Wahl, 1957). In the upper Midwest, anticyclones composed mainly of *mP* air trailing weak cold fronts become frequent in late August, and function to bring to an end the showery rains so typical of summer. By early autumn the Canadian cold source has waxed sufficiently to cause the average polar front on frequent occasions to stagnate in its location across the upper Midwest, bringing to that region abundant rains of the more steady variety. This situation operates to produce the second and later of the warm-season rainfall peaks, resembling in weather and precipitation that of June.

Table 1 and Figure 25 in Bryson and Lahey (1958) point up the change from frequent dry cold fronts in August to the more numerous warm fronts and mobile disturbances, accompanied by prolonged and heavy rains, typical of September. Early in September, in the relatively meridional circula-

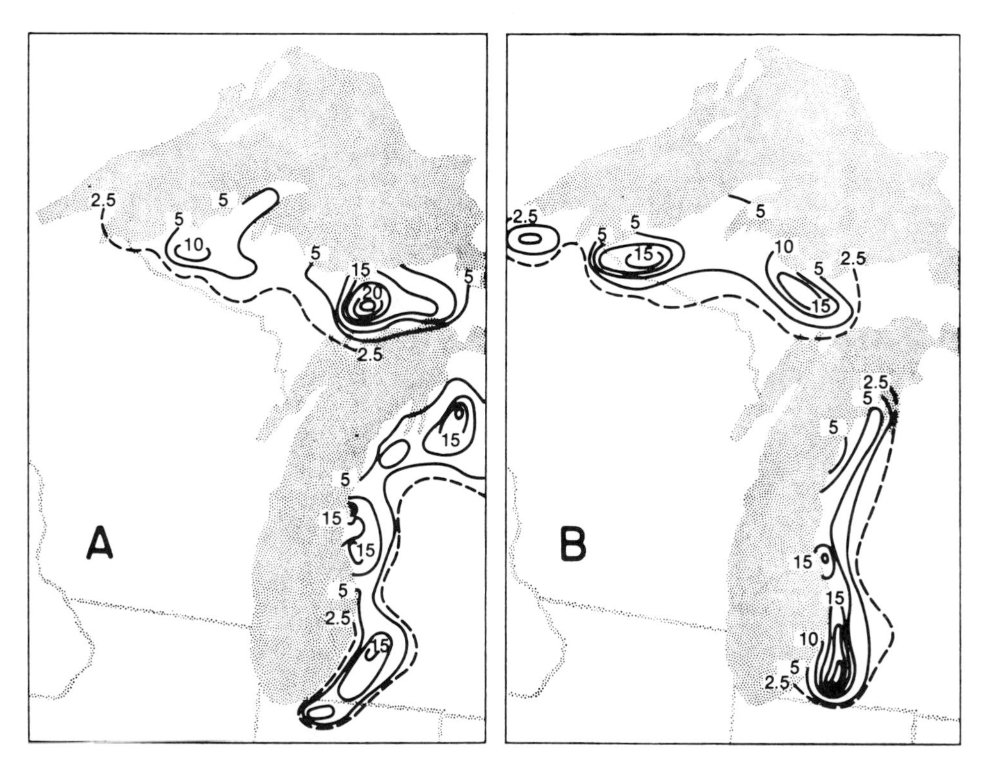

Fig. 18.11. Precipitation patterns (in mm) in the vicinity of Lake Superior and Lake Michigan during a February 1948 cold spell when the whole region was under the control of a rather uniform mass of cold arctic air. No fronts were present. During the first part of the period (A) a strong outbreak of arctic air prevailed and northwesterly winds were strong. During the second part (B) the northerly current was weaker. (After Petterssen and Calabrese, 1959.)

tion that continues to prevail, it is still the moist tropical air from the Gulf of Mexico that continues to be drawn into the disturbances and frontal systems of the upper Midwest. But as the circulation becomes increasingly zonal in late September, the warm-air source shifts westward, and to an increasing degree air is drawn from the southwestern dry lands instead of from the Gulf of Mexico. In addition, late September sees an increase in the number of migrating anticyclones over the upper Midwest. All of these changes operate to diminish the late September rains, and usher in the drier, much cherished Indian Summer weather so characteristic of October.

From a somewhat different point of view, one may consider the modest midsummer dip in the annual rainfall profile to be the positive feature requiring explanation, which in turn results in the two rainfall peaks of June and September. In other words, this is a normal annual rainfall profile for the interior of a middle-latitude continent whose customary midsummer apex has suffered a slight-to-modest collapse. To be sure, throughout most summers in the Midwest that region is dominated by a flow of warm, southerly, humid air arriving from the Gulf of Mexico. Frontal passages are common and these produce the moderate rainfalls of July and August. But it is also true that in these same months rainfall amounts vary considerably from year to year. The somewhat reduced average rainfalls of midsummer months, and also the dry spells within normal summers, appear to be related to those occasions when a zonal circulation pattern prevails, and dry air from the western interior invades the Midwest. This occasional zonal circulation pattern is scarcely to be observed on the average streamline map for July or August. But Borchert (1950) has shown that in abnormally dry summers in the North American Grassland region, and extending even into the Midwest, the normal southerly flow of air from the Gulf of Mexico fails, and is replaced by a zonal westerly flow from the hot and dry interior.

Local Effects of the Great Lakes upon Precipitation.—Although not all of the Great Lakes lie within the boundaries of Subtype 3d, it appears appropriate to include general comments on some of their climatic effects at this point. Considering the extensive area of these water bodies, their influence on the precipitation of their marginal lands seems remarkably small (Eshleman, 1921). Their chief effect appears to be upon snowfall, there being a somewhat greater fall in the immediate vicinity of the lakes than elsewhere, and more especially downwind from a lake, or along the eastern and southern sides on occasions of strong northerly and westerly winds in the cooler months.

Northern Upper Michigan on the southern side of Lake Superior, northwestern and western Lower Michigan lying to the southeast and east of Lake Michigan, peninsular Ontario in Canada, and those parts of New York State and Pennsylvania lying adjacent to Lakes Erie and Ontario, all show an excess of snowfall over that received by surrounding areas farther removed from the influence of the lakes. Thus, Grandhaven on the eastern side of Lake Michigan has 250 mm more snow than Milwaukee on the western side. Wiggin (1950) recognizes two kinds of snows in the vicinity of the lakes, (1) cyclonic snows which are widespread and general to the area, since the Great Lakes are a preferred region for cyclogenesis and cyclonic activity during the colder months, and (2) lake snows which are directly attributable to the lake influence and are markedly more local in character. The latter are chiefly the result of the warming and humidifying effects on cold dry *cP* air as it moves over the relatively warmer water in fall, winter, and spring (Fig. 18.11). Most

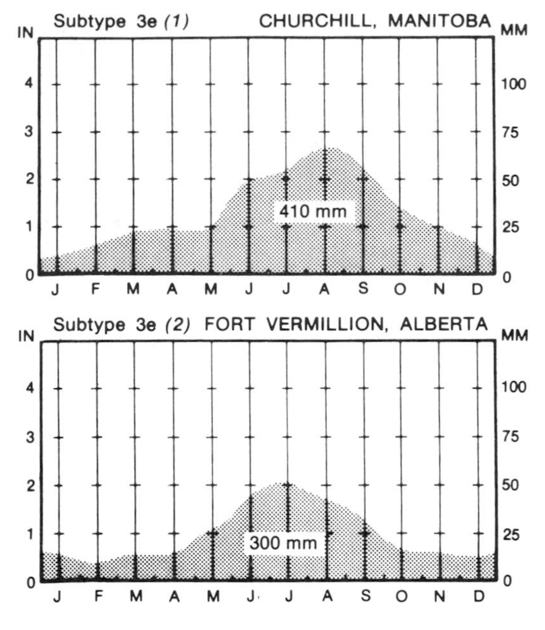

Fig. 18.12. Annual march of precipitation in Churchill, Manitoba, and Fort Vermillion, Alberta.

of this precipitation is dropped in the form of heavy snow showers or snow bursts over the lakes and along the downwind coasts in a narrow strip roughly a score of miles in width. The warm lake source has the effect of lifting the inversion at the top of the turbulent layer from 1 km to about 2.5 km, while cyclonic curvature is induced in the cold air mass resulting from decreased surface friction as the cold air moves from a land to a water surface (Eichmeier, 1951; Petterssen and Calabrese, 1959). The warming effect of the water surface also has a tendency to produce a shallow thermal low which favors vertical upward movement, and as the humidified air comes onshore it is further lifted by orographic and frictional effects.

It is difficult to determine to what extent the influence of the lakes actually affects the shape of the annual precipitation profile. There is some evidence that where the marine influence is strong it tends to slightly depress the total rainfall of summer and increase that of fall and early winter. Thus, in western New York State fall and winter not uncommonly are the seasons of maximum precipitation, December–January at Buffalo, and October–November–December at Oswego. On the first harmonic phase-angle chart (Horn, Bryson, and Lowry, 1957) there is a strong packing of the isochrones on the east and south shores of Lakes Erie and Ontario where there is an abrupt change from a late fall maximum immediately along the coast to a late summer maximum farther inland. It is not entirely clear whether the fall maximum on the south side of these lakes can be attributed to lake-induced snows, although it may be associated with the early cold outbreaks of the year.

THE LA PORTE, INDIANA, RAINFALL ANOMALY.—There may exist in the vicinity of La Porte, Indiana, a local, annual rainfall anomaly (Harman and Elton, 1971). Annual precipitation totals and frequencies, during the period from about the mid-1920s to around 1960, were higher at La Porte and vicinity by 10± mm than elsewhere in northwestern Indiana. This differential was especially marked in summer. At first it was suggested that this rainfall anomaly might have arisen from the increased heat and condensation nuclei being added to the local atmosphere by the expanding urban-industrial complex at the southern end of Lake Michigan. But both the data and conclusions of the original study have been questioned. Indeed, the apparent fading of the anomaly since about 1960 has raised doubts even about its validity. If it did, or does, exist, then Harman and Elton suggest that the added summer rainfall of La Porte may show the effects of a local, afternoon lake-breeze front in the presence of a westerly flow from across Lake Michigan. Passage of the lake-breeze front usually coincides with a change in wind direction, and a decrease in temperature and relative humidity. Such a local front would have the effect of increasing convective overturning.

At this stage what is needed for a better understanding of the influence of the lakes is a more penetrating analysis of their dynamic effects on polar air masses, and more especially their modifying influence on particular synoptic situations that have resulted in heavy local snowfalls.

Subtype 3e.—Because of the meager data available no attempt has been made to regionalize the extensive subarctic and arctic continental region of Anglo-America. Most of its shows a relatively strong summer maximum, the cold anticyclonic winters being relatively dry (Fig. 18.12). Precipitation is dominantly cyclonic in origin.

Weatherwise, there may be recognized a twofold subdivision of the northern lands, a relatively quieter west and a more stormy east. This results from the facts of perturbation sources, and also because humid maritime air can enter much more readily on the Atlantic flank than on the Pacific. The western Arctic-subarctic, including much of interior Alaska, the Yukon, the Mackenzie Valley, and the mainland east to the 100° or 105° meridian, is drier and has less disturbed weather. Of the two main streams of cyclones, one from the Pacific and the other from the interior of the continent, it is the first stream only which affects the western half. Moreover, the western cyclones from the Pacific are weak, for they are only the remnants aloft of coastal cyclones that have crossed the mountains and whose lower parts have been disrupted. Their precipitation is light both in summer and in winter.

The stormier and wetter east is affected by both the regenerated Pacific storms as well as those from the interior, and moreover, these are operating in air which has a higher humidity content. Here the weather is more disturbed, with gales, heavier snowfall, and high windchill characterizing the winter season. All of the eastern half has over 250 mm of annual precipitation, and the region east of Hudson Bay has 500 to 1,000 mm. Over 1,000 mm of precipitation characterize southwest Greenland.

A notable feature of those subarctic localities lying along the windward side of open water in winter is the tendency toward a fall maximum in precipitation. Thus southeastern Baffin and the Ungava coast have heavy fall snows as does the western side of the Labrador Peninsula facing Hudson Bay. As the cold surges of polar air cross the open water they are greatly warmed and humidified, resulting in instability snow showers from dense cumulus clouds on the windward shores (Hare, 1951). These effects disappear later in the winter as the areas of open water are greatly reduced or disappear entirely.

Weather and precipitation on the interior of the Greenland Ice Cap have been a controversial question. The surface winds are prevalently radial and outward, to be sure, but this is not the result of a permanent anticyclone but only downslope gravity flow from the ice plateau. Such radial winds are especially prevalent during a quiet period of general circulation. Often upslope winds prevail at times of cyclone passage. The evidence is strong that precipitation on the ice cap is cyclonic and orographic in origin, the latter chiefly on the flanks. Rapid fluctuations in pressure, wind direction and speed, and cloud types indicate a typical disturbed cyclonic climate. Probably the sea-level pressure systems do not cross the ice cap, but high-level perturbations overlying the surface cyclones do cross the ice plateau and subsequently regenerate as surface cyclones on the lee side (Dorsey, 1945).

CHAPTER 19

Regional Rainfall Types: The South and East

TYPE 4. THE OHIO VALLEY REGION

Around the southern and eastern margins of the great interior region (Type 3) with its relatively simple rainfall profile, characterized by a warm-season maximum and a cold-season minimum, are a number of transitional and pure types, most of which are more complicated in character. The broader patterns of profile modification, from the interior toward the Atlantic Ocean and Gulf of Mexico, are as follows. North of about 40°N precipitation increases eastward and becomes less seasonally accented, so that along the Atlantic margins winter precipitation may equal, or in places even slightly exceed, that of summer. Because the precipitation of the cooler months is greater than in the interior, the annual profile tends to be flatter. Southward and eastward from the interior toward the Gulf of Mexico and the subtropical Atlantic there is likewise an increase in the annual rainfall, but the profiles are variable and complicated. Along the Gulf and subtropical Atlantic margins of the continent the profiles resemble those of the interior in that there is a strong maximum in the warm season. But in contrast to the interior, winters here are relatively wet and a secondary maximum in early spring is fairly characteristic. Between the interior and the subtropical coasts with their summer maximum rainfall, is a region of variable and complex profiles, over most of whose area winter rainfall exceeds that of summer.

Type 4, focused in the Ohio River drainage basin, is less a discrete type of annual rainfall variation than it is a collection of variable transition types. As a rule the 1,100 to 1,300 mm of precipitation are well distributed throughout the year, but usually there is some accent upon the warmer months (Fig. 19.1). Multiple small rises and falls in the profile are characteristic. The summer maximum becomes somewhat better defined toward the east and south with increasing proximity to the subtropical Atlantic coast.

TYPE 5. THE SUBTROPICAL INTERIOR

That this type has been recognized, although variously named, by a number of writers testifies to its distinctive character. By Greely and Ward it was labeled Tennessee type, by Henry it was designated as Southern Appalachian and Tennessee type, while more recently Crowe has assigned to it the name, Mississippi Lowlands type. Its anomalous character lies in the fact that here in an inland subtropical location on the leeward side of a great continent is an extensive region where the maximum rainfall coincides with the cooler months. Both to the north and west as well as to the south and east a summer maximum prevails. Moreover, in comparable latitudinal and geographical locations on the other continents, not only is a cool-season maximum absent, but the summer maximum is ordinarily strong, especially so in China. There can be no doubt that this area in subtropical southeastern United States has an exceptional rainfall profile.

But within this extensive region extending from eastern Texas to the western Carolinas, where the total winter rainfall exceeds that of summer, it is not true that winter is everywhere the season of maximum. Actually winter and spring are nearly on a par over extensive areas in terms of rainfall amounts. Employing 30-year averages (1920–1950) of rainfall, the ratios of winter to summer, and also of winter to spring, were determined and plotted on a map and isolines drawn. The resulting map shows that within a more extensive region where winter rainfall exceeds that of summer, there is a more restricted, but still large, core area where winter exceeds not only summer but also spring, and consequently is the wettest season (Fig. 19.2). Usually the winter excess over spring is not great, but in a very restricted area the ratio may rise to 130 percent. Surrounding the core area where winter is rainier than any other season, but still lying within the more extensive region where winter's rainfall exceeds summer's, is a belt of variable width where spring is the rainiest season.

This unique rainfall region with its cool-season maximum is approximately delimited also by a line bounding the area where more than 50 percent of the year's rainfall occurs in the cooler half-year, or where more than 35 percent falls in the four months December to March inclusive (Fig. 19.3). It is likewise conspicuous on the phase-angle map of

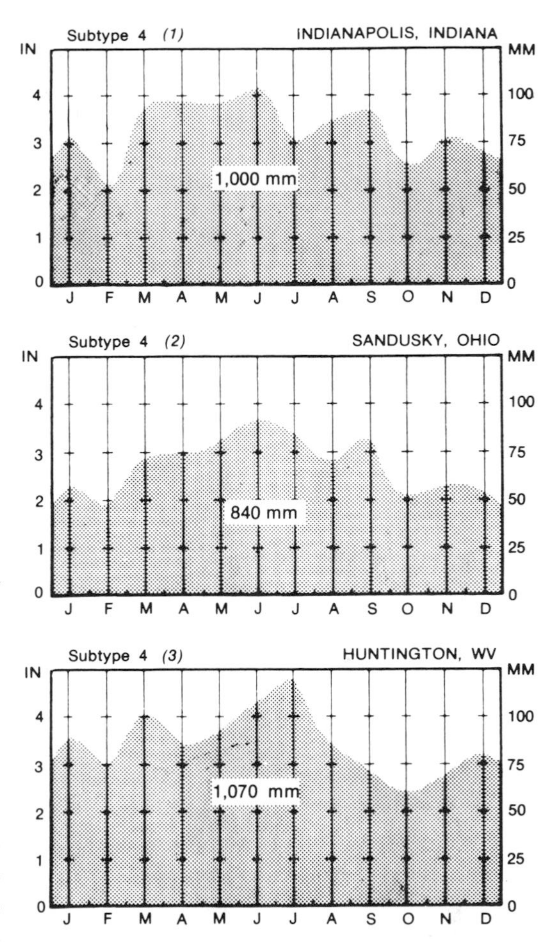

Fig. 19.1. Annual march of rainfall in Indianapolis, Indiana; Sandusky, Ohio; and Huntington, West Virginia.

the first harmonic as defined by the isochrones representing a February–March maximum of rainfall (Horn, Bryson, and Lowry, 1957). A close packing of the isochrones to the northwest and southeast of the core area represents the rapid transition to a summer maximum in both directions. In combination these several maps, although based upon different criteria, which result in somewhat contrasting boundaries, serve to establish the approximate location of a cool-season rainfall type in the interior subtropics of eastern United States. The representative rainfall profiles for different parts of this ex-

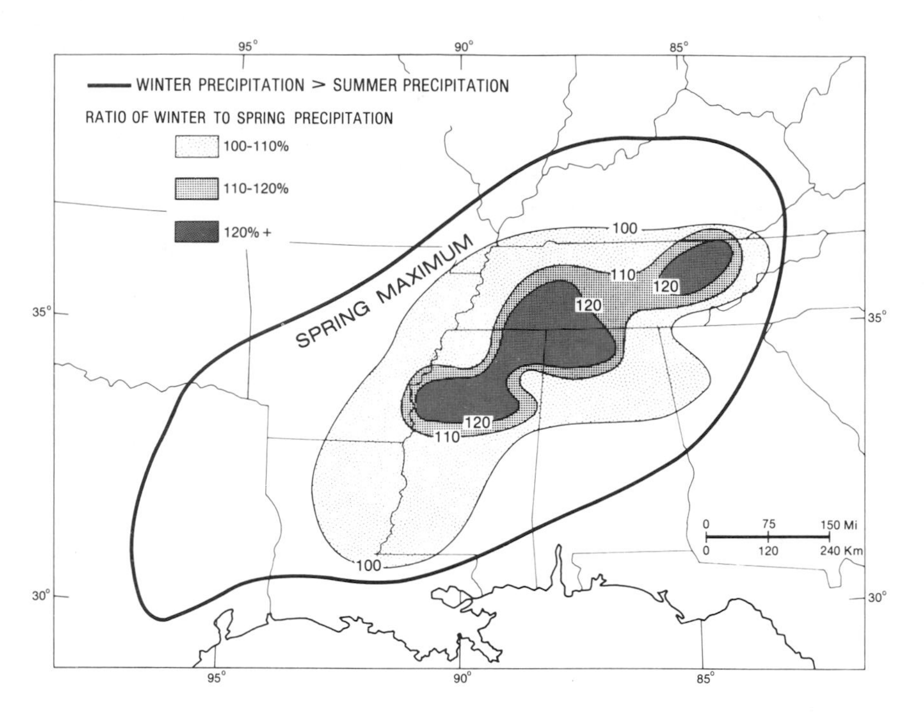

Fig. 19.2. Showing the location of an extensive region in the subtropical southeastern United States where winter precipitation exceeds that of summer. Winter and spring are the wettest seasons, with winter exceeding spring in the central part of the region thus defined.

tensive area are by no means strongly similar, and usually they are complex.

Four subtypes are recognized. In Subtype 5a, the core area, the principal maximum is in winter and early spring, with March commonly the wettest month (Fig. 19.4). A secondary peak is usually evident in midsummer. There is no dry season but the most pronounced minimum is in fall with a secondary in May-June. This annual rainfall profile is not unlike that of the Gulf coast just to the south, except that the latter's strong primary summer maximum here is distinctly reduced. Some control appears to dampen the summer rain-making processes going northward along a line from Pensacola, Florida, to Chattanooga, Tennessee. Subtype 5b is similar to 5a except that the summer secondary peak is weak or even absent. Maxima are in spring and in winter with a principal minimum in fall-summer, and a brief secondary in February (Fig. 19.5). Subtype 5c, transitional in character, shows a primary late-spring (May) maximum and a secondary in early fall. Minima

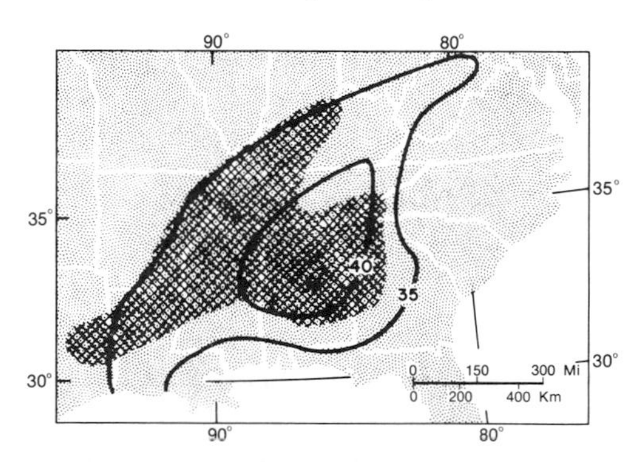

Fig. 19.3. Within the heavily shaded area less than 50 percent of the annual precipitation occurs in the warmer half year, April–September inclusive. Numerals on the solid line indicate the percentage of the annual precipitation occurring in the four months December to March inclusive. (After Kincer, 1922, and Visher, 1954.)

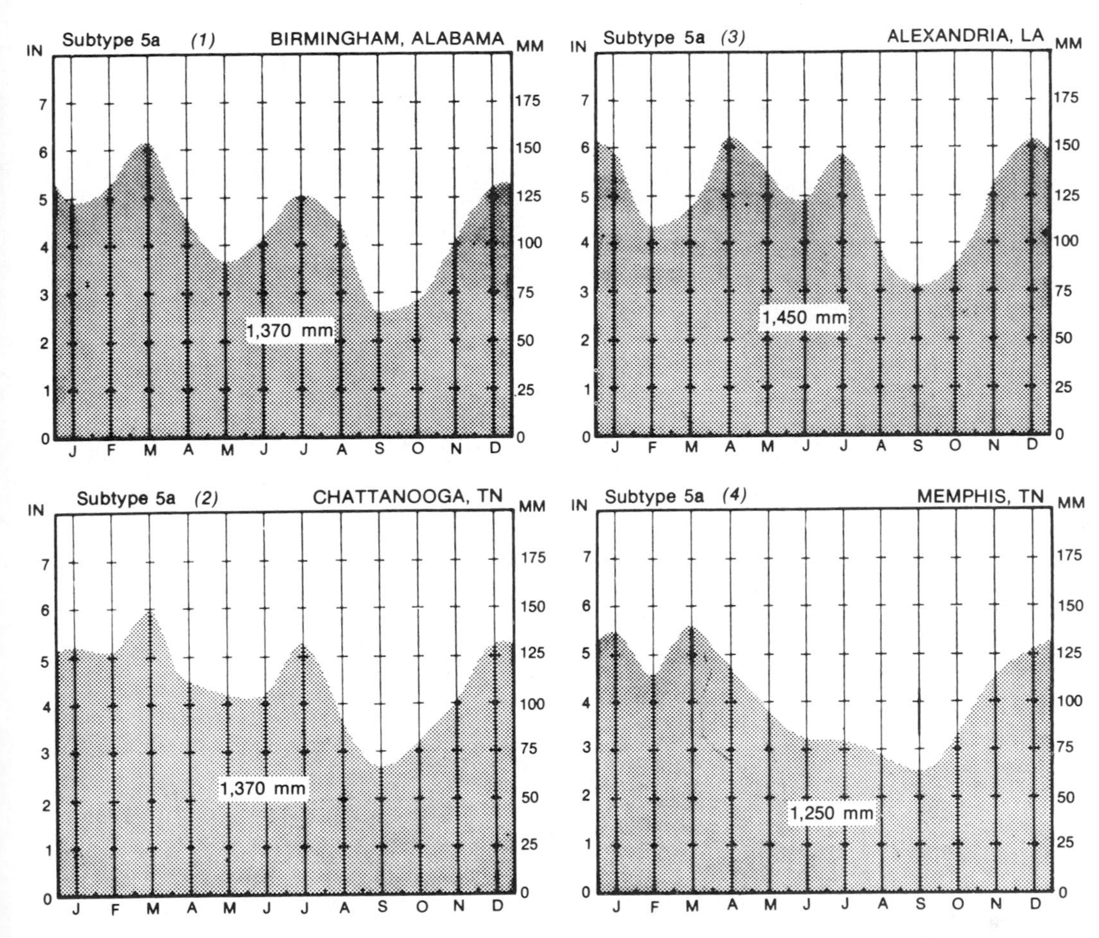

Fig. 19.4. Annual march of precipitation in Birmingham, Alabama; Chattanooga, Tennessee; Alexandria, Louisiana; and Memphis, Tennessee.

occur both in midsummer and midwinter (Fig. 19.6). Compared with 5a and 5b, winter rainfall shows a large decline.

In the explanatory section to follow attention will be focused on the core area 5a, characterized by a primary winter-early spring maximum, and a secondary in midsummer. Much of this region is coincident with the area of heaviest winter rainfall in the United States east of the Rocky Mountains. Subtypes 5b and 5c lie within the area of closely packed isochrones on the first harmonic phase-angle map and emphatically are transitional types in the zone of rapid change toward the interior region of dominantly summer rainfall.

Several fundamental questions require answers. (1) What causes such a rapid decline in summer precipitation inland from the northern Gulf coast toward Subtype 5a, so that what was a strong July-August primary maximum at Pensacola on the coast with about 200 mm of rainfall in each of these two summer months, has become only a secondary maximum a few hundred kilometers inland with the consequence that at Chattanooga these same months record only 100 to 130 mm? (2) Why is the winter-spring pre-

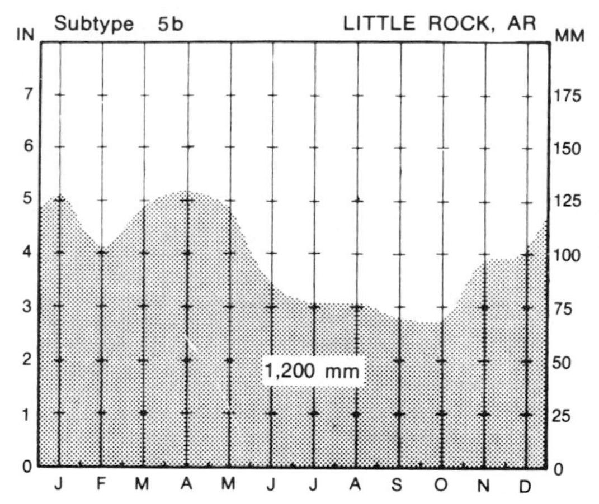

Fig. 19.5. Annual march of precipitation in Little Rock, Arkansas.

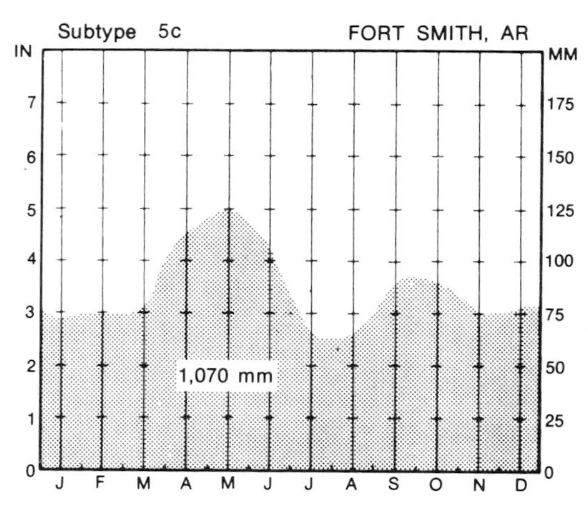

Fig. 19.6. Annual march of precipitation in Fort Smith, Arkansas.

cipitation so heavy in this subtropical southeastern region, with no decrease inland from the coast? Thus, while the isohyets are closely spaced and parallel to the Gulf coast in summer, representing a rapid decline in rainfall inland, such is not the case in winter and spring, the interior having quite as much rainfall as the coast at these seasons, if not more. (3) What are the causes for a primary minimum in September-October and a lesser minimum in late spring?

One reason for the decline in summer rainfall inland may be found in the warm-season location of the mid-troposphere troughs and ridges and their associated wet and dry tongues of air as described by Wexler and Namias (1938; Wexler, 1943). According to their analysis the eastern Gulf coast and Florida in July-August are located underneath a trough and its deep moist tongue of air which provide a favorable environment for heavy convective rainfall (Fig. 18.4). But in a northwesterly direction from the Gulf coast the effects of the moist tongue become progressively weaker and those of the mid-continent pressure ridge and its dry air stream increase. Throughout July-August most of the region included within Type 5 lies on the fluctuating southeastern margins of the mid-continent dry tongue and experiences its damping effects upon convective overturning. The relative positions of the wet and dry tongues over southern United States in summer are reflected in the rapid decline in summer rainfall inland from the Gulf, and a similar decline in the number of summer thunderstorms (Wexler, 1943) (Figs. 18.5, 18.6).

Computed divergence by the Bellamy method reveals significant contrasts between the coastal margins and the subtropical interior in July. In two triangles within Subtype 5a that were employed, one in the west and another in the east, the one showed only weak convergence at lower levels but strong convergence at levels over 4,600 m, while in the other triangle there was positive divergence throughout the air column, but weaker at the surface than aloft. Such a vertical structure does not represent as favorable a condition for convective overturning as does that over the Florida Gulf coast where there exists stronger convergence at all levels up to about 3,000 m.

Auxiliary causes for the strong decline in summer thunderstorms and summer rainfall inland from the coast may be the sharp decline in the effects of afternoon sea-breeze convergence away from the coast, and likewise the rapid deterioration inland of tropical disturbances that advance westward along the northern Gulf coast in the easterly circulation of the warm season (Riehl, 1947). Hosler (1956) indicates that the easterly wave is a dominant factor in the climatology of the whole Gulf coast from Florida to Texas in summer, but the rainfall effects of this disturbance decline both westward and inland.

The heavy winter-early spring rainfall characteristic of subtropical southeastern United States is of the general widespread type associated with traveling depressions. Such disturbances are at their maximum frequency and intensity over the Gulf states in the period December through March. Not only are the depressions numerous and well developed at this time, but because of the proximity of the moisture source provided by the Gulf of Mexico, the individual winter disturbance in the southeast yields more precipitation than its counterpart farther north and therefore farther removed from the primary source of *mT* air. It is these storms which result in the southeast being not only the region of heaviest winter rainfall in the United States east of the Rockies, but also the area of greatest frequency of heavy prolonged winter rains and of the heaviest rainfall on each winter day having precipitation (Visher, 1954). Commonly appearing first in the Texas region, these cool-season disturbances follow a northeastward route which concentrates their rainfall along the northern Gulf coast and in the region included in Subtype 5a just to the north. Because the storms take a route to the northeast instead of eastward to cross the south Atlantic seaboard, Florida and the coastal region south of Cape Hatteras feel their effects less than do the northern Gulf coast and the subtropical interior, and as a consequence have less winter rainfall. Westward from Louisiana the winter rains diminish sharply, for here the depressions are still in their incipient stages and have not had the benefit of a long continued invasion of humid Gulf air.

The frequency of well-developed depressions in these subtropical latitudes in the cooler months is related to a number of circumstances. Surface temperature gradients are unusually steep over the region and the oscillating polar front is positioned in the same vicinity (Brunnschweiler, 1952). Isotachs of the mean resultant geostrophic winds at the 500 mb level indicate that the core area of highest velocity winds (the jet stream) in North America is positioned over southeastern United States from December through March, with the farthest southward displacement of the jet occurring in March, which is characteristically the month of heaviest precipitation (Lahey et al., 1958). Significantly the alignment of the isotachs of the resultant geostrophic winds in March at the 500 mb level is such that the area of highest wind velocity extends in a SW-NE direction from the Texas Gulf coast to about Cape Hatteras. It is clear that the concentration of winter depressions and their heavy rainfall closely coincide with the position of the jet stream aloft, suggesting that the jet plays a significant role in originating and steering the perturbations. Corroborating the evidence above is the fact that on the 500 mb wind maps there is a high standard deviation of the daily value of the south-north component of the geostrophic wind during winter and early spring (Fig. 19.7). This strong north-south mixing is in keeping with the concept of the cool months as the period of maximum cyclonic and frontal activity.

Fall, the period of least rainfall, sees the retreat of the polar front, major storm tracks, and jet stream to their northernmost position and hence farthest removed from the subtropical southeast. On the 500 mb charts, resultant geostrophic winds over southeastern United States in fall have speeds only one-fifth to one-third as great as they are in

March, while the standard deviation of the daily value of the north-south component of the geostrophic wind is more than halved, indicating a weakened cyclonic control (Fig. 19.7).

From an analysis of the 5-day surface charts it can be observed that fall, the period of the primary rainfall minimum, is characterized by a westward extension of the Atlantic high with frequently a separate and distinct cell centered over region 5a. In wetter July, by contrast, the Atlantic high barely reaches into the same area.

Tabulations of the frequency of occurrence of certain synoptic types occurring within region 5a, as observed on the daily surface synoptic chart, show significant correlations with wet and dry seasons (see Table 19.1). Thus, there is a maximum frequency of highs and a minimum of fronts during September, the driest month. By contrast, March, the wettest month, experiences a minimum of highs, a strong maximum of warm fronts, and about the same number of "all fronts" as January, another wet month. July, the wettest month of summer, holds an intermediate position in number of highs and fronts between wet March and relatively dry September.

Type 6. The Subtropical Oceanic Margins, Atlantic and Gulf

Along the subtropical Atlantic and Gulf margins of the United States from southern New Jersey in the northeast to about the Texas-Louisiana boundary in the southwest, an abundant annual rainfall of 1,100 to 1,500 mm is concentrated in the warm season. In spite of important intraregional variations, the primary summer maximum is common throughout. Thus, summer, the season of only a secondary maximum in the subtropical interior (Subtype 5a), becomes the season of dominant rainfall along the oceanic margins. The annual profile of rain-

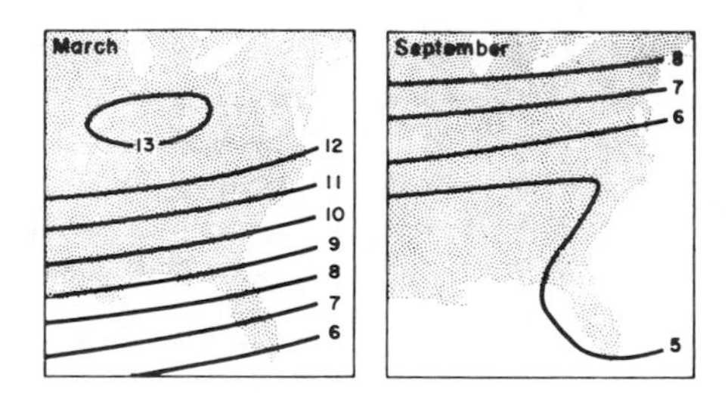

Fig. 19.7. Standard deviation of the daily value of the south-north component of the geostrophic wind in m/sec at the 500 mb level in the eastern United States. (After Lahey et al., 1958.)

fall occurrence, or rainy days, likewise shows a primary summer maximum, but it is not as emphatic as in the profile of rainfall amounts, indicating that the showery convective rainfall of summer produces somewhat more rainfall per rainy day than do the more general cyclonic rains of winter. No other part of the country has such abundant warm-season precipitation, while winter is likewise wet.

Because of its very southerly position this region has a marginal location between the easterly flow of the tropics and the westerly flow of the middle latitudes, and experiences the weather features of both. Westerly control dominates a greater share of the time, however. This is prevailingly the case in the cooler seasons, but even in summer, except in peninsular Florida and the southern tip of Texas, the westerly days are two and one-half to three times as numerous as are the easterly or tropical days (Riehl, 1947). Over

Table 19.1.
Frequency of occurrence of certain synoptic types over the state of Tennessee for the period 1925–1935, expressed in percentage of the total number of days

	Highs	All fronts	Warm fronts
July	59	32	20
September	80	17	8
January	32	41	16
March	42	40	27

Florida the activity of the westerlies and their disturbances diminishes rapidly in May and June, the upper trough which earlier was located east of the 80–85th meridians shifts westward into the Gulf of Mexico, and the wind aloft backs from northwest to southwest or southeast. Coincident with these changes the summer rainy season of maximum thunderstorm activity begins. The reverse windshift from south to northwest occurs over Florida in late September and October as the cooling land permits the westerly flow to move southward. Concurrently the main summer rainy season in Florida comes to an end.

Throughout the summer period the weather of this subtropical region is conditioned by the presence of a deep current of humid tropical air aloft associated with an upper trough which is positioned over the area (Wexler, 1943; Riehl, 1947) (Fig. 18.4). Most of the abundant precipitation is of the shower type, frequently accompanied by thunder and lightning, Florida and the northern Gulf coast possessing the distinction of having one of the highest frequencies of thunderstorm occurrence for any part of the earth. A marked diurnal periodicity characterizes the shower activity, daylight hours, and especially those of the afternoon, being favored. This fact immediately suggests that daytime surface heating is an important factor in their genesis. But while insolational heating is an essential part of the convective process, it is, on the other hand, not the only one. There are numerous days during which there is strong surface heating of the deeply humidified *mT* Gulf air when showers and thunderstorms are lacking. In addition to insolational heating, dynamically induced low-level horizontal convergence is likewise an essential element in thunderstorm formation. This may be provided by directional or speed convergence associated with the sea breeze, which similarly is diurnal in its occurrence, or by disturbances of tropical or middle-latitude origin. Thus, there is superimposed on the diurnal pattern of shower and thunderstorm occurrence associated with daytime heating and sea breeze, still another of an aperiodic character related to various synoptic patterns which produce sustained convergence and divergence situations sometimes extending over several days. Apparently some of the disturbances inducing convergence are so weak that present-day methods of measurement are not capable of detecting them.

Subtype 6a includes chiefly the northern Gulf coast, excluding peninsular Florida, as far west as the Texas-Louisiana boundary. Here there prevails a biannual rainfall profile characterized by a primary maximum in summer (July-August) and a secondary maximum in the cooler months, which rather consistently reaches a peak in March (Fig. 19.8). Inland from the coast the summer maximum decreases in prominence so that not a few stations have a March peak as high as that of summer. One element favoring the summer maximum is the positioning over the extreme southeast of a trough aloft, together with its tongue of deep, moist, southerly air, as described by Wexler and Namias (1938), and elaborated in an earlier section (Figs. 18.4, 18.5, 18.6). This situation creates a general environment favorable for strong shower activity in neutral or unstable Gulf air during the hours of maximum surface heating. That this summer maximum is strongest near the coast and declines inland confirms the concept of the sea breeze as an important element making for low-level convergence during the warmer hours. The sea breeze appears to cause a modest maximum of convergence in the lower 600 m of atmosphere in the late afternoon, although it is not so strong here as in peninsular Florida. It is likewise along the littoral that the rainfall effects of summer tropical disturbances, especially easterly waves, are concentrated. The secondary maximum in winter-early

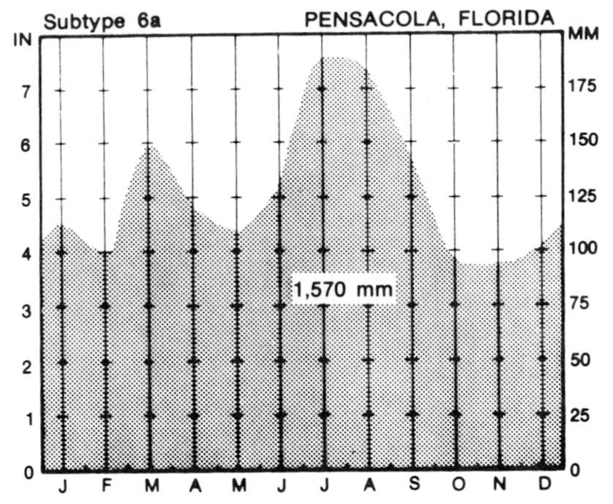

Fig. 19.8. Annual march of precipitation in Pensacola, Florida.

spring reflects the operation of controls similar to those that produce a primary maximum at the same period in the interior subtropical region just to the north (Subtype 5a) which has been described previously.

Subtype 6b, confined largely to peninsular Florida, has an annual profile which contrasts with that of Subtype 6a in important respects: the biannual feature has largely disappeared; a much larger percentage of the year's total rainfall is concentrated in the months June to September inclusive; and winter rainfall has declined in relative and absolute importance (Fig. 19.9). The broad, summer rainfall peak characteristically remains at the high level of 180–200 mm a month over a period of about one-third of the year, while along the northern Gulf coast the summer peak is distinctly narrower. At Orlando, Florida, the four warm months account for two-thirds of the year's total rainfall. In this respect it is worthy of note that the shift in fall from a tropical to a middle-latitude circulation occurs a month or more later in Florida than it does along the Gulf and subtropical Atlantic coasts farther north, so that in Florida rains continue with summer vigor throughout September. Over a period of 15 years Florida experienced in September a total of 40 deep lows and hurricanes, while the northern Gulf coast was affected by only 11 and the Atlantic coast south of Delaware Bay by only 8. The three winter months provide only 10 to 15 percent of the year's rainfall. Moreover, the March peak has largely disappeared and is obvious in rudimentary form only in extreme northern Florida.

It is to the greatly increased thunderstorm activity in peninsular Florida that the region's unusually heavy summer precipitation is largely due. In interior Florida thunderstorm frequency reaches a maximum for any part of the United States and possibly for the entire earth. Byers ascribes this phenomenon to the special conditions prevailing over the peninsula which feels the effects of strong low-level convergence resulting from the afternoon sea breeze moving into the peninsula from both east and west (Byers and Rodebush, 1948). Such a two-sided sea breeze acts to produce a maximum of convergence in the afternoon over interior locations, resulting in a higher moisture content aloft, which in turn stimulates the vertical growth of convective clouds. Computed

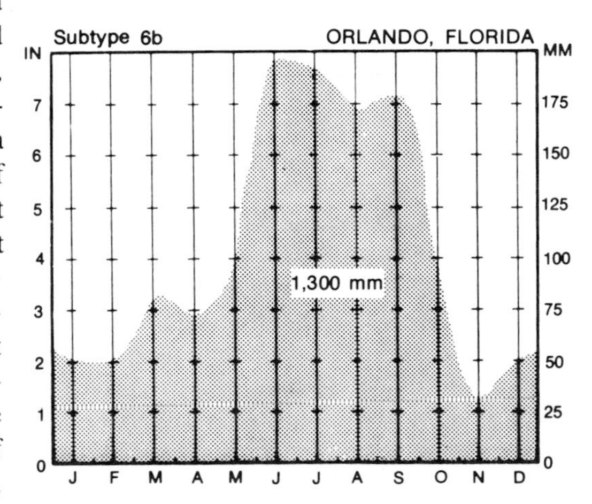

Fig. 19.9. Annual march of precipitation in Orlando, Florida.

convergence following the Bellamy method indicates that a strong afternoon convergence prevails up to an altitude of 1,100 to 1,200 m, with a maximum reached in the late afternoon when the effects of the double sea breeze should be strongest. By contrast, the single sea breeze along the northern Gulf coast is able to produce only a weaker and shallower convergence which results in appreciably fewer thunderstorms and less summer rainfall.

It has been pointed out by Day (1953), however, that the very numerous thunderstorms at Miami on the southeast coast of Florida cannot be attributed to a double sea-breeze convergence. As a consequence the Miami area does not have as many afternoon thunderstorms as the interior, but on the other hand it does have a larger percentage which occur during night and morning hours. Also, the southeast coast has days with little or no rain when the diurnal precipitation pattern of the interior remains unchanged. This fact suggests a modification of the genetic concept of interior thunderstorms as described by Byers and Rodebush (1948), and seems to indicate a mechanism which combines sea-breeze effects with those of traveling synoptic features, some of which are impossible by present methods to detect. At Miami, in addition to a diurnal oscillation of convergence during the summer months, it is found that wet-day curves of convergence show an exaggeration of the normal diurnal curves, with an excess of convergence during the afternoon hours. But there likewise exist patterns of convergence-divergence of a larger scale extending over as much as a 5–7 day cycle.

The decreased winter preciptation in peninsular Florida as compared with that of the northern Gulf coast, may be attributed to the fact that the former's more southerly location causes it to be less influenced by the westerly circulation of winter. Those well-developed winter and early-spring depressions moving east and northeast from the Texas area, and delivering heavy rainfall to the northern Gulf coast and the subtropical interior, are too distant from peninsular Florida to produce similar amounts of rainfall there. Most of the Texas disturbances as they move away from their region of origin normally follow a route which bends to the northeast, so that the extreme southeast in Florida tends to escape their principal effects. Thus, the frequency of January cyclones in Florida is only one-third what it is in western Louisiana.

Subtype 6c, characteristic of Virginia and the Carolinas east of the Appalachians, has an annual rainfall profile which resembles that of Subtype 6a, except that there is somewhat less winter precipitation and the secondary March peak is lacking or present only in rudimentary form. A summer maximum is striking, but it is a sharper and narrower peak than that of Florida, while its winter rainfall is more abundant (Fig. 19.10).

The origin of the summer maximum is related to the same causes as were described earlier for the northern Gulf coast (Subtype 6a), for the upper-level trough and its deep current of humid tropical air similarly control the summer weather of this region. Sea-breeze effects are likewise present, stimulating convective activity.

That winter-early spring precipitation is in excess of that in peninsular Florida reflects the closer proximity of Subtype 6c to the routes of winter storms moving along a northeasterly course from the Texas area of origin. On the other hand, the decreased winter, and especially March, precipitation, compared with that in Subtypes 6a and 5a, appears to be a consequence of the latter areas lying more directly in the paths of the winter lows from the southwest.

Subtype 6d, whose location is Middle Atlantic, is transitional between the subtropical

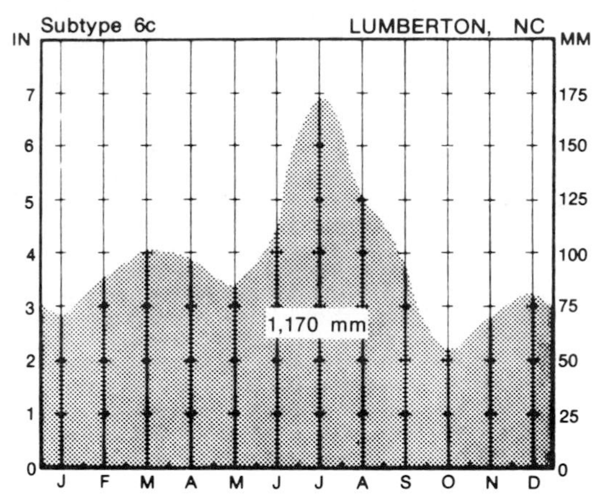

Fig. 19.10. Annual march of precipitation in Lumberton, North Carolina.

Atlantic region with its strong summer maximum, and the New England region where seasonal accent is largely lacking. In 6d the summer maximum is still present, but is far less conspicuous than it is in the other subdivisions of Type 6 farther south. Total annual precipitation is likewise somewhat less (Fig. 19.11). The waning summer maximum reflects the fact that in the hot months the Middle Atlantic lies north and west of the upper trough and its associated deep stream of humid southerly air, which is so influential in creating an environment favoring summer precipitation in the subtropical regions farther south. Thunderstorms are only one-third to one-half as numerous as they are along the Gulf and subtropical Atlantic coasts, while the tropical storms of late summer are also fewer and less effective rainbringers.

Type 7. The Eastern Region

This type, including New England and the Atlantic provinces of Canada, together with parts of New York State and the provinces of Quebec and Ontario, is characterized by moderate precipitation amounts well distributed throughout the year and usually without marked seasonal accent. It is the latter feature, viz., an annual precipitation profile lacking conspicuous seasonal peaks and depressions, and which in marine locations even exhibits a modest winter maximum, that chiefly distinguishes this type from the neighboring types (Fig. 19.12). No other extensive region in Anglo-America manifests such seasonal monotony in its annual precipitation profiles. Even more noteworthy is the striking contrast of this type's station profiles, some of which show a cool-season precipitation concentration, with the representative profiles characteristic of the eastern continental margins of Asia in similar latitudes, where there is a pronounced summer maximum and winter minimum. In the latter area the much stronger and persistent winter anticyclone has the effect of markedly lowering the winter precipitation. Other than in Type 5, Type 7 is the only one east of the Rockies in which there are fairly extensive areas where winter is wetter than summer.

An explanation of the moderately uniform seasonal distribution of rainfall characteristic of most parts of Type 7 is to be found in the annual march of the weather elements. Thunderstorm days here reach a minimum

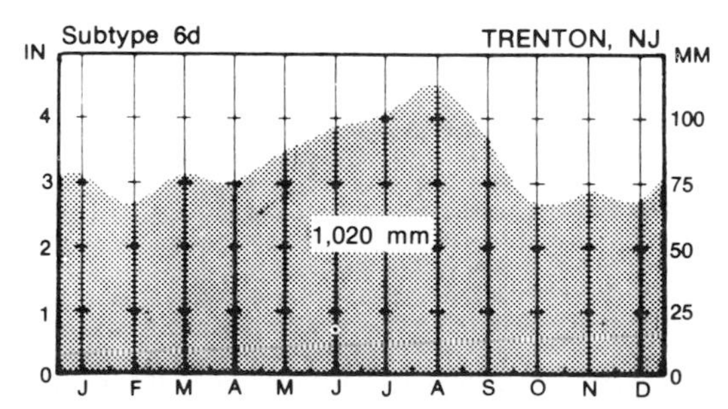

Fig. 19.11. Annual march of precipitation in Trenton, New Jersey.

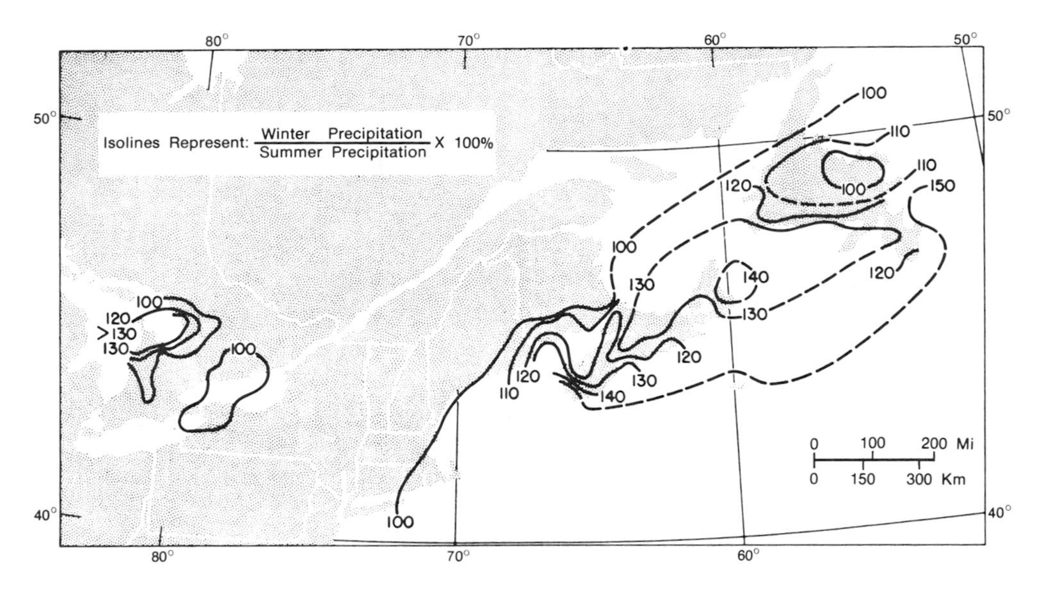

Fig. 19.12. Showing the location of areas in the Atlantic seaboard and Great Lakes regions where winter precipitation exceeds that of summer.

for any part of Anglo-America east of the Rocky Mountains in similar latitudes, eastern New England recording fewer than 20 a year. Northerly location, cool Atlantic waters offshore, a deeply indented coastline, and possibly the marine influence of the Great Lakes in summer, all act to inhibit the convective processes and consequently to reduce the amount of summer shower precipitation. Thus, the warm-season rainfall of maritime New England is no greater than that of the Great Plains of the Dakotas, Kansas, and Nebraska where the total annual precipitation is only half as much. As compared with the interior of the country, a much larger proportion of the annual precipitation of New England comes in the cooler seasons and is derived from cyclonic disturbances, which likewise continue to be relatively numerous even in summer. In this respect it is climatically significant that storm tracks which have their origins in almost any latitude of the far west or the interior tend to converge over the Great Lakes–New England area, so that there is a marked bunching of the tracks over this northeastern part of the United States and adjacent parts of Canada. It is for this reason that cool-season precipitation is relatively abundant, cloudiness high, and snowfall relatively great.

The Atlantic and northern Gulf of Mexico coasts of Anglo-America and the adjacent oceans comprise one of the two most favored regions of winter cyclogenesis in the Northern Hemisphere. This is especially true of that section of the Atlantic coast and ocean between Cape Hatteras and New England. In winter this region is proverbially stormy and becomes the locus of a major winter cyclone track. Here the generation of cyclones is often so rapid that what is in the beginning a weak weather disturbance may be transformed into a major storm within 12 to 24 hours. Winter in this region is characterized by great temperature contrasts between a warm sea, where the Gulf Stream parallels the coast, and a cold continent. During outbreaks of cold *CP* air there takes place a huge transfer of sensible heat at the air-sea interface. And it has been shown that there is a maximum transfer of sensible heat from warm ocean to frigid atmosphere and a release of latent heat just before and during rapid cyclogenesis (Min and Horn, 1974).

Subtype 7a, including most of New England and reaching inland along the Mohawk depression to the Lake Ontario lowlands, and including much of peninsular Ontario in Canada, has an annual rainfall profile which is conspicuously flat (Fig. 19.13). A slight cool-season maximum is observable along the extreme Atlantic margins of New England but this feature quickly disappears inland. Likewise along the eastern shores of Lakes Huron and Ontario and of Georgian Bay it is common for winter precipitation to exceed that of summer, the ratio rising to 120–130 percent in a few restricted spots.

Subtype 7b includes the most maritime sections of Atlantic Canada, where the excess of winter precipitation over that of summer is relatively marked, more so than along the New England coast farther south (Figs. 19.12, 19.14). Limited areas along Georgian Bay in the Ontario Peninsula would also qualify for inclusion in Subtype 7b. In parts of southeastern Nova Scotia and southeastern Newfoundland the ratio of winter to summer precipitation reaches values in excess of 130 percent. Obviously the winterfall storms, especially those moving northeastward along the coast, bring heavier winter precipitation to the more exposed coastal sections than to the interior. Throughout those parts of the maritime northeast where winter's precipitation exceeds summer's, as often as not fall is as wet as or wetter than winter.

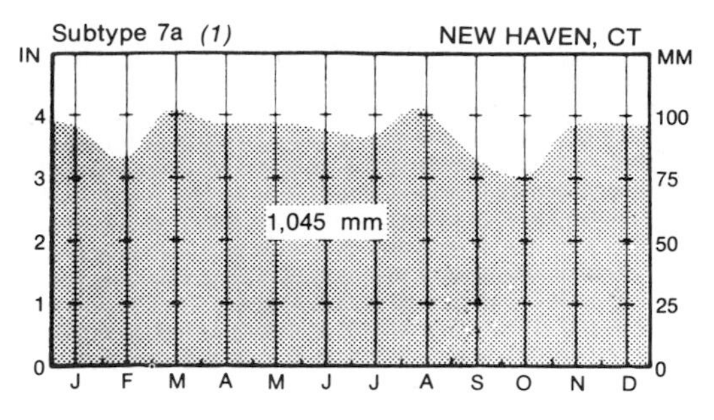

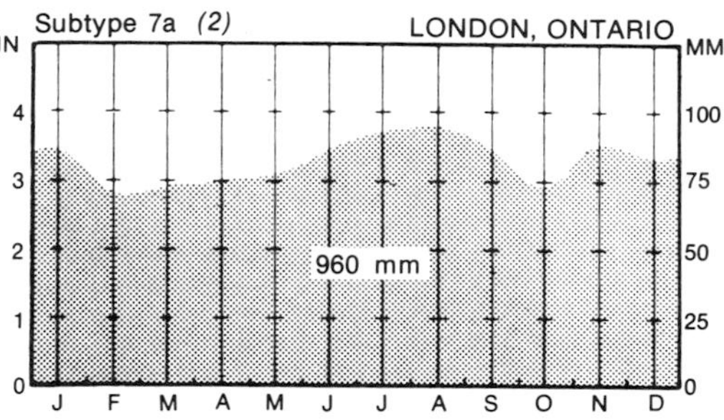

Fig. 19.13. Annual march of precipitation in New Haven, Connecticut, and London, Ontario.

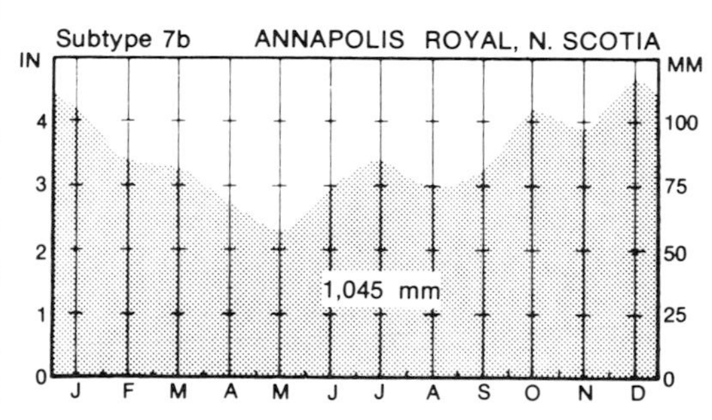

Fig. 19.14. Annual march of precipitation in Annapolis Royal, Nova Scotia.

Subtype 7c, located in northern Newfoundland and easternmost Labrador peninsula, has annual precipitation profiles which bear considerable resemblance to those of Subtype 7a, where seasonal uniformity is conspicuous (Fig. 19.15). Local profile peculiarities are numerous, however.

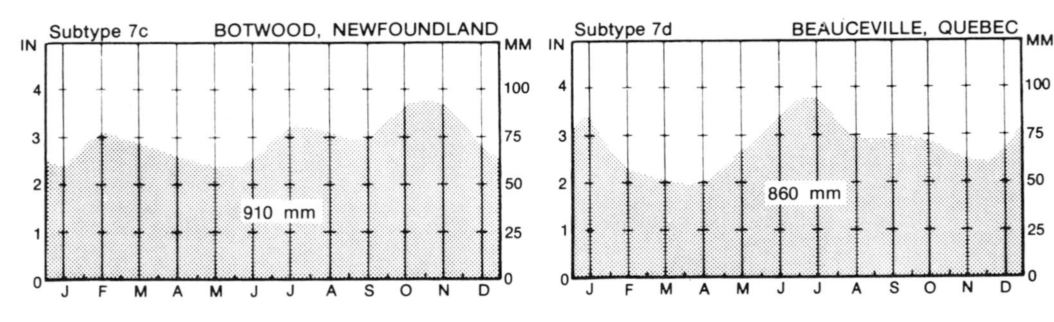

Fig. 19.15. Annual march of precipitation in Botwood, Newfoundland.

Fig. 19.16. Annual march of precipitation in Beauceville, Quebec.

Subtype 7d, mostly interior in location, is distinctly transitional, a majority of its station profiles exhibiting a warm-season precipitation maximum, but with not a few of them likewise manifesting an uptrend in winter resulting in a secondary maximum at that time (Fig. 19.16). The region included within Subtype 7d could, with almost equal logic, be included as a subtype within the great interior region designated as Type 3.

REFERENCE MATTER

Appendix

Köppen's Classification of Climates

Köppen recognizes five main groups of world climate which are intended to correspond with five principal vegetation groups. The five climatic groups, each designated by a capital letter, are as follows: *A*, tropical rainy climates with no cool season; *B*, dry climates; *C*, middle-latitude rainy climates with mild winters; *D*, middle-latitude rainy climates with severe winters; and *E*, polar climates with no warm season. Each of these in turn is subdivided into climatic types based upon the seasonal distribution of rainfall or the degree of dryness or cold. The small letters *f, s,* and *w* indicate the seasonableness of precipitation; no dry season (*f*); dry season in summer (*s*); dry season in winter (*w*). The capital letters *S* and *W* are employed to designate the two subdivisions of dry climate; semiarid, or steppe (*S*), and arid, or desert (*W*). Capital letters *T* and *F* are similarly employed to designate the two subdivisions of polar climate; tundra (*T*), and icecap (*F*). Table I.1 shows the Köppen scheme of five main climatic groups and eleven climatic types. The parentheses indicate that the combinations *As* and *Ds* rarely occur, and for this reason they are not recognized as among the principal climatic types.

A Climates

A = Tropical rainy climates; temperature of the coolest month above 18°C (64.4°F). With monthly temperatures lower than 18° C certain sensitive tropical plants do not thrive. Within the *A* group of climates two main types are recognized: one has adequate precipitation throughout the year, while the other includes a distinctly dry season which affects vegetation adversely.

Af = Tropical wet climate; *f*: rainfall of the driest month is at least 6 cm. Within this climate there is a minimum of seasonal variation in temperature and precipitation, both remaining high throughout the year.

Aw = Tropical wet-and-dry climate; *w*: distinct dry season in low-sun period or winter. A marked seasonal rhythm of rainfall characterizes *Aw* climates; at least 1 month must have less than 6 cm. Temperature is similar to that in *Af*.

Table I.1. Köppen's climatic groups and types

Climatic group	Symbol	Dry period	Degrees of dryness or cold	
Tropical rainy climates	*A*	*f (s) w*		
Dry climates	*B*		*S*	*W*
Mild temperate rainy climates	*C*	*f s w*		
Cold snow-forest climates	*D*	*f (s) w*		
Polar climates	*E*		*T*	*F*

Other small letters used with *A* climates are as follows:

m (monsoon) = short dry season, but with total rainfall so great that ground remains sufficiently wet throughout the year to support rainforest. *Am* is intermediate between *Af* and *Aw*, resembling *Af* in amount of precipitation and *Aw* in seasonal distribution.

w′ = rainfall maximum in autumn.

w″ = two distinct rainfall maxima separated by two dry seasons.

s = dry season during high-sun period (rare).

i = range of temperature between warmest and coldest months less than 5°C (9°F).

g = Ganges types of annual march of temperature; hottest month comes before the solstice and the summer rainy season.

B Climates

B = Dry climates in which there is an excess of evaporation over precipitation. No surplus of water remains, therefore, to maintain a constant groundwater level so that permanent streams cannot *originate* within *B* climates. There are two main subdivisions of *B* climates, the arid, or desert, type *BW* (*W* from the German word, *Wüste,* meaning "desert"), and the semiarid or steppe, type *BS* (*S* from the word *steppe*, meaning dry grassland).

BW = Arid climate or desert

BS = Semiarid climate or steppe.

Other small letters used with *B* climates are as follows:

h (heiss) = average annual temperature over 18°C (64.4°F). *BWh* and *BSh* therefore are low-latitude, or tropical, deserts and steppes.

k (kalt) = average annual temperature under 18°C (64.4°F). *BWk* and *BSk* therefore are middle-latitude, or cold, deserts and steppes.

k′ = temperature of the warmest month under 18°C (64.4°F).

s = summer drought; at least three times as much rain in the wettest winter month as in the driest summer month (see footnote under *Cs* climate).

w = winter drought; at least ten times as much rain in the wettest summer month as in the driest winter month (see footnote under *Cw* climate).

n (Nebel) = frequent fog. *BWn* and *BSn* climates are usually found along littorals paralleled by cool ocean currents.

C Climates (Mesothermal)

C = Mild temperate rainy climates; average temperature of coldest month 18°C (64.4°F) but above −3°C (26.6°F); average temperature of warmest month over 10°C (50°F). The average monthly temperature of −3°C (26.6°F) for the coldest month supposedly roughly coincides with the equatorward limit of frozen ground and a snow cover lasting for a month or more. Within the *C* group of climates three contrasting rainfall regimes are the basis for recognition of three principal climatic types: the *f* type with no dry season; the *w* type with a dry winter; and the *s* type with a dry summer.

Cf = no distinct dry season; difference between the rainiest and driest months is less than for *w* and *s*, and the driest month of summer receives more than 3 cm (1.2 in).

Cw = Winter dry; at least ten times as much rain in the wettest month of summer as in the driest month of winter.[1] This type of climate has two characteristic locations: (1) elevated sites in the low latitudes where altitude reduces the temperature of the *Aw* climates which prevail in the adjacent lowlands, and (2) mild middle-latitude monsoon lands of southeastern Asia, particularly northern India and southern China.

Cs = Summer dry; at least three times as much rain in the wettest month of winter as in the driest month of summer, and the driest month of summer receives less than 3 cm.[2]

[1] Alternative definition: 70 percent or more of the average annual rainfall is received in the warmer 6 months.

[2] Alternative definition: 70 percent or more of the average annual rainfall is received in the winter 6 months.

Other small letters used with *C* climates are as follows:

a = hot summer, average temperature of warmest month over 22°C (71.6°F).

b = cool summer; average temperature of warmest month under 22°C (71.6°F).

c = cool short summer; less than 4 months over 10°C (50°F).

i = same as in *A* climates.

g = same as in *A* climates.

x = rainfall maximum in late spring or early summer; drier in late summer.

n = same as in *B* climates.

D CLIMATES (MICROTHERMAL)

D = Cold snow-forest climates; average temperature of coldest month below −3°C (26.6°F), average temperature of warmest month above 10°C (50°F). The average temperature of 10°C for the warmest month approximately coincides with the poleward limits of forest. *D* climates are characterized by frozen ground and a snow cover of several months' duration. Two principal subdivisions of the *D* group are recognized: the one, *Df*, with no dry season, and the other, *Dw*, with dry season in winter.

Df = Cold climate with humid winters.

Dw = Cold climate with dry winters; characteristic of northeastern Asia, where the winter anticyclone is well developed.

Other small letters used with *D* climates are as follows:

d = average temperature of coldest month below −38°C (−36.4°F).

f, *s*, *w*, *a*, *b*, and *c* are the same as in *C* climates.

E CLIMATES

E = Polar climates; average temperature of the warmest month below 10°C (50°F). In the higher latitudes, once the temperatures are well below freezing and the ground frozen it makes little difference to plant life how cold it gets. Rather, it is the intensity and duration of a season of warmth which is critical. For this reason a warm-month isotherm is employed as the poleward boundary of *E* climates. Two climatic subdivisions are recognized: one, *ET*, in which there is a brief growing season and a meager vegetation cover, and the other, *EF*, in which there is perpetual frost and no vegetation.

ET = Tundra climate; average temperature of warmest month below 10°C (50°F) but above 0°C (32°F).

EF = Perpetual frost; average temperature of all months below 0°C (32°F). Such climates persist only over the permanent icecaps.

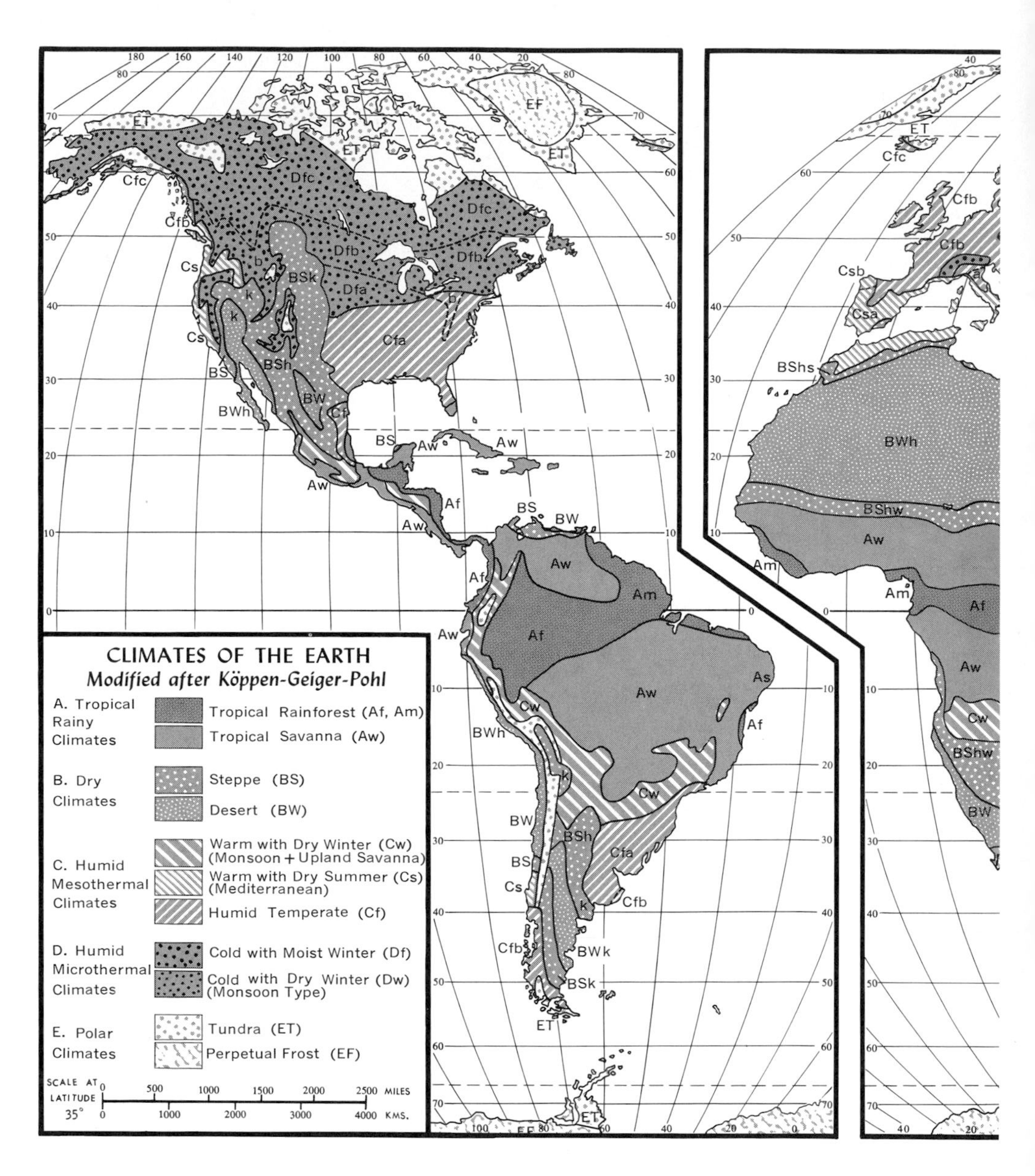

I.1 Köppen system of world climates

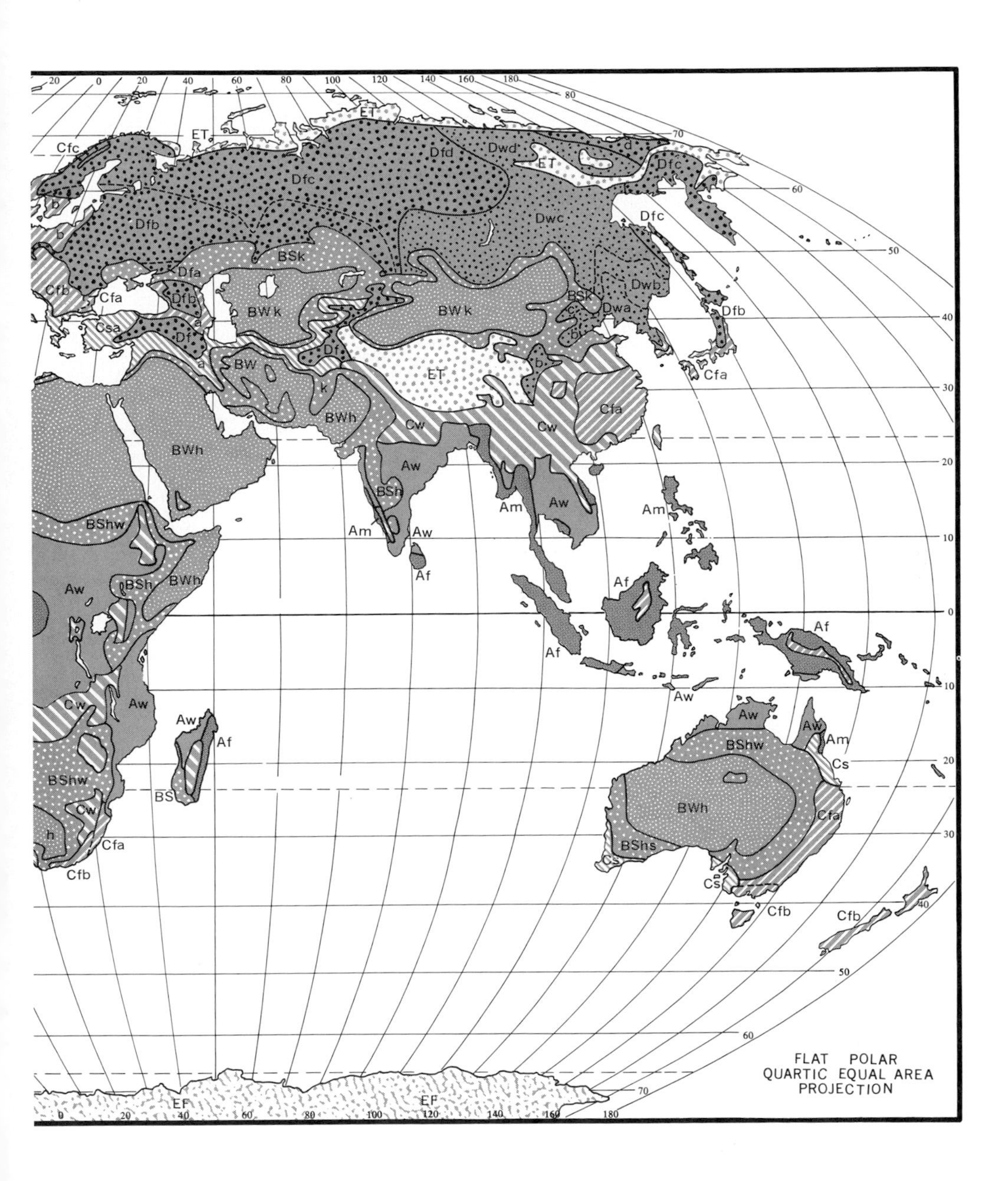
ET
Cfc
Dfd
Dwd
Dfc
Dfb
Dwc
BSk
Dfa
Cfb
Cfa
Dwb
Dwa
BWk
Csa
Df
BW
ET
BWh
Cw
Cfa
Aw
BSh
Am
Af
BShw
Cs
BShs
Cfb
BS
EF
FLAT POLAR
QUARTIC EQUAL AREA
PROJECTION

References

Alexander, Charles S. 1958. The geography of Margarita and adjacent islands, Venzuela. Univ. Calif. Publ. Geograph., 12, No. 2: 85–192.

Alpert, Leo. 1948. Notes on the areal distribution of annual mean rainfall over the tropical eastern Pacific Ocean. Bull. Am. Meteorol. Soc., 29: 38–41.

Alt, E. 1932. Klimakunde von Mittel- und Südeuropa. *In*, Handbuch der Klimatologie, W. Köppen and R. Geiger, Eds., Band 3, Teil m. Gebrüder Borntraeger, Berlin.

Amiran, D., and A. Wilson, Eds. 1973. Coastal deserts, their natural and human environments. Univ. of Arizona Press, Tucson.

Akiyama, Tokako. 1975. Southerly transversal moisture flux into the extremely heavy rainfall zone in the bai-u season. Jour. Meteorol. Soc. Japan, 53: 304–15.

Aperçus sur la climatologie de l'A.E.F. 1956. Monógraphies de la méteorologie nationale, Paris, No. 1: 1–4.

Arias, Alfonso Contreras. 1958. Bosquejo climatologica. *In*, Los recursos naturales del sureste y su aprovechamiento, II Parte, Tome 2, Instituto Mexicano de Recursós Naturales Renovables, A.C.: 95–158.

Arroyo, Elías Almeyda. 1955. Chilean temperature anomalies. Geograph. Rev., 45: 419–22.

Asai, T., and E. Fukui, 1977. The short dry period in mid-summer. *In*, The Climate of Japan, E. Fukui, Ed., Developments in Atmospheric Science, no. 8, Kandansha Ltd., Tokyo and Elsevier, New York: 103–12.

Asakura, T. 1957. A synoptic study on the seasonal change of the general circulation from summer to "Shurin" in 1951. Jour. Meteorol. Soc. Japan, 35: 278–87.

———. 1971*a*. Transport and source of water vapor in the northern hemisphere and monsoon Asia. *In*, Water balance in monsoon Asia, M. Yoshino, Ed., Univ. of Hawaii Press, Honolulu: 27–51.

———. 1971*b*. Distribution and variation of cloudiness and precipitable water during the rainy season over monsoon Asia. *In*, Water balance of monsoon Asia, M. Yoshino, Ed., Univ. of Hawaii Press, Honolulu: 131–51.

Asnani, G. C. 1968. The equatorial cell in the general circulation. Jour. Atmos. Sci., 25: 133–34.

Atlas Pluviométrico do Brasil (1914–1938). 1948. Ministerio da Agrucutura, Rio de Janeiro. Boletim No. 5.

Australian Bureau of Meteorology. 1930. Australian rainfall in district averages. Melbourne. Bulletin No. 23.

Bargman, D. J., Ed. 1959. Tropical meteorology in Africa. Symposium on tropical meteorology, Nairobi, 1959. Munitalp Foundation, Nairobi.

Bauer, G. 1935. Luftzirkulation und Niederschlagsverhältnisse in Vorderasien. Gerlands Beitr. Geophys., 45: 384.

Baum, Werner A., and Laurence B. Smith. 1952. Semimonthly mean sea-level pressure maps for the Mediterranean area. Florida State Univ., Dept. of Meteorol., Sci. Repts., No. 2; Also,

Contract AF (122)–466, Geophysics Res. Div., Air Force Cambridge Research Center, Cambridge, Mass.

Bebber, W. J. van. 1882. Typische Witterungserscheinungen. Arch. deut. Seewarte, 5: 3 (1886) 9: 2.

Beelitz, Paul. 1932. Die Haupttypen des jährlichen Ganges der Niederschläge in Europa. Inaugural dissertation. Berlin.

Benton, George S., and Mariano A. Estoque. 1954. Water-vapor transfer over the North American continent. J. Meteorol., 11: 469–71.

Bergeron, Tor. 1949. The problem of artificial control of rainfall on the globe, II. The coastal orographic maxima of precipitation in autumn and winter. Tellus, 1: 15–32.

Bernardes, Lysia M. C. 1951. Tipos de clima do Brasil. Bol. Geográf., 9: 988–97.

Berry, F. A., E. Bollay, and Norman R. Beers, Eds., 1945. Handbook of meteorology. McGraw-Hill, New York.

Biel, Erwin R. 1944. Climatology of the Mediterranean area. Univ. of Chicago, Inst. of Meteorology, Misc. Repts., No. 13: 129–30.

———. 1945. Weather and climate of China, parts A. and B. U.S. Army, Weather Division, Hdq. Army Air Forces, Report 890: 22–24; 104.

Birkeland, B. J., and N. J. Föyn. 1932. Klima von Nordwesteuropa. *In*, Handbuch der Klimatologie, W. Köppen and R. Geiger, Eds., Band 3, Teil 1. Gebrüder Borntraeger, Berlin.

Bjerknes, J. A. 1960. A Possible response of the atmospheric Hadley circulation to equatorial anomalies of ocean temperatures. Tellus, 18: 820–29.

———. 1966. Survey of El Niño 1957–1958 in its relation to tropical Pacific meteorology. Inter-Am. Trop. Tuna Comm. Bull., 12, No. 2: 25–86.

Blair, Thomas A. 1942. Climatology, general and regional. Prentice-Hall, New York.

Bleeker, W., and M. J. Andre. 1951. On the diurnal variation of precipitation, particularly over central U.S.A., and its relation to large-scale orographic circulation systems. Quart. J. Roy. Meteorol. Soc., 77: 260–71.

Blume, Helmut. 1962. Beiträge zur Klimatologie Westindiens. Erdkunde, 4, 16: 271–89.

Boffi, Jorge Alberto. 1949. Effect of the Andes mountains on the general circulation over the southern part of South America. Bull. Am. Meteorol. Soc., 30: 242–47.

Borchert, John. 1950. The Climate of the central North American grassland. Ann. Assoc. Am. Geographs., 40: 1–39.

———. 1971. The desert bowl in the 1970's. Ann. Assoc. Am. Geographs., 81: 1–22.

Bose, B. L. 1957. The nor'wester and the lower level convergence. Indian J. Meteorol. Geophys., 8: 391–98.

Boss, G. 1954. Klimastudie über Angola. Meteorol. Rundschau, 7: 10–13.

Botts, Adelbert K. 1930. The rainfall of Salvador. Monthly Weather Rev., 58: 459–66.

Boyden, M. W., and Earle C. Fowler. 1943. The weather disturbances of French North Africa. *In*, A report of synoptic conditions in the Mediterranean area. Prepared for the U. S. Army for limited distribution by the Univ. of Chicago, Dept. of Meteorol., 114–27.

Braak, C. 1931. Klimakunde von Hinterindien und Insulinde. *In*, Handbuch der Klimatologie, W. Köppen and R. Geiger, Eds., Band 4, Teil r. Gebrüder Borntraeger, Berlin.

Brooks, C. E. P. 1946. Annual recurrences of weather: "singularities." Weather, 1: 107–13; 130–34.

———. 1954. The English climate. English Universities Press, London.

———, and S. T. A. Mirrlees. 1932. A study of the atmospheric circulation over tropical Africa. Great Britain Meteorol. Office, Geophys. Mem. No. 55.

Brückner, Werner. 1951. Weather observations in Colombia. Weather, 6: 54–58.

Brunnschweiler, Dieter H. 1952. The geographic distribution of air masses in North America. Vierteljahrschr. Naturforsch. Ges. Zurich, 97: 43–44.

Bryson, Reid A. 1957*a*. Fourier analysis of the annual march of precipitation in Australia. Univ. of Arizona, Inst. of Atmos. Physics, Tech. Repts. on the Meteorol. and Climatol. of Arid Regions, No. 5.

———. 1957*b*. The annual march of precipitation in Arizona, New Mexico, and Northwestern Mexico. Univ. of Ariz., Inst. of Atmo-

spheric Physics, Tech. Repts. on the Meteorol. and Climatol. of Arid Regions, No. 6.
———. 1966. Air masses, streamlines, and the boreal forest. Geog. Bull., 8: 228–269.
———, and D. A. Baerreis. 1967. Possibilities of major climatic modification and their implications: Northwest India, a case study. Bull. Am. Meteorol. Soc., 48: 136–42.
———, and F. K. Hare, Eds. 1976. The climates of North America. World survey of climatology, vol. 11, Elsevier, New York.
———, and Peter M. Kuhn. 1961. Stress-differential induced divergence with application to littoral precipitation. Erdkunde, 15: 287–94.
———, and James F. Lahey. 1955. Factors in Wisconsin rainfall. Univ. of Wisconsin, Dept. of Meteorol., 26 pp.
———, and ———. 1958. The march of the seasons. Univ. of Wisconsin, Dept. of Meteorol., Sci. Repts., No. 12: 1–41.
———, and William P. Lowry. 1955*a*. Synoptic climatology of the Arizona summer monsoon. Univ. of Wisconsin, Dept. of Meteorol., Sci. Repts., No. 1.
———, and ———. 1955*b*. Synoptic climatology of the Arizona summer precipitation singularity. Bull. Am. Meteorol. Soc., 36: 329–39.
Bultot, F. 1952. Sur le caractere organise de la pluie au Congo Belge. Bureau Climatologique. Publications de l'institut national pour l'étude agronomique du Congo Belge, Communication No. 6.
Burbridge, F. E. 1951. The modification of continental polar air over Hudson Bay. Quart. J. Roy. Meteorol. Soc., 77: 365–74.
Burgos, J. J., and A. L. Vidal. 1951. The climates of the Argentine Republic according to the new Thornthwaite classification. Ann. Assoc. Am. Geographs., 41: 237–63.
Byers, Horace R., and Harriet R. Rodebush. 1948. Causes of thunderstorms of the Florida Peninsula. J. Meteorol., 5: 275–80.
Calef, Wesley. 1950. Interdiurnal variability of temperature extremes in the United States. Bull. Am. Meteorol. Soc., 31: 300–302.
———, et al. 1957. Winter weather type frequencies, northern Great Plains. U.S. Army, Environmental Protection Research Division, Hdq. Quartermaster Research and Engineering Division Command, Natick, Mass., Tech. Rept., EP–64.
Carter, Douglas B. 1954. Climates of Africa and India according to Thornthwaite's 1948 classification. Johns Hopkins Univ. Laboratory of Climatol., Publ. Climatol., 7, No. 4: 453–74.
Caviedes, César N. 1973*a*. Secas and El Niño: two simultaneous climatical hazards in South America. Proc. Assoc. Amer. Geogs., 5: 44–49.
———. 1973*b*. The climatic profile of the North Chilean Desert at latitude 20°S. *In*, Coastal deserts, their natural and human environments, D. Amiran and A. Wilson, Eds., Univ. of Arizona Press, Tucson: 115–21.
Chakravarti, A. K. 1968. Summer rainfall in India. Atmosphere, 6: 87–114.
———. 1976. Precipitation deficiency patterns in the Canadian prairies, 1921–1970. Prairie Forum, 1: 95–110.
Chakravortty, K. C., and S. C. Basu. 1957. The influence of western disturbances on the weather over Northeast India in monsoon months. Indian J. Meteorol. Geophys., 8: 261–72.
Chancellor, Justus. 1946. West African weather patterns. M. A. thesis, Univ. of Chicago.
Chang, Jen-hu. 1962. Comparative climatology of the tropical western margins of the northern oceans. Ann. Assoc. Am. Geographs., 52: 221–27.
———. 1967. The Indian summer monsoon. Geograph. Rev., 57: 372–96.
———. 1972. Atmospheric circulation systems and climates. The Oriental Publishing Co., Honolulu. 274–97.
Chapman, John D. 1952. The climate of British Columbia. *In*, Trans. British Columbia natural resources conf., 5th Conf.: 8–37.
Chaudhury, A. M. 1950. On the vertical distribution of wind and temperature over Indo-Pakistan along the meridian 76°E in winter. Tellus, 2: 56–62.
Chen, Kuo-yen. 1974. Water balance and water resources in China. Geophys. Mag, 37: 95–122.
Cheng, Chwen-shu. 1949. Chinese synoptic weather patterns. Shanghai Observatory.
Chu, Co-Ching. 1939. Southeast monsoon and rainfall in China. J. Chinese Geograph. Soc., 1: 1–27. *Also in*, 1954. Collected sci. papers,

meteorol., 1919–1949. Academia Sinica, Peking: 475–93.

Chu, P. H. 1950. The frontal waves in spring in China. J. Chinese Geophys. Soc., 2: 131–43.

Connor, A. J. 1938. The climates of North America, Canada. *In*, Handbuch der Klimatologie, W. Köppen and R. Geiger, Eds., Band 2, Teil j. Gerbrüder Borntraeger, Berlin: 332–424.

Conrad, V. 1943. The climate of the Mediterranean region. Bull. Am. Meteorol. Soc., 24: 134.

———. 1946. Usual formulas of continentality and their limits of validity. Trans. Am. Geophys. Union, 27: 663–64.

Costa, José Rutllant. 1978. Paleoclimatology, oceanography, El Niño, and climate-related resources of the arid zones in Peru and northern Chile. *In*, Exploring the earth's driest climate, H. H. Lettau and Katharina Lettau, Eds., Univ. of Wisconsin, Center for Climatic Research, Inst. for Environmental Studies, Report 101: 163–81.

Coyle, J. R. 1940. A practical analysis of weather along the east coast of South America. Meteorological Service, Panair do Brasil, Rio de Janeiro.

———. 1950. Notes concerning the analysis of South American weather maps. Meteorological Service, Panair do Brasil, Rio de Janeiro. Mimeographed.

———. No date, *a*. Anticyclones and anticyclonic circulations about the tropical regions of Brasil. Meteorological Service, Panair do Brasil, Rio de Janeiro. Mimeographed.

———. No date, *b*. Cold front slopes in Brasil frequently transform to overrunning slopes. Meteorological Service, Panair do Brasil, Rio de Janeiro. Unpublished.

———. No date, *c*. Notes on the movement of fronts along the Brasilian coast and into the southeasterly trades. Meteorological Service, Panair do Brasil, Rio de Janeiro. Unpublished.

———. No date, *d*. Seepage northward from the westerlies. Meteorological Service, Panair do Brasil, Rio de Janeiro. Unpublished.

———. No date, *e*. Weather on the direct route Rio-Manaus. Meteorological Service, Panair do Brasil, Rio de Janeiro. Mimeographed.

———. No date, *f*. A quasi-desert within a tropical rain belt. *In*, A series of papers on the weather of South America. Navaer U.S. Navy Reprint, Pan American Airways, Rio de Janeiro: 18–22.

———. No date, *g*. Injection of cold air masses into the westerlies of the Southern Hemisphere in the region of the South American continent. *In*, A series of papers on the weather of South America. Navaer U.S. Navy Reprint, Pan American Airways, Rio de Janeiro: 1–8.

———. No date, *h*. Movement of cold air masses over South America. *In*, A series of papers on the weather of South America. Navaer U.S. Navy Reprint, Pan American Airways, Rio de Janeiro.

Currie, R. 1953. Upwelling in the Benguela current. Nature, London: 407–500.

Das, P. K. 1962. Mean vertical motion and nonadiabatic heat sources over India during the monsoon. Tellus, 14: 212–20.

———. 1968. The monsoons. National Book Trust, New Delhi.

Day, Stanley. 1953. Horizontal convergence and the occurrence of summer precipitation at Miami, Florida. Monthly Weather Rev., 81: 155–61.

Deppermann, C. E. 1941. On the occurrence of dry, stable maritime air in equatorial regions. Bull. Am. Meteorol. Soc., 22: 143–49.

Desai, B. N. 1950. Mechanism of nor'westers of Bengal. Indian J. Meteorol. Geophys., 1: 74–76.

———. 1951. On the development and structure of monsoon depressions in India. Mem. Indian Meteorol. Dept., 28, Part 5.

———, and P. Koteswaram. 1951. Air masses and fronts in the monsoon depressions in India. Indian J. Meteorol. Geophys., 2: 250–65.

———, and S. Mal. 1938. Thunderstorms of Bengal. Gerlands Beitr. Geophys., 53: 285–304.

Dey, B. 1976. The variability of summer precipitation in the Canadian prairies. Albertan Geographer, 12: 16–25.

———. 1977. The summer monsoon in South Asia. Geog. Rev. India, 39: 266–81.

———, and A. Chakravarti. 1976. A synoptic climatological analysis of summer dry spells in the Canadian prairies. Great Plains-Rocky

Mountains Geog. Jour., 5: 30–46.
Doberitz, R. 1967. Zum Küstenklima von Peru. Deutscher Wetterdienst, Einzelveröff. Seewetter Amt., No. 59.
D'Ooge, Charles L. 1955. Continentality in the western United States. Bull. Am. Meteorol. Soc., 36: 175–77.
Dorsey, Herbert G., Jr. 1945. Some meteorological aspects of the Greenland Ice Cap. J. Meteorol., 2: 135–42.
Duvergé, Pierre. 1949. Principes de météorologie dynamique et types de temps à Madagascar. Publication du Service Météorologique Madagascar, Tananarive, No. 13.
East Africa Royal Commission 1953–1955 Report. 1955. London, Cmd 9975.
Eichmeier, A. H. 1951. Snowfalls—Paul Bunyan style. Weatherwise, 4: 124–27.
Eldridge, R. H. 1957. A synoptic study of West African disturbance lines. Quart. J. Roy. Meteorol. Soc., 83: 303–14.
El-Fandy, M. G. 1946. Barometric lows of Cyprus. Quart. J. Roy. Meteorol. Soc., 72: 291–306.
Eshleman, C. H. 1921. Do the Great Lakes diminish rainfall in the crop growing season? Monthly Weather Rev., 49: 500–502.
Ferdon, Edwin N. 1950. Studies in Ecuadorian geography. Monogr. Sch. Am. Res. No. 15: 35–76.
Findlater, J. 1969. A major low-level air current near the Indian Ocean during the northern summer. Quart. J. Roy. Meteorol. Soc., 95: 362–80.
Fleming, Richard H. 1941. A contribution to the oceanography of the Central America region. Proc. Pacific Sci. Congr., Pacific Sci. Assoc., 6th Congr., 1939, 3: 167–75.
Flohn, Hermann. 1948. Zur Kenntnis des jährlichen Ablaufs der Witterung im Mittelmeergebiet. Geofis. Pura e Appl., 13: 184–87.
———. 1949. Klimatologische Homologien (Tropische Orkane und Zyklonentätigkeit in der Westdriftzone). Meteorol. Rundschau, 2: 198–292.
———. 1950*a*. Ablauf und Struktur des ostasiatischen Sommermonsuns. Ber. Deut. Wetterdienstes in der U.S. Zone, No. 18: 21–33.
———. 1950*b*. Studien zur allgemeinen Zirkulation der Atmosphäre. Ber. Deut. Wetterdienstes in der U.S. Zone, No. 18: 5–52.
———. 1951. Passatzirkulation und äquatoriale Westwindzone. Arch. Meteorol. Geophys. Bioklimatol., Ser. B, 3: 3–15.
———. 1953. Tropical circulation patterns. World Meteorological Organization, Commission for Aerology, First Session. Toronto, Scientific Paper No. 5.
———, Ed., 1954. Witterung und Klima in Mitteleuropa. S. Hirzel, Zurich.
———. 1956*a*. Der indische Sommermonsun als Glied der planetarischen Zirkulation der Atmosphäre. Ber. Deut. Wetterdienstes in der U.S. Zone, No. 22: 134–39.
———. 1956*b*. Investigations on the general atmospheric circulation, especially in lower latitudes. *In*, I.U.G.G. Assoc. of Meteorol., Rome, Sept. 1954. Sci. Proc., Butterworth, London: 431–42.
———. 1956*c*. Zur Kenntnis der aequatorialen Westwindzone ueber Angola. Miscelânea Geofísica Publicada Pelo Serviço Meteorológico de Angola em Comemoraçâo do X Aniversário do Serviço Meteorológico Nacional, Luanda: 25–28.
———. 1957. Large-scale aspects of the "summer monsoon" in South and East Asia. Jour. Meteorol. Soc. Japan, 75th Anniversary Volume: 180–86.
———. 1960. Equatorial westerlies over Africa, their extension and significance. *In*, Tropical meteorology in Africa. Symposium on Tropical Meteorology, Nairobi, 1959. D. J. Bargman, Ed., Munitalp Foundation, Nairobi: 266–67.
———. 1965. Studies in the meteorology of tropical Africa. Bonner Meteorol. Abhandl., 5: 1–57.
———. 1967. Dry equatorial zones and asymmetry of the global atmospheric circulation. Bonner Meteorol. Abhandl., 7: 3–7.
———. 1968. Contributions to a meteorology of the Tibetan Highlands. Colorado State Univ., Dept of Atmospheric Science, Fort Collins, Colo., Atmospheric Science Paper No. 130.
———, Ed. 1969. Climate and weather in Central Europe. B. V. de G. Walden, trans. McGraw-Hill, New York.
———. 1970*a*. Elements of a climatology of the Indo-Pakistan subcontinent. Bonner. Meteorol.

Abhandl., 14: 5–28.

———. 1970*b*. Elements of a synoptic climatology of the Indo-Pakistan subcontinent. *In*, Studies in the climatology of South Asia, U. Schweinfurth, H. Flohn, and M. Domrös, Eds., South Asia Institute, Dept. of Geography, Franz Steiner Verlag, Wiesbaden: 3–5.

———. 1971. Tropical circulation patterns. Bonner Meteorol. Abhandl., 15.

———. 1972. Investigations of equatorial upwelling and its climatic role. *In*, Studies in physical oceanography, Arnold L. Gordon, Ed., Gordon and Breach Science Publishers, New York: 93–102.

———, and K. Hinkelmann. 1952. Äquatoriale Zirkulationsanomalien und ihre klimatische auswirkung. Ber. Deut. Wetterdienstes in der U.S. Zone, No. 42: 114–21.

———, and Josef Huttary. 1950. Über die Bedeutung der V^b-lagen für das Niederschlagsregime Mitteleuropas. Meteorol. Rundschau, 3: 167–70.

———, and H. Oeckel. 1956. Water vapour flux during the summer rains over Japan and Korea. Geophys. Mag., 27: 527–32.

Fobes, Charles B. 1954. Continentality in New England. Bull. Am. Meteorol. Soc., 35: 197; 207.

Foley, J. C. 1956. 500 mb. contour patterns associated with the occurrences of widespread rains in Australia. Australian Meteorol. Mag., 13: 1–18.

Forsdyke, A. G. 1944. Synoptic analysis in the western Indian Ocean. East African Meteorol. Dept., Memoirs, 2, No. 3; 1–11.

———. 1949. Weather forecasting in tropical regions. Great Britain Meteorol. Office, Geophys. Mem., 10, No. 82.

Fraselle, E. 1947. Introduction a l'etude de l'atmosphere Congolaise, Inst. roy. Colonial belge Mém., 16, Fasc. 3.

Freeman, John C., Jr. 1950. The wind field of the equatorial East Pacific as a Prandtl-Meyer expansion. Bull. Am. Meteorol. Soc., 31: 303–4.

Freise, Friedrich. 1938. The drought region of northeastern Brazil. Geograph. Rev., 28: 363–78.

Frost, R. 1953. Upper air circulation in low latitudes in relation to certain climatological discontinuities. Great Britain Meteorol. Office, Professional Note No. 107: 1–25.

Fukui, E. 1964. A short dry period of mid-summer in Japan. (English abstract) Geog. Rev. Japan, 37: 531–47.

———, Ed. 1977. The climate of Japan. Developments in Atmospheric Science, No. 8, Kandansha Ltd., Tokyo ard Elsevier, New York.

Gabites, F. J. 1943. Weather analysis in the tropical South Pacific. New Zealand Meteorol. Office, Misc. Meteorol. Notes, Ser. A, No. 1.

Garbell, Maurice A. 1947. Tropical and equatorial Meteorology. Pitman, New York.

Garnier, B. J. 1946. The climates of New Zealand: according to Thornthwaite's classification. Ann. Assoc. Am. Geographs., 36: 64–65; 134–41; 151–77.

———. 1950. The seasonal climates of New Zealand. *In*, New Zealand weather and climate, B. J. Garnier, Ed., Special Publication of the New Zealand Geograph. Soc., Misc. Ser., No. 1: 105–39.

———. 1958. The climate of New Zealand: a geographic survey. E. Arnold, London.

Gehrke, Willis T. No date. The coastal steppe of Gold Coast and Togo. Unpublished seminar report, Univ. of Wisconsin.

Gentilli, J. 1952. Climatology of the Central Pacific. Proc. Pacific Sci. Congr., 7th Congr., Wellington, 3: 92–100.

George, P. A. 1956. Effects of off-shore vortices on rainfall along the west coast of India. Indian J. Meteorol. Geophys., 7: 225–40.

George, Joseph J., and Paul M. Wolff. 1953. Cyclogenesis along the east coast of Asia. Bureau of Aeronautics, Project AROWA, Technical Report Task 13.

Gichuiya, S. N. 1970. Easterly disturbances in the south-east monsoon. *In*, Proc. of the symposium on tropical meteorology, June 2–11, 1970. Univ. of Hawaii, Hawaii Institute of Geophysics, Honolulu, 13: 1–7.

Gierloff-Emden, Hans-Günter. 1959. Der Humboldtstrom und die pazifischen Landschaften seines Wirkungsbereiches. Petermanns Geograph. Mitt., 103: 1–17.

Gleeson, Thomas A. 1954. Cyclogenesis in the Mediterranean region. Arch. Meteorol. Geophys. Bioklimatol., Ser. A, 6: 165–71.

Glover, J., P. Robinson, and J. P. Henderson. 1954. Provisional maps of the reliability of an-

nual rainfall in East Africa. Quart. J. Roy. Meteorol. Soc., 80: 602–9.
Gordon, Arnold L., Ed. 1972. Studies in physical oceanography. Gordon and Breach Science Publishers, New York.
Graves, Maurice E., Erwin Schweigger, and Jorge P. Valdivia. 1955. Un año anormal. Boletín Científico de la Compañía Administradora del Guano, Lima: 2.
Great Britain Air Ministry. 1937. Weather in the Mediterranean. vol. 1, General Information. Meteorol. Office: 98–99.
Great Britain Meteorological Office. 1944*a*. Weather in the Indian Ocean. vol. 2, Local information. Part 2, The Gulf of Aden and West Arabian Sea to longitude 60°E., 451 (b) 2.
———. 1944*b*. A pilot's primer of West African weather. M. O. 469: 5–11.
———. 1948. Aviation meteorology repts., No. 1.
———. 1949. Weather on the west coast of tropical africa. M. O. 492: 131–40.
Gregory, Stanley. 1954. Climatic classification and climatic change. Erdkunde, 8: 246–52.
Griffiths, J. F. 1959. The variability of annual rainfall in East Africa. Bull. Am. Meteorol. Soc., 40: 361–62.
———. 1972. The horn of Africa. *In*, The climates of Africa, J. F. Griffiths, Ed., World survey of climatology, vol. 10, Elsevier, New York: 133–65.
———, Ed. 1972. The climates of Africa. World survey of climatology, vol. 10, Elsevier, New York.
Grove, Edward L. 1959. The Cuban desert: a climatological study. Los Angeles State College. Unpublished.
Gunther, E. R. 1936*a*. A report on oceanographical investigations in the Peru coastal current. "Discovery" Reports, 13: 107–276.
———. 1936*b*. Variations in behaviour of the Peru coastal current—with an historical introduction. Geograph. J., 88: 37–65.
Guppy, H. B. 1906. Observations of a naturalist in the Pacific between 1896 and 1899. vol. 2, Macmillan, London.
Gutnick, Murray. 1958. Climatology of the trade-wind inversion in the Caribbean. Bull. Am. Meteorol. Soc., 39: 410–20.
Hales, J. 1974. Southwestern United States summer monsoon source—Gulf of Mexico or Pacific Ocean. Jour. Applied Meteorol., 43: 331.
Hamilton, R. A., and J. W. Archbold. 1945. Meteorology of Nigeria and adjacent territory. Quart. J. Roy. Meteorol. Soc., 71: 238–44.
Hare, F. K. 1943. Atlas lee depressions and their significance for sirocco. Great Britain Meteorol. Office, Synoptic Div., Tech. Mem., No. 43.
———. 1944. The crachin. Great Britain Meteorol. Office, Tech. Mem., No. 87.
———. 1951. Some climatological problems of the Arctic and Subarctic. *In*, Compendium of meteorology, Thomas F. Malone, Ed., American Meteorological Society, Boston: 952–63.
———. 1956. The climate of the American northlands. Book I, The dynamic north. U.S. Navy, Technical Asst. to the Chief of Naval Operations for Polar Projects (OP–O3A3).
———, and Margaret R. Montgomery. 1949. Ice, open water, and winter climate in the eastern Arctic of North America: part I. Arctic, 2: 79–89.
Harman, J. R., and Wallace M. Elton. 1971. The La Porte, Indiana, precipitation anomaly. Ann. Assoc. Am. Geographs., 61: 468–80.
Hastenrath, Stefan. 1976. Variations in low-latitude circulation and extreme climatic events in the tropical Americas. Jour. Atmos. Sci., 33: 202–15.
———, and Leon Heller. 1977. Dynamics of climatic hazards in Northeast Brazil. Quart. J. Roy. Meteorol. Soc., 103: 77–92.
Haurwitz, Bernhard, and James M. Austin. 1944. Climatology. McGraw-Hill, New York: 176–211.
Hawaii Institute of Geophysics. 1970. Proceedings of the symposium on tropical meteorology, Honolulu, Hawaii, June 2–11, 1970. Univ. of Hawaii, Honolulu.
Hellmann, Gustav. 1928. Die Trockengebiete Europas und deren Ursachen. Z. Gesellschaft Erdkunde zu Berlin, 353–58.
Henderson, J. P. 1949. Some aspects of climate in Uganda. East African Meteorol. Dept., Memoirs, 2, No. 5: 2–15.
Hess, Paul, and Helmuth Brezowsky. 1952. Katalog der *Grosswetterlagen* Europas. Ber. Deut. Wetterdienstes in der U.S. Zone, No. 33.
Hogan, J., and J. V. Maher. 1943. A paper on the

Australian cold fronts over the Coral Sea and its environs. Australian Meteorol. Service, Research Repts., Ser. 10.

Holdridge, L. R. 1947. The forests of western and central Ecuador. U.S. Dept. of Agriculture, Forest Service. Multilithed pamphlet.

Holm, Karl-Friedrich. 1953. Die thermische Kontinentalität in Europa während der "Normalperiode" 1901–30 und ihre Schwankung. Petermanns Geograph. Mitt., 97: 26–30.

Horn, Lyle H., and Reid A. Bryson. 1960. Harmonic analysis of the annual march of precipitation over the United States. Ann. Assoc. Am. Geographs., 50: 157–71.

———, ———, and William P. Lowry. 1957. An objective precipitation climatology of the United States. Univ. of Wisconsin, Dept. of Meteorol., Sci. Repts., No. 6: 1–34.

Hosler, Charles R. 1956. A study of easterly waves in the Gulf of Mexico. Bull. Am. Meteorol. Soc., 37: 101–7.

H. R. J. 1954. Weather in Sierra Leone. Geograph. J., 120: 124–27.

Hsü, Shu-Ying. 1958. Water-vapor transfer and water balance over eastern China. Acta Meteorologica Sinica, Translation Emm-65-29. Air Force Cambridge Research Laboratories, Bedford, Mass., 29, No. 1: 33–42.

Hubbard, J. H. 1954. A note on the rainfall of Accra, Gold Coast. Geograph. Studies, 1, No. 1: 64–75.

———. 1956. Daily weather at Achimora, near Accra, Gold Coast. Geograph. Studies, 3, No. 1: 56–63.

Hunt, H. A., G. Taylor, and E. T. Quayle. 1913. Climate and weather of Australia. Minister of State for Home Affairs, Melbourne.

Hunter, Robert S. No date. Correlation of monsoon rain of Northeast Brasil with polar pressure waves and polar front frontolysis. Panair do Brasil, Rio de Janeiro. Unpublished.

Huttary, Josef. 1950. Die Verteilung der Niederschläge auf die Jahreszeiten im Mittelmeergebiet. Meteorol. Rundschau, 3: 111–19.

Industrial Meteorology Association. 1948. The climatographic atlas of Japan.

Instituto Mexicano de Recursós Naturales Renovables, A. C. Los recursós naturales del sureste y su aprovechamiento. II Parte, Tome 2.

Jackson, S. P. 1951. A preliminary study of the atmospheric circulation over South Africa. Ph.D. thesis, Univ. of London.

———. 1952. Atmospheric circulation over South Africa. S. African Geograph. J., 34: 48–59.

James, Preston E. 1952. Observations on the physical geography of Northeast Brazil. Ann. Assoc. Am. Geographs., 42: 153–76.

Japanese National Commission for UNESCO. 1955. Proceedings of the UNESCO symposium on typhoons, Tokyo, November 9–12, 1954.

Jeandidier, G. 1956. Orages et mousson sur le bassin du Congo. Publ. serv. météorol. Afrique Équatorial Franç., Brazzaville, No. 3: 1–9.

Jefferson, Mark. 1921. The rainfall of Chile. Am. Geograph. Soc. Research Ser., No. 7.

John, I. G., and F. Kenneth Hare. No date. Winter circulation over Burma, Thailand, and Indo-China. Great Britain Meteorol. Office, Synoptic Tech. Mem., No. 120.

Johnson, A. M. 1976. The climate of Peru, Bolivia and Ecuador. *In*, Climates of Central and South America, W. Schwerdtfeger, Ed., World survey of climatology, vol. 12, Elsevier, New York: 147–218.

Johnson, D. H. 1962. Rain in East Africa. Quart. J. Roy. Meteorol. Soc., 88, No. 375: 1–19.

———, and H. T. Mörth. 1959. Forecasting research in East Africa. *In*, Tropical meteorology in Africa. Symposium on tropical meteorology, Nairobi, 1959. D. J. Bargman, Ed., Multialp Foundation, Nairobi: 56–137. (Also in, 1961. East Africa Meteorology Dept., Memoirs, 3, No. 9.)

Jong, P. C. 1950. The Kunming quasi-stationary front. J. Chinese Geophys. Soc., 2: 87–103.

Jurwitz, Louis R. 1953. Arizona's two-season rainfall pattern. Weatherwise, 6: 96–99.

Karelsky, S. 1956. Classification of the surface circulation in the Australasian region. Australian Bureau of Meteorol., Meteorol. Study No. 8: 1–36.

Katow, K. 1957. Analysis of tropospheric structure during the baiu season 1954. Geophys. Inst. Tokyo Univ., Collected Meteorol. Papers, 7, No. 92.

Kendrew, W. G. 1953. The climates of the continents. 4th ed., Clarendon, Oxford.

Kerr, D. P. 1951. The summer-dry climate of Georgia Basin, British Columbia. Trans. Roy. Can. Inst., 29: 28–29.

Kerr, I. S. No date. Seasonal variations of weather types in New Zealand. New Zealand Meteorol. Office, Circular Note No. 25.

Kidson, E. 1931. The annual variation of rainfall in New Zealand. New Zealand J. Sci. Technol., 12: 268–71.

———. 1932. Climatology of New Zealand. *In*, Handbuch der Klimatologie, W. Köppen and R. Geiger, Eds., Band 4, Teil s. Gebrüder Borntraeger, Berlin.

———. 1950. The elements of New Zealand's climate. *In*, New Zealand weather and climate, B. J. Garnier, Ed., Special Publication of the New Zealand Geograph. Soc., Misc. Ser., No. 1: 44–83.

Kimura, J. 1966. The beginning and the end of the Shurin season in Japan. Geog. Rep. Tokyo Metropolitan Univ., 1: 113–18.

Kincer, J. B. 1922. Precipitation and humidity, atlas of American agriculture. Part II, Climate. Washington, D.C.

Klute, Fritz, Ed. 1930. Handbuch der geographischen Wissenschaft. vol. 4, Südamerika. Akademische Verlagsgesellschaft Athenaion, Wildpark-Potsdam.

Knoch, K. 1930. Klimakunde von Südamerika. *In*, Handbuch der Klimatologie, W. Köppen and R. Geiger, Eds., Band 2, Teil b. Gerbrüder Borntraeger, Berlin.

———. 1951. Klima von Europa. *In*, World atlas of epidemic diseases, E. Rodenwaldt and H. J. Jusatz, Eds., Falk-Verlag, Hamburg.

———, and A. Schulze. 1956. Precipitation, temperature and sultriness in Africa. *In*, World atlas of epidemic diseases, E. Rodenwaldt and H. J. Jusatz, Eds., Falk-Verlag, Hamburg.

Kopec, R. 1965. Continentality around the Great Lakes. Bull. Am. Meteorol. Soc., 46: 54–57.

Köppen, W. 1923. Die Klimate der Erde. Metzger and Wittig, Berlin and Leipzig.

———, and R. Geiger, Eds. 1930–1939. Handbuch der Klimatologie. Gebrüder Borntraeger, Berlin.

Koteswaram, P. 1958. The easterly jet stream in the tropics. Tellus, 10: 43–57.

———, and C. A. George. 1958. On the formation of monsoon depressions in the Bay of Bengal. Indian J. Meteorol. Geophys., 9: 9–22.

———, and V. Srinivasan. 1958. Thunderstorms over Gangetic West Bengal in the pre-monsoon season and the synoptic factors favourable for their formation. Indian J. Meteorol. Geophys., 9: 301–12.

Kuepper, William. 1968. Rainfall deficiencies on the plateau of Tanzania. Ph.D. thesis, Univ. of Wisconsin.

———. No date. Eastern Africa and the Indian Ocean monsoons. Unpublished.

Kurashima, A. 1968. Studies of the summer monsoons in East Asia based on dynamic concepts. Geophys. Mag. (Tokyo), 34: 145–235.

———, and Y. Hiranuma. 1971. Synoptic and climatological study of the upper moist tongue extending from Southeast Asia to East Asia. *In*, Water balance of monsoon Asia, M. Yoshino, Ed., Univ. of Hawaii Press, Honolulu: 153–69.

Lahey, James F. 1958. On the origin of the dry climate in northern South America and the southern Caribbean. Ph.D. dissertation, Univ. of Wisconsin. *Also*, Air Force Cambridge Research Center, ARDC, Contract 19–604992, AFCRC–TN58–218, Geophysics Research Directorate. Scientific Report No. 10.

———. 1973. The origin of the dry climate in northern South America and the southern Caribbean. *In*, Coastal deserts, D. Amiran and A. Wilson, Eds., Univ. of Arizona Press, Tucson: 75–90.

———, Reid Bryson, and Eberhard Wahl. 1957. Atlas of five-day normal sea-level pressure charts. Univ. of Wisconsin, Dept. of Meteorol., Sci. Repts., No. 7. *Also*, U.S. Air Force Contract, AF19 (604)–992.

———, et al. 1958. Atlas of 500 mb wind characteristics for the Northern Hemisphere. Univ. of Wisconsin, Dept. of Meteorol., Sci. Repts., No. 1. *Also*, U. S. Air Force Contract, AF19(604)–2278.

Lamb, H. H. 1950. Types and spells of weather around the year in the British Isles: annual trends, seasonal structure of the year, singularities. Quart. J. Roy. Meteorol. Soc., 76:

393–498.

———. 1953. British weather around the year. 1953. Weather, 8: 131–36; 176–82.

———. 1959. The southern westerlies: a preliminary survey; main characteristics and apparent associations. Quart. J. Roy. Meteorol. Soc., 85: 1–23.

Lamb, Peter. 1978. Case studies of tropical Atlantic surface circulation patterns during recent sub-Saharan weather anomalies. Monthly Weather Rev., 106: 482–91.

Landsberg, H. 1944. A climatic study of cloudiness over Japan. Univ. of Chicago, Dept. of Meteorol., Misc. Repts., No. 15: 6–83.

La Seur, N. E. 1964. Synoptic models in the tropics. *In*, Proc. of the symposium on tropical meteorology, Rotoura, New Zealand, 1963. New Zealand Meteorological Service, Wellington: 319–28.

Lauscher, Friedrich. 1954. Dynamische Klimaskizze von Österreich. *In*, Witterung und Klima in Mitteleuropa, Hermann Flohn, Ed., S. Hirzel, Zurich: 145–58.

Lautensach, Hermann. 1951. Die Niederschlagshöhen auf der Iberischen Halbinsel. Petermanns Geograph. Mitt., 95: 145–60.

Lee, Byong-Sul. 1974. A synoptic study of the early summer and autumn rainy season in Korea and East Asia. Geog. Reports Tokyo Metropolitan Univ., 9: 79–96.

Lee, John. 1937. The cause of scarcity of precipitation on the islands along the Chinese coast. Academy Sinica, Mem. Nat. Research Inst. Meteorol., No. 9: 43–53.

Lembke, Herbert. 1940. Eine neue Karte des Jahresniederschlages im westlichen Vorderasien. Petermanns Geograph. Mitt., 86: 217–25.

Lettau, H. 1976. Dynamic and energetic factors which cause and limit aridity along South America's Pacific coast. *In*, Climates of Central and South America, W. Schwerdtfeger, Ed., World survey of climatology, vol. 12, Elsevier, New York: Appendix I, 188–92.

———. 1978. Explaining the world's driest climate. *In*, Exploring the earth's driest climate, H. H. Lettau and Katharina Lettau, Eds., Univ. of Wisconsin, Center for Climatic Research, Inst. for Environmental Studies, Report 101: 182–248.

———, and Katharina Lettau, Eds. 1978. Exploring the earth's driest climate. Univ. of Wisconsin, Center for Climatic Research, Inst. for Environmental Studies, Report 101.

Logan, R. 1960. The central Namib Desert, South West Africa. National Academy of Sciences/National Research Council, Washington, D.C.: Publication 758. (O.N.R. Field Research Program, Report 9.)

Loon, H. van. 1955. Mean air-temperature over the southern oceans. Notos, 4: 292–308.

Lu, A. 1945. The winter frontology of China. Bull. Am. Meteorol. Soc., 26: 309–14.

Ludwig, Hildegard. 1953. Regionale Typen im Jahresgang der Niederschläge in Vorderindien und ihre Beziehung zu Landschaftsgrundlagen. Abhandl., Gebiet der Auslandskunde, Band 57, Reihe C. Also in, Naturwissenschaften, Band 16.

Lydolph, Paul E. 1955. A comparative analysis of the dry western littorals. Ph.D. thesis, Univ. of Wisconsin.

———. 1957. A comparative analysis of the dry western littorals. Ann. Assoc. Am. Geographs., 47: 213–30.

———. 1973. On the causes of aridity along a selected group of coasts. *In*, Coastal deserts, their natural and human environments, D. Amiran and A. Wilson, Eds., Univ. of Arizona Press, Tuscon: 67–72.

———, Ed. 1977. Climates of the Soviet Union. World survey of climatology, vol. 7, Elsevier, New York.

McRae, J. N. 1955. A case of cyclogenesis associated with an upper cut off low. Australian Meteorol. Mag., 11: 36–46.

Maede, Hans. 1951. Strömungsdivergenz als Ursache für die Niederschlagsarmut der südlichen Ostseeküste. Z. Meteorol., 5: 26–30.

———. 1954. Die Regenwetterlagen an der südlichen Ostsee. Abhandl. Meteorol. u. Hydrol. Dienstes Deut. Demokratischen Republik, 4, No. 3: 14–18.

Malone, Thomas F., Ed. 1951. Compendium of meteorology. American Meteorological Society, Boston.

Malurkar, S. L. 1950. Notes on analysis of weather of India and neighborhood. Mem. Indian Meteorol. Dept., 28, Part 4.

———. 1956. A brief report of a discussion on "western disturbances" held at New Delhi. Indian J. Meteorol. Geophys., 7, No. 1: 1–6.

Manley, Gordon. 1945. The effective rate of altitudinal change in temperate Atlantic climates. Geograph. Rev., 34: 408–17.

———. 1952. Climate and the British scene. Collins, London.

Markham, C. 1975. Twenty-six-year cyclical distribution of drought and flood in Ceará, Brazil. Professional Geographer, 27: 454–56.

Marvin, C. F. 1919. Are irregularities in the annual march of temperature persistent? Monthly Weather Rev., 47: 544–45.

Matsumoto, S., H. Itoo, and Akio Arakawa. 1954. An aerological study on the pre-summer rainy season in Japan. Jour. Meteorol. Soc. Japan, 32: 85–95.

Maung, Po E. 1955. Storms and depressions over burma which have their origin in the typhoons of the Pacific Ocean and the South China Sea during 1949–53. *In*, Proc. of the UNESCO symposium on typhoons, Tokyo, November 9–12, 1954. Japanese National Commission for UNESCO, Tokyo.

Mbele-Mbong, Samuel. 1974. Rainfall in West Central Africa. Colorado State Univ., Dept. of Atmospheric Science, Fort Collins, Colo., Atmospheric Science Paper No. 222.

Means, Lynn B. 1946. The nocturnal maximum occurrence of thunderstorm in the midwestern states. Univ. of Chicago, Dept. of Meteorol., Misc. Repts., No. 16: 1–37.

Meigs, Peveril. 1955, Hurricanes and rain in Southern Baja California, Mexico. Ann. Assoc. Am. Geographs., 45: 202.

Miller, A. 1976. The climate of Chile. *In*, Climates of Central and South America, W. Schwerdtfeger, Ed., World survey of climatology, vol. 12, Elsevier, New York: 113–45.

Miller, James E., and Homer T. Mantis. 1947. Extratropical cyclogenesis in the Pacific coastal region of Asia. J. Meteorol., 4: 29–34.

Min, Kyung D., and Lyle H. Horn. 1974. The generation of available potential energy by sensible heating along the east coast of Asia and North America. Jour. Meteorol. Soc. Japan, 52: 204–17.

Mintz, Yale, and Gordon Dean. 1952. The observed mean field of motion of the atmosphere. Geophys. Research Papers, No. 17, Air Force Cambridge Research Center, Bedford, Mass.

Mohri, Keitaro. 1953. On the fields of wind and temperature over Japan and adjacent waters during winter of 1950–51. Tellus, 5: 340–47.

Mooley, D. A. 1957. The role of western disturbances in the production of weather over India during different seasons. Indian J. Meteorol. Geophys., 8: 253–60.

———. 1975. Climatology of the Asian summer monsoon rainfall. Geog. Rev. India, 37: 7–20 (1976) 38: 1–5; 282–92; 321–29 (1977) 39: 1–11.

Mörth, H. T. 1963. Five years of pressure analyses over tropical Africa. *In*, Proc. of the symposium on tropical meteorology, Rotoura, New Zealand, 1963. Wellington: 329–46.

———. 1970. A study of the areal and temporal distribution of rainfall in East Africa. *In*, Proc. of the symposium on tropical meteorology, June 2–11, 1970. Univ. of Hawaii, Hawaii Institue of Geophysics, Honolulu.

Mosiño Alemán, Pedro A., and Enriqueta Garcia. 1976. The climate of Mexico. *In*, Climates of North America, Reid Bryson and F. K. Hare, Eds., World survey of climatology, vol. 11, Elsevier, New York: 345–404.

Murakami, M. 1976. Analysis of summer monsoon fluctuations over India. Jour. Meteorol. Soc. Japan, 54: 15–31.

Murakami, T. 1951. On the study of the change of the upper westerlies in the last stage of Baiu season (rainy season in Japan). Jour. Meteorol. Soc. Japan, 29: 175.

———. 1959. The general circulation and water-vapor balance over the Far East during the rainy season. Geophys. Mag. (Tokyo), 29, No. 2: 131–71.

———, Y. Arai, and K. Tomatsu. 1962. On the rainy season in early autumn. Jour. Meteorol. Soc. Japan, 40: 330–49.

Murphy, Robert Cushman. 1923. The oceanography of the Peruvian littoral. Geograph. Rev., 13: 64–85.

———. 1926. Oceanic and climatic phenomena along the west coast of South America during 1925. Geograph. Rev., 16: 26–54.

———. 1939. The littoral of Pacific Colombia

and Ecuador. Geograph. Rev., 29: 2–33.

Musk, L. F. 1976. Rainfall variability and the Walker cell in the equatorial Pacific Ocean. Weather, 31: 34–47.

Nakamura, K. 1967. Climate of East Africa as related to equatorial westerlies. Geographical Reports of Tokyo Metropolitan Univ., July: 49–69.

Nakamura, S., and E. Arai. 1957. An upper air wind analysis during the later stage of "bai-u" in 1952. (in Japanese; English summary.) J. Aerological Observatory at Tateno, 6, No. 1: 9–16.

Namias, Jerome. 1972. Influence of Northern Hemisphere general circulation on drought in N.E. Brazil. Tellus, 24: 336–42.

National Academy of Sciences. 1977. Plan for U.S. participation in the monsoon experiment. Washington, D.C.

Navarro, J. 1950. Les grains du Nord Est et le régime des pluies sur la Côte du Golfe de Guinée. Cited in, R. J. Harrison Church, West Africa. Longmans, Green, and Co., London, 1957.

Newton, C. W. 1951. Note on the mechanism of nor'westers of Bengal. Indian J. Meteorol. Geophys., 2: 48–50.

New Zealand Meteorological Service. Proceedings of the symposium on tropical meteorology, Rotoura, New Zealand, 1963. 1964. Wellington.

Nunn, Roscoe. 1927. The "January thaw." Monthly Weather Rev., 55: 20–21.

Ojo, Oyediran. 1977. The climates of West Africa. Heinemann, London.

Oliver, Mildred B. 1956. Some aspects of the rainfall in the Kenyan Highlands. Geograf. Ann., 38: 102–11.

———, and Earl C. Fowler. The weather disturbances of French North Africa and the western Mediterranean. Univ. of Chicago, Library of the Dept. of Meteorol. Unpublished.

Oosawa, K. 1951. A normal broad-weather cycle in the "baiu," the rainy season in Japan. Papers Meteorol. Geophys. (Tokyo), 2: 45–51.

Otani, T. 1954. Converging line of the northeast trade wind and converging belt of the tropical air current. Geophys. Mag., (Tokyo) 25: 64–73.

Palmer, C. E. 1942. Synoptic analysis over the southern oceans. New Zealand Meteorol. Office, Professional Note No. 1.

———. 1952. Tropical meteorology. Quart. J. Roy. Meteorol. Soc. 78: 126–64.

Pant, P. S., and E. M. Rwandsya. 1971. Climates of East Africa. East African Meteorolog. Dept., Tech. Mem. 18.

Patton, Clyde Perry. 1956. Climatology of summer fogs in the San Francisco Bay area. Univ. Calif. Publ. Geograph., 10, No. 3: 113–200.

Petterssen, S., and P. A. Calabrese. 1959. On some weather influences due to warming of the air by the Great Lakes in winter. Jour. Am. Meteorol. Soc., 16: 646–52.

Pfalz, Richard. 1943. Die Regenverteilung in Belgisch-Kongo. Petermanns Geograph. Mitt., 89: 263–68.

Philippson, A. 1947. Griechenlands zwei Seiten. Erdkunde, 1: 144–62.

Pincock, G. L. 1951. An analysis of precipitation on the British Columbia coast. Meteorol. Division, Vancouver District, Local Forecast Study No. 9: 1–7.

Pisharoty, P. R. 1963. Monsoon pulses. *In*, Proc. of the symposium on tropical meteorology, Rotoura, New Zealand, 1963. Wellington: 373–79.

———, and B. N. Desai. 1956. "Western disturbances" and Indian weather. Indian J. Meteorol. Geophys., 7: 333–38.

Portig, Wilfried H. 1959. Air masses in Central America. Bull. Am. Meteorol. Soc., 40: 301–4.

———. 1965. Central American rainfall. Geograph. Rev., 55: 68–90.

———. 1976. The climate of Central America. *In*, Climates of Central and South America, W. Schwerdtfeger, Ed., World survey of climatology, vol. 12, Elsevier, New York: 405–78.

Potts, Allan. 1971. Application of harmonic analysis to the study of East African rainfall data. Jour. Tropical Geog., 33: 31–42.

Prager, Erwin. 1952. Der Niederschlag auf See und an der Dünenflachküste. Ann. Meteorol., 5: 259–67.

Prohaska, F. J. 1973. New evidence of the climatic controls along the Peruvian coast. *In*, Coastal deserts, their natural and human environments, D. Amiran and A. Wilson, Eds., Univ. of Arizona Press, Tuscon: 91–107.

———. 1976. The climate of Argentina, Paraguay and Uruguay. *In*, Climates of Central and South America, W. Schwerdtfeger, Ed., World survey of climatology, vol. 12, Elsevier, New York: 13–112.

Putnins, P., and Nina A. Stepanova. 1956. The dynamic north. Book I, Climate of the Eurasion Northlands. U.S. Navy, Technical Asst. to the Chief of Naval Operations for Polar Projects (OP–O3A3).

Pyke, C. 1972. Some meteorological aspects of the seasonal distribution of precipitation in the Western United States and Baja California. Univ. of Calif., Water Resources Center, Davis, Calif., Contribution No. 139.

Queiroz, Dario X. 1955. Variabilidade das chuvas em Angola. Serviço Meteorológico de Angola, Luanda.

Quinn, W. H. 1974. Monitoring and predicting El Niño invasions. Jour. Applied Meteorol., 13: 825–30.

Rahmatullan, M. 1952. Synoptic aspects of the monsoon circulation and rainfall over Indo-Pakistan. J. Meteorol., 9: 176–79.

Rainteau, P. 1955. Étude de l'evolution du temps sur le bassin du Congo (5°N–5°S) de 1er au 15 Mars, 1954. Publ. serv. météorol. Afrique Équatoriale Franç., Brazzaville, No. 3: 2–4.

Ramage, C. S. 1951. Analysis and forecasting of summer weather over and in the neighborhood of South China. J. Meteorol., 8: 289–92.

———. 1952*a*. Diurnal variation of summer rainfall over East China, Korea and Japan. J. Meteorol., 9: 83–86.

———. 1952*b*. Relationship of the general circulation to normal weather over Southern Asia and the western Pacific during the cool season. J. Meteorol., 9: 403–6.

———. 1952*c*. Variation of rainfall over South China through the wet season. Bull. Am. Meteorol. Soc., 33: 308–11.

———. 1954. Non-frontal crachin and the cool season clouds of the China Seas. Bull. Am. Meteorol. Soc., 35: 404–11.

———. 1955. The cool-season tropical disturbances of Southeast Asia. J. Meteorol., 12: 252–62.

———. 1966. The summer atmospheric circulation over the Arabian Sea. Jour. Atmos. Sci., 23: 144–50.

———. 1971. Monsoon meteorology. Academic Press, New York.

———. 1975. Preliminary discussion of the meteorology of the 1972–73 El Niño. Bull. Am. Meteorol. Soc., 56: 234–42.

Ramanathan, K. R. 1955. On upper tropospheric easterlies and the travel of monsoon and post-monsoon storms and depressions. *In*, Proc. of the UNESCO Symposium on typhoons, November 9–12, 1954. Japanese National Commission for UNESCO.

Ramaswamy, G. R. 1956. On the sub-tropical jet stream and its role in the development of large-scale convection. Tellus, 8: 27–36.

Ramos, R. P. L. 1974. Precipitation characteristics in the Northeast Brazil dry region. Colorado State Univ., Dept. of Atmospheric Science, Fort Collins, Colo., Atmospheric Science Paper No. 224.

———. 1975. Precipitation characteristics in the Northeast Brazil dry region. Jour. Geophys. Research, 80: 1665–78.

Rao, P. R. Krishna, and P. Jagannathan. 1953. A study of the northeast monsoon rainfall of Tamilnad. Indian J. Meteorol. Geophys., 4: 22–44.

Rao, Y. P., and B. N. Desai. 1970. The Indian summer monsoon. *In*, Proc. of the symposium on tropical meteorology, June 2–11, 1970. Univ. of Hawaii, Hawaii Institute of Geophysics, Honolulu: J-V-1-6.

Ratcliffe, R. A. S. 1950. Upper air analysis and tropical forecasting. Great Britain Meteorol. Office, Meteorol. Repts., No. 8: 1–18.

Ratisbona, L. R. 1976. The climate of Brazil. *In*, Climates of Central and South America, W. Schwerdtfeger, Ed., World survey of climatology, vol. 12, Elsevier, New York: 219–93.

Reichel, Eberhard. 1949. Die Niederschlagshäufigkeit im Mittelmeergebiet. Meteorol. Rundschau, 2: 129–42.

Rex, Daniel F. 1950. Blocking action in the middle troposphere and its effect upon regional climate. Tellus, 2: 196–211; 275–301.

———. 1951. The effect of atlantic blocking action upon European climate. Tellus, 3: 100–111.

Riehl, Herbert. 1947. Subtropical flow patterns in summer. Univ. of Chicago, Dept. of Meteorol., Misc. Repts., No. 22: 6–10, 49–50.

———. 1950. On the role of the tropics in the

general circulation of the atmosphere. Tellus, 2: 1–17.

———. 1954. Tropical meteorology. McGraw-Hill, New York.

———. No date. Waves in the easterlies and the polar front in the tropics. Univ. of Chicago, Dept. of Meteorol., Misc. Repts., No. 17: 40–53.

———, et al. 1954. The jet stream. Am. Meteorol. Soc. Publ., Meteorol. Monographs, 2, No. 7: 1–100.

Rodenwalt, E., and H. J. Jusatz, Eds. 1956. World atlas of epidemic diseases. Falk-Verlag, Hamburg.

Rubin, Morton J. 1956. The associated precipitation and circulation patterns over southern Africa. Notos, 5: 53–63.

———, and Harry van Loon. 1954. Aspects of the circulation of the Southern Hemisphere. J. Meteorol., 11: 68–76.

Rudloff, W. 1956. Pacific cold fronts and the weather at Lima. Mimeographed sheets prepared at the Lima weather station.

Rudolph, William E. 1953. Weather cycles on the South American west coast. Geograph. Rev., 43: 565–66.

Rupprecht, E. 1970. A quantitative investigation of the aridity of the desert of Thar. Bonner Meteorol. Abhandl., 14: 81–99.

Sabbagh, M., and Reid Bryson. 1962. An objective precipitation climatology of Canada. Univ. of Wisconsin, Dept. of Meteorol., Tech. Report, No. 8: 1–33.

Saha, K. 1970. On the nature and origin of the double intertropical front. *In*, Proc. of the symposium on tropical meteorology, August, 1970. Univ. of Hawaii, Amer. Meteorol. Soc., Honolulu: section F-II.

Saito, N. 1966. A preliminary study of the summer monsoon of southern and eastern Asia. Jour. Meteorol. Soc. Japan, 44: 44–49.

Saito, Ren-ichi. 1959. The climate of Japan and her meteorological disasters. *In*, Proc. I.G.U. regional conf. Japan, August 28–September 3, 1957, Science Council of Japan. Tokyo: 173–83.

Sanders, R. A. 1953. Blocking highs over the eastern North Atlantic Ocean and western Europe. Monthly Weather Rev., 81: 67–73.

Sanderson, Robert. 1954. Notes on the climate of Indochina. Weatherwise, 7: 56–69.

Sands, R. D. 1959. A study in the regional climatology of Mexico with precipitation as the correlative factor. Ph.D. thesis, Clark Univ., Worcester, Mass.

Sandström, J. W. 1926. Über den Einfluss des Golfstromes auf die Wintertemperatur in Europa. Meteorol. Zeit., 43: 410–11.

Sangster, Wayne E. 1958. An investigation of nighttime thunderstorms in the central United States. Univ. of Chicago, Dept. of Meteorol., Tech. Repts., No. 5. Also, (AFCRC) TN–58–211, Contract AF19 (604)–2179.

Sansom, H. W. 1954. The climate of East Africa. East African Meteorol. Dept., Memoirs, 3, No. 2.

Sapper, K. 1932. Klimakunde von Mittelamerika. *In*, Handbuch der Klimatologie, W. Köppen and R. Geiger, Eds., Band 2, Teil h. Gebrüder Borntraeger, Berlin: 19–47.

Sawyer, J. S. 1947. The structure of the intertropical front over N.W. India during the S.W. monsoon. Quart. J. Roy. Meteorol. Soc., 73: 346–69.

———. 1952. Memorandum on the intertropical front. Great Britain Meteorol. Office, Meteorol. Repts., 10: 1–14.

Schedler, A. 1924. Die Zirkulation im Nordatlantischen Ozean und den anliegenden Teilen der Kontinente, dargestellt durch Häufigkeitswerte der Zyklonen. Ann. Hydrog. u. Maritimen Meteorol., 52: 1–14.

Schmidt, F. H. 1949. On the theory of small disturbances in equatorial regions. J. Meteorol., 6: 427–28.

———, and J. H. A. Ferguson. 1951. Rainfall types based on wet and dry period ratios for Indonesia with western New Guinee. Republik Indonesia. Djawatan Meteorol. dan Geofis., Verhandelingen No. 42.

Schmidt, R. D. 1952. Die Niederschlagsverteilung im andinen Kolumbien. Bonner Geograph. Abhandl., 9: 99–119.

Schneider-Carius, Karl. 1947–1948. Die Etesien. Meteorol. Rundschau, 1: 464–70.

Schott, Gerhard. 1932. The Humboldt current in relation to land and sea conditions on the Peruvian coast. Geography, 17: 87–98.

———. 1935. Geographie des Indischen und Stillen Ozeans. C. Boysen, Hamburg, Tafel 19.

Schröder, Rudolf. 1955. Die Verteilung der Regenzeiten im nördlichen tropischen Amerika. Petermanns Geograph. Mitt., 99: 263–69.

Schüepp, Max. 1954. Witterungsklimatologie der Schweiz. *In*, Witterung und Klima in Mitteleuropa, Hermann Flohn, Ed., S. Hirzel, Zurich: 159–67.

Schulze, Alfred. 1956. Eine Methode zur Erfassung von Jahresgängen mit praktischer Anwendung auf Lufttemperatur und Niederschlagsmenge in Europa. Petermanns Geograph. Mitt., C: 34–39.

Schulze, B. R. 1958. The climate of South Africa according to Thornthwaite's rational classification. S. African Geograph. J., 40: 31–53.

Schupelius, G. 1976. Monsoon rains over West Africa. Tellus, 28: 533–37.

Schütte, K. 1968. Untersuchungen zur Meteorologie und Klimatologie des El Niño Phänomens in Ecuador und Nordperu. Bonner Meteorol. Abhandl., 9.

Schweigger, Erwin. 1947. El litoral peruano. Compana Administradora del Guano, Lima.

———. 1949. Der Perustrom nach zwölfjährigen Beobachtungen. Erdkunde, 3: 121–32; 229–41.

———. 1959. Die Westküste Südamerikas im Bereich des Peru-Stroms. Kaysersche Verlagsbuchhandlung, Heidelberg-Munich.

Schweinfurth, U., H. Flohn, and M. Domrös, Eds. 1970. Studies in the climatology of South Asia. South Asia Institute, Dept. of Geography, Franz Steiner Verlag, Wiesbaden.

Schwerdtfeger, W. 1976. High thunderstorm frequency over the subtropical Andes during summer. *In*, Climates of Central and South America, W. Schwerdtfeger, Ed., World survey of climatology, vol. 12, Elsevier, New York: 192–95.

———, Ed. 1976. The climates of Central and South America. World survey of climatology, vol. 12, Elsevier, New York.

Science Council of Japan. Proceedings of the I. G. U. regional conference Japan, August 28–September 3, 1957. 1959. Tokyo.

Scully, E. C. 1951. Notes on synoptic analysis and forecasting of Mediterranean summer weather. Bull. Am. Meteorol. Soc., 32: 163–65.

Seelye, C. J. 1950. Rainfall and its variability over the central and southwestern Pacific. New Zealand Meteorol. Service, Meteorol. Office Note No. 35.

Sekiguchi, Takeshi. 1952. The rainfall distribution in the Pacific region. Proc. Pacific sci. congr., 7th Congr., Wellington, 3: 101–2.

Sekeguti, T., and H. Tamiya. 1968. A climatology of autumnal rains in Japan. Geog. Rev. Japan, 41: 258–79.

Serra, Adalberto. 1941. The general circulation over South America. Bull. Am. Meteorol. Soc., 22: 173–78.

———. 1946. As sêcas do nordeste. Translation by L. de N. Friburg. Ministério da Agricultura, Servício de Meteorologia, Brazil.

Servício de Meteorologia, Republica de Venezuela. 1957. Atlas Climatologico Provisional (Periodo 1951–55).

Servício Meteorologico Nacional. 1958. Régimen de las precipitaciónes en la Patagonia. Buenos Aires. Typed sheets.

Shanbhag, G. Y. 1956. The climates of India and its vicinity according to a new method of classification. Indian Geograph. J., 31: 1–25.

Sinha, K. L. 1952. An analysis of the space distribution of rainfall in India and Pakistan. Indian J. Meteorol. Geophys., 3, No. 1: 1–16.

Sircar, N. C. Rai. 1956. A climatological study of storms and depressions in the Bay of Bengal. Indian J. Meteorol. Geophys., 7: 157–60.

Snow, J. W. 1976. The climate of northern South America. *In*, Climates of Central and South America, W. Schwerdtfeger, Ed., World survey of climatology, vol. 12, Elsevier, New York: 370–76.

Solot, Samuel B. 1943. The meteorology of Central Africa. Report of weather research center, 19th weather region, Accra, Gold Coast, November: 20–24.

———. 1950. General circulation over the Anglo-Egyptian Sudan and adjacent regions. Bull. Am. Meteorol. Soc., 31: 85–94.

Staff Members of the Institute of Meteorology. 1944. A report on synoptic conditions in the Mediterranean area. Univ. of Chicago, Inst. of Meteorol., Misc. Repts., No. 14.

Staff Members of the Section of Synoptic and Dynamic Meteorology, Institute of Geophysics and Meteorology, Academia Sinica, Peking. 1957. On the general circulation over Eastern Asia. Tellus, 9: 432–46.

———. 1958. On the general circulation over Eastern Asia. Tellus, 10: 58–75; 299–312.

Stupart, R. F. 1905. The Canadian climate. Report of the eighth international geographic congress, 1904. U.S. Government Printing Office, Washington, D.C.: 294–307.

Suda, K., and T. Asakura. 1955. A study on the unusual "baiu" season in 1954 by means of Northern Hemisphere upper air mean charts. Jour. Meteorol. Soc. Japan, 33: 233–44.

Sumner, Alfred R. 1953. Standard deviation of mean monthly temperatures in Anglo-America. Geograph. Rev., 43: 50–59.

Sung, Shio-wang. 1931. The extratropical cyclones of East China and their characteristics. Mem. Nat. Research Inst. Meteorol. (China), No. 3.

Supan, Alexander. 1927. Grundzüge der physischen Erdkunde. 7th ed., vol. 1, Gross-Oktav, Berlin.

Svenson, H. K. 1946. The vegetation of the coast of Ecuador and Peru and its relation to the Galapagos Islands. Am. J. Botany, 33: 394–498.

Sverdrup, H. U. 1942. Oceanography for meteorologists. Prentice-Hall, New York: 189–93.

Swan, A. D. 1958. The West African monsoon. Ghana Meteorol. Dept., Accra, Dept. Note No. 8.

Taljaard, J. J. 1953. The mean circulation in the lower troposphere over southern Africa. S. African Geograph. J., 35: 33–45.

———. 1955. Stable stratification in the atmosphere over southern Africa. Notos, 4, No. 3: 217–30.

———, and T. E. W. Schumann. 1940. Upper-air temperatures and humidities at Walvis Bay, South West Africa. Bull. Am. Meteorol. Soc., 21: 293–96.

Thompson, B. W. 1951. An essay on the general circulation of the atmosphere over South-East Asia and the West Pacific. Quart. J. Roy. Meteorol. Soc., 77: 569–97.

———. 1957*a*. Some reflections on equatorial and tropical forecasting. East African Meteorol. Dept., Tech. Mem., No. 7.

———. 1957*b*. The diurnal variation of precipitation in British East Africa. East African Meteorol. Dept., Tech. Mem., No. 8.

———. 1965. The climate of Africa. Oxford Univ. Press, New York.

Thornthwaite, C. Warren. 1933. The climate of the earth. Geog. Rev. (Amer. Geog. Soc.), 23, No. 3: 433–40.

———. 1948, An approach toward a rational classification of climate. Geogr. Rev., 38, No. 1: 55–94.

Tomsett, J. E. 1969. Average monthly and annual rainfall maps of East Africa. East Africa Meteorol. Dept., Tech. Mem., No. 14.

Transactions of the British Columbia Natural Resources Conference, 5th Conf. 1952.

Troll, Carl. 1930. Die tropischen Andenländer. *In*, Handbuch der geographischen Wissenschaft, Fritz Klute, Ed., vol. 4, Südamerika. Akademische Verlagsgesellschaft Athenaion Wildpark-Potsdam: 398–99.

Troup, A. J. 1956. An aerological study of the "meridional front" in western Australia. Australian Meteorol. Mag., 14: 1–20.

Tümertekin, Erol. 1955. The relationship between the wheat growing period and dry months in Turkey. Rev. Geograph. Inst. Univ. Istanbul, No. 2.

Tyner, R. V. 1951. Polar outbreaks into the Okanagan Valley and southern coast of British Columbia. Meteorol. Division, Vancouver District, Local Forecast Study No. 3.

Union of South Africa Meteorological Office. 1941. Atmospheric pressure and weather charts.

Union of South Africa Meteorological Services of the Royal Navy and the South African Air Force. 1944. vol. 2, Weather on the coasts of southern Africa.

Union of South Africa Weather Bureau. 1957. Climate of South Africa. Part 4, Rainfall maps. W.B. 22.

U.S. Army Air Force Hdq., Weather Research Center. 1942. Climate and weather of Southeastern Asia. Part I, India, Burma, and Southern China. 5, No. 3.

U.S. Army, Environmental Protection Division, Hdq. Quartermaster Research and Development Command. 1955. The standard deviation as a measure of variability of monthly mean temperature in the North Hemisphere. Tech.

Rept. EP–16.
U.S. Hydrographic Office. 1945. Weather summary of Brazil. U.S. Government Printing Office, Washington, D.C., No. 527.
U.S. Weather Bureau. 1938. Atlas of climatic charts of the oceans. U.S. Government Printing Office, Washington, D.C. W.B. 1247.
———. 1947. Thunderstorm rainfall. U.S. Government Printing Office, Hydrometeorological Report. No. 5.
———. 1949. The thunderstorm, report of the thunderstorm project. U.S. Government Printing Office, Washington D.C.
———. 1957. Principle tracks and mean frequencies of cyclones and anticyclones in the Northern Hemisphere. U.S. Government Printing Office, Washington, D.C.
Van Lingen, Lieut. M. S. 1945. The coastal low of the south and south east coasts of the Union of South Africa. South Africa Air Force, Meteorological Section, Tech. Notes, Pretoria, No. 33. Mimeographed.
Villmow, Jack Richard. 1955. The nature and origin of the Canadian dry belt. Ph.D. thesis, Univ. of Wisconsin.
———. 1956. The nature and origin of the Canadian dry belt. Ann. Assoc. Am. Geographs, 46: 211–32.
Visher, S. S. 1954. Climatic atlas of the United States. Harvard University Press: Cambridge.
Vowinckel, E. 1953. Zyklonenbahnen und zyklogenetische Gebiete der Suedhalbkugel. Notos, 2: 28–39.
———. 1955*a*. Beitrag zur Witterungsklimatologie Südafrikas. Arch. Meteorol. Geophys. Bioklimatol., Ser. B, Band 7, Heft 1: 11–31.
———. 1955*b*. Southern Hemisphere weather map analysis: five-year mean pressures; Part II. Notos, 4: 204–16.
———. 1956. Ein Beitrag zur Witterungsklimatologie des suedlichen Mozambiquekanals. Miscelânea Geofísica Publicada Pelo Serviço Meteorológico de Angola em Comemoração do X Aniversário do Serviço Meteorológico Nacional, Luanda: 63–86.
Vuorela, Lauri A. 1953. Some results from the Finnish Atlantic expedition, 1939. Ann. Acad. Sci. Fennicae, Ser. A.
Wagner, A. 1931. Zur Aerologie des indischen Monsuns. Gerlands Beitr. Geophys, 30: 196–238.
Wagner, M., and E. Rupprecht. 1975. Materialien zur Entwicklung des indischen Sommermonsuns. Bonner. Meteorol. Abhandl., 23: 1–29.
Wahl, Eberhard. 1952. The January thaw in New England. Bull. Am. Meteorol. Soc., 33: 380–86.
———. 1953. Singularities in the general circulation. J. Meteorol., 10: 42–45.
———. 1954. A weather singularity over the U.S. in October. Bull. Am. Meteorol. Soc., 35: 351–56.
Waibel, Leo. 1922. Winterregen in Deutsch-Südwest-Afrika. Abhandl. Gebiet der Auslandskunde, Bank 9, Reihe C.
Waite, Paul J. 1958. September rains in the upper midwest. Unpublished.
Walker, H. O. 1958. The monsoon in West Africa. Ghana Meteorol Dept., Accra, Dept. Note No. 9.
Walker, Sir Gilbert T., and J. C. Kamesvara Rav. 1925. Rainfall types in India in the cold weather period Dec. 1 to March 15. Mem. Indian Meteorol. Dept., 24: Part 11.
Ward, Robert De C. 1925. The climates of the United States. Ginn & Co.: Boston.
———, and Charles F. Brooks. 1934. Climatology of the West Indies. *In*, Handbuch der Klimatologie, W. Köppen and R. Geiger, Eds., Band 2, Teil a, Gebrüder Borntraeger, Berlin.
———, ———, and A. J. Connor. 1930–39. The climates of North America. *In*, Handbuch der Klimatologie, W. Köppen and R. Geiger, Eds., Band 2, Teil j, Gebrüder Borntraeger, Berlin.
Watts, I. E. M. 1945. Forecasting New Zealand weather. New Zealand Geograph., 1: 119–38.
———. 1947. The relations of New Zealand weather and climate: an analysis of the westerlies. New Zealand Geograph., 3: 115–36.
———. 1955. Equatorial weather, with particular reference to Southeast Asia. University of London Press, London.
Webb, Kempton E. 1954. The climate of northeast Brazil according to the Thornthwaite classification. M.A. thesis, Syracuse Univ.
Weickmann, L. 1922. Luftdruck und Winde im

östlichen Mittelmeergebiet. Habilitationsschrift, 99.
Wellington, John H. 1955. Southern Africa, a geographical study. vol. 1, Physical Geography. Cambridge: Cambridge University Press.
West, Robert C. 1957. The Pacific lowlands of Colombia, a Negroid area of the American tropics. Louisiana State Univ. Press, Baton Rouge.
Wexler, Harry. 1943. Some aspects of dynamic anticyclogenesis. Univ. of Chicago, Inst. of Meteorol., Misc. Repts., No. 8: 3–13.
———, and J. Namias. 1938. Mean monthly isentropic charts and their relation to departures of summer rainfall. Trans. Am. Geophys. Union, 19: 164–70.
Wiggin, B. L. 1950. Great snows of the Great Lakes. Weatherwise, 3: 123–26.
Wilhelmy, Herbert. 1953. Die pazifische Küstenebene Kolumbiens. Deut. Geographentag Essen, 96–100.
Williams, Philip, Jr. 1948. The variation of the time of maximum precipitation along the west coast of North America. Bull. Am. Meteorol. Soc., 29: 143–45.
Wise, Colin G. 1944. Climatic anomalies on the Accra Plain. Geography, 29: 35–38.
Wyrtki, Klaus. 1956. The rainfall over the Indonesian waters. Republik Indonesia. Djawatan Meteorol. dan Geofis., Verhandelingen No. 49.
———, et al. 1976. Predicting and observing El Niño. Science, 191: 343–46.
Yao, C. S. 1940. On the origin of the depressions in southern China. Bull. Am. Meteorol. Soc., 21: 351–55.
Yasui, H., K. Shimakawa, U. Morino, and T. Yamada. 1955. Blocking activity and Bai-u phenomenon in the Far East. (In Japanese; English summary.) J. Meteorol. Research (Japan), 7: 213–22.
Yeh, Tu-Cheng. 1950. The circulation of the high troposphere over China in the winter of 1945–46. Tellus, 2: 173–83.
Yin, Maung Tun. 1949. A synoptic-aerologic study of the onset of the summer monsoon over India and Burma. J. Meteorol., 6: 393–400.
Yoshino, M. 1963. Four stages of the rainy season in early summer over East Asia. Jour. Meteorol. Soc. Japan, Part 1, 43: 231–45; Part 2, 44: 209–17.
———. 1965. Frontal zones and precipitation distribution in the rainy season over East Asia. Geog. Rev. Japan, 38: 14–28.
———. 1971. Water balance problems in monsoon Asia from the viewpoint of climatology. *In*, Water balance of monsoon Asia, M. Yoshino, Ed., Univ. of Hawaii Press, Honolulu: 3–23.
———, Ed. 1971. Water balance of monsoon Asia: a climatological approach. Univ. of Hawaii Press, Honolulu.
———. 1977. Bai-u, the rainy season in early summer. *In*, The climate of Japan, E. Fukui, Ed., Developments in Atmospheric Science, No. 8, Kandansha Ltd., and Elsevier, New York: 85–101.
———. 1977. Dynamic and synoptic aspects of Japan's climate. *In*, The climate of Japan, E. Fukui, Ed., Elsevier, New York: 29–63.
———. 1977. The winter monsoon. *In*, The climate of Japan, E. Fukui, Ed., Developments in Atmospheric Science, No. 8, Kandansha Ltd., and Elsevier, New York: 65–84.
Zimmerschied, Wilhelm. 1947–1948. Über typische Wetterlagen der Iberischen Halbinsel. Meteorol. Rundschau, 1: 405–7.

Index

Accra. *See* Ghana

Africa: circulation patterns, 105–6; climatic pattern, 106; lack of cordillera, 105; moisture regions, 108; northern, *ITC* in, 109; problem climates, 106–7; subtropical, compared with subtropical South America, 107–8; subtropical anticyclones, 105–6. *See also* East Africa; Southern Africa

Amazon Valley: *Af*, *Am*, and *Aw* climates in, 63–65; atmospheric disturbances, 64–65; climatic peculiarities, 63–64; *friagem*, 64

Andes: effects on subtropical anticyclones, 12

Anglo-America, general: Arctic-subarctic, 282–83; Canadian "Dry Belt," 290–91; chinook effects, 286–88; climatic controls, 281–82; compared climatically with Eurasia, 281–82; eastern regions, 285–88; Great Plains dry region, 289; Gulf of Mexico as a moisture source, 289–90; interdiurnal temperature variability in, 284–86; interior region of, 311–24; intermontane region of, 304–10; January Thaw in, 288; nocturnal summer rainfall in, 293–96; Ohio Valley region, 326; Pacific coast region, 298–304; Prairie Wedge, 291–93; precipitation amounts, 289–93; rainfall regions of, 298; regional types of annual precipitation variation, 297–325; seasonal rainfall, 296; seasonal temperatures, 282–83; subarctic precipitation of, 325; subtropical interior region, 326–32; subtropical oceanic margins, 332–36; temperature continentality, 283–84; temperature singularities, 288–89; temperature variability, 284–88; water vapor transport across Pacific and Gulf coastlines, 289; variability of monthly mean temperatures, 284–88; west Gulf borderlands and northeast Mexico, 289–90

Anglo-America, regional

—Eastern region: cyclogenetic regions, 336–39; rainfall subtypes, 338–39; small annual range of precipitation, 336; storm tracks, 337; winter maximum precipitation, 336–38

—Interior région: cyclogenetic regions, 311; annual precipitation variation, 311; early summer rainfall maximum in northern Great Plains, 314–16; effect of Great Lakes on precipitation, 323–24; subarctic parts, 321–23; twinned summer rainfall maximum in Texas section, 312–14; twinned summer rainfall maximum in Upper Mississippi Valley region, 316–21; weather of Greenland Ice Cap, 325

—Intermontane region: biannual rainfall variation in, 305; July–August rainfall maximum in Arizona, 305–7; June rainfall maximum in Washington-Oregon, 308–10; moisture sources, 306–7; precipitation singularity in Arizona, 307; winter secondary precipitation maximum in Arizona, 306

—Ohio Valley region, 326

—Pacific Coast region: amounts of precipitation, 302–4; Baja California rainfall regimes, 302; cool coastal current of, 300; distinctive precipitation features of, 298–99; dry-summer climate displaced poleward in, 299–300; low-rainfall areas in, 302–3; origin of rainfall in, 303; rainfall subtypes of, 303–4; San Francisco Bay area, 303–4; seasonal southward progression of rainfall maximum, 300–302; time of precipitation maximum a function of latitude in, 301

—Subtropical interior region, 326–32; cool-season precipitation maximum in, 327; cyclonic storms in, 331; positioning of jet stream over, 331; rainfall anomaly in, 326–27; rainfall subtypes, 328–29; resultant geostrophic winds of March and September, 331–32; seasonal synoptic types, 332

—Subtropic oceanic margins region: Atlantic Ocean and Gulf of Mexico, 332–36; double sea breeze effects in, 334–35; diurnal periodicity of rainfall in, 333; rainfall subtypes of, 333–36; sea breeze convergence in, 334–35; seasonal circulations in, 332–33, 334; synoptic types, 332–35; thunderstorms in, 332–36; tropical disturbances in, 332–33. *See also* Florida

Angola, 131–33; coastal rainfall gradients of, 115–16; coastal rainfall variability of, 132–33; dry coast of,

130–33; effects of cold water on, 105, 131–32; *ITC* in, 131
Argentina, 45–46
Ascension Island: compared with Galapagos Islands, 168; dry climate of, 168
Asia. *See also* East Asia; Indian subcontinent; South Asia; Southeast Asia; Southwest Asia
—Eastern: summer transport of water vapor over, 209
—Eastern-Southern: equatorial southwesterlies in, 175–77; seasonal circulation patterns of, 173–77; summer circulation of, 175–77; wind discontinuities and fronts in, 173–77; winter circulation of, 174–75
Australia, 87–93; climatic pattern of, 87–88; cold fronts in, 89; easterly waves in, 90–91; *ITC* in, 88; meridional fronts in, 88–90; problem climates of, 91–93; tropical disturbances in, 90–91; typhoons in, 90–91; warm fronts, 89; weather element in, 88–90; westerly disturbances in, 88–90

Baiu: in China, 224–25; in Japan, 227–29
Baja California: rainfall regime, 302
Benguela cool current, 167–68
Blocking action by anticyclone, effects of: on cyclone routes in Europe, 239–40; on jet stream in Europe, 239; on seasonal precipitation in western Europe, 241; on temperatures in Europe, 240–42; on temperatures in the Mediterranean Basin, 242
Brazil: coastal *As* region, 50–56; cold fronts in, 48–50, 63–64; easterly waves in, 53; eastern, inversion in, 12; easternmost, 12, 55; effects of South Atlantic anticyclone on, 48, 55; equatorial westerlies in, 53–55; *ITC* in, 12; northeastern anticyclonic circulation in, 12; positioning of *ITC* over, 12
—Northeast, 56–63; causes of wet and dry years in, 50–51; coastal subsidence in, 59–60; equatorial westerlies in, 53–55; *ITC* in, 53–55, 58–59; location of, 56; reasons for drought, 60–63; seasonal rainfall in, 57; terrain in, 57
Brazilian climates: Amazon Basin, 63–65; controls of, 47–48; effects of South Atlantic anticyclone on, 47–48

"Callao Painter," 36–37
Canaries current, 106
Caribbean region, 66–83; aberrant climates, 76–83; dry parts of, 66–73, 75–76; lack of *ITC* in, 74; *Nortes*, 76; temperature inversions in, 78–80; Venezuela dry littoral, 66–73. *See also* Middle America
Chile: desert (*BW*) climate in, 23–78; subtropical dry-summer (*Cs*) climate, 39; tundra (*Ft*) climate, 39–40
Chilean-Peruvian desert, 23–38; anticyclonic circulation, 24, 31–32; cloud and rainfall, 28–31; Ekman drift effects, 25, 34–35; El Niño effects, 26, 35–38; fog, 29–30; *garúa*, 30–31; intense aridity of, 23, 31–35; *ITC* in, 37–38; Lettau's hypothesis, 34–35; Peru Coastal Current, 24–25; sea breeze effects, 26, 31–35; stratus cloud deck, 27; temperature, 26–28; temperature inversions, 27–28; thunderstorms in subtropical Andes, 35; upwelled cool coastal water, 24–25
China: air masses in, 208–9, 210–11; Baiu rainy season in, 224–25; *crachin* in, 222; cyclogenesis in, 214–18; diurnal periodicity of summer precipitation in, 225; easterly waves in, 217; fall precipitation in, 225; jet stream location in, 207; latitudinal precipitation gradients in, 220–21; Maiu rainy season in, 210, 224; moisture sources for, in summer, 209–13; positioning of winter jets over, 207–9; precipitation features of, 220–25; reduced rainfall of southeast coast of, 221; seasonal circulation, 207–13; seasonal precipitation of, 221–23; spring precipitation in, 222–23; summer disturbances in, 217–18; summer precipitation bimodal maximum in, 224–25; summer precipitation in, 223–25; thunderstorms in, 218–19; tropical storms in, 217–18; water vapor flux over, in summer, 209; water vapor gradient, 211; weather systems, 214–18; winter air masses of, 208–9; winter disturbances in, 214–17; winter fronts in, 208–9, 214–17; winter precipitation as related to subtropical jet stream, 222; winter precipitation of, 221–22; winter temperatures of, 219–20. *See also* East Asia
Colombia, 13–19; annual march of rainfall, 16–17; atmospheric circulation, 18–19; causes for extreme wetness, 18–19; climates of, 16; diurnal march of rainfall, 17–18; location of *ITC*, 18; night rains, 17–18; winds, 18–19
Congo Basin, 125–31; atmospheric disturbances of, 126–30; circulation features of, 125–26; climatic arrangement in, 130; climatic peculiarities of, 130–31; Cuvette Central, 130–31; line squalls of, 127–30; organized convection in, 127; stability features of southwesterlies in, 125–27, 130; subhumid coast of, south of equator, 125, 131–32; surges in the southwesterlies, 127
Crachin: in China, 222; in Southeast Asia, 206

Drizzle-mist: in coastal Ecuador, 21–22; in Peru, 29–31

East Africa
—North: bimodal annual rainfall profile, 149; causes of drought in, 150; climatic peculiarities of, 149–55; coastal fogs of, 154–55; coastal subsidence in, 154–55; cool coastal waters of, 154; Flohn's explanation of Somali's dryness, 150, 154; high-sun cir-

culation patterns of, 152–55; *ITC* in, 152–54; low-sun circulation patterns of, 150–52; structure of summer atmosphere in, 152–55
—South: cold-front disturbances of, 157–58; effects of Madagascar on rainfall of, 156–57; origin of rainfall deficiency in, 156–57; rainfall distribution in, 155; rainfall reliability, 156; tropical storms of, 157
—Tropical: annual march of rainfall, 148–49; causes for general dryness, 139–43; circulation patterns, 139–43; cold fronts, 126–27; diffluent character of both monsoons, 122–23; diffluent effects of Great Rift heat low, 142; "dry wedge," 139; equatorial westerlies, 140; fog and mist, 154–55; meridional flow of monsoon, 140–43; mobile weather disturbances, 146–48; moisture sources, 122; monsoon circulations, 121–22; monsoon surges, 146–47; rainfall distribution, 134–39; random thermal convection, 146; southerly monsoon, 140; stability of monsoons, 140–43; weather systems (large scale), 144–46
East Asia: circulation features, 207–31; climatic features of, 207; cold waves of, 208–9; cyclogenetic regions of, 214–18; easterly waves in, 217; effects of Tibetan lee convergence zone on cyclones, 214–16; fronts, 208–9; jet streams in, 207; Maiu front, 208–11; Maiu rains, 210, 213; moisture sources of, in summer, 209, 213; monsoons, 210–14; precipitation features of, 220–32; seasonal circulation features of, 207–14; Shurin rainfall maximum, 230–31; spring weather, 216–17; summer circulation features, 209; summer disturbances in, 217–18; temperature features of, 219–20; thunderstorms in, 218–19; Tibetan lee convergence zone of, 207, 214–16; tropical storms in, 217–18; unusual climatic features of, 219–31; water vapor flux in summer, 209; water vapor gradient, 211; winter air masses and fronts in, 208–9; winter weather, types of, 214–16. *See also* China; Japan
Ecuador: drizzle of, 21–22; drought controls in, 22; dry littoral of, 20–22; fog of, 22; *verano-invierno* in, 22
El Niño: current, 26; effects of, 35–38; relation to Northeast Brazil droughts, 58–59; wet years in north Peru, 35–38
Equatorial Pacific dry belt, 97–102; boundaries of, 97–98; Cromwell current, 101; double *ITC* structure in, 100–101; Hadley cell in, 100; relationship to cool water, 100–101
Europe: arrangements of climates in, 236–37; climatic controls of, 235; climatic features of, 235–36; summer rainfall minimum, 236; winter rainfall maximum, 236. *See also* Mediterranean Basin; Western-Central Europe; Western Europe

Florida: broad summer rainfall maximum of, 333–34; double sea breeze effect in, 334; hurricanes in, 334; thunderstorms in, 332–35
Friagem, 63–64

Galapagos Islands, 32–33, 97–98
Garúa: in coastal Ecuador, 21–22; in Chilean-Peruvian desert, 29–31
Ghana-Togo dry littoral, 120–24; annual rainfall profile of, 118; anomalous features of, 121; coastal divergence along, 121–22; effect of cool water on, 123; intensified summer drought of, 123–24; relation of, to disturbance lines, 122
Great Lakes: effects of, on precipitation, 323–24
Great Plains: origin of rainfall deficiency on, 289–91
Greenland: as an air-mass source region, 282–83
Grosswetterlagen: in Europe, 251–54
Guinea: August secondary rainfall minimum of, 122–23; circulation patterns of, 112; climatic peculiarities of, 116–24; disturbance lines of, 114–16; Ghana dry littoral, 120–24; harmattan, 114; monsoons, 111; seasonal reversal of winds on, 111; southerly disturbances of, 115–16; temperature inversions, 118; vertical structure of air over, 118–19; weather disturbances of, 114–16; weather zones of, 112–14; winter rainfall of, 119

Harmattan, 110–11
Honduras, 80

Indian subcontinent: active monsoon, 187; "andhis" in, 181–82; break in the monsoon, 187–88; annual march of rainfall in, 196–98; cool-season rainfall along southeast coast of, 196; cool-season weather of, 177–80, 197–98; disappearance of westerly jet in summer in, 184–85; disturbances in summer monsoon, 184–90; dry areas of, 193–96; easterly jet of summer in, 178; fall disturbances on southeast coast, 191–92; fall weather of, 191–92; fall-winter rainfall maximum on southeast coast of, 196–97; hurricanes in, 191; jet streams in, 179, 180; moisture regions, 193; monsoon circulation, 173–77; monsoon depressions, 189–90; nor'westers in, 181–82; onset of summer monsoon, 183–85; onset and withdrawal of monsoon, 191; origin of drought in the Deccan, 195–96; origin of drought in northwestern parts of, 193–95; regions of rainfall deficiency, 193–98; secondary cool-season rainfall maximum of, 180–81; spring rainfall distribution in, 179, 181–82; spring temperature maximum in, 192; spring weather of, 180–83; summer circulation in, 175–77; summer monsoon of, 183–90; summer weather of, 183–90; surges in summer monsoon, 188–89; Tamilnad coast, fall-winter rainfall maximum of, 197; temperature

peculiarities of, 192; Thar desert, 194–95; Tibet's role in monsoon, 184–85; unusual climatic features of, 192–98; weather of, 177–92; westerly troughs, 190; western disturbances in, 178–80; winter monsoon of, 177–78; winter rainfall distribution in, 177–80
Invierno: in Ecuador, 22

Japan: annual precipitation features of, 224–32; Baiu, rainy season in, 228–30; diurnal periodicity of summer rainfall in, 231; jet stream location in, 226, 228–29; maps of seasonal rainfall, 231–32; positioning of summer jets over, 228; precipitation contrasts of, with China, 225–26; precipitation features of, 225–32; seasonal precipitation features of, 227–32; Shurin, rainy period of, 230–31; summer precipitation bimodal maximum in, 227–31; temperature features of, 220; thunderstorms in, 232; water vapor flux in summer, 180; winter precipitation in, 232. *See also* East Asia

Köppen classification of climate, 343–45

La Porte, Indiana: rainfall anomaly, 324–25

Mediterranean Basin: annual rainfall profiles of, 242–43; Black Sea borderlands of, 274–75; cold-front precipitation in, 265–67; cyclogenesis in, 257–63; distribution of pure *Cs* climatic type in, 255, 271; distribution of seasonal rainfall frequency in, 268–71; distribution of seasonal rainfall amounts in, 271–73; effect of blocking action on weather and temperature of, 258; effect of trough-and-ridge steering in, 258–60; extensive development of *Cs* climate in, 255, 269; geographical features, 255–56; high-index situations in, 257; instability showers in, 265; local winds of, 257; low-index situations in, 257–61; modified *Cs* climate in, 255; northern, bimodal annual rainfall profile of, 273–74; northwest flow pattern in, 258–61; origin of precipitation in, 265–67; polar air invasions and cyclogenesis in, 257–60; precipitation associated with polar air invasions of, 265–67; precipitation features of, 265–76; rainfall contrasts between east and west coasts, 267; rainfall frequency associated with wind directions in, 265–67; seasonal circulations of, 256–57; seasonal weather in, 263; southwest steering and flow pattern in, 260–61; stable type of precipitation in, 265–67; summer disturbances in, 262–63; temperature contrasts between east and west coasts, 256; temperature features of, 264–65; thunderstorms in, 267; tracks of disturbances in, 263–65; warm-front precipitation in, 265–67; weather types of, 257–65; winter convergence in, 256

Mediterranean Sea: climatic and weather effects of, 255–56; effect of, on extensive *Cs* development, 255; effect of, on invading air masses, 255, 257–64
Meridional fronts: in Australia, 89–90
Mexico: subhumid northeastern littoral of, 81–83; Yucatan dry region, 80–81
Middle America, 73–83; asymmetry of annual rainfall profile in, 78–80; circulation and synoptic patterns of, 74–76; climatic anomalies, 76–83; climatic arrangements in, 73–75; dry climates in, 76–77; Honduras winter-rainfall maximum region, 80; problem climates of, 76–83; midsummer secondary rainfall minimum of, 77–78; temporal variations in rainfall, 76–77. *See also* Caribbean region
Monsoon depressions: in India, 189–90
Monsoons: Halley's theory of, 173–74

Namib Desert, 167–69
New Zealand: meridional fronts in, 93; problem climates of, 95–97; synoptic types of, 93–95; tropical cyclones in, 93–94
North America. *See* Anglo-America
Northeast Brazil. *See* Brazil, Northeast

Pakistan. *See* Indian subcontinent
Patagonia: dry climate, 41–45; cause of rainfall increase inland, 45; circulation patterns of, 41–42; climatic anomalies of, 41–42; effect of Andes on, 43–44; origin of drought in, 43–45; seasonal rainfall in, 42–43
Peruvian-Chilean desert. *See* Chilean-Peruvian desert
Peru Coastal Current, 24–26, 32; aridifying effects of, 32–33
Peru oceanic current, 24–25
Prairie Wedge, 291–93

Sahara: harmattan wind, 110–11; easterly jet in, 110; effect of cool water on, 107; southern, origin of rainfall in, 110; structure of *ITC* in, 109; unusual latitudinal breadth of, 107–108; weather zones, 112–14
Sahel (Africa), 110
Sêcas: in Northeast Brazil, 58–62
Shurin: rainy period in Japan, 230–31
Somalia, 149–52, 154–55
South America: Atlantic coast, 41–65; circulation features of, 11–13; climatic pattern of, 12–13; cold fronts in, 48–50; eastern, subtropical climates of, 45–46; meridional exchange of air over, 12; Pacific coast, 13–40; positioning of *ITC* in, 12; problem climates in, 12–13
Southeast Asia, tropical: air stream convergences in, 199–200; annual rainfall variations in, 204–6; areas of low-sun (winter) rainfall maximum in, 205–6; circulation features of, 199–202; climatic peculiarities of, 202–6; cold winters of Tonkin Delta in, 206; *cra-*

chin weather type, 206; equatorial disturbances in, 200–1; line squalls in, 201; movements of *ITC* in, 199; subhumid Mandalay basin of, 204; subhumid-semiarid areas in, 202–4; summer-dry areas in, 202; tropical disturbances in, 201–2
Southern Africa: absence of monsoon circulation in, 160–61; annual variation of rainfall in, 159–60; causes of dry climates in, 165–67; climatic peculiarities of, 165–69; cold fronts in, 166; extensiveness of dry climates in, 165–67; Limpopo, dry areas in, 166–67; Namib Desert, 167–69; origin of summer rainfall maximum in, 164–65; origin of wetter east side in, 165; rainfall distribution in, 159; seasonal circulation patterns in, 160–61; seasonal rainfall, 159–60; weather types, 161–65
Southern Caribbean. *See* Caribbean region
Southern Hemisphere: Antarctic cold surges in, 11; atmospheric circulation in, 11–13; cyclogenesis in, 11; *ITC* shifted northward in, 12; subtropical belt of high pressure in, 11–12
Southwest Asia, 275–78; etesians in, 277–78; Indian Ocean monsoon in, 277; summer circulation in, 277; summer precipitation in, 277–78; winter circulation in, 275–77; winter precipitation in, 276
Sudan, 111–16; circulation patterns of, 111–14; disturbance lines in, 114–15; Sahel in, 110; southerly disturbances in, 115; structure of *ITC* in, 109; weather disturbances of, 114–16; weather zones of, 112–14

Thar Desert, 194–96
Texas: subhumid littoral of, 313–15

Venezuela dry littoral, 66–73; causes for, 67–73; cool water along, 69; description of, 66–67; seasonal synoptic patterns of, 73; terrain effects, 67–69
Veranillo, 77
Verano: in Ecuador, 22

Western-Central Europe: annual range of precipitation in, 247; climatic boundaries (Köppen) of, 236–37; climatic effects of anticyclonic blocking action, 238–42; cool-season maximum of precipitation in, 247–48; effects of anticyclonic blocking on cyclone routes in, 239–41; effects of anticyclonic blocking on precipitation in, 241–42; effects of anticyclonic blocking on temperatures in, 241–42; *Grosswetterlagen* of, 251–54; modest lowland precipitation of, 243–45; precipitation features of, 242–47; *Schafskälte*, 251; seasonal isotherms in, 237–38; seasonal periodicity of blocking action in, 239; seasonal precipitation characteristics, 247–54; singularities, 249–54; small annual range of rainfall, 247; slow decline of rainfall inland in, 242–43; temperature anomalies in, 237–38; temperature features of, 237–42; weather-type climatology of, 250–54. *See also* Western Europe
Western Europe: coastal precipitation features of, 245–49; coastal rainfall maximum in, 245–47; cool uplands in, 238; cyclone routes, 244–45; effects of blocking action on seasonal precipitation in, 250; effects of blocking action on temperature, 238–42; effects of large-scale weather situations on seasonal precipitation in, 250–51; fall precipitation maximum in, 248–50; lack of latitudinal progression of monthly rainfall maximum, 247–48; modest lowland rainfall, 243–45; spring precipitation maximum in, 248–50; steep vertical temperature gradients in, 238; unusual features of coastal rainfall, 245–47. *See also* Western-Central Europe
World pattern of climates, 3–5
World pattern of seasonal rainfall concentration, 4

Yucatan: dry northern coast of, 80–81

DESIGNED BY RON FENDEL
COMPOSED BY GRAPHIC COMPOSITION, INC.
ATHENS, GEORGIA
MANUFACTURED BY THE NORTH CENTRAL PUBLISHING CO.
ST. PAUL, MINNESOTA
TEXT IS SET IN TIMES ROMAN
DISPLAY LINES IN BODONI

Library of Congress Cataloging in Publication Data
Trewartha, Glenn Thomas, 1896–
The Earth's problem climates.
Bibliography: pp. 349-366
Includes index.
1. Climatology.
I. Title.
QC981.T648 1980 551.6 80–5120
ISBN 0–299–08230–X